T0093161

Electronics

ALL-IN-ONE

3rd Edition

by Doug Lowe

for dummies®
A Wiley Brand

Electronics All-in-One For Dummies®, 3rd Edition

Published by: **John Wiley & Sons, Inc.,** 111 River Street, Hoboken, NJ 07030-5774, www.wiley.com

Copyright © 2022 by John Wiley & Sons, Inc., Hoboken, New Jersey

Published simultaneously in Canada

No part of this publication may be reproduced, stored in a retrieval system or transmitted in any form or by any means, electronic, mechanical, photocopying, recording, scanning or otherwise, except as permitted under Sections 107 or 108 of the 1976 United States Copyright Act, without the prior written permission of the Publisher. Requests to the Publisher for permission should be addressed to the Permissions Department, John Wiley & Sons, Inc., 111 River Street, Hoboken, NJ 07030, (201) 748-6011, fax (201) 748-6008, or online at http://www.wiley.com/go/permissions.

Trademarks: Wiley, For Dummies, the Dummies Man logo, Dummies.com, Making Everything Easier, and related trade dress are trademarks or registered trademarks of John Wiley & Sons, Inc. and may not be used without written permission. All other trademarks are the property of their respective owners. John Wiley & Sons, Inc. is not associated with any product or vendor mentioned in this book.

LIMIT OF LIABILITY/DISCLAIMER OF WARRANTY: THE PUBLISHER AND THE AUTHOR MAKE NO REPRESENTATIONS OR WARRANTIES WITH RESPECT TO THE ACCURACY OR COMPLETENESS OF THE CONTENTS OF THIS WORK AND SPECIFICALLY DISCLAIM ALL WARRANTIES, INCLUDING WITHOUT LIMITATION WARRANTIES OF FITNESS FOR A PARTICULAR PURPOSE. NO WARRANTY MAY BE CREATED OR EXTENDED BY SALES OR PROMOTIONAL MATERIALS. THE ADVICE AND STRATEGIES CONTAINED HEREIN MAY NOT BE SUITABLE FOR EVERY SITUATION. THIS WORK IS SOLD WITH THE UNDERSTANDING THAT THE PUBLISHER IS NOT ENGAGED IN RENDERING LEGAL, ACCOUNTING, OR OTHER PROFESSIONAL SERVICES. IF PROFESSIONAL ASSISTANCE IS REQUIRED, THE SERVICES OF A COMPETENT PROFESSIONAL PERSON SHOULD BE SOUGHT. NEITHER THE PUBLISHER NOR THE AUTHOR SHALL BE LIABLE FOR DAMAGES ARISING HEREFROM. THE FACT THAT AN ORGANIZATION OR WEBSITE IS REFERRED TO IN THIS WORK AS A CITATION AND/OR A POTENTIAL SOURCE OF FURTHER INFORMATION DOES NOT MEAN THAT THE AUTHOR OR THE PUBLISHER ENDORSES THE INFORMATION THE ORGANIZATION OR WEBSITE MAY PROVIDE OR RECOMMENDATIONS IT MAY MAKE. FURTHER, READERS SHOULD BE AWARE THAT INTERNET WEBSITES LISTED IN THIS WORK MAY HAVE CHANGED OR DISAPPEARED BETWEEN WHEN THIS WORK WAS WRITTEN AND WHEN IT IS READ.

For general information on our other products and services, please contact our Customer Care Department within the U.S. at 877-762-2974, outside the U.S. at 317-572-3993, or fax 317-572-4002. For technical support, please visit https://hub.wiley.com/community/support/dummies.

Wiley publishes in a variety of print and electronic formats and by print-on-demand. Some material included with standard print versions of this book may not be included in e-books or in print-on-demand. If this book refers to media such as a CD or DVD that is not included in the version you purchased, you may download this material at http://booksupport.wiley.com. For more information about Wiley products, visit www.wiley.com.

Library of Congress Control Number: 2022932613

ISBN 978-1-119-82211-0 (pbk); ISBN 978-1-119-82212-7 (ebk); ISBN 978-1-119-82213-4 (ebk)

SKY10069712_031424

Contents at a Glance

Table of Contents

Introduction

Welcome to the amazing world of electronics!

Ever since I was a kid, I've been fascinated with electronics. When I was about 10 years old, my dad bought me an electronic experimenter's kit from the local RadioShack store. I still have it; it's pictured here.

I have incredible memories of evenings spent with my dad, wiring the sample circuits to make squawking police sirens, flashing lights, a radio receiver, and even a telegraph machine.

The best part was dreaming that when I grew up, I'd have a job in the field of electronics, that someday I'd understand exactly how those resistors, capacitors, inductors, transistors, and integrated circuits actually worked, and I'd use that knowledge to design televisions or computers or communication satellites.

Well, that dream didn't come true. Instead, I went into a closely related field: computer programming. But my love of electronics never died, and I've spent the last 40 years or so experimenting with electronics as a hobbyist.

This book is an introduction to electronics for people who have always been fascinated by electronics but didn't make a career out of it. In these pages, you'll find clear and concise explanations of the most important concepts that form the basis of all electronic devices, concepts such as the nature of electricity (if you think you really know what it is, you're kidding yourself); the difference between voltage, amperage, and wattage; and how basic components such as resistors, capacitors, diodes, and transistors work.

Not only will you gain an appreciation for the electronic devices that are a part of everyday life, but you'll also learn how to build simple circuits that will not only impress your friends but may actually be useful!

About This Book

Electronics All-in-One For Dummies, 3rd Edition, is intended to be a reference for the most important topics you need to know when you dabble in building your own electronic circuits. It's a big book made up of eight smaller books, which we at the home office like to call *minibooks*. Each of these minibooks covers the basics of one key topic for working with electronics, such as circuit building techniques, how electronic components like diodes and transistors work, or using integrated circuits.

This book doesn't pretend to be a comprehensive reference for every detail on every possible topic related to electronics. Instead, it shows you how to get up and running fast so that you have more time to do the things you really want to do. Designed using the easy-to-follow For Dummies format, this book helps you get the information you need without laboring to find it.

Whenever one big thing is made up of several smaller things, confusion is always a possibility. That's why this book is designed with multiple access points to help you find what you want. At the beginning of the book is a detailed table of contents that covers the entire book. Then each minibook begins with a minitable of contents that shows you at a miniglance what chapters are included in that minibook. Useful running footers appear at the bottom of each page to point out the topic discussed on that page, and handy thumb tabs run down the side of the pages to help you find each minibook quickly. Finally, a comprehensive index lets you find information anywhere in the entire book.

This isn't the kind of book you pick up and read from start to finish as if it were a cheap novel. If I ever see you reading it at the beach, I'll kick sand in your face. Beaches are for reading romance novels or murder mysteries, not electronics books. Although you could read this book straight through from start to finish, this book is designed like a reference book, the kind of book you can pick up, open to just about any page, and start reading.

You don't have to memorize anything in this book. It's a "need-to-know" book: You pick it up when you need to know something. Need a reminder on how to calculate the correct load resistor for an LED circuit? Pick up the book. Can't remember the pinouts for a 555 timer IC? Pick up the book. After you find what you need, put the book down and get on with your life.

You can find a total of 54 projects strewn throughout this book's chapters. You'll find a plethora of simple projects you can build to demonstrate the operation of typical circuits. For example, in the chapter on transistors, you'll find several simple projects that demonstrate common uses for transistors, such as driving an LED, creating an oscillator, or inverting an input.

I suggest you build each of the projects as you read the chapters. Reading about electronics circuits is one thing, but to understand how a circuit works, you really need to build it and see it in operation. Most of the projects are simple enough that you can build them in 20 to 30 minutes, assuming you have the parts on hand.

If you are lucky enough to have a store that carries electronic components in your community, you're in luck! If you want to build one of the projects on a Saturday afternoon, you can buzz over to your local electronics store, pick up the parts you'll need, take them home, and build the circuit.

Of course, you can also purchase the components you need at any other store that stocks electronic hobbyist components, and you can find many sources for purchasing the parts online.

Finally, most of the electronic circuits described in this book are perfectly safe: They run from common AA or 9 V batteries and therefore don't work with voltages large enough to hurt you.

However, you'll occasionally come across circuits that work with higher voltages, which can be dangerous. Any project that involves line voltage (that is, that you plug into an electrical outlet) should be considered potentially dangerous and handled with the utmost care. In addition, even battery-powered circuits that use large capacitors can build up charges that can deliver a potentially painful shock.

When you work with electronics, you'll also encounter dangers other than those posed by electricity. Soldering irons are hot and can burn you. Wire cutters are sharp and can cut you. And there are plenty of small parts that can fall on the floor and find themselves in the mouths of kids or pets.

Safety is an important enough topic that I've devoted a chapter to it in Book 1. I strongly urge you to read Book 1, Chapter 4 *before* you build anything.

Please be careful! The projects that are presented in Book 8 all work directly with line-level voltage and should be considered dangerous. You must exercise great care if you decide to build any of those projects, as a single mistake could kill you or someone else. Those projects are offered as educational prototypes that are designed to be operated only within the safe confines of your workbench, where you can control the power connections so that no one is exposed to dangerous voltages.

Foolish Assumptions

Throughout this book, I make very few assumptions about what you may know about the subject of electronics. I certainly don't assume that you've ever taken a class on electronics, have ever assembled a circuit, or are well versed in advanced science or math.

In fact, there are really very few things I do assume:

> » **You're curious about the fascinating world of electronics.** For example, if you've ever wondered how a radio works or what makes a computer possible, this book is for you.

> » **You like to build things.** The best way to learn *about* electronics is to *do* electronics. This book has plenty of simple projects for you to build and back your knowledge up with first hand experience.

> » **You have a space to work and some basic tools.** You'll need at least a small workspace and basic tools such as a screwdriver and wire cutters.

> » **You can afford to spend a little money to get the parts you need.** Although a few of the projects later in the book require that you purchase items that may cost as much as a hundred dollars or more, most of the components you need can be purchased for just a few dollars.

Icons Used in This Book

Like any *For Dummies* book, this one is chock-full of helpful icons that draw your attention to items of particular importance. You find the following icons throughout this book:

TIP

Pay special attention to this icon; it lets you know that some particularly useful tidbit is at hand.

TECHNICAL STUFF

Hold it — overly technical stuff is just around the corner. Obviously, because this is an electronics book, almost every paragraph of the entire book could get this icon. So I reserve it for those paragraphs that go into greater depth, down into explaining how something works under the covers — probably deeper than you really need to know to use a feature, but often enlightening. You also sometimes find this icon when I want to illustrate a point with an example that uses some electronics gadget that hasn't been covered so far in the book but is covered later. In those cases, the icon is just a reminder that you shouldn't get bogged down in the details of the illustration and should instead focus on the larger point.

WARNING

Danger, Will Robinson! This icon highlights information that may help you avert disaster. You should definitely pay attention to the warning icons because they will let you know about potential safety hazards.

REMEMBER

Did I tell you about the memory course I took?

Beyond the Book

In addition to the material in the print or e-book you're reading right now, this product also comes with some access-anywhere goodies on the web. Check out the free Cheat Sheet for some safety rules to follow, a list of electronic resistor color codes, and more. To get this Cheat Sheet, simply go to www.dummies.com and type **Electronics All-in-One For Dummies Cheat Sheet** in the Search box.

Where to Go from Here

Yes, you *can* get there from here. With this book in hand, you're ready to plow right into the exciting hobby of electronics. Browse through the table of contents and decide where you want to start. Be bold! Be courageous! Be adventurous! And above all, have fun!

1
Getting Started
with Electronics

Contents at a Glance

Chapter **1**

Welcome to Electronics

I thought it would be fun to start this book with a story, so please bear with me. In January of 1880, Thomas Edison filed a patent for a new type of device that created light by passing an electric current through a carbon-coated filament contained in a sealed glass tube. In other words, Edison invented the light bulb. (Students of history will tell you that Edison didn't really invent the light bulb; he just improved on previous ideas. But that's not the point of the story.)

Edison's light bulb patent was approved, but he still had a lot of work to do before he could begin manufacturing a commercially viable light bulb. The biggest problem with his design was that the lamps dimmed the more you used them. This was because when the carbon-coated filament inside the bulb got hot, it shed little particles of carbon, which stuck to the inside of the glass. These particles resulted in a black coating on the inside of the bulb, which obstructed the light.

Edison and his team of engineers tried desperately to discover a way to prevent this shedding of carbon. One day, someone on his team noticed that the black carbon came off of just one end of the filament, not both ends. The team thought that maybe some type of electric charge was coming out of the filament. To test this theory, they introduced a third wire into the lamp to see if it could catch some of this electric charge.

It did. They soon discovered that an electric current flowed from the heated filament to this third wire, and that the hotter the filament got, the more electric current flowed. This discovery, which came to be known as *the Edison Effect*, marks the beginning of technology known as *electronics*. The device, which Edison patented on November 15, 1883, is the world's first electronic device.

When Edison patented his device in 1883, he had no idea what it would lead to. Now, nearly 140 years later, it's hard to imagine a world without electronics. Electronic devices are everywhere. There are more television sets in the United States than there are people. No one uses film to take pictures anymore; cameras have become electronic devices. And you rarely see a teenager anymore without headphones in their ears.

Without electronics, life would be very different.

Have you ever wondered what makes these electronic devices tick? In this chapter, I lay some important groundwork that will help the rest of this book make sense. I examine the bits and pieces that make up the most common types of electronic devices, and take a look at the basic concept that underlies all of electronics: electricity.

I promise I won't bore you too much with tedious or complicated physics concepts, but I must warn you from the start: In order to learn how electronics works at a level that will let you begin to design and build your own electronic devices, you need to have at least a basic idea of what electricity is. Not just what it does, but what it actually *is*. So put on your thinking cap and get started.

What Is Electricity?

Before you can understand even the simplest concepts of electronics, you must first understand what electricity is. After all, the whole purpose of electronics is to get electricity to do useful and interesting things.

The concept of electricity is both familiar and mysterious. We all know what electricity is or at least have a rough idea based on practical experience. In particular, consider these points:

>> We are very familiar with the electricity that flows through wires like water flows through a pipe. That electricity comes from power plants that burn coal, catch the wind, absorb sunlight, or harness nuclear reactions. It travels from the power plants to our houses in big cables hung high in the air or buried in

the ground. Once it gets to our houses, it travels through wires through the walls until it gets to electrical outlets. From there, we plug in power cords to get the electricity into the electrical devices we depend on every day, such as ovens and toasters and vacuum cleaners.

» We know, because the electric company bills us for it every month, that electricity isn't free. If we don't pay the bill, the electric company turns off our electricity. Thus, we know that electricity is valuable.

» We know that electricity can be stored in batteries, which contain a limited amount of electricity that can be used up. When the batteries die, all their electricity is gone.

» We know that some kinds of batteries, like the ones in our cellphones, are *rechargeable*, which means that when they've been drained of all their electricity, more electricity can be put back into them by plugging them into a charger, which transfers electricity from an electrical outlet into the battery. Rechargeable batteries can be filled and drained over and over again, but eventually they lose their ability to be recharged — and you have to replace them with new batteries. (Or, in the case of your iPhone, you have to buy a whole new phone.)

» We also know that electricity is the stuff that makes lightning strike in a thunderstorm. In grade school, we were taught that Ben Franklin discovered this by conducting an experiment involving a kite and a key, which we should not attempt to repeat at home.

» We know that electricity can be measured in *volts*. Household electricity is 120 volts (abbreviated 120 V). Flashlight batteries are 1.5 volts. Car batteries are 12 volts.

» We also know that electricity can be measured in *watts*. Traditional incandescent light bulbs are typically 60, 75, or 100 watts (abbreviated 100 W). Microwave ovens and hair dryers are 1,000 or 1,200 watts. The more watts, the brighter the light or the faster your pizza reheats and your hair dries.

» Except that we also know that some technologies do more work for a given wattage. Thus, a 20-watt CFL bulb and a 12-watt LED bulb both produce as much light as a 75-watt incandescent bulb.

» We also may know that there's a third way to measure electricity, called *amps*. A typical household electrical outlet is 15 amps (abbreviated 15 A).

» The truth is, most of us don't really know the difference between volts, watts, and amps. (Don't worry; by the time you finish Chapter 2 of this minibook, you will!)

>> We know that there's a special kind of electricity called *static electricity* that just sort of hangs around in the air but can be transferred to us by dragging our feet on a carpet, rubbing a balloon against our hairy arms, or forgetting to put an antistatic sheet in the dryer.

REMEMBER

>> And finally, we know that electricity can be very dangerous. In fact, dangerous enough that for almost 100 years electricity was used to administer the death penalty. Every year, hundreds of people die in the United States from accidental electrocutions.

But Really, What Is Electricity?

In the previous section, I list several ideas most of us have about electricity based on everyday experience. But the reality of electricity is something very different. Chapter 2 of this minibook is devoted to a deeper look at the nature of electricity, but for the purposes of this chapter, I want to start by introducing you to three very basic concepts of electricity: namely, *electric charge*, *electric current*, and *electric circuit*.

>> *Electric charge* refers to a fundamental property of matter that even physicists as smart as Neil deGrasse Tyson don't totally understand. Suffice it to say that two of the tiny particles that make up atoms — protons and electrons — are the bearers of electric charge. There are two types of charge: *positive* and *negative*. Protons have positive charge, electrons have negative charge.

Electric charge is one of the basic forces of nature that hold the universe together. Positive and negative charges are irresistibly attracted to each other. Thus, the attraction of negatively charged electrons to positively charged protons hold atoms together.

If an atom has the same number of protons as it has electrons, the positive charge of the protons balances out the negative charge of the electrons, and the atom itself has no overall charge.

However, if an atom loses one of its electrons, the atom will have an extra proton, which gives the atom a net positive charge. When an atom has a net positive charge, it goes looking for an electron to restore its balanced charge.

Similarly, if an atom somehow picks up an extra electron, the atom has a net negative charge. When this happens, the atom goes looking for a way to get rid of the extra electron to once again restore balance.

TECHNICAL STUFF

Okay, technically, atoms don't really go "looking" for anything. They don't have eyes, and they don't have minds that are troubled when they're short an electron or have a few too many. However, the natural attraction of negative to positive charges causes atoms that are short an electron to be attracted to atoms that are long an electron. When they find each other, something almost magic happens. . . . The atom with the extra electron gives its electron to the atom that's missing an electron. Thus, the charge represented by the electron moves from one atom to another, which brings us to the second important concept . . .

>> *Electric current* refers to the flow of the electric charge carried by electrons as they jump from atom to atom. Electric current is a very familiar concept: When you turn on a light switch, electric current flows from the switch through the wire to the light, and the room is instantly illuminated.

Electric current flows more easily in some types of atoms than in others. Atoms that let current flow easily are called *conductors*, whereas atoms that don't let current flow easily are called *insulators*.

Electrical wires are made of both conductors and insulators, as illustrated in Figure 1-1. Inside the wire is a conductor, such as copper or aluminum. The conductor provides a channel for the electric current to flow through. Surrounding the conductor is an outer layer of insulator, such as plastic or rubber.

The insulator serves two purposes. First, it prevents you from touching the wire when current is flowing, thus preventing you from being the recipient of a nasty shock. But just as importantly, the insulator prevents the conductor inside the wire from touching the conductor inside a nearby wire. If the conductors were allowed to touch, the result would be a *short circuit*, which brings us to the third important concept . . .

>> An *electric circuit* is a closed loop made of conductors and other electrical elements through which electric current can flow. For example, Figure 1-2 shows a very simple electrical circuit that consists of three elements: a battery, a lamp, and an electrical wire that connects the two.

The circuit shown in Figure 1-2 is, as I already said, very simple. Circuits can get much more complex, consisting of dozens, hundreds, or even thousands or millions of separate components, all connected with conductors in precisely orchestrated ways so that each component can do its bit to contribute to the overall purpose of the circuit. But all circuits must obey the basic principle of a closed loop.

REMEMBER

All circuits must create a closed loop that provides a complete path from the source of voltage (in this case, the battery) through the various components that make up the circuit (in this case, the lamp) and back to the source (again, the battery).

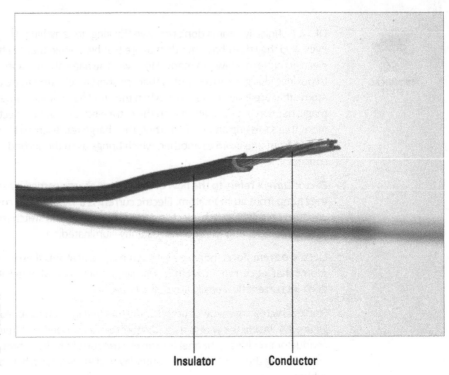

FIGURE 1-1:
An insulated
wire consists
of a conductor
surrounded by an
insulator.

Insulator   Conductor

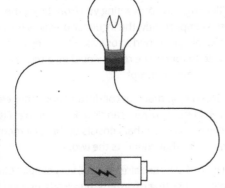

FIGURE 1-2:
A simple
electrical circuit
consisting of a
battery, a lamp,
and some wire.

What Is Electronics?

One of the reasons I started this chapter with the history lesson about Thomas Edison was to point out that when the whole field of electronics was invented in 1883, electrical devices had already been around for at least 100 years. For example:

>> Benjamin Franklin was flying kites in thunderstorms more than 100 years before.

>> The first electric batteries were invented by a fellow named Alessandro Volta in 1800. Volta's contribution is so important that the common term *volt* is named for him. (There is some archeological evidence that the ancient Parthian Empire may have invented the electric battery in the second century BC, but if so we don't know what they used their batteries for, and their invention was forgotten for 2,000 years.)

>> The electric telegraph was invented in the 1830s and popularized in America by Samuel Morse, who invented the famous Morse code used to encode the alphabet and numerals into a series of short and long clicks that could be transmitted via telegraph. In 1866, a telegraph cable was laid across the Atlantic Ocean, allowing instantaneous communication between the United States and Europe.

>> Contrary to popular belief, Benjamin Franklin wasn't the first to fly a kite in a thunderstorm. In 1752, he published a paper outlining his idea. Then he let a few other people try it first. After they survived, he tried the experiment himself and wound up getting all the credit. Benjamin Franklin was not only very *smart*; he was also very *wise*.

All of these devices and many other common devices still in use today, such as light bulbs, vacuum cleaners, and toasters, are known as *electrical devices*. So what exactly is the difference between electrical devices and *electronic devices*?

The answer lies in how devices manipulate electricity to do their work. Electrical devices take the energy of electric current and transform it in simple ways into some other form of energy — most often light, heat, or motion. For example, light bulbs turn electrical energy into light so that you can stay up late at night reading this book. The heating elements in a toaster turn electrical energy into heat so that you can burn your toast. And the motor in your vacuum cleaner turns electrical energy into motion that drives a pump that sucks the burnt toast crumbs out of your carpet.

In contrast, electronic devices do much more. Instead of just converting electrical energy into heat, light, or motion, electronic devices are designed to manipulate the electrical current itself to coax it into doing interesting and useful things.

That very first electronic device invented in 1883 by Thomas Edison manipulated the electric current passing through a light bulb in a way that let Edison create a device that could monitor the voltage being provided to an electrical circuit and automatically increase or decrease the voltage if it became too low or too high.

TIP

Don't worry if you aren't certain what the term *voltage* means at this point. You learn about voltage in Chapter 2 of this minibook.

One of the most common things that electronic devices do is manipulate electric current in a way that adds meaningful information to the current. For example, audio electronic devices add sound information to an electric current so that you can listen to music or talk on a cellphone. And video devices add images to an electric current so you can watch great movies like *Office Space*, *Ferris Bueller's Day Off*, or *The Princess Bride* over and over again until you know every line by heart.

Keep in mind that the distinction between electric and electronic devices is a bit blurry. What used to be simple electrical devices now often include some electronic components in them. For example, your toaster may contain an electronic thermostat that attempts to keep the heat at just the right temperature to make perfect toast. (It will probably still burn your toast, but at least it tries not to.) And even the most complicated electronic devices have simple electrical components in them. For example, although your TV remote control is a pretty complicated little electronic device, it contains batteries, which are simple electrical devices.

What Can You Do with Electronics?

The amazing thing about electronics is that it's being used today to do things that weren't even imaginable just a few years ago. And of course, that means that in just a few years we'll have electronic devices that haven't even been thought up yet.

That being said, the following sections give a very brief overview of some of the basic things you can do with electronics.

Making noise

One of the most common applications for electronics is making noise. Often in the form of music, though the distinction between *noise* and *music* is often debatable. Electronic devices that make noise are often referred to as *audio devices*. These devices convert sound waves to electrical current, and then store, amplify, and otherwise manipulate the current, and eventually convert the current back to sound waves you can hear.

Most audio devices have these three parts:

>> A *source*, which is the input into the system. The source can be a microphone, which is a device that converts sound waves into an electrical signal. The

subtle fluctuations in the sound waves are translated into subtle fluctuations in the electrical signal. Thus, the electrical signal that comes from the source contains audio information.

The source may also be a recorded form of the sound, such as sound recorded on a CD or in an MP3.

>> An *amplifier*, which converts the small electrical signal that comes from the source into a much larger electrical signal that, when sent to the speaker or headphones, can be heard.

Some amplifiers are small, as they need to boost the signal only enough to be heard by a single listener wearing headphones. Other amplifiers are huge, as they need to boost the signal enough so that 80,000 people can hear, for example, a famous singer forget the words to "The Star Spangled Banner."

>> *Speakers*, which convert electrical current into sound you can hear. Speakers may be huge, or they may be small enough to fit in your ear.

Making light

Another common use of electronics is to produce light. The simplest electronic light circuits are LEDs, which are the electronic equivalent of a light bulb.

TECHNICAL STUFF

LED stands for *light-emitting diode*, but that won't be on the test, at least not for this chapter. However, it will definitely be on the test for Book 2, Chapter 5, where you learn how to work with LEDs.

Video electronic devices are designed to create not just simple points of light but complete images that you can look at. The most obvious examples are television sets, which can provide hours and hours of entertainment and ask for so little in return — just a few of your brain cells.

Some types of electronic devices work with light that you can't see. The most common are TV remote controls, which send infrared light to your television set whenever you push a button. (That is, assuming you can find the remote control.) The electronics inside the remote control manipulate the infrared light in a way that sends information from the remote control to the TV, telling it to turn up the volume, change channels, or turn off the power. (You learn how to work with infrared devices in Book 4, Chapter 4.)

Transmitting to the world

Radio refers to the transmission of information without wires. Originally, radio was used as a wireless form of telegraph, broadcasting nothing more than audible clicks. Next, radio was used to transmit sound. In fact, to this day the term *radio* is usually associated with audio-only transmissions, either in the form of music or the ever-popular talk radio. However, the transmission of video information — in other words, television — is also a form of radio, as are wireless networking, cordless phones, and cellphones.

You learn much more about radio electronics in Book 4, Chapter 3.

Computing

One of the most important applications of electronics in the last 50 years has been the development of computer technology. In just a few short decades, computers have gone from simple calculating machines to machines that can beat humans at games we once thought humans were the masters of, such as chess, *Jeopardy!*, and Go.

Computers are the most advanced form of a whole field of electronics known as *digital electronics*, which is concerned with manipulating data in the binary language of zeros and ones. You learn plenty about digital electronics in Books 5 through 8.

Looking inside Electronic Devices

Have you ever taken apart an electronic device that no longer works, like an old clock radio or VHS tape player, just to see what it looks like on the inside?

I just took some hefty server computers to an e-waste recycler. You can bet that before I did, I opened them up to see what they looked like on the inside. And I removed a couple of the more interesting pieces to keep on my shelf. (Call me weird if you want. Some people collect teacups, some people collect spoons from around the world, and some people collect shot glasses. I collect old computer CPUs.)

Inside most electronic devices, you'll find a *circuit board* (or circuit *card*; it's all the same), which is a flat, thin board that has electronic gizmos mounted on it. In most cases, one side of the circuit board is populated with tiny devices that look like little buildings. These are the components that make up the electric circuit: the resistors, capacitors, diodes, transistors, and integrated circuits that do the work that the circuit is destined to do. The other side is painted with little lines of silver or copper that look like streets. These are the conductors that connect all the components so that they can work together.

An electronic circuit board looks like a little city! For example, have a look at the typical circuit board pictured in Figure 1-3. The top of the card is shown on the top; it has a variety of common electronic components. The underbelly of the circuit card is shown on the bottom; it has the typical silver streaks of conductors that connect the components topside so that they can perform useful work.

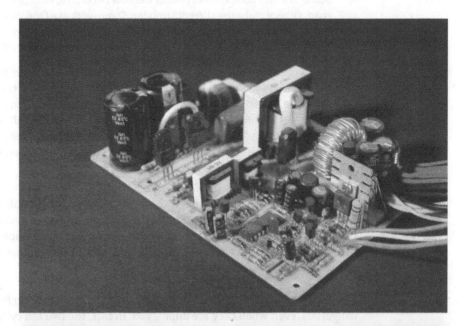

FIGURE 1-3:
A typical
electronic
circuit board.

Here's the essence of what's going on with these two sides of the circuit card:

>> The *component* side of the card — the side with the little buildings — holds a collection of electronic components whose sole purpose in life is to bend, turn, and twist electric current to get it to do interesting and useful things. Some of those components restrict the flow of current, like speed bumps on a road. Others make the current stronger. Some work like One-Way street signs that allow current to flow in only one direction. Still others try to smooth out any ripples or variations in the current, resulting in smoother traffic flow.

>> The *circuit* side of the card — the side with the roads — provides the conductive pathways for the electric current to flow from one component to the other in a certain order. The whole trick of designing and building electronic circuits is to connect all the components together in just the right way so that the current that flows out of one component is passed on to the next component. The circuit side of the board is what lets the components work together in a coordinated way.

WARNING

Okay, I couldn't even get through the first chapter of this book without having to give you the first of many warnings about the dangers of working with electronics. So here it goes: Do not under any circumstances plunge carelessly into the disassembly of old electronic circuits until you're certain that you know what you're doing.

The little components on a circuit card, such as the one shown in Figure 1-3 can be dangerous, even when they are unplugged. In fact, the two tall cylindrical components near the back edge of this circuit card are called *capacitors*. They can contain stored electrical energy that can deliver a powerful — even fatal — shock long after you've unplugged the power cord. Please see Chapter 4 of this minibook before you begin disassembling anything!

IN THIS CHAPTER

» Unraveling the deep mysteries of the matter and energy (well, only a little bit)

» Learning about three important aspects of electric circuits: current, voltage, and power

» Looking at the difference between direct and alternating current

» Learning your first electrical equation (don't worry; it's simple)

Chapter **2**

Understanding Electricity

Frankly, the title of this chapter is a bit ambitious. Before you can do much that's very interesting with electronics, you need to have a basic understanding of what electricity is and how it works, but unfortunately, understanding electricity is a tall order. Don't let this discourage or dissuade you: Even the smartest physicists in the world don't really understand it.

At the start of this chapter, you examine the very nature of electricity: what it is and what causes it. This first part of the chapter will remind you of a seventh- or eighth-grade science class, as you delve into the insides of atoms and learn about protons, neutrons, and electrons.

The second part of this chapter introduces you to three things you have to know about electricity if you want to design and build circuits: current, voltage, and power — the Manny, Moe, and Jack of electricity. Or if you prefer, the Huey, Dewey, and Louie, or the bacon, lettuce, and tomato, or the — you get the idea.

Pondering the Wonder of Electricity

The exact nature of electricity is one of the core mysteries of the universe. Although we don't really know exactly what electricity is, we do know a lot about what it does and how it behaves.

Strange as it may sound, your understanding of electricity will improve right away if you avoid using the term *electricity* to describe it. That's because the term *electricity* isn't very precise. We use the word *electricity* to refer to any of several different but related things. Each has a more precise name, such as *electric charge*, *electric current*, *electric energy*, *electric field*, and so on. All these things are commonly called *electricity*.

Electricity isn't so much a specific thing, but a phenomenon that has many different faces. So to avoid confusion, I try to avoid the word *electricity* in the rest of this book. Instead, I use more precise terms such as *charge* or *current*.

I really hate to use the word *phenomenon* here because it sounds so scientific. I feel like I should wear a bow tie whenever I say the word *phenomenon*. I consulted my thesaurus to see if there was a simpler word I could use instead. None of the suggestions really seemed to fit, but the one that came closest was *wonder*.

WHERE DID THE WORD *ELECTRICITY* COME FROM?

Do you remember the movie *Jurassic Park*, where scientists discovered the DNA of dinosaurs locked inside bits of amber? Amber is fossilized tree resin, and it played a key role in the history of our knowledge about electricity.

Since the days of the ancient Greeks, people have known that if you rubbed sticks of amber with fur, the amber could then be used to raise the hair on your head and that lightweight objects like feathers would stick to it. They had no idea why this happened, but they knew that it did happen.

The Greek word for amber is *elektron*. The Latin version of the word was *electricus*.

At the beginning of the seventeenth century, an English scientist named William Gilbert began to study electricity. He used these ancient words to describe the phenomena he was investigating, including the Latin term *electricus*. The influence of Gilbert's book, which was written in Latin, led to the word *electricity* in the English language.

Wonder isn't a bad substitute. When it comes down to it, the phenomenon we call electricity is pretty amazing. It really does qualify as one of the great wonders of the universe.

Remember the so-called "Seven Wonders of the Ancient World," which included the Great Pyramid of Giza, the Hanging Gardens of Babylon, the Temple of Artemis at Ephesus, the Statue of Zeus at Olympia, the Mausoleum of Halicarnassus, the Colossus of Rhodes, and the Lighthouse of Alexandria? If we were to make a list called "The Seven Wonders of the Universe," I suppose it would have to include Matter, Gravity, Time, Light, Life, Pizza, and Electricity.

Looking for Electricity

One of the most amazing things about electricity is that it is, literally, everywhere. By that I don't mean that electricity is commonplace or plentiful, or even that the universe has an abundant supply of electricity. Instead, what I mean is that electricity is a fundamental part of everything.

To get an idea of what I mean, consider a common misconception about electric current. Most of us think that wires carry electricity from place to place. When we plug in a vacuum cleaner and turn on the switch, we believe that electricity enters the vacuum cleaner's power cord at the electrical outlet, travels through the wire to the vacuum cleaner, and then turns the motor to make the vacuum cleaner suck up dirt and grime and dog hair. But that's not the case. The truth is that the electricity was already in the wire. The electricity is always in the wire, even when the vacuum cleaner is turned off or the power cord isn't plugged in. That's because electricity is a fundamental part of the copper atoms that make up the wire inside the power cord. Electricity is also a fundamental part of the atoms that make up the rubber insulation that protects you from being electrocuted when you touch the power cord. And it's a fundamental part of the atoms that make up the tips of your finger which the rubber keeps from touching the wires.

In short, electricity is a fundamental part of the atoms that make up all matter. So, to understand what electricity is, we must first look at atoms.

Peering Inside Atoms

As you probably learned in grade school, all matter is made up of unbelievably tiny bits that are called *atoms*. They're so tiny that the period at the end of this sentence contains several trillion of them.

It's hard for us to comprehend numbers as large as trillions. For the sake of comparison, suppose you could enlarge the period at the end of this sentence until it was about the size of Texas. Then, each atom would be about the size of — you guessed it — the period at the end of this sentence.

The word *atom* comes from an ancient Greek fellow named Democritus. Contrary to what you might expect, the word *atom* doesn't mean "really small." Rather, it means "undividable." Atoms are the smallest part of matter that can't be divided without changing it to a different kind of matter. In other words, if you divide an atom of a particular element, the resulting pieces are no longer the same thing.

For example, suppose you have a handful of some basic element such as copper and you cut it in half. You now have two pieces of copper. Toss one of them aside, and cut the other one in half. Again, you have two pieces of copper. You can keep doing this, dividing your piece of copper into ever smaller halves. But eventually, you'll get to the point where your piece of copper consists of just a single copper atom.

If you try to cut that single atom of copper in half, the resulting pieces will not be copper. Instead, you'll have a collection of the basic particles that make up atoms. There are three such particles, called neutrons, protons, and electrons.

The neutrons and protons in each atom are clumped together in the middle of the atom, in what is called the *nucleus*. The electrons spin around the outside of the atom.

When I first learned about atoms as a kid, I was taught that the electrons orbit around the nucleus much like planets orbit around the sun in our solar system. Even today, kids are taught this. School children are still being taught to create models of atoms using Styrofoam balls and wires, like the one shown in Figure 2-1.

That turns out to be a really bad analogy. Instead, the electrons whiz around the nucleus in a cloud that's called, appropriately enough, the *electron cloud*. Electron clouds have weird shapes and properties, and strangely enough, it's next to impossible to figure out exactly where in its cloud an electron actually is at any given moment.

FIGURE 2-1:
A common model
of an atom.

Examining the Elements

Several times in this chapter, I use the term *element* without explaining it. So here's the deal: An element is a specific type of atom, defined by the number of protons in its nucleus. For example, hydrogen atoms have just one proton in the nucleus, an atom with two protons in the nucleus is helium, atoms with three protons are called lithium, and so on.

The number of protons in the nucleus of an atom is called the *atomic number*. Thus, the atomic number of hydrogen is 1, the atomic number of helium is 2, lithium is 3, and so on. Copper — an element that plays an important role in electronics — is atomic number 29. Thus, it has 29 protons in its nucleus.

What about neutrons, the other particle found in the nucleus of an atom? Neutrons are extremely important to chemists and physicists. But they don't really play that big of a role in the way electric current works, so we can safely ignore them in this chapter. Suffice it to say that in addition to protons, the nucleus of each atom (except hydrogen) contains neutrons. In most cases, there are a few more neutrons than protons.

The third particle that makes up atoms is the electron. Electrons are what we're most interested in when we work with electricity because they are the source of electric current. They're unbelievably small; a single electron is about 200,000 times smaller than a proton. To gain some perspective on that, if a single electron were the size of the period at the end of this sentence, a proton would be about the size of a football field.

REMEMBER

Atoms usually have the same number of electrons as protons, and thus an atom of the element copper has 29 protons in a nucleus that is orbited by 29 electrons. When an atom picks up an extra electron or finds itself short of an electron, things get interesting because of a special property of protons and electrons called *charge*, which I explain in the next section.

Minding Your Charges

Two of the three particles that make up atoms — electrons and protons — have a very interesting characteristic called *electric charge*. Charge can be one of two *polarities*: negative or positive. Electrons have a *negative* polarity, while protons have a *positive* polarity.

The most important thing to know about charge is that opposite charges attract and similar charges repel. Negative attracts positive and positive attracts negative, but negative repels negative and positive repels positive.

As a result, electrons and protons are attracted to each other, but electrons repel other electrons and protons repel other protons.

The attraction between protons and electrons is what holds the electrons and the protons of an atom together. This attraction causes the electrons to stay in their orbits around the protons in the nucleus.

Here are a few more enlightening details about charge:

» Charge is a property of one of the fundamental forces of nature known as *electromagnetism*. The other three forces are *gravity*, the *strong force*, and the *weak force*.

» As I say in the previous section, an atom normally has the same number of electrons as protons. This is because the electromagnetic force causes each proton to attract exactly one electron. When the number of protons and electrons is equal, the atom itself has no net charge. It is then said to be *neutral*.

However, it's possible for an atom to pick up an extra electron. When it does, the atom has a net negative charge because of the extra electron. It's also possible for an atom to lose an electron, which causes the atom to have a net positive charge because it has more protons than electrons.

TECHNICAL
STUFF

>> If you've been paying attention, you may have wondered how it can be that the nucleus of an atom can stay together if it consists of two or more protons that have positive charges. After all, don't like charges repel? Yes they do, but the electrical repellent force is overcome by a much more powerful force called, for lack of a better term, the *strong force*. Thus, the strong force holds protons (and neutrons) together in spite of the protons' natural tendency to avoid each other.

>> The strong force doesn't affect electrons, so you never see electrons clumped together the way protons do in the nucleus of an atom. The electrons in an atom stay well away from each other.

>> If one were so inclined, one might liken the strong force to the patriotic force that binds the citizens of a nation together in spite of their differences. It's this force that keeps a country together in spite of the fact that its political parties seem to hate each other. Let's hope the strong force remains strong.

Conductors and Insulators

Some elements don't hold on to their outermost electrons very tightly. These elements frequently lose electrons or pick up extra electrons, and so they frequently get bumped off of neutral and become either negatively or positively charged. Such elements are called *conductors*. The best conductors are the metals silver, copper, and aluminum.

Other elements hold on to their electrons tightly. In these elements, it's hard to pry loose an electron or force another electron in. These elements almost always stay neutral. They're called *insulators*.

In a conductor, electrons are constantly skipping around between nearby atoms. An electron jumps out of one atom — call it Atom A — into a nearby atom, which I'll call Atom B. This creates a net positive charge in Atom A and a net negative charge in Atom B. But almost immediately, an electron will jump out of another nearby atom – call it Atom C — into Atom A. Thus, Atom A again becomes neutral, and now Atom C is negative.

This skipping around of electrons in a conductor happens constantly. Atoms are in perpetual turmoil, giving and receiving electrons and constantly cycling their net charges from positive to neutral to negative and back to positive.

Ordinarily, this movement of electrons is completely random. One electron might jump left, but another one jumps right. One goes up, another goes down. One goes east, the other goes west. The net effect is that although all the electrons are moving, collectively they aren't going anywhere. They're like Keystone Kops, running around aimlessly in every direction, bumping into each other, falling down, picking themselves back up, and then running around some more. When this randomness stops and the Keystone Kops get organized, the result is electric current, as explained in the next section.

Understanding Current

Electric current is what happens when the random exchange of electrons that occurs constantly in a conductor becomes organized and begins to move in the same direction.

When current flows through a conductor such as a copper wire, all those electrons that were previously moving about randomly get together and start moving in the same direction. A very interesting effect then happens: The electrons transfer their electromagnetic force through the wire almost instantaneously. The electrons themselves all move relatively slowly — on the order of a few millimeters a second. But as each electron leaves an atom and joins another atom, that second atom immediately loses an electron to a third atom, which immediately loses an electron to the fourth atom, and so on trillions upon trillions of times.

The result is that even though the individual electrons move slowly, the current itself moves at nearly the speed of light. Thus, when you flip a light switch, the light turns on almost immediately, no matter how much distance separates the light switch from the light.

Here are a few additional points that may help you understand the nature of current:

>> One way to illustrate this principle is to line up 15 balls on a pool table in a perfectly straight line, as shown in Figure 2-2. If you hit the cue ball on one end of the line, the ball on the opposite end of the line will almost immediately move. The other balls will move a little, but not much (assuming you line them up straight and strike the cue ball straight).

FIGURE 2-2:
Electrons transfer
current through a
wire much like a
row of pool balls
transfers motion.

This is similar to what happens with electric current. Although each electron moves slowly, the ripple effect as each atom loses and gains an electron is lightning fast (literally!).

**TECHNICAL
STUFF**

>> It's no coincidence that moving water is also called *current*. Many of the early scientists who explored the nature of electricity believed that electricity was a type of fluid, and that it flowed in wires in much the same way that water flows in a river.

>> The strength of an electric current is measured with a unit called the *ampere*, sometimes used in the short form *amp* or abbreviated *A*. The ampere is nothing more than a measurement of how many charge carriers (in most cases, electrons) flow past a certain point in one second. One ampere is equal to 6,240,000,000,000,000,000 electrons per second. That's 6,240 quadrillion electrons per second. (That's a huge number, but remember that electrons are incredibly small. To give it some perspective, though, imagine that each electron weighed the same as an average grain of sand. If that were true, one amp of current would be equivalent to the movement of nearly 350 tons of sand per second.)

>> Most electric incandescent light bulbs have about one amp of current flowing through them when they are turned on. A hair dryer uses about 12 amps.

>> Current in electronic circuits is usually much smaller than current in electrical devices like light bulbs and hair dryers. The current in an electronic circuit is often measured in thousandths of amps, or *milliamps* (abbreviated *mA*).

>> Current is often represented by the letter *I* in electrical equations. The *I* stands for *intensity*.

Understanding Voltage

In its natural state, the electrons in a conductor such as copper freely move from atom to atom but in a completely random way. To get them to move together in one direction, all you have to do is give them a push. The technical term for this push is *electromotive force*, abbreviated *EMF*, or sometimes simply *E*. But you know it more commonly as *voltage*.

A voltage is nothing more than a difference in charge between two places. For example, suppose you have a small clump of metal whose atoms have an abundance of negatively charged atoms and another clump of metal whose atoms have an abundance of positively charged atoms. In other words, the first clump has too many electrons and the second clump has too few. A voltage exists between those two clumps. If you connect those two clumps with a conductor such as a copper wire, you create what is called a *circuit* through which electric current will flow.

This current continues to flow until all the extra negative charges on the negative side of the circuit have moved to the positive side. When that has happened, both sides of the circuit become electrically neutral and the current stops flowing.

Here are some additional points to ponder concerning voltage:

>> Whenever there's a difference in charge between two locations, there's a possibility that a current will flow between the two locations if those locations are connected by a conductor. Because of this possibility, the term *potential* is often used to describe voltage. Without voltage, there can be no current. Thus, voltage creates the potential for a current to flow.

>> If current can be compared to the flow of water through a hose, voltage can be compared to water pressure at the faucet. It's water pressure that causes the water to flow in the hose.

>> Voltage is measured using a unit called, naturally, the *volt*, usually abbreviated *V.* The voltage that's available in a standard electrical outlet in the United States is about 117 V. The voltage available in a flashlight battery is about 1.5 V. A car battery provides about 12 V.

>> You can find out how much voltage exists between two points by using a device known as a *voltmeter*, which has two wire *test leads* that you can touch to different points in a circuit to measure the voltage between those points. Figure 2-3 shows a typical voltmeter. (Actually, the meter shown in the figure is technically called a *multimeter* because it can measure things other than voltage. For more information about using a voltmeter, see Chapter 8 of this minibook.)

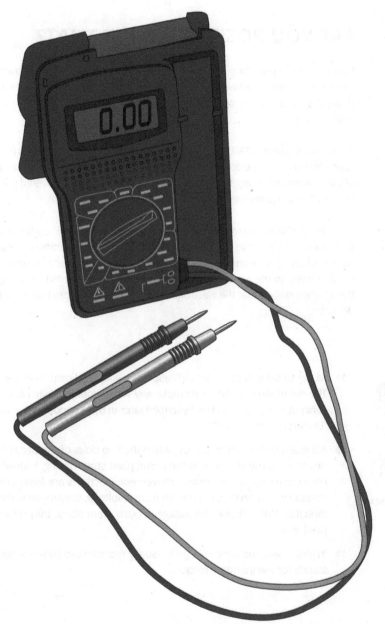

FIGURE 2-3:
You can use a
multimeter like
this to measure
voltage.

» Voltages can be considered positive or negative but only when compared with some reference point. For example, in a flashlight battery, the voltage at the positive terminal is +1.5 V relative to the negative terminal. The voltage at the negative terminal is –1.5 V relative to the positive terminal.

ARE YOU POSITIVE ABOUT THAT?

For the first 150 years or so of serious research into the nature of electricity, scientists had electric current backward: They thought that electric current was the flow of positive charges and that electric current flowed from the positive side of a circuit to the negative side.

It wasn't until around 1900 that scientists began to unravel the structure of atoms. They soon figured out that electrons have a negative charge, and current is actually the flow of these negatively charged electrons. In other words, they discovered that current flows in the opposite direction from what they had long thought.

Old ideas die hard, and to this day most people think of electric current as flowing from positive to negative. This concept of current flow is sometimes called *conventional current*. Modern electronic circuits are almost always described in terms of conventional current, so the assumption is that current flows from positive to negative, even though the reality is that the electrons in the circuit are actually flowing in the opposite direction.

TECHNICAL STUFF

REMEMBER

» I'd like to tell you the exact definition of a volt, but I can't — at least not yet. The definition of a volt won't make any sense until you learn about the concept of *power*, which is described later in this chapter in the section titled "Understanding Power."

» Although current stops flowing when the two sides of the circuit have been neutralized, the electrons in the circuit don't stop moving. Instead, they simply revert to their natural random movement. Electrons are always moving in a conductor. When they get a push from a voltage, they move in the same direction. When there's no voltage to push them along, they move about randomly.

» In electrical equations, voltage is usually represented by the letter E, which stands for *electromotive force*.

Comparing Direct and Alternating Current

An electric current that flows continuously in a single direction is called a *direct current*, or *DC*. The electrons in a wire carrying direct current move slowly, but eventually they travel from one end of the wire to the other because they keep plodding along in the same direction.

The voltage in a direct-current circuit must be constant, or at least relatively constant, to keep the current flowing in a single direction. Thus, the voltage provided by a flashlight battery remains steady at about 1.5 V. The positive end of the battery is always positive relative to the negative end, and the negative end of the battery is always negative relative to the positive end. This constancy is what pushes the electrons in a single direction.

Another common type of current is called *alternating current*, abbreviated *AC*. In an alternating-current circuit, voltage periodically reverses itself. When the voltage reverses, so does the direction of the current flow. In the most common form of alternating current, used in most power distribution systems throughout the world, the voltage reverses itself either 50 or 60 times per second, depending on the country. In the United States, the voltage is reversed 60 times per second.

Alternating current is used in nearly all the world's power distribution systems for the simple reason that AC current is much more efficient when it's transmitted through wires over long distances. All electric currents lose power when they flow for long distances, but AC circuits lose much less power than DC circuits.

The electrons in an AC circuit don't really move along with the current flow. Instead, they sort of sit and wiggle back and forth. They move one direction for 1/60th of a second and then turn around and go the other direction for 1/60th of a second. The net effect is that they don't really go anywhere.

For your further enlightenment, here are some additional interesting and useful facts concerning alternating current:

>> A popular toy called *Newton's Cradle* might help you understand how alternating current works. The toy consists of a series of metal balls hung by string from a frame, such that the balls are just touching each other in a straight line, as shown in Figure 2-4. If you pull the ball on one end of the line away from the other balls and then release it, that ball swings back to the line of balls, hits the one on the end, and instantly propels the ball on the other end of the line away from the group. This ball swings up for a bit, and then it turns around and swings back down to strike the group from the other end, which then pushes the first ball away from the group. This alternating motion, back and forth, continues for an amazingly long time if the toy is carefully constructed.

Alternating current works in much the same way. The electrons initially move in one direction but then reverse themselves and move in the other direction. The back and forth movement of the electrons in the circuit continues as long as the voltage continues to reverse itself.

FIGURE 2-4:
Newton's Cradle
works much
like alternating
current.

TIP

>> If you want to see a Newton's Cradle in action, go to YouTube and search for *Newton's Cradle*.

>> The reversal of voltage in a typical alternating current circuit isn't instantaneous. Instead, the voltage swings smoothly from one polarity to the other. Thus, the voltage in an AC circuit is constantly changing. It starts out at zero, then increases in the positive direction for a bit until it reaches its maximum positive voltage, and then it decreases until it gets back to zero. At that point, it increases in the negative direction until it reaches its maximum negative voltage, at which time it decreases again until it gets back to zero. Then the whole cycle repeats itself.

>> The fact that the amount of voltage in an AC circuit is always changing turns out to be incredibly useful. You learn why in Book 4, Chapter 1 when I give you a deeper look at alternating current.

Understanding Power

At the start of this chapter, I mention the three key concepts you need to know about electricity before you can start to work with your own circuits. The first two — current and voltage — are described earlier in this chapter. To recap, *current* is the organized flow of electric charges through a conductor, and *voltage* is the driving force that pushes electric charges to create current.

The third piece of the puzzle is called *power* (abbreviated P in equations). Simply put, power is the work done by an electric circuit. Electric current, in and of itself, isn't all that useful. It becomes useful only when the energy carried by an electric current is converted into some other form of energy, such as heat, light, sound, or radio waves. For example, in an incandescent light bulb, voltage pushes current through a *filament*, which converts the energy carried by the current into heat and light.

Power is measured in units called *watts* (abbreviated W). The definition of one watt is simple: One watt is the amount of work done by a circuit in which one ampere of current is driven by one volt.

This relationship lends itself to a simple equation. I promised myself when I started this book that I would use as few equations as possible, but I knew I'd have to include at least some of the basic equations. Fortunately, this one is pretty simple:

$$P = E \times I$$

In other words, *power* (P) equals *voltage* (E) times *current* (I).

TIP

To use the equation correctly, you must make sure that you measure power, voltage, and current using their standard units: watts, volts, and amperes. For example, suppose you have a light bulb connected to a 10-volt power supply, and one-tenth of an ampere is flowing through the light bulb. To calculate the wattage of the light bulb, you use the $P = E \times I$ formula like this:

$$P = 10\ V \times 0.1\ A = 1\ W$$

Thus, the light bulb is doing 1 watt of work.

Often, you know the voltage and the wattage of the circuit and you want to use those values to determine the amount of current flowing through the circuit. You can do that by turning the equation around, like this:

$$I = \frac{P}{E}$$

For example, if you want to determine how much current flows through a lamp with a 100-watt light bulb when it's plugged into a 117-volt electrical outlet, use the formula like this:

$$I = \frac{100 \text{ W}}{117 \text{ V}} = 0.855 \text{ A}$$

Thus, the current through the circuit is 0.855 amperes.

Here are some final thoughts concerning the concept of power:

>> The term *dissipate* is often used in association with power. As the energy carried by an electric current is converted into another form such as heat or light, the circuit is said to dissipate power.

>> Did you notice that current and voltage are represented by the letters *I* and *E*, not the letters *C* or *V* as you might expect, but power is represented by the letter *P*? Sometimes you wonder if the people who make the rules are just trying to confuse everyone.

Maybe the following table will help you keep things sorted out:

Concept	Abbreviation	Unit
Current	I	Amp (A)
Voltage	E or EMF	Volt (V)
Power	P	Watt (W)

>> Earlier in this chapter, in the section "Understanding Voltage," I say that I can't define "one volt" until you know what power is. Now that you know, you can see that the definition of a volt is simple: One volt is the amount of electromotive force (EMF) necessary to do one watt of work at one ampere of current.

>> Calculating the power dissipated by a circuit is often a very important part of circuit design. That's because electrical components such as resistors, transistors, capacitors, and integrated circuits all have maximum power ratings. For example, the most common type of resistor can dissipate at most ¼ watt. If you use a ¼-watt resistor in a circuit that dissipates more than ¼ watt of power, you run the risk of burning up the resistor.

WARNING

Chapter 3

Creating Your Mad-Scientist Lab

I loved to watch Frankenstein movies as a kid. My favorite scenes were always the ones where Dr. Frankenstein went into his laboratory. Those laboratories were filled with the most amazing and exotic electrical gadgets ever seen. The mad doctor's assistant, Igor, would throw a giant knife switch at just the right moment, and sparks flew, and the music rose to a crescendo, and the creature jerked to life, and the crazy doctor yelled, "It's ALIVE!"

The best Frankenstein movie ever made is still the original 1931 *Frankenstein*, directed by James Whale and starring Boris Karloff. The second-best Franken-stein movie ever made is the 1974 *Young Frankenstein*, directed by Mel Brooks and starring Gene Wilder. Both have great laboratory scenes.

In fact, did you know that the laboratory in *Young Frankenstein* uses the very same props that were used in the original 1931 classic? The genius who created those props was Kenneth Strickfaden, one of the pioneers of Hollywood special effects. Strickfaden kept the original Frankenstein props in his garage for decades. When Mel Brooks asked if he could borrow the props for *Young Frankenstein*, Strickfaden was happy to oblige.

You don't need an elaborate mad-scientist laboratory like the ones in the Frankenstein movies to build basic electronic circuits. However, you will need to build yourself a more modest workplace, and you'll need to equip it with a basic set of tools as well as some basic electronic components to work with.

However, no matter how modest your work area is, you can still call it your mad-scientist lab. After all, most of your friends will think you're a bit crazy and a bit of a genius when you start building your own electronic gadgets.

In this chapter, I introduce you to the stuff you need to acquire before you can start building electronic circuits. You don't have to buy everything all at once, of course. You can get started with just a simple collection of tools and a small space to work in. As you get more advanced in your electronic skills, you can acquire additional tools and equipment as your needs change.

Setting Up Your Mad-Scientist Lab

First, you must create a good place to work. You can build a fancy workbench in your garage or in a spare room, but if you don't have that much space, you can set up an ad-hoc mad-scientist lab just about anywhere. All you need is a place to set up a small workbench and a chair.

I do most of my electronics work in a spare room in my home, which also doubles as a display area for some of the Halloween props I've built over the years for my haunted house. Thus, as Figure 3-1 shows, my Mad-Scientist Lab really is a mad-scientist lab!

Here are the essential ingredients of any good work area for electronic tinkering:

>> **Adequate space:** You'll need to have adequate space for your work. When you're just getting started, your work area can be small — maybe just 2 or 3 feet in the corner of the garage. But as your skills progress, you'll need more space.

WARNING

It's very important that the location you choose for your work area is secure, especially if you have young children around. Your work area will be filled with perils — things that can cause shocks, burns, and cuts, as well as things that under no circumstances should be ingested. Little hands are incredibly curious, and children are prone to put anything they don't recognize in their mouths. So be sure to keep everything safely out of reach, ideally behind a locked door.

FIGURE 3-1:
My Mad-Scientist
Lab really is a
mad-scientist lab!

>> **Good lighting:** The ideal lighting should be overhead instead of from the side or behind you. If possible, purchase an inexpensive fluorescent shop light and hang it directly over your work area. If your chosen spot doesn't allow you to hang lights from overhead, the next best bet is a desk lamp that swings overhead, to bring light directly over your work.

>> **A solid workbench:** Initially, you can get by with something as simple as a card table or a small folding table. Eventually, though, you'll want something more permanent and substantial. You can make yourself an excellent workbench from an old door set atop a pair of old file cabinets, or you can hit the yard sales on a Saturday morning in search of an inexpensive but sturdy office desk.

TIP

If your only option for your workbench is your kitchen table, go to your local big box hardware store and buy a 24-inch square piece of ⅝-inch plywood. This will serve as a good solid work surface and protect your kitchen table until you can acquire a real workbench.

>> **Comfortable seating:** If your workbench is a folding table or desk, the best seating is a good office chair. However, many workbenches stand 4 to 6 inches taller than desk height. This allows you to work comfortably while standing. If your workbench is tall, you'll need to get a seat of the correct height. You can purchase a bench stool from a hardware store, or you can shop the yard sales for a cheap bar stool.

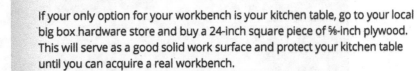

>> **Plenty of electricity:** You will obviously need a source of electricity nearby as you build electronic projects. A standard 15-amp electrical outlet will provide enough current capacity, but it probably won't provide enough electrical outlets for your needs. The easiest way to meet that need is to purchase several multi-outlet power strips and place them in convenient locations behind or on either side of your work area.

>> **Plenty of storage:** You'll need a place to store your tools, supplies, and components. The ideal storage for hand tools is a small sheet of pegboard mounted on the wall right behind your workbench. Then, you can use hooks to hang your tools within easy reach. For larger tools, such as a drill or saw, built-in cabinets are best.

For small parts, multicompartment storage boxes, shown in Figure 3-2, are best. I suggest you get one or two of these to store all the little components such as resistors, diodes, capacitors, transistors, and so on. If you get two boxes, get one that has a few larger compartments and another that has a greater number of smaller compartments.

TIP

It's also a good idea to keep a few small, shallow storage bins handy. These are especially useful for storing parts for the project you're working on. It helps to keep your parts together in a shallow bin rather than having them scattered loose all over your work area.

FIGURE 3-2:
Multicompartment
storage boxes
are ideal for
storing small
components.

Equipping Your Mad-Scientist Lab

Like any hobby, electronics has its own special tools and supplies. Fortunately, you don't need to run out and buy everything all at once. But the more involved you get with the hobby, the more you will want to invest in a wide variety of quality tools and supplies. The following sections outline some of the essential stuff you'll need at your disposal.

Basic hand tools

For starters, you'll need a basic set of hand tools, similar to the assortment shown in Figure 3-3. Specifically, you'll need these items:

>> **Screwdrivers:** Most electronic work is relatively small, so you don't need huge, heavy-duty screwdrivers. But you should get yourself a good assortment of small- and medium-sized screwdrivers, both flat-blade and Phillips head.

A set of *jeweler's screwdrivers* is sometimes very useful. The swiveling knob on the top of each one makes it easy to hold the screwdriver in a precise position while turning the blade.

FIGURE 3-3:
Basic hand
tools you'll
want to have.

>> **Pliers:** You will occasionally use standard flat-nosed pliers, but for most electronic work, you'll depend on *needle-nose pliers* instead, which are especially adept at working with wires — bending and twisting them, pushing them through holes, and so on. Most needle-nose pliers also have a cutting edge that lets you use them as wire cutters.

Get a small set of needle-nose pliers with thin jaws for working with small parts, and a larger set for bigger jobs.

>> **Wire cutters:** Although you can use needle-nose pliers to cut wire, you'll want a few different types of wire cutters at your disposal as well. Get something heavy-duty for cutting thick wire, and something smaller for cutting small wire or component leads.

>> **Wire strippers:** Figure 3-4 shows two pieces of wire that I *stripped* (removed the insulation from). I stripped the one on top with a set of wire cutters and the one on the bottom with a set of *wire strippers.* Notice the crimping in the one at the top, at the spot where the insulation ends? That was caused by using just a bit too much pressure on the wire cutters. That crimp has created a weak spot in the wire that may eventually break.

YOU GET WHAT YOU PAY FOR

When it comes to tools, the old mantra "you get what you pay for" is generally true. Good tools that are manufactured from the best materials and with the best quality fetch a premium price. Cheap tools are, well, cheap. The price range can be substantial. You can easily spend $20 or $25 on decent wire cutters, or you can buy cheap ones for $3 or $4.

There are two main drawbacks to cheap tools. First, they don't last. The business end of a cheap tool wears out very fast. Each time you cut a wire with a pair of $4 wire cutters, you ding the cutting blade a bit. Pretty soon, the cutters can barely cut through the wire. The second drawback of cheap tools is a consequence of the first: When tools wear out, they tend to damage the materials you use them on. For example, tightening a screw with a badly worn screwdriver can strip the screw. Likewise, attempting to loosen a tight nut with a worn-out wrench can strip the nut.

There are a few situations in which I would endorse spending money on cheap tools. One is as a way of getting started in this fascinating hobby as inexpensively as possible. You can always start with cheap tools, and then replace them one by one with more expensive tools as your experience, confidence, love of the hobby, and budget increases. Another good reason to buy cheap tools is if you're absentminded (like me) and tend to lose things. There's not much point in buying expensive tools if you're going to have to replace them every few months because you keep losing them!

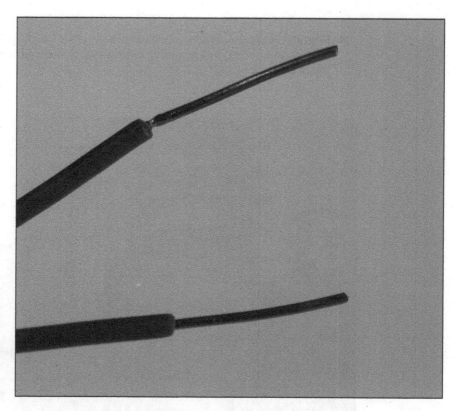

FIGURE 3-4:
The wire on the
top was stripped
with wire cutters;
the one on the
bottom was
stripped with a
wire stripper.

TIP

To avoid damaging your wires when you strip them, I suggest you purchase an inexpensive (under $10) wire stripping tool. You'll thank me later.

Magnifying glasses

One of the most helpful items you can have in your tool arsenal is a good magnifying glass. After all, electronic stuff is small. Resistors, diodes, and transistors are downright tiny.

Actually, I suggest you have at least three types of magnifying glasses on hand:

» A **handheld magnifying glass** to inspect solder joints, read the labels on small components, and so on.

» A **magnifying glass mounted on a base** so that you can hold your work behind the glass. The best mounted glasses combine a light with the magnifying glass, so the object you're magnifying is bright.

>> **Magnifying goggles,** which provide completely hands-free magnifying for delicate work. Ideally, the goggles should have lights mounted on them (see Figure 3-5).

FIGURE 3-5: The author modeling his favorite magnifying headgear.

Third hands and hobby vises

A *third hand* is a common tool amongst hobbyists. It's a small stand that has a couple of clips that you use to hold your work, thus freeing up your hands to do delicate work. Most third-hand tools also include a magnifying glass. Figure 3-6 shows an inexpensive third-hand tool holding a circuit card.

The most common use for a third hand in electronics is soldering. You use the clips to hold the parts you want to solder, positioned behind the magnifying glass so that you can get a good look.

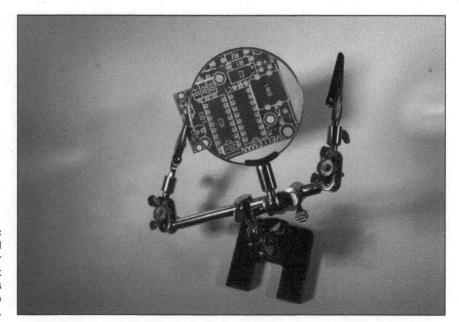

FIGURE 3-6:
A third hand
can hold your
stuff so that
both your hands
are free to do
the work.

TIP

Although the magnifying glass on the third hand is helpful, it does tend to get in the way of the work. It can be awkward to maneuver your soldering iron and solder behind the magnifying glass. For this reason, I often remove the magnifying glass from the third hand and use my favorite magnifying headgear instead.

The third hand is often helpful for assembling small projects, but it lacks the sturdiness required for larger projects. Eventually you'll want to invest in a small hobby vise such as the one shown in Figure 3-7. This one is made by PanaVise (www.panavise.com).

Here are a few things to look for in a hobby vise:

>> **Mount:** Get a vise that has a base with the proper type of workbench mount. There are three common types of mounts:

- *Bolt mount:* The base has holes through which you can pass bolts or screws to attach the vise to your workbench. This is the most stable type of mount, but it requires that you put holes in your workbench.

- *Clamp mount:* The base has a clamp that you can tighten to fix the base to the top and bottom of your workbench. Clamp mounts are pretty stable but can be placed only near the edge of your workbench.

- *Vacuum mount:* The base has a rubber seal and lever that you pull to create a vacuum between the seal and the workbench top. Vacuum mounts are the most portable but work well only when the top of your workbench is smooth.

FIGURE 3-7:
A hobby vise.

>> **Movement:** Get a vise that has plenty of movement so that you can swivel your work into a variety of different working positions. Make sure that when you lock the swivel mount into position, it stays put. You don't want your work sliding around while you are trying to solder on it.

>> **Protection:** Make sure the vise jaws have a rubber coating to protect your work.

Soldering iron

Soldering is one of the basic techniques used to assemble electronic circuits. The purpose of soldering is to make a permanent connection between two conductors — usually between two wires or between a wire and a conducting surface on a printed circuit board.

The basic technique of soldering is to physically connect the two pieces to be soldered, and then heat them with a soldering iron until they are hot enough to melt *solder* (a special metal made of lead and tin that has a low melting point), then apply the solder to the heated parts so that it melts and flows over the parts.

Once the solder has flowed over the two conductors, you remove the soldering iron. As the solder cools, it hardens and bonds the two conductors together.

You learn all about soldering in Chapter 7 of this minibook. For now, suffice it to say that you need three things for successful soldering:

>> **Soldering iron:** A little hand-held tool that heats up enough to melt solder. An inexpensive soldering iron from RadioShack or another electronics parts store is just fine to get started with. As you get more involved with electronics, you'll want to invest in a better soldering iron that has more precise temperature control and is internally grounded.

>> **Solder:** The soft metal that melts to form a bond between the conductors.

>> **Soldering iron stand:** To set your soldering iron on when you aren't solder-ing. Some soldering irons come with stands, but the cheapest ones don't. Figure 3-8 shows a soldering iron that comes with a stand. You can purchase this type of soldering iron from RadioShack for about $25.

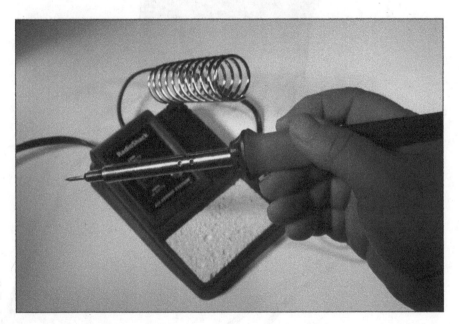

FIGURE 3-8:
A soldering iron with a stand.

Multimeter

In Chapter 2 of this minibook, you learn that you can measure voltage with a volt-meter. You can also use meters to measure many other quantities that are impor-tant in electronics. Besides voltage, the two most common measurements you'll need to make are current and resistance.

Rather than use three different meters to take these measurements, it's common to use a single instrument called a *multimeter*. Figure 3-9 shows a typical multimeter purchased from RadioShack for about $20.

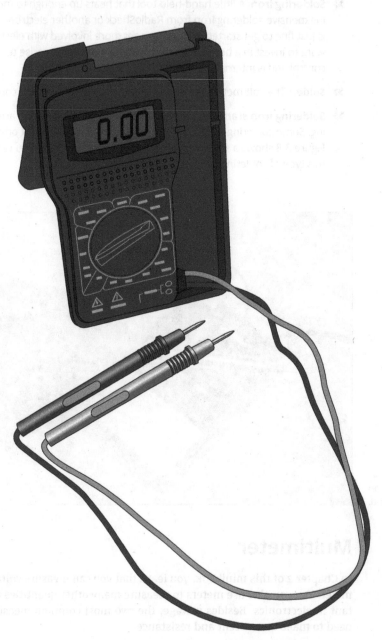

FIGURE 3-9:
An inexpensive
multimeter.

Solderless breadboard

A *solderless breadboard* — usually just called a *breadboard* — is a must for experimenting with circuit layouts. A breadboard is a board that has holes in which you can insert either wires or electronic components such as resistors, capacitors, transistors, and so on to create a complete electronic circuit without any soldering. When you're finished with the circuit, you can take it apart, and then reuse the breadboard and the wires and components to create a completely different circuit.

Figure 3-10 shows a typical breadboard, this one purchased from RadioShack for about $20. You can purchase less expensive breadboards that are smaller, but this one (a little bigger than 7 x 4 inches) is large enough for all the circuits presented in this book.

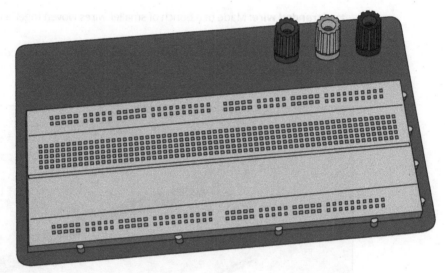

FIGURE 3-10: A solderless breadboard.

What makes breadboards so useful is that the holes in the board are actually solderless connectors that are internally connected to one another in a specific, well-understood pattern. Once you get the hang of working with a breadboard, you'll have no trouble understanding how it works.

Throughout the course of this book, I show you how to create dozens of different circuits on a breadboard. As a result, you'll want to invest in at least one. I suggest you get one similar to the one shown in Figure 3-10, plus one or two other, smaller breadboards. That way, you won't always have to take one circuit apart to build another.

You can learn more about working with solderless breadboards in Chapter 6 of this minibook.

Wire

One of the most important items to have on hand in your lab is *wire*, which is simply a length of a conductor, usually made out of copper but sometimes made of aluminum or some other metal. The conductor is usually covered with an outer layer of insulation. In most wire, the insulation is made of polyethylene, which is the same stuff used to make plastic bags.

Wire comes in these two basic types:

>> **Solid wire:** Made from a single piece of metal

>> **Stranded wire:** Made of a bunch of smaller wires woven together

Figure 3-11 shows both types of wire with the insulation stripped back so you can see the difference.

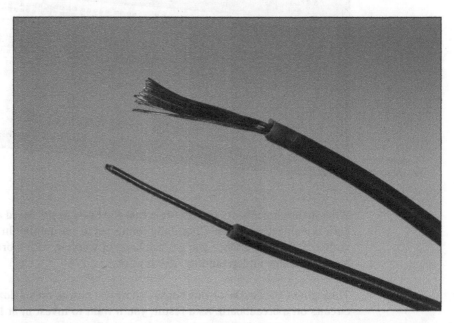

FIGURE 3-11:
Solid and
stranded wire.

For most purposes in this book, you'll want to work with solid wire because it's easier to insert into breadboard holes and other types of terminal connections. Solid wire is also easier to solder. When you try to solder stranded wire, inevitably one of the tiny strands gets separated from the rest of the strands, which can create the potential for a short circuit.

On the other hand, stranded wire is more flexible than solid wire. If you bend a solid wire enough times, you'll eventually break it. For this reason, wires that are frequently moved are usually stranded.

Wire comes in a variety of sizes, which are specified by the wire's *gauge*, and is generally coiled in or on the packaging. Strangely, the larger the gauge number, the smaller the wire. For most electronics projects, you'll want 20- or 22-gauge wire. You'll need to use large wires (usually 14 or 16 gauge) when working with household electrical power.

Finally, you may have noticed that the insulation around a wire comes in different colors. The color doesn't have any effect on how the wire performs, but it's common to use different colors to indicate the purpose of the wire. For example, in DC circuits it's common to use red wire for positive voltage connections and black wire for negative connections.

To get started, I suggest you purchase a variety of wires — at least four rolls: 20-gauge solid, 20-gauge stranded, 22-gauge solid, and 22-gauge stranded. If you can find an assortment of colors, all the better.

TIP

In addition to wires on rolls, you may also want to pick up *jumper wires*, which are precut, stripped, and bent for use with solderless breadboards. Figure 3-12 shows an assortment I bought at RadioShack for about $6.

Batteries

Don't forget the batteries! Most of the circuits covered in this book use either AA or 9-volt batteries, so you'll want to stock up.

If you want, you can use rechargeable batteries. They cost more initially, but you don't have to replace them when they lose their charge. If you use rechargeables, you'll also need a battery charger.

To connect the batteries to the circuits, you'll want to get several AA battery holders. Get one that holds two batteries and another that holds four. You should also get a couple of 9-volt battery clips. These holders and clips are shown in Figure 3-13.

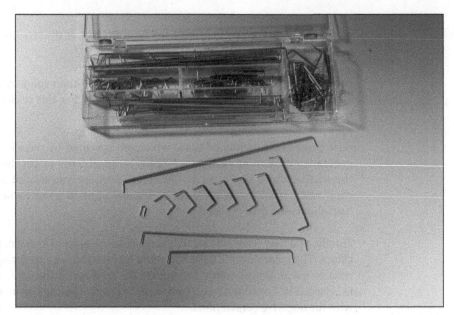

FIGURE 3-12:
Jumper wires
for working with
a solderless
breadboard.

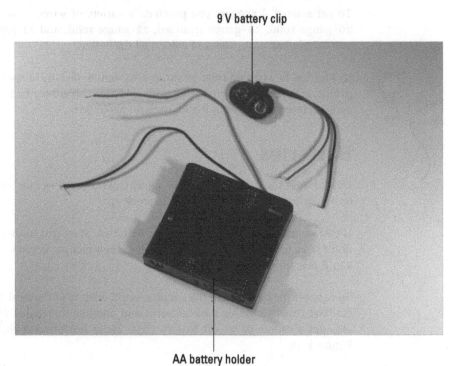

9 V battery clip

FIGURE 3-13:
Battery holders
will help deliver
power to your
circuits.

AA battery holder

Other things to stock up on

Besides all the stuff I've listed so far, here are a few other items you may need from time to time:

TIP

>> **Electrical tape:** Get a roll or two of plain black electrician's tape. You'll use it mostly to wrap around temporary connections to hold them together and keep them from shorting out.

Do yourself a favor, and don't waste time with cheap electrical tape. There's a huge difference in quality. In my opinion, far and away the best electrical tape is Scotch Super 33+ Vinyl Electrical Tape. It's about twice as expensive as bargain-basement tape, but you won't regret spending a few extra dollars.

>> **Compressed air:** A small can of compressed air can come in handy to blow dust off an old circuit board or component.

>> **Cable ties:** Also called *zip ties*, these little plastic ties are handy for temporarily (or permanently) holding wires and other things together.

>> **Jumper clips:** These are short (typically 12 or 18 inches) wires that have *alligator clips* attached on either end, as shown in Figure 3-14. You'll use them to make quick connections between components for testing purposes.

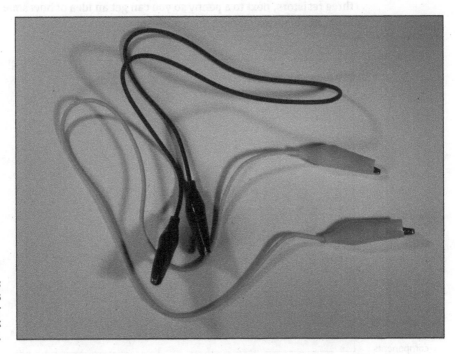

FIGURE 3-14:
Jumper clips are great for making quick connections.

Stocking up on Basic Electronic Components

Besides all the tools and supplies I've described so far in this chapter, you'll also need to gather a collection of inexpensive electronic components to get you started with your circuits. You don't have to buy everything all at once, but you'll want to gather at least the basic parts before you go much farther in this book.

Unfortunately, most communities don't have a store where you can buy these components in person. (If you're lucky enough to have a specialty electronics store in your community, purchase your parts from them to help these local businesses stay open!) If you don't have a local supplier, you can buy the parts online from Amazon (www.amazon.com), Jameco (www.jameco.com), RadioShack (www.radioshack.com), or any other electronic parts distributor.

Resistors

A *resistor* is a component that resists the flow of current. It's one of the most basic components used in electronic circuits; in fact, you won't find a single circuit anywhere in this book that doesn't have at least one resistor. Figure 3-15 shows three resistors, next to a penny so you can get an idea of how small they are. You'll learn all about resistors in Book 2, Chapter 2.

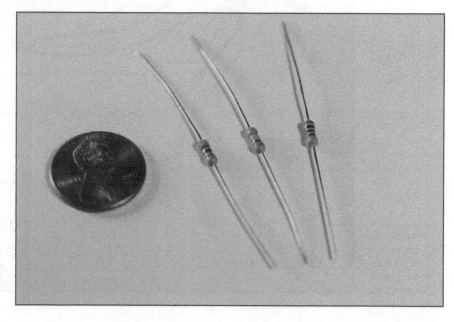

FIGURE 3-15:
Resistors are one of the most commonly used circuit components.

Resistors come in a variety of *resistance values* (how much they resist current, measured in units called ohms and designated by the symbol Ω) and *power ratings* (how much power they can handle without burning up, measured in watts).

All the circuits in this book can use resistors rated for one-half watt. You'll need a wide variety of resistance values. I recommend you buy at least 10 each of the following 12 resistances:

470 Ω	4.7 kΩ	47 kΩ	470 kΩ
1 kΩ	10 kΩ	100 kΩ	1 MΩ
2.2 kΩ	22 kΩ	33 kΩ	220 kΩ

TIP

You can save money by purchasing a package that contains a large assortment of resistors. For example, most suppliers sell a package that contains an assortment of 500 or so resistors — at least 10 of all values listed here, plus a few others, for about $15.

Capacitors

Next to resistors, capacitors are probably the second most commonly used component in electronic circuits. A *capacitor* is a device that can temporarily store an electric charge. You learn all about capacitors in Book 2, Chapter 3. Figure 3-16 shows some capacitors.

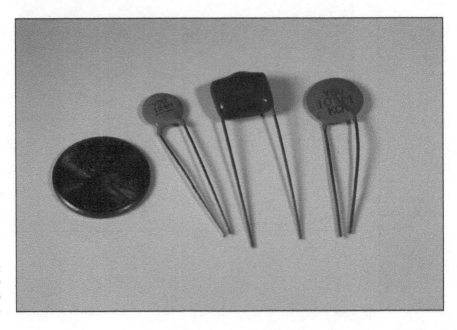

FIGURE 3-16:
Capacitors come in many shapes and sizes.

Capacitors come in several different varieties, the two most common being *ceramic disk* and *electrolytic*. The amount of capacitance of a given capacitor is usually measured in *microfarads*, abbreviated µF. As a starting assortment of capacitors, I suggest you get at least five each of the following capacitors:

>> **Ceramic disk:** 0.01 µF and 0.1 µF

>> **Electrolytic:** 1 µF, 10 µF, 100 µF, 220 µF, and 470 µF

TIP

As with resistors, you can purchase kits with a wide variety of assorted capacitor values. For about $15, you can get 500 or more capacitors in a range of useful values.

Diodes

A *diode* is a device that lets current flow in only one direction. Figure 3-17 shows an assortment of various types of diodes.

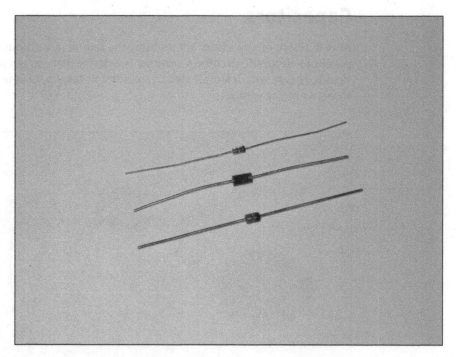

FIGURE 3-17:
An assortment
of diodes.

A diode has two terminals, called the *anode* and the *cathode*. Current will flow through the diode only when positive voltage is applied to the anode and negative voltage to the cathode. If these voltages are reversed, current will not flow.

You learn all about diodes in Book 2, Chapter 5. For now, I suggest you get at least five of the basic diodes known as the 1N4001 (the middle one in Figure 3-17). You should be able to find these at any electronics parts distributor.

Light-emitting diodes

A *light-emitting diode* (or *LED*) is a special type of diode that emits light when current passes through it. You learn about LEDs in Book 2, Chapter 5. Although there are many different types of LEDs available, I suggest you get started by purchasing at least five red diodes. Figure 3-18 shows a typical red LED.

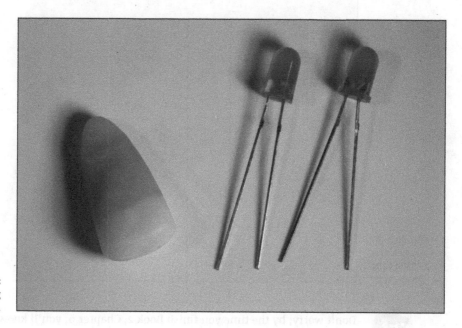

FIGURE 3-18:
Light-emitting
diodes.

Transistors

A *transistor* is a three-terminal device in which a voltage applied to one of the terminals (called the *base*) can control current that flows across the other two terminals (called the *collector* and the *emitter*). The transistor is one of the most important devices in electronics, and I cover it in detail in Book 2, Chapter 6. For now, you can just get a few simple 2N3904 NPN transistors, shown in Figure 3-19, to have on hand.

FIGURE 3-19: A look at a 2N3904 NPN transistor.

TECHNICAL STUFF

Don't worry; by the time you finish Book 2, Chapter 6, you'll know what the designation *2N3904 NPN* means.

Integrated circuits

An *integrated circuit* is a special component that contains an entire electronic circuit, complete with transistors, diodes, and other elements, all photographically etched onto a tiny piece of silicon. Integrated circuits are the building blocks of modern electronic devices such as computers and cellphones.

You learn how to work with some basic integrated circuits in Book 3. To get started, you'll want to pick up a few each of at least two different types of integrated circuits: a 555 timer and an LM741 op-amp. These chips are shown in Figure 3-20.

FIGURE 3-20: Two popular integrated circuits: A 555 timer and an LM741 op-amp.

One Last Thing

There is, of course, one last thing your mad-scientist lab will need to make it complete. That one last thing is a sign that properly warns your friends and family that you are indeed a mad scientist. You have my express permission to photocopy the sign, shown in Figure 3-21, and place it in a prominent spot near your workbench.

FIGURE 3-21:
Make sure your
friends and
family are
properly warned.

Chapter **4**

Staying Safe

When I was a kid, I helped a good friend named Barry who built a Tesla coil. By "helped," I mean that I hung out in his garage and watched while he meticulously wrapped thousands of turns of bare copper wire around a huge glass milk bottle, painted it with dozens of coats of lacquer, and polished the brass ball that attached to the very top of the coil. I'm quite certain he couldn't have done it without me.

When it was done, we plugged it in and marveled at what it could do. Sparks flew at random a foot or two into the air from the ball at the top of the coil. If you held a crowbar in one hand, you could draw a spark several feet from the ball to the crowbar. The current coming off the ball flew through the air and into the crowbar, and then passed through our bodies and into the ground. You could also light up a fluorescent light tube simply by holding the tube in your hand within a few feet of the coil.

To this day, I cannot believe Barry's parents let him build it. I know my parents wouldn't have let me build one. My mom was kind of like the mom in *A Christmas Story,* who wouldn't let her son Ralphie have a Red Ryder BB gun (the one with a compass in the stock and this thing that tells time) because "you'll shoot your eye out."

That was my mom. No Tesla coils for me. Too dangerous.

None of the electronic projects described in this book are anywhere near as dangerous as a Tesla coil. In fact, most of them pose no threat at all. That being said, it's important to remember that whenever you're working with electricity, you're working with something that's potentially very dangerous.

The possibility of electric shock is always present whenever you work with electricity, but there are other potential dangers as well. You probably won't shoot your eye out, but if you're not careful, you might start a fire or otherwise injure yourself or someone else.

The purpose of this chapter, then, is to keep you safe while you experiment with electronics. Please read it well, and please heed every bit of advice I give here.

Facing the Realities of Electrical Dangers

There's no escaping the simple fact that an electric shock, if strong enough, can kill you. So whenever you work with electricity, you must be sure to take every precaution you can to avoid being the recipient of a shock strong enough to do damage.

In the United States, somewhere between 500 and 1,000 people die every year from accidental electrocutions. Many of those are industrial or weather-related accidents in which people come into contact with downed power lines. But many of them are completely avoidable accidents that happen in the home. In the sections that follow, I give you specific guidelines for avoiding accidental electrocution.

Household electrical current can kill you!

Too many people are under the false impression that the 120 volts of alternating current running through household electrical wires isn't enough to kill. So let's start by getting one fact straight:

WARNING

The electricity in your home wiring system is more than strong enough to kill you.

You're exposed to household electrical current primarily in two places: in electrical outlets and in the lamp sockets within light fixtures. As a result, you should be extra careful whenever you plug or unplug something into or from an electrical outlet, and you should be careful whenever you change a light bulb. Specifically, you should follow these precautions:

>> **Never change a light bulb when the light is turned on.** If the light is controlled by a switch, turn the switch off. If the light isn't controlled by a switch, unplug the light from the wall outlet.

>> **If an extension cord becomes frayed or damaged in any way, discard it.** When the insulation begins to rub off of an extension cord, the shock hazard is very real.

>> **Never perform electrical wiring work while the circuit is energized.** If you insist on changing your own light switches or electrical outlets, *always* turn off the power to the circuit by turning off the circuit breaker that controls the circuit before you begin. Many people die every year because they think that they can be careful enough to safely work with live power.

IS IT TRUE THAT CURRENT, NOT VOLTAGE, KILLS?

There is an old adage that "it's the current that kills, not the voltage." Although this statement may be technically true, it's also dangerously misleading. In fact, it stems from a fundamental misunderstanding of what current and voltage are. It can cause you to take dangerous risks if you don't understand the relationship between current and voltage.

The danger from electric shock occurs as current passes through vital parts of your body — specifically, your heart. It takes only a few milliamperes of current to stop your heart. At somewhere around 10 mA, muscles seize up, making it impossible to let go if you're holding a live wire. At around 15 mA, the muscles in your chest can seize up, making it impossible to breath. And at around 60 mA, your heart can stop. It takes only a few moments of exposure for these effects to occur.

So yes, it is current passing through your body that can kill you.

But current is inseparable from voltage. Current can't happen without voltage, and all other things being equal, the greater the voltage, the greater the current. As a result, it's very difficult to receive a lethal shock from three volts even if you're dripping wet and standing on bare concrete. But under those conditions, 30 volts may be enough to create a painful and damaging shock.

Saying "it's the current that kills, not the voltage" is kind of like saying "it's lack of oxygen, not water" that causes drowning. Although it may be technically true, isn't it the water that causes the lack of oxygen?

>> **Never work on an AC-powered appliance when it has power applied.** Simply turning the appliance off isn't enough to be safe. If the appliance has a power cord, unplug it before you work on it. If it doesn't have a power cord, turn off the power to the appliance by throwing the circuit breaker on your home's electrical panel.

>> **Take extra precautions when you're working with your own AC circuits.** In Book 4, I tell you more about working with AC circuits; I say more about AC safety then.

Even relatively small voltages can hurt you

Most of the projects in this book work with AA batteries, usually two or four of them tied together to produce a total of three or six volts. That's not enough voltage to do serious harm. Even if you do get a shock with three or six volts, you will probably barely feel it.

However, it's possible to injure yourself with voltages even as low as three or six volts. If you accidentally create a short circuit between the two poles of a battery, a lot of current will flow very fast. This will very likely cause the wire connecting the two ends of the battery to get very hot, and the battery itself may also heat up. The heat may be enough to inflict a nasty burn.

WARNING

If the racing current goes unchecked, there's also the possibility that the battery will explode. Trust me; you don't want to be nearby if that happens. You really don't want to make a trip to the emergency room to have fragments of an exploded battery removed from your eyes.

As a result of this danger, you should take the following precautions when working with the battery-powered circuits described in this book:

>> **Don't connect power to the circuit until the circuit is completely finished and you've reviewed your work to ensure that everything is connected properly.**

>> **Don't leave your circuits unattended when they're connected to power.** Always remove the batteries before you walk away from your workbench.

>> **Periodically touch the batteries with your finger to make sure they aren't hot.** If they're getting warm, remove the batteries and recheck your circuit to make sure you haven't made a wiring mistake.

>> **If you smell anything burning, remove the batteries and recheck your circuit.**

STAYING SAFE BY STAYING DRY

We've all seen murders committed on TV crime dramas by throwing a plugged-in electrical appliance such as a hair dryer into a bathtub while the victim was taking a bath. I've always wondered how often that really happens, and how likely it's fatal. For example, how quickly would the circuit breaker kick in and cut power to the hair dryer? Would the special GFCI-protection devices required in all modern bathrooms work as designed and cut power to the hair dryer in time?

I've never wanted to conduct an experiment to actually find out — nor should you, under any circumstances. Water and electricity are a very bad combination because water is an excellent conductor of electricity, and it flows everywhere.

Strictly speaking, pure uncontaminated water is actually an insulator. But pure water is very rare. Most water is filled with contaminates, and those contaminants turn the water into an excellent conductor. Thus, it's true that you should avoid water when working with electrical current. Here are a few tips for staying safe by staying dry:

- **Make sure the floor is dry.** Don't work on electronic or electrical devices in an area where the floor is wet.

- **Beware of high humidity, especially if it condenses into moisture on your projects.**

- **Dry your hands before working with electrical current.** Even a small amount of sweat on your hands can lower your body's natural resistance and accentuate the danger of electrical shocks from lower voltages.

>> **Always wear protective eyewear to protect yourself against exploding batteries.** (Under the right circumstances, other components can explode as well!)

Sometimes voltage hides in unexpected places

One of the biggest shock risks in electronics comes from voltages that you didn't expect to be present. It's easy enough to keep your eye on the voltages that you know about, such as in your power supply or batteries, but some electronic circuits are designed to amplify voltages. So even though your circuit runs on 6-volt batteries, there may be much larger voltages at specific points within your circuit.

In addition, some electrical devices can actually store an electric charge long after the power from your circuit has been disconnected. The most notorious device with this characteristic is the *capacitor*, which alternately builds up and then releases electrical charges. Thus, you should be wary of any circuit that contains capacitors — especially if the capacitors are large. Common ceramic-disk capacitors, which are typically smaller than a tiddlywink, don't store much charge. However, if your circuit has capacitors the size of batteries, you should be very careful when working around them. Such capacitors can hold large charges long after the power has been cut off.

Here are some safety points concerning capacitors:

WARNING

>> **One of the most common places to find large capacitors is in the power-supply circuit.** Any electronic device that plugs into a household electrical outlet has a power-supply circuit that may contain a large capacitor. Be very careful around these capacitors. In fact, if the power-supply circuit is inside its own enclosed box, *don't open the box*. Instead, replace the entire power supply if you suspect it's bad.

>> **Another common place to find high-voltage capacitors is in a flash camera.** Even though the battery may be just 1.5 V, the capacitor that drives the flash unit may well be holding a charge of 300 V or more.

>> **Before working on a circuit that contains a capacitor, always discharge the capacitor first.** You can discharge small capacitors by shorting out their leads with the blade of a screwdriver. Make sure you touch only the insulated handle of the screwdriver while you short out the leads, and don't touch any other part of the circuit with your free hand.

>> **Larger capacitors should be discharged by connecting their leads to a lamp or a large resistor.** The easiest way to do this is to wire up a lamp holder to a pair of alligator clips, screw a lamp into the lamp holder, then carefully connect the clips to the capacitor leads. If the capacitor is holding a charge, the lamp will glow for a moment as the capacitor discharges through the lamp.

WARNING

>> **If you don't feel completely confident in what you're doing where large capacitors are concerned, walk away from the project.**

Other Ways to Stay Safe

Electric shock isn't the only danger you'll encounter when you work with electronics. The following paragraphs summarize a few of the other risks you may be exposed to and describes the precautions you should take to minimize those risks:

>> **Soldering poses an obvious fire hazard.** If your soldering iron is hot enough to melt solder, it's also hot enough to ignite combustible materials such as paper, wire insulation, and so on. Therefore:

- *Always be aware of when your soldering iron is on.* Don't plug it in until you need it, and unplug it when you're finished soldering.

- *Never set a hot soldering iron down directly on your workbench.* Instead, get a soldering iron holder to safely hold the soldering iron while it's hot. Figure 4-1 shows a soldering iron resting in a simple stand. As you can see, this stand keeps the business end of the soldering iron safely elevated away from the work surface.

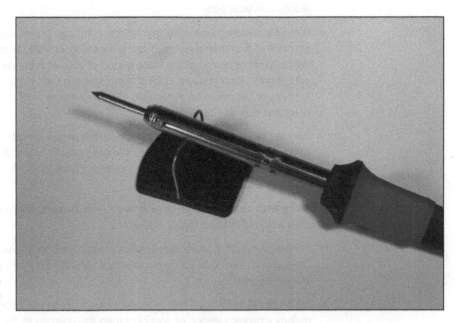

FIGURE 4-1:
A soldering iron resting on a stand.

- *Give your soldered joints a few minutes to cool down before you handle them.*

- *Watch out for the soldering iron's electrical cord.* Obviously, you want to avoid burning the cord with the soldering iron. As ridiculous as it sounds, I did this myself once when I carelessly set the soldering iron aside, directly on top of its own power cord. Fortunately, I noticed my mistake before the soldering iron melted much of the power cord's insulation.

  Make sure the soldering iron's power cord is placed safely away from your stuff so that you won't bump it as you work, knocking it out of its stand and perhaps causing a burn.

Staying Safe

- *Be sure to wear eye protection when you solder.* As solder melts, it occasionally boils and splatters little globules of hot solder through the air. You really don't want molten metal anywhere near your eyes.

» **Electronics — and especially soldering — can also create a chemical hazard.** When you solder, small amounts of lead are released into the air. Therefore:

- *Always work in a well-ventilated place.*

- *Wash your hands after you work with solder or any other electronic components before you touch your face, mouth, nose, or eyes.* Small amounts of lead and potentially other toxic substances are bound to get on your hands. It's best to wash them frequently to keep whatever gunk they pick up from getting into your body.

- *Keep your soldering tools away from children.* Young children and pets love to stick things in their mouths. If you leave solder or little electronic parts like resistors or diodes sitting loose on top of your workbench, your kids or pets may decide to make a meal of them, so keep such things safely stored in boxes or cabinets and, if possible, keep your entire work area safely off-limits and behind closed doors.

- *Don't get into the habit of sticking parts into your mouth to hold them while you're working.* As crazy as it sounds, I've seen people hold a dozen resistors in their mouth while soldering each one into a printed circuit board. That's definitely a bad idea.

» **Working with sharp tools such as knives, wire cutters, and power drills creates a risk of cutting injury.** Therefore:

- *Think before you cut.* Make sure you know exactly where you want to make the cut, and make sure you know exactly where all your fingers are before you start the cut.

- *Let the tool do the work.* Don't apply excessive force to coerce a tool into making a bigger, deeper, or wider cut than it's designed to do.

- *Keep your tools sharp.* Working with dull tools causes you to use extra force, which often results in the tool slipping and finding itself lodged in your finger.

- *Remove jewelry such as rings, wristwatches, and long dangling necklaces before you start — especially if you're working with power tools.*

- *Wear safety goggles whenever you're cutting, sawing, or drilling.* Little pieces of the work or blade can easily break off and hit you in the face. Add bits of insulation, copper wire, and broken drill bits to the growing list of things you don't want in your eyes.

Keeping Safety Equipment on Hand

In spite of every precaution you might take, accidents are bound to happen as you work with electronics. Other than preventing an accident from happening in the first place, the best strategy for dealing with an accident is to be prepared for it, so I recommend you keep the following items nearby whenever you're working with electronics:

>> **Fire extinguisher:** So you can quickly put out any fire that might start before it gets out of hand.

>> **First-aid kit:** For treating small cuts and abrasions as well as small burns. The kit should include bandages, antibacterial creams or sprays, and burn ointments.

>> **Phone:** So that you can call for assistance in case something goes really wrong.

>> **Friend:** If your project works with household current (120 volts), a friend can help in case you get shocked.

Protecting Your Stuff from Static Discharges

Static electricity — more properly called *electrostatic charge* — results when electric charges (that is, voltage) builds up in the absence of a circuit that allows current to flow. Your own body is frequently the carrier of static charge, which can be created by a variety of causes. The most common is friction that results from simple things such as walking across a carpet. Your clothes can also pick up static charge, and usually do when you toss them around in a clothes dryer.

Static charge accumulated in your body usually discharges itself over time. However, if you touch a conductor — such as a brass doorknob — while you're charged up, the charge will dissipate itself quickly in an annoying shock.

If the conductor happens to be a sensitive electronic component such as a transistor or an integrated circuit rather than a brass doorknob, the discharge can be more than annoying; it can fry the innards of the component, rendering it useless for your projects. For this reason, it's wise to protect your stuff from static discharge when you work on your electronic projects. The easiest way to do that is to make sure you're properly discharged before you start your work. If you have a metal workbench or a large metal tool such as a drill press or grinder near your workbench, simply reach out and touch it after you've settled in to your seat and before you begin your work.

A more reliable way to protect your gear from static discharge is to wear a special *antistatic wristband* on one wrist, as shown in Figure 4-2. Wear the wristband tightly so that it's in good solid contact with your skin all the way around your wrist. Then, plug the alligator clip into a metal surface such as your workbench frame or that nearby drill press.

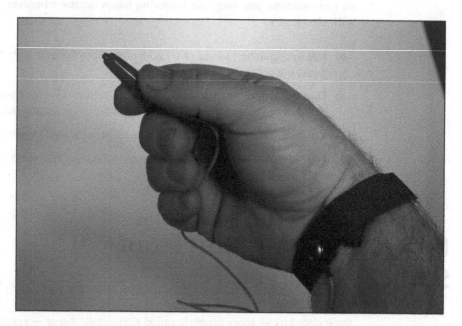

FIGURE 4-2:
An antistatic
wristband.

TIP

For best results, the alligator clip on your antistatic wristband should be connected to a proper *earth ground*. To create a proper earth ground, clamp a long length of wire to a metal water pipe. The wire should be long enough to reach from the pipe to your workbench. Carefully route the wire from the pipe to your workbench, strip off an inch or so of insulation, and staple or clamp the wire to the workbench, leaving the stripped end free so you can attach the alligator clip from your antistatic wristband to it. (Note that this technique works only if the building uses metal pipes throughout. If the building uses plastic pipe, the water-pipe won't provide a proper ground.)

WARNING

An often-recommended way to connect the wristband to an earth ground is to connect it to the ground receptacle of a properly grounded electrical outlet. I'm definitely not a fan of this method, as the key to its operation lies in the term "properly grounded electrical outlet." All it takes is one stupid wiring mistake, or one wire shaken loose by a sonic boom or a mild earthquake, and suddenly that ground wire might not be a ground wire anymore — it might be energized. Call me paranoid if you wish, but there's no way I can recommend strapping a conductor around your wrist and then plugging it into an electrical outlet.

IN THIS CHAPTER

» Examining how schematic diagrams provide a road map for electronic circuits

» Looking at the most commonly used component symbols

» Noting how voltage supply and common ground circuits are often drawn

» Seeing how components are typically labeled

Chapter 5

Reading Schematic Diagrams

I love maps. I think I've kept every map I've used on every trip I've taken. I have big maps of entire countries and states, maps of cities, walking maps, maps of parks and museums, and even subway maps. My favorite maps are topographical maps of the areas where I've gone on weeklong backpacking trips. These maps show not only the routes I've hiked, but also elevation lines that represent every painful uphill step I've carried my 50-pound backpack up.

Without maps, we'd be lost. We'd never get to our destinations because we wouldn't know where the roads are. Think of all the sights we'd miss along the way!

Electronics has its own form of maps. They're called *schematic diagrams*. They show how all the different parts that make up an electronic circuit are connected.

Just as maps use symbols to represent features like cities, bridges, and railroads, schematic diagrams use special symbols to represent the different parts of a circuit, such as batteries, resistors, and diodes, and like maps, schematic diagrams have conventions that almost always are used. For example, positive voltages are almost always shown at the top of a schematic diagram, just as north is almost always shown at the top of a map.

In this chapter, you learn about the symbols used in schematic diagrams and the conventions used to draw them.

Introducing a Simple Schematic Diagram

I've read a lot of computer programming books in my day, and I've written a few too. In a computer programming book, the first complete computer program usually shown is a program called Hello World, a program that simply displays the text "Hello World!" on a screen, and then quits. It's pretty much the simplest possible computer program that can be written. It doesn't do anything useful, but it's a great starting point for learning how to write computer programs.

Figure 5-1 shows a schematic diagram that is the electronic equivalent of the Hello World program. This diagram is about the simplest schematic diagram possible that actually does something: It lights a lamp, thus announcing to the world that a circuit is indeed working.

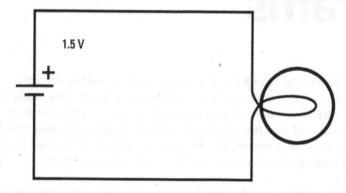

FIGURE 5-1:
A simple schematic of a circuit that lights a lamp.

1.5 V

This diagram contains two symbols representing the two components in the circuit: a 1.5 V battery and an incandescent lamp. The lines that connect the two components represent conductors, which could be actual wires or traces of copper in a printed circuit board.

In the circuit depicted in this schematic, the positive side of the battery is connected to one lead from the lamp, and the other lead from the lamp is connected to the negative side of the battery. Once these connections are made, current will flow from the battery to the lamp, through the lamp's filament to produce light, and then back to the battery.

Schematic diagrams always depict *conventional current flow*, which, as you learn in Chapter 2 of this minibook, means that current flows from positive to negative. Thus, the current flows from the positive terminal of the battery through the lamp and then back to the negative terminal of the battery.

REMEMBER

In reality, conventional current flow is the opposite of the actual flow of electrons through the circuit. The negative side of the battery has an excess of negatively charged particles (extra electrons) whereas the positive side has an excess of positively charged particles (missing electrons). Thus, the electric charge flows through the conductor from the negative side of the battery, through the lamp, and back to the positive side. (For more about the difference between real current flow and conventional current flow, see Chapter 2 of this minibook.)

As it passes through the lamp, the resistance of the lamp's filament causes the current to heat the filament, which in turn causes the filament to emit visible light.

Laying Out a Circuit

REMEMBER

One of the most important things to realize about a schematic diagram is that the arrangement of components in the diagram doesn't necessarily correspond to the physical arrangement of parts in the circuit when you actually build the circuit.

For example, the circuit shown in Figure 5-1 shows the battery on the left side of the circuit and the lamp on the right. It also shows the battery oriented so that the positive terminal is at the top and the negative terminal is at the bottom. However, that doesn't mean the circuit would actually have to be built that way. If you want, you could put the lamp on the left and the battery on the right, or you could put the battery at the top and the lamp on the bottom.

The physical arrangement of the circuit doesn't matter as long as the component connections remain the same as shown in the schematic. Thus, in this example, no matter how you physically arrange the components, you must connect the positive terminal of the battery to one lead of the lamp and the negative terminal to the other lead.

Because there are only two components and two conductors in the circuit shown in Figure 5-1, it would be pretty hard to mess up the connections. However, in a more complicated circuit with perhaps dozens of components and dozens of connections, laying out the circuit and making sure that all the connections exactly match the connections indicated in the schematic can be a challenge. Each connection must be checked carefully to make sure it's correct.

To Connect or Not to Connect

One of the goals when laying out a schematic circuit diagram is to keep the diagram as simple as possible. However, the lines in all but the simplest of schematic diagrams will at some places need to cross over each other. When they do, it's vital that you can tell whether the lines that cross represent actual connections (also called *junctions*) between the conductors or whether the lines cross over each other but don't actually connect.

Unfortunately, there isn't one clear and universally used standard that dictates how to indicate whether crossed lines represent a junction. Figure 5-2 shows some of the ways for showing crossed wires with or without junctions.

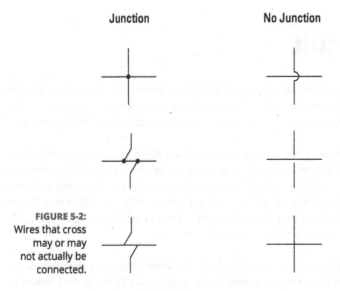

Junction   No Junction

FIGURE 5-2:
Wires that cross
may or may
not actually be
connected.

The three examples on the left side of Figure 5-2 show how junctions are indicated. The example at the top left shows the most common way to indicate a junction: by placing a conspicuous dot at the point where the wires cross. Any time you see a dot where two lines intersect, you know that the two lines form a junction.

In the two junction styles shown in the middle-left and bottom-left examples in Figure 5-2, the vertical lines are angled to avoid coming together at the same spot on the horizontal line. With or without the dot, junctions are clearly indicated in both of these examples.

The three examples on the right side of Figure 5-2 show how lines that cross but don't connect to form junctions are most commonly shown. In the top two examples, one line "hops" over the other, and one of the lines is broken at the spot where it crosses the other.

The example in the bottom-right corner of Figure 5-2 is a bit ambiguous. Here, the lines cross each other. However, there's no hop or break to indicate that no junction is present, nor is there a dot to indicate that a junction should be present. So is there a junction here or not? The answer is, in most cases, no. You can usually assume that a junction is *not* present when lines cross but there's no dot. However, you should examine the rest of the diagram to make sure. If you find other places in the diagram where nonjunctions are indicated by a hop or a break, the crossed lines without the hop or break may indeed indicate a junction.

TIP

To avoid ambiguity altogether, the schematic diagrams in this book always use a dot to indicate a junction and a hop to indicate a nonjunction. You'll never see lines simply cross without a hop or a dot.

Looking at Commonly Used Symbols

The circuit shown in Figure 5-1 has just two components: a battery and a lamp. Most electronic circuits will have additional components. There are hundreds of different types of electronic components, and each has its own unique schematic diagram symbol. Fortunately, you need to know only a few basic symbols to get started. These symbols are summarized in Table 5-1. (Note that when used in an actual circuit diagram, the symbols are often rotated.)

Figure 5-3 shows a schematic diagram that includes several of these components. Don't worry — you don't need to understand this diagram right now. I just want you to get an idea of what real-world schematic diagrams look like and how to read them.

TABLE 5-1

Common Symbols for Schematic Diagrams

Symbol	Description
	Battery
	Capacitor
	Diode
	Ground
	Inductor (coil)
	Lamp
	Light-emitting diode
	Resistor
	Source voltage connection
	Speaker
	Switch
	Transformer
	Transistor (NPN)
	Transistor (PNP)
	Variable resistor (potentiometer)

As you can see, the circuit shown in Figure 5-3 contains six components. Working from left to right, they are:

» 6 V battery

» NPN transistor

» Resistor

» Capacitor

» PNP transistor (at the top right)

» Light-emitting diode (at the bottom right)

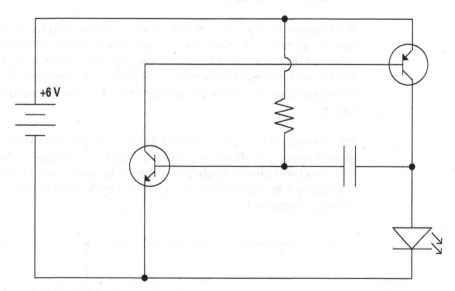

FIGURE 5-3:
A typical
schematic
diagram.

Throughout the course of this book, I use these and other symbols in the schematic diagrams that describe the circuits. Whenever I use a symbol for the first time, I explain what it is and how it works.

Simplifying Ground and Power Connections

In many electronic circuits, the distribution of voltage connections is one of the most complicated aspects of the circuit. For example, about half of the connections in the schematic diagram shown in Figure 5-3 are used to connect the resistor, transistors, and the LED to either the positive or negative terminal of the battery.

In a more complicated circuit, there can be dozens or even hundreds of power connections. If all the lines representing those connections had to be drawn to the positive or negative side of the battery symbol, schematic diagrams would quickly be overwhelmed by the power connections.

Most circuits have a common path by which current returns to its source. In the case of Figure 5-3, it's the conductor at the very bottom of the diagram that collects current from the LED and the resistor and returns it to the battery. This conductor is necessary to complete the circuit so that current can flow in a complete loop from the battery through the various components and then back to the battery.

This common return path is often called the *ground*, and can be replaced by the ground symbol, as shown in Table 5-1. Figure 5-4 shows a schematic diagram that uses three ground symbols to indicate the path by which current returns to the battery. The circuit shown in Figure 5-4 is identical in function to the circuit shown in Figure 5-3.

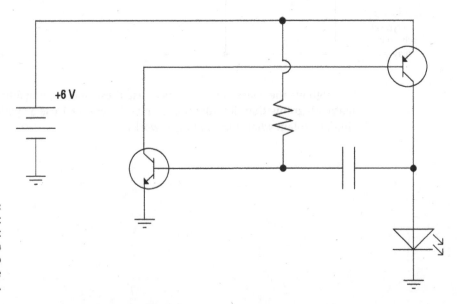

FIGURE 5-4: A schematic diagram that uses a common ground to complete the circuit.

In addition to a common ground path, most circuits also have a common voltage path. In the case of the circuit shown in Figures 5-3 and 5-4, the common voltage path goes from the battery to the resistor and on to the second transistor. This conductor can be replaced by symbols representing voltage sources that appear wherever voltage is required in a circuit.

The symbol for a voltage source is either an open circle or an arrow. The quantity of voltage is always indicated next to the circle or arrow. When a voltage source symbol is used in a schematic diagram, the symbol for the battery (or other power source if the circuit isn't powered by a battery) is omitted. Instead, the presence of voltage source symbols implies that voltage is provided by some means, either by a battery or by some other device such as a solar cell or a power supply plugged into an electrical outlet.

Figure 5-5 shows a schematic diagram for the same circuit that is shown in Figures 5-3 and 5-4, but with voltage source symbols instead of a battery symbol. As you can see, +6 V is required in two places in the circuit: at the resistor and at the second transistor. This circuit is functionally identical to the circuits shown in Figures 5-3 and 5-4.

TIP

Although the circuit shown in Figure 5-5 has a positive voltage source and the ground is negative, this isn't always the case. You can also use the voltage source symbol to refer to negative voltage. In that case, the ground actually carries positive voltage back to the source.

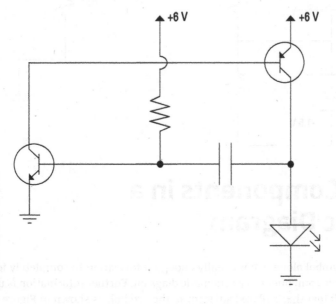

FIGURE 5-5: A schematic diagram that uses a common ground to complete the circuit, with voltage source symbols.

In some cases, a circuit may require both positive and negative voltages at different places within the circuit. Remember from Chapter 2 of this minibook that voltages are always measured with respect to two points in a circuit. Thus, voltages are always relative. For example, the positive pole of a AAA battery is +1.5 V relative to the negative pole. At the same time, the negative pole of the battery is −1.5 V relative to the positive pole.

Now suppose you connect two AAA batteries end to end. Then, the voltage at the positive terminal of the first battery will be +3 V relative to the voltage at the negative terminal of the second battery. But, the voltage at the positive pole of the first battery will be +1.5 V relative to the point between the batteries, and the voltage at the negative pole of the second battery will be −1.5 V relative to the point between the batteries.

Figure 5-6 shows how this arrangement might be drawn in a schematic diagram, with a pair of resistors connected across each battery to the middle point. The diagram on the left shows the batteries and connections to them. The diagram on the right shows the same circuit using ground and voltage source symbols instead.

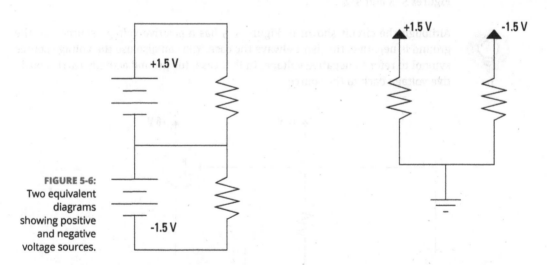

FIGURE 5-6:
Two equivalent diagrams showing positive and negative voltage sources.

Labeling Components in a Schematic Diagram

A symbol alone is not usually enough information to completely identify an electronic component in a schematic diagram. Further information is usually included with text that's placed adjacent to the symbol, as shown in Figure 5-7. This additional information usually includes the following:

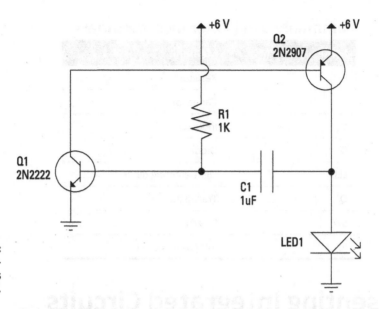

FIGURE 5-7:
A schematic dia-
gram with parts
labeled.

>> **Reference identifier:** Each component is usually labeled with a letter that designates the type of component followed by a number that helps identify each component of the same type. For example, if a circuit has four resistors, the resistors are identified as R1, R2, R3, and R4. The most commonly used letters are shown in Table 5-2.

>> **Value or part number:** For components such as resistors and capacitors, the value is given in ohms (for resistors) and microfarads (for capacitors). Thus, a 470 Ω resistor would have the number 470 next to it, and a 100 µF capacitor would have the number 100 next to it.

The letters k and M are used to denote thousands and millions. For example, a 10,000 Ω resistor is identified as 10k in a schematic.

Components such as diodes, transistors, and integrated circuits don't have values; instead, they have manufacturer's part numbers. Thus, you might find a part number such as 1N4001 (for a diode), 2N2222 (for a transistor), or 555 (for an integrated circuit, IC) next to one of these components.

In some cases, the value or part number is omitted from the schematic diagram itself and instead included in a separate parts list that identifies the value or part number of each referenced part that appears in the schematic. Then, to find the value or part number of a particular component, you look up the component by its reference identifier in the parts list.

TABLE 5-2

Commonly Used Reference Identifiers

Letter	Meaning
R	Resistor
C	Capacitor
L	Inductor
D	Diode
LED	Light-emitting diode
Q	Transistor
SW	Switch
IC	Integrated circuit

Representing Integrated Circuits in a Schematic Diagram

One important symbol that isn't shown in Table 5-1 is the symbol for an IC (integrated circuit). ICs are small assemblies that usually have multiple leads, called *pins*, which connect to various parts of the circuit contained within the assembly. Some ICs have as few as six or eight pins; others have dozens or even hundreds. These pins are numbered, beginning with pin 1. Each pin in an IC has a distinct purpose, so connecting to the correct pins in your circuit is vital to the circuit's proper operation. If you connect to the wrong pins, your circuit won't work, and you may damage the IC.

The most common way to depict an integrated circuit in a schematic diagram is as a simple rectangle with leads coming out of it to depict the various pins. The arrangement of the pins in the schematic diagram doesn't necessarily correspond to the physical arrangement of pins on the IC itself. Instead, the pins are positioned to provide for the simplest circuit paths in the diagram. The pins in the diagram are numbered to indicate the correct pin to use.

For example, Figure 5-8 shows a schematic diagram that uses a popular IC called a 555 timer IC to make an LED flash. The 555 has eight pins, and you can see that the schematic calls for connections on all eight. However, the pins in the diagram are arranged in a manner that simplifies the connections to be made to the pins. In an actual 555 IC, the pins are arranged in numerical order on either side of the IC, with pins 1 through 4 on one side and pins 5 through 8 on the other side.

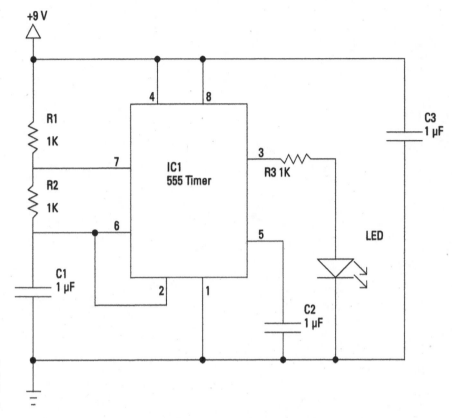

FIGURE 5-8:
A circuit that uses
an integrated
circuit.

TIP

Don't worry about any of the details of the operation of this circuit. You learn how
it works in Book 3, Chapter 2. My only purpose for including it here is so you can
see how integrated circuits are depicted in a schematic diagram.

IN THIS CHAPTER

» **Fleshing out an idea for an electronic project**

» **Creating a workable circuit design**

» **Building a prototype on a solderless breadboard**

» **Creating a permanent circuit on a printed circuit board**

» **Finishing the project by putting everything into a suitable enclosure**

Chapter **6**

Building Projects

Yogi Berra is alleged to have said, "In theory, there is no difference between theory and practice. But in practice, there is."

Much of this book is theoretical — how electric current works, how individual electronic components like resistors, capacitors, and transistors work, how digital logic works, and so on.

But the heart of electronics is building things. The reason for learning all the theory is so that you can practice the art by actually building circuits and putting them to use.

Throughout this book, I back up theoretical explanations about how various types of electronic components work with simple construction projects you can build to demonstrate the theory in actual use. In this chapter, you learn the basic construction techniques needed to build these projects.

Specifically, you learn how to create a prototype of a circuit using a handy device called a *solderless breadboard*. Then, you learn several techniques for creating a more permanent version of the circuit, in which the components and all the circuit's interconnections are soldered together on a circuit board. Finally, you learn how to enclose your circuit board in a project box or other enclosure.

In this chapter, I walk you through the process of building a fairly sophisticated electronics project. Although you're welcome to do so if you wish, I don't expect you to actually build the project as you read this chapter. Instead, I simply want you to gain an appreciation for the process of building a nontrivial project from start to finish.

Looking at the Process of Building an Electronic Project

Electronic projects such as the ones you learn about in this book typically follow this predictable sequence of general steps from start to finish:

1. **Decide what you want to build.**

   Before you can design or build an electronic project, you must have a solid idea in mind for what you expect the project to do, what you want it to look like, and how human beings will interact with it.

2. **Design the circuit.**

   After you've settled on what you want to build, you need to design an electronic circuit that gets the job done. The end result of this step is a schematic diagram.

3. **Build a prototype.**

   Before you invest the time and materials needed to build a permanent circuit, it's a good idea to first build a *prototype*, which lets you quickly test the circuit to make sure it works. Usually, you build the prototype on a solderless breadboard.

4. **Build a permanent circuit.**

   When your prototype is working, you can build a permanent version of the circuit. Usually, you build the permanent version by soldering components onto a printed circuit board.

5. **Finish the project.**

   To finish the project, you mount the circuit board along with any other necessary components such as batteries, switches, or light-emitting diodes in a suitable enclosure.

The remaining sections in this chapter describe each of these steps in greater detail.

Envisioning Your Project

Before you get lost in the details of designing and building your project, you should step back and look at the big picture. First, you need to make sure you have a solid idea for your project. Why do you want to build it? What will it do, who will use it, and why?

For example, every year I like to build something to scare trick-or-treaters on Halloween. A few years ago, I built a giant jack-in-the-box that pops up and screams when people walk up to it. The box was made of plywood, and the pop-up mechanism that made the door open and the scary clown pop up was driven by compressed air. Figure 6-1 shows the finished contraption. Trust me; I scared a lot of kids and more than a few adults with it.

I knew right away that I'd need some type of electronic circuit to control the jack-in-the-box. At first, I wasn't sure exactly what type of circuit I'd need, but I knew I needed a circuit of some kind.

FIGURE 6-1:
One of my scarier electronics projects.

When you have a general idea for a project, you can flesh out the details. You'll need to answer questions like these:

>> What will its *user interface* be? That is, how will a person work with the device to get it to do what it's supposed to do?

>> Will it be a stand-alone device, or will it interact with other devices?

>> Will it be powered by batteries, or will it plug into a wall outlet to get its power? Or will it be solar powered?

>> How big will it be? Does it need to be small enough to hold in your hand or fit in your pocket? Or will it sit on a shelf?

The jack-in-the-box Halloween prop is a fairly complicated project — too complicated to use as an illustration this early in the book. So, here's a simpler project: an electronic decision maker. Have you ever resorted to tossing a coin to make a difficult decision? For this project, you create an electronic version of a coin toss. Instead of flipping a coin into the air to see if it lands heads or tails, you build an electronic device that does the coin toss. That way, you can make decisions even when you're penniless.

The specifications for the coin-toss project are as follows:

>> The device will have two LED indicators to indicate heads and tails.

>> It will also have two small metal contacts, which the user can touch with their finger. When the user touches both of the posts, the LEDs start flashing, alternating back and forth, much like a coin flips end over end when you toss it into the air.

>> When the user removes his finger from the two metal contacts, one of the two lights will stay lit, indicating whether the result of the coin toss is heads or tails. Which light stays lit will be essentially random.

>> To conserve battery life, the device will have an on/off push button. The user must depress the push button to make the device work; when the button is released, the device is turned off.

>> The device will be battery powered and contained in an enclosure small enough to hold in your hand.

TIP

As you flesh out the details for your project, you may want to start drawing diagrams to show how it will look. Figure 6-2 shows a hand-drawn sketch I created for the electronic coin tosser.

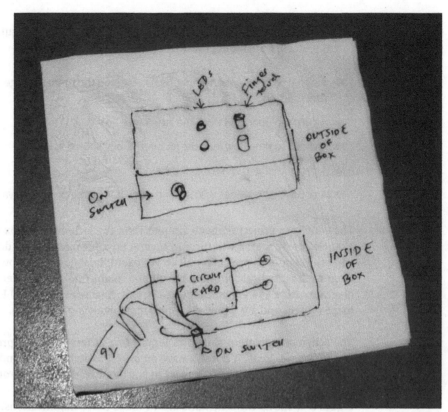

FIGURE 6-2:
A hand-drawn
sketch for an
electronic coin
tosser.

Designing Your Circuit

After you have an idea for a project, the next step is to design a circuit that meets the project's needs. At first, you'll find it very difficult to design your own circuits, so you'll turn to books like this one or to the Internet to find other people's circuit designs. With a bit of Google searching, you can probably find a schematic diagram that's very close to what your project needs.

In many cases, you won't be able to find exactly the circuit you're looking for. You may find a circuit that's close, but you may need to make minor modifications to make the circuit fit your project's needs. At first, making modifications to a circuit may seem beyond your abilities. But as you gain experience, you'll find yourself tweaking circuits all the time to fit specific applications.

One helpful strategy for designing circuits is to break complex requirements down into simpler parts. For example, consider the pop-up jack-in-the-box Halloween prop I mention earlier. The complete circuit for this project required several different elements, including these:

>> A circuit to detect when someone has entered the room to trigger the prop's action

>> A circuit to open and close the jack-in-the-box

>> A circuit to time how long the jack-in-the-box should stay open

>> A circuit that plays a screaming sound

>> A circuit that provides a 30-second delay before the prop is activated again

The coin-toss project is much simpler than the jack-in-the-box project. In fact, a quick Google search will turn up several possible circuits that do almost exactly what the coin-toss project requires. For example, Figure 6-3 shows the schematic diagram for a typical coin-toss circuit you might find on the Internet. This circuit diagram uses a 555 Timer integrated circuit, four resistors, two LEDs, one capacitor, a switch, and a 9 V power supply (most likely a 9 V battery).

The schematic diagram shown in Figure 6-3 differs from our project's needs in just two ways. First, it doesn't have an on/off switch. And second, it uses a push button instead of the user's fingers to start and stop the LEDs from flashing.

Figure 6-4 shows the schematic after I made those modifications. As you can see, I added a push-button switch that must be pressed to provide the +9 V voltage needed to run the circuit, and I replaced the push button that was in the original schematic with two open terminals. When the user touches these two terminals, the resistance of their finger completes the circuit.

REMEMBER

Please don't worry at all if you don't understand how the circuit depicted in Figure 6-4 works. I wouldn't expect you to at this point in the book! Understanding how a circuit works and building that circuit are two entirely different things; you can (and probably will) build plenty of circuits whose operation you don't understand. The only thing you should focus on at this point is how the schematic diagram indicates the various connections between the parts in the circuit. You learn the details of how this circuit works in Book 3, Chapter 2.

One final step you might want to consider when designing a circuit is to create a final version of the schematic diagram that indicates what components will be mounted on your final circuit board and what components won't be on the circuit board. This diagram will come in handy later when you're ready to create the circuit board that will become the permanent home of your circuit.

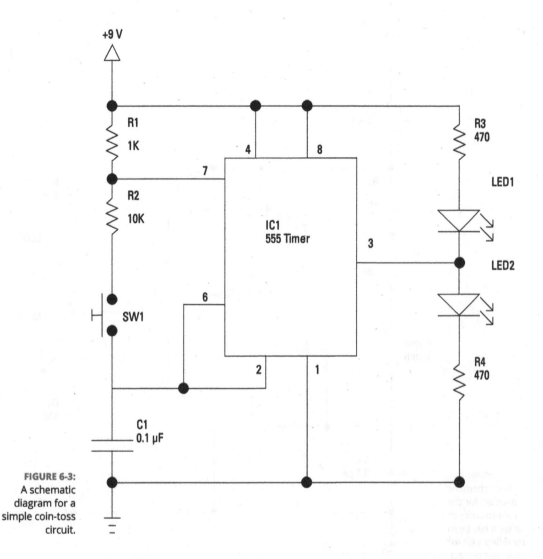

FIGURE 6-3:
A schematic
diagram for a
simple coin-toss
circuit.

For example, Figure 6-5 shows a version of the coin-toss circuit that uses a dashed line to delineate the items that won't be mounted on the circuit board: the battery power supply (that is, the +9 V voltage source and the ground), the push-button power switch, the two metal finger contacts, and the two LEDs. Instead, they'll be mounted separately within the project box. Thus, the circuit board will need to hold only six components: the 555 timer integrated circuit, the four resistors, and the capacitor.

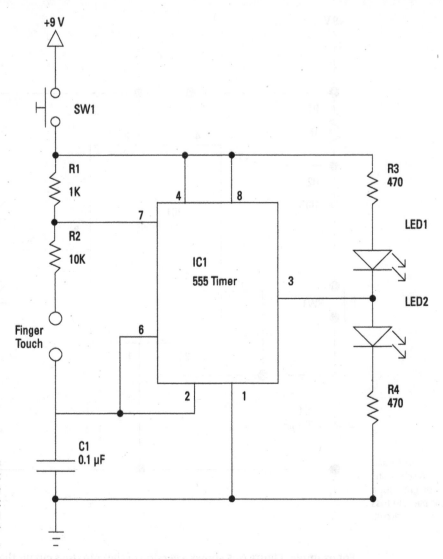

+9 V

SW1

R1
1K

R2
10K

7

4   8

IC1

555 Timer

3

6

2   1

Finger
Touch

C1
0.1 µF

R3
470

LED1

LED2

R4
470

FIGURE 6-4:
The schematic
diagram for the
coin-toss circuit
after it has been
modified a bit for
our project.

After you've completed your circuit design, you'll want to compile a list of all the parts you'll need to build the circuit. Then, you can rummage through your parts bin to figure out what parts you already have at your disposal and what parts you'll need to purchase. Here's a list of the components you'll need to build the coin-toss circuit:

Part ID	Description
R1	1 kΩ, ¼ W resistor
R2	10 kΩ, ¼ W resistor
R3	470 Ω, ¼ W resistor
R4	470 Ω, ¼ W resistor
C1	0.1 µF capacitor
LED1	5 mm red LED
LED2	5 mm green LED
IC1	555 timer IC
SW1	Momentary-contact, normally open push button

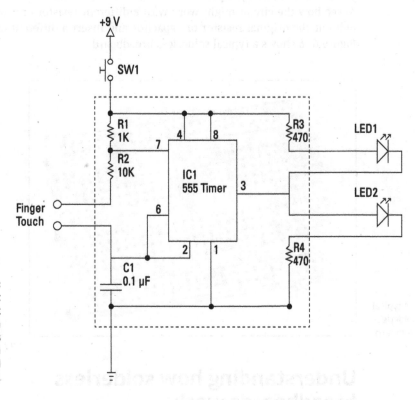

FIGURE 6-5:
A schematic
diagram that
indicates which
components
are on the main
circuit board and
which aren't.

Prototyping Your Circuit on a Solderless Breadboard

Before you commit your circuit to a permanent circuit board, you want to make sure it works. The easiest way to do that is to build the circuit on a *solderless breadboard.* The solderless breadboard lets you quickly assemble the components of your circuit without soldering anything. Instead, you just push the bare wire leads of the various components you need into the holes on the breadboard and then use *jumper wires* to connect the components together.

The beauty of working with a solderless breadboard is that if the circuit doesn't work the way you expect it to, you can make changes to the circuit simply by pulling components or jumper wires out and inserting new ones in their place. If you discover that your schematic diagram is missing an important connection, you can add another jumper wire to create the missing connection or, if you want to see how the circuit might work with a different resistor or capacitor, you can pull out the original resistor or capacitor and insert a different one in its place. Figure 6-6 shows a typical solderless breadboard.

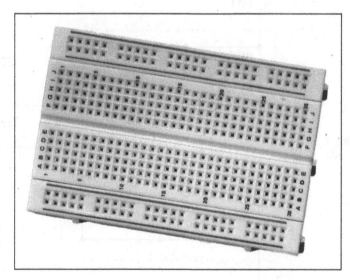

FIGURE 6-6:
A typical
solderless
breadboard.

Understanding how solderless breadboards work

Although many different manufacturers make solderless breadboards, they all work pretty much the same way. The board consists of several hundred little holes called *contact holes* that are spaced $\frac{1}{10}$ inch apart. This is a convenient spacing because it also happens to be the standard spacing for the pins that come out of the bottom or sides of most integrated circuits. Thus, you can insert all the pins of even a large integrated circuit directly into a solderless breadboard.

Beneath the plastic surface of the solderless breadboard, the contact holes are connected to one another inside the breadboard. These connections are made according to a specific pattern that's designed to make it easy to construct even complicated circuits. Figure 6-7 shows how this pattern works.

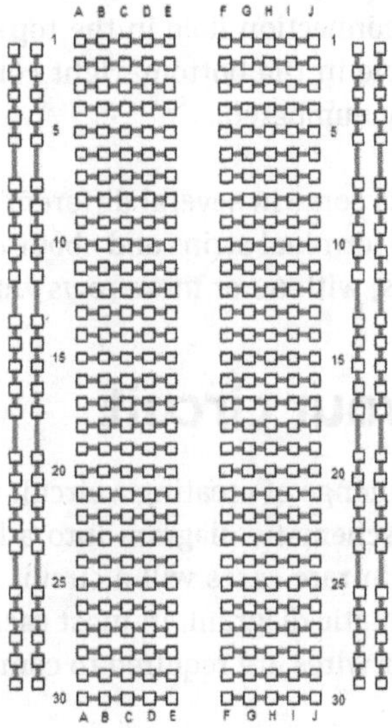

FIGURE 6-7:
The contact holes in typical solderless breadboards are internally connected following this pattern.

The holes in the middle portion of a solderless breadboard are connected in groups of five that are called *terminal strips*. These terminal strips are arranged in two groups, with a long open slot between the two groups, like a little ditch. It is in these holes that you will connect components such as resistors, capacitors, diodes, and integrated circuits.

It's important to note that the rows of holes are not connected across the ditch. Thus, each row comprises two electrically separate terminal strips: one that connects the holes labeled A through E, the other connecting the holes labeled F through J.

TIP

The breadboard is designed so that integrated circuits can be placed over the top of the ditch, with the pins on either side of the integrated circuit pushed into the holes on either side of the ditch.

The holes on the outside edges of the breadboard are called *bus strips*. There are two bus strips on either side of the breadboard. For most circuits, you will use the bus strips on one side of the breadboard for the voltage source and use the bus strips on the other side of the board for the ground circuit.

Most solderless breadboards use numbers and letters to designate the individual connection holes in the terminal strips. In Figure 6-7, the rows are labeled with numbers from 1 through 30, and the columns are identified with the letters A through J. Thus, the connection hole in the top-left corner of the terminal strip area is A1, and the hole in the bottom-right corner is J30. The holes in the bus strips are not typically numbered.

Solderless breadboards come in several different sizes. Small breadboards usually have about 30 rows of terminal strips and about 400 holes altogether. But you can get larger breadboards, with 60 or more rows with 800 or more holes.

Laying out your circuit

The most difficult challenge of creating a circuit on a solderless breadboard is the task of translating a schematic diagram into a layout that can be assembled on the breadboard. Only in rare cases will a circuit assembled on a breadboard look like the circuit's schematic diagram. In most cases, the components are arranged differently and jumper wires are required to connect the components together.

REMEMBER

The key when assembling a circuit on a solderless breadboard is to ensure that every connection represented in the schematic diagram is faithfully re-created on the breadboard. For example, the schematic diagram in Figure 6-4 indicates that pin 1 of the 555 timer IC must be connected to ground. Thus, when you build the circuit on a breadboard, you must ensure that this connection is properly made.

One of the first challenges you face when building a circuit on a breadboard is connecting the pins on an integrated circuit. In a schematic diagram, the pin connections on an integrated circuit are rarely drawn in numerical order. For example, in the schematic diagram shown in Figure 6-4, the pin connections on the 555 timer IC are listed in this order, going counterclockwise from the top left: 7, 6, 2, 1, 3, 8, and 4. (Pin 5 is not used.)

But the pins on an actual 555 timer IC chip are arranged in numerical order starting at the top-left corner of the chip, as shown in Figure 6-8. Notice also that there are pins on the left and right side of the chip but none on the top or bottom. (The dot imprinted on the top of the chip is used to identify pin 1.)

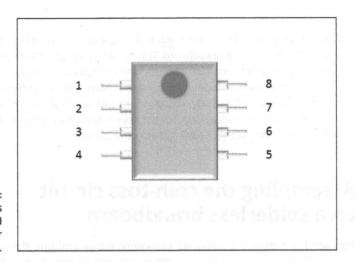

FIGURE 6-8:
How the pins
are numbered
on a 555 timer
integrated circuit.

You'll have to use your wits to re-create a circuit represented by a schematic dia-gram on a solderless breadboard. Here are some pointers to get you started:

>> Start by designating the top row of bus strips as the positive power supply and the bottom row as the ground. Connect your battery connector to holes in one end of these bus strips, but don't yet connect the battery; it's never a good idea to apply power to your circuit before you've finished assembling it.

>> Next, insert any ICs required for the circuit. Insert them so that they straddle the ditch in the middle of the terminal rows and, if your circuit has more than one IC, orient them all the same. You'll only confuse yourself if pin 1 is on the bottom-left corner of some of your ICs and on the top-right corner of others.

TIP

>> Each pin of each IC is connected to a terminal strip that has four additional connection holes. Thus, you can connect as many as four additional compo-nents or jumper wires to each pin. If your circuit requires more than four component connections to a single pin, use a jumper wire to extend the pin's terminal strip to an unused row anywhere on the breadboard.

>> Use jumper wires to connect the voltage source and ground pins for each IC to the nearest available connection hole in the voltage and ground buses.

>> Now work your way around the rest of the pins for each IC, connecting each component as needed. If one end of a component connects to an IC pin and the other end connects to either the voltage source or ground, plug one end of the component into an available connection hole on the terminal strip for the IC pin and plug the other end into the nearest available connection hole on either the voltage supply or ground buses.

>> If you want, you can trim the leads of the various components so that the parts fit closer to the breadboard. This results in a breadboard circuit that's neater, and with fewer bare leads sticking up high above the breadboard, the likelihood of leads accidentally coming in contact with each other and creating a short-circuit are less likely. I usually don't trim the leads, however, unless the circuit is complex enough that I'm not able to keep the component leads away from each other without cutting them down to size.

Assembling the coin-toss circuit on a solderless breadboard

This section presents a complete procedure for assembling the coin-toss circuit on a small solderless breadboard. When you get all your materials together, you should be able to complete this project in about an hour.

All the parts required to build this prototype circuit can be purchased from RadioShack, or you can order them online from any electronic parts supplier. For your convenience, here is a complete list of the parts you'll need to build this prototype circuit, along with the RadioShack catalog part numbers:

Quantity	Description
1	Small solderless breadboard
1	Solderless breadboard jumper wire kit
1	LM555 timer IC
1	1 kΩ, ¼ W resistor (5 per package)
1	10 kΩ, ¼ W resistor (5 per package)
2	470 Ω, ¼ W resistor (5 per package)
1	0.1 µF polyester film capacitor
1	Red LED 5 mm
1	Green LED 5 mm
1	9 V battery snap connector
1	9 V battery

You can build this circuit using equivalent parts from any supplier. So if you already have equivalent parts on hand, you don't need to run out to RadioShack and purchase them just for the sake of spending money.

You won't need many tools for this project. You can probably assemble it without any tools at all, but you may want to keep your wire cutters, wire strippers, and tweezers handy.

The steps that follow identify specific holes in the terminal strip area of the breadboard using numbers and letters. If you're using a different breadboard than the one listed in the parts list, you might encounter a different numbering system. If so, you can refer to Figure 6–7 to translate the numbers given in the steps for the breadboard you're using.

After you have everything you need, follow these steps to assemble the circuit:

Follow these three steps to insert the IC and connect it to power.

1. **Insert the 555 timer IC.**

   Take a close look at the 555 timer IC. On the top, notice a small dot in one corner; this dot marks the location of pin 1. Carefully insert the leads of the 555 timer IC into the breadboard near the middle of the board, inserting pin 1 into hole E14 and pin 8 in hole F14. The IC will straddle the groove that runs down the center of the board.

2. **Connect pin 1 of the 555 timer IC to the ground bus.**

   Insert one end of a small jumper wire into hole A14 and the other end into the nearest available hole in the bottommost bus strip.

3. **Connect pin 8 of the 555 timer IC to the +9 V bus.**

   Insert one end of a small jumper wire into hole J14 and the other end into the nearest available hole in the topmost bus strip.

4. **Connect pins 2 and 6 of the 555 timer together.**

   Insert one end of a small jumper wire into hole C15 and the other end in hole H16. The jumper wire will reach over the top of the 555 timer chip.

   Figure 6-9 shows what the breadboard looks like after these three steps.

The next five steps connect the LEDs and resistors R3 and R4. The LEDs will use the terminal strips in rows 19 and 21.

1. **Connect pin 3 of the IC to row 19.**

   Insert one end of a short jumper wire in hole C16 and the other end into hole C19.

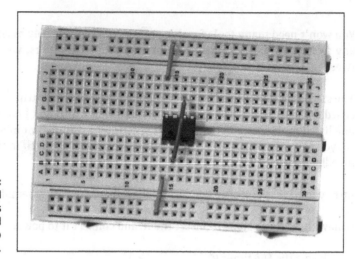

FIGURE 6-9:
The breadboard
after the IC has
been inserted
and connected to
the power buses.

2. **Connect the two segments of row 19.**

   Insert one end of a short jumper wire into hole E19 and the other end into hole F19. This jumper wire bridges the gap between the two terminal strips in row 19, effectively making them a single terminal strip.

3. **Insert the red LED.**

   If you look carefully at the red LED, you'll see that one lead is a bit shorter than the other. This short lead is called the *cathode.* The longer lead is called the *anode.* Insert the cathode (shorter lead) into hole D21. Then, insert the anode (longer lead) into hole D19.

4. **Insert the green LED.**

   The green LED also has a short cathode lead and a longer anode lead. Insert the anode (long) lead in hole G21 and the cathode (short) lead in hole G19.

   TECHNICAL STUFF

   Note that the leads of the two LEDs are installed reversed from one another: the red LED's anode and the green LED's cathode are inserted into row 19, while the red LED's cathode and the green LED's anode are inserted into row 21. There's a very good reason for this, but I wouldn't expect you to understand it yet even if I tried to explain it. So for now, take it on faith that you must install the two LEDs reversed like this for the circuit to work. (You learn more about LEDs, cathodes, and anodes in Book 2, Chapter 5.)

5. **Insert resistors R3 and R4.**

   Both of these resistors are 470 Ω. You can identify these resistors by looking at the three color strips painted on the resistors — they're yellow, purple, and brown. Insert one end of the first resistor in hole B21 and the other end in the nearest available hole in the bottommost bus strip (the ground bus). Then, insert one end of the other resistor in hole I21 and the other end in the nearest available hole in the topmost bus strip (the +9 V bus).

Figure 6-10 shows what the breadboard looks like after these steps.

FIGURE 6-10:
The breadboard after the LEDs have been connected.

The next five steps connect the finger–touch circuit that lets the user activate the coin toss by touching the two metal contacts. For the purposes of this prototype, you connect one end of a pair of jumper wires to the circuit and leave the other ends protruding from the end of the breadboard. Touching the bare ends of these wires with your fingers will simulate touching the metal contacts that you use in the final version of the circuit. The two jumper wires will be inserted into holes in row 9.

1. **Insert resistor R1 from pin 7 of the IC to the +9 V bus.**

   Resistor R1 is the 1 kΩ resistor, which should be connected between pin 7 of the IC and the +9 V bus. This resistor has stripes in the following sequence: brown, black, and red. Insert one end of this resistor into hole J15 and the other end into the nearest available hole in the topmost bus strip.

2. **Insert capacitor C1 from pin 2 of the IC to the ground bus.**

   Insert one lead of the capacitor (it doesn't matter which) into hole B15, and then insert the other into the nearest available slot in the bottommost bus strip.

3. **Insert resistor R2 from pin 7 of the IC to one of the metal contacts.**

This resistor is the 10 kΩ. It should be connected between pin 7 of the IC and one of the metal contacts that the user will touch with his finger to activate the coin-toss action. This resistor has the following sequence of color stripes: brown, black, and orange. Insert one end of it into hole H15 and the other end into hole H9.

4. **Connect a jumper wire from pin 2 of the IC to the other metal contact.**

Insert one end of a short jumper wire into hole B15 and the other end into hole B9.

5. **Insert the two jumper wires that simulate the metal contacts.**

Pick out a couple of jumper wires long enough to reach from row 9 and dangle an inch or so over the edge of the breadboard. Insert one end of these wires into holes E9 and F9 and leave the other ends free. Separate the ends of the two jumper wires to make sure that they're not touching; they should be about ½ inch apart.

Figure 6-11 shows what the breadboard looks like after these steps.

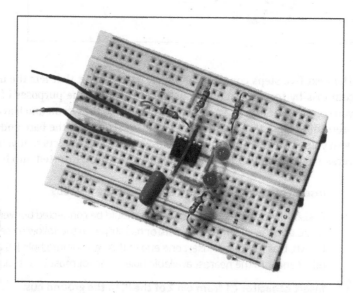

FIGURE 6-11:
The breadboard after the finger contact jumpers have been connected.

The remaining two steps complete the circuit by connecting the power supply.

1. **Connect the battery snap connector.**

The leads on the battery snap connector use stranded rather than solid wire, so you'll need to prepare them a bit before you insert them into the breadboard.

a. Use your wire strippers to strip off about ½ inch of insulation from the end of both leads.

b. Use your fingers to twist the leads as tightly as you can, so that no individual strands are protruding from the very tip of the wire.

c. Insert the red lead into the last hole of the topmost row and insert the black lead into the last hole of the bottommost row.

2. Connect the 9 V battery to the snap connector.

The red LED should immediately light up. (If not, see the troubleshooting tips in the next section.)

You can now test the circuit by touching both of the two free jumper wires. Pinch them both between your thumb and index finger, but don't let the wires actually touch each other. The resistance in your skin will conduct enough current to complete the circuit, and the LEDs will start alternately flashing red, green, red, green, and so on. They will continue to flash until you let go of the jumper wires. Then, one or the other will stay lit. When you touch the wires again, the flashing will resume.

Figure 6-12 shows the completed circuit in operation.

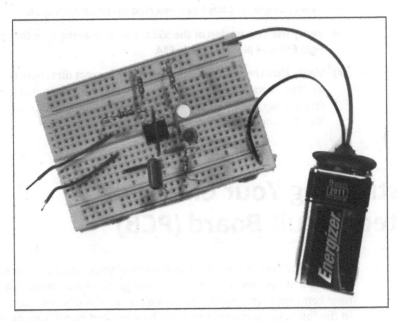

FIGURE 6-12:
The prototype of the coin tosser in operation.

Notice that if you squeeze the wires tightly, the rate at which the LEDs flash increases. If you squeeze tight enough, the LEDs will flash so fast that both will appear to be solid on. The LEDs are still flashing alternately, but they're flashing faster than your eye's ability to discern the difference, so they appear to be on constantly.

What if it doesn't work?

If your circuit doesn't work, there are a number of troubleshooting steps you can take to find out why and correct the problem. Here are some helpful troubleshooting tips:

>> **Examine all the component leads to make sure none of them are touching each other.** If any of the leads are touching, gently adjust them so that they're not touching.

>> **Make sure that the circuit is getting power.** Use your multimeter to test the battery voltage (see Chapter 8 of this minibook for information on how to do that), and verify that the leads from the battery snap connector are inserted properly into the solderless breadboard.

>> **Carefully double-check your wiring to make sure that every jumper and every component has been inserted in the correct spot.**

>> **Verify the orientation of the 555 timer IC, making sure that pin 1 is in hole E14 and pin 8 is in hole F14.**

>> **Verify that the LEDs are inserted in the correct direction.** For the red LED, the short lead (the cathode) goes in D21, and the long lead (the anode) goes in D19. For the green LED, the short lead (the cathode) goes in G19, and the long lead (the anode) goes in G21.

Constructing Your Circuit on a Printed Circuit Board (PCB)

When you're satisfied with the operation of your circuit, the next step is to build a permanent version of the circuit. Although there are several ways to do that, the most common is to construct the circuit on a *printed circuit board*, also called a *PCB*. In the following sections, you learn how printed circuit boards work and how to assemble the coin-toss circuit on a PCB.

TIP

Note that assembling a circuit on a PCB requires that you know how to solder. To learn that all-important skill, see Chapter 7 of this minibook.

Understanding how printed circuit boards work

A printed circuit board is made from a layer of insulating material such as plastic or some similar material. Copper circuit paths are bonded to one side of the board. The circuit paths consist of *traces*, which are like the wires that connect components, and *pads*, which are small circles of copper that the component leads can be soldered to. Figure 6-13 shows a typical printed circuit board.

FIGURE 6-13:
A printed circuit board.

There are two basic styles of printed circuit boards available:

>> **Through-hole:** A PCB in which the copper circuits are on one side of the board and the components are installed on the opposite side. In a through-hole PCB, small holes (usually $\frac{1}{16}$ inch in diameter) are drilled through the board at the center of the copper pads. Components are mounted to the blank side of the board by passing their leads through the holes and soldering the leads to the copper pads on the other side of the board. When the solder joint is completed, any excess wire lead is trimmed away.

>> **Surface-mount:** A PCB in which the components are installed on the same side of the board as the copper circuits. No holes are drilled.

Building Projects

Surface-mount PCBs are easier for large-scale automated circuit assembly. However, they're much more difficult to work with as a hobbyist because the components tend to be smaller and the leads are closer together. Thus, all the PCBs used in this book are of the through-hole variety.

Using a preprinted PCB

The easiest way to work with printed circuit boards is to purchase a preprinted board from an electronics parts supplier. You'll find a wide variety of preprinted PCBs at online distributors such as Hobby Engineering (www.hobbyengineering.com), Jameco Electronics (www.jameco.com), or All Electronics (www.allelectronics.com).

As you can see, preprinted PCBs come in a wide variety of shapes and sizes. The most useful, from a hobbyist's point of view, are the ones that mimic the terminal-strip and bus-strip layout of a solderless breadboard. For example, Figure 6-14 shows a PCB that has 550 holes laid out in a standard breadboard arrangement. With a preprinted PCB that has a breadboard layout, you can transfer your breadboard prototype circuit to the PCB without having to come up with an entirely new layout.

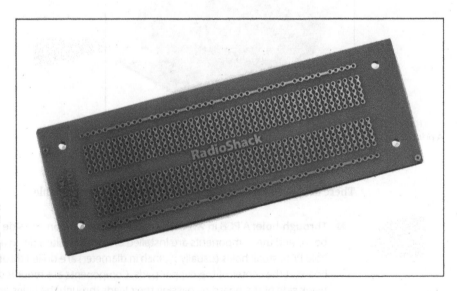

FIGURE 6-14:
A preprinted
PCB that uses a
standard
breadboard
layout.

Some preprinted circuit boards have layouts that are similar to standard breadboard layouts, but not identical. So check carefully before you build; you may have to make minor adjustments to your circuit layout to accommodate the PCB you're using.

TIP

If necessary, you can cut a larger PCB to a smaller size. One way to cut a PCB is to score it on both sides with a heavy-duty utility knife, and then snap it at the score. Another way is to cut it with a rotary tool such as a Dremel.

CREATING A CUSTOM PCB

As you get more advanced in your electronics skills, you may find that you want to create your own custom-printed circuit boards rather than force your circuit designs to fit the limited variety of preprinted circuit boards that are available. Although it isn't a trivial or inexpensive process, it's possible to make your own printed circuit boards customized for your circuit.

Here are the basic steps for creating your own printed circuit boards:

1. **Purchase a blank PCB. The entire surface of one side of this board will be completely coated with copper.**

2. **Create a *mask* on the copper surface that indicates the circuit layout.**

   There are several ways to do this. For simple circuits, you can simply hand-draw the circuit onto the copper using a special pen designed for this purpose. For more complicated circuits, you can purchase special stickers that are shaped like pads and common traces and place the stickers directly on the copper. Or you can design the circuit on your computer using any graphics drawing software, print the design on special paper, and transfer the design to the copper using a hot iron.

3. **Etch the board by dipping it into a special chemical that eats away all the copper that isn't covered by your mask.**

   This is a nasty process that needs to be done outside in a well-ventilated area while wearing gloves, a face mask, and goggles. When the etching is finished, all the copper that wasn't covered by the mask will be gone.

4. **Wash the board to get all that nasty copper etching solution off.**

5. **Scrub away the mask to reveal the beautiful copper circuit pattern left behind.**

6. **Drill holes in the center of each pad, and then assemble your circuit.**

Note that there are several companies on the Internet that will make small printed circuit boards for you. It isn't cheap, but the price per board comes down dramatically if you have them make more than just one. For example, PCBWay (www.pcbway.com) will make 4-inch-square circuit boards for about $3 per board if you order ten.

Building the coin-toss circuit on a PCB

This section presents a complete procedure for building the coin-toss circuit on a small preprinted PCB. When you get all your materials together, you should be able to complete this project in about an hour.

All the parts required to build this prototype circuit can be purchased from your local RadioShack. Or, you can order them online from any electronic parts supplier. For your convenience, here is a complete list of the parts you'll need to build this prototype circuit, along with the RadioShack catalog part numbers:

Quantity	Description
1	General-purpose, dual-printed circuit board
1	LM555 timer IC
1	1 kΩ, ¼ W resistor (5 per package)
1	10 kΩ, ¼ W resistor (5 per package)
2	470 Ω, ¼ W resistor (5 per package)
1	0.1 µF polyester film capacitor
1	Red LED 5 mm
1	Green LED 5 mm
1	Normally open, momentary-contact push button
1	9 V battery snap connector
1	9 V battery

You will also need about a foot each of 22-gauge solid insulated wire and 22-gauge stranded insulated wire. The color doesn't matter.

Note: This list is similar to the list I give earlier in this chapter for building the coin-toss circuit on a solderless breadboard. If you have the parts from that project, you can reuse them here.

Figure 6-15 shows the layout of the preprinted circuit board. Before we start building the circuit, study the layout of this board for a moment to familiarize yourself with it. As you can see, this board doesn't contain bus strips like those found on a breadboard. However, the overall layout of the board is similar to the layout of the terminal strips on a breadboard. The center portion of the board contains a total of 20 terminal strips, 10 on each side of the ditch. Each strip has three holes but is also connected to a second strip of two holes along the edge of the board. Thus, each strip effectively has five holes.

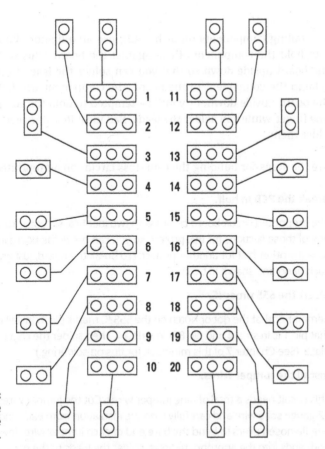

FIGURE 6-15:
The layout for the
PCB used in the
coin-toss circuit.

The strips aren't numbered on the board, but I've numbered them in Figure 6-15. I use the numbers 1 through 10 to number the strips on the left side of the board and the numbers 11 through 20 to number the strips on the right. In the instructions that follow, I use these numbers to indicate which holes to attach components or jumper leads to. *Note:* To keep things simple, I just specify the terminal strip number and leave it up to you to decide which of the five holes in the strip to use.

REMEMBER

The numbers used in the PCB layout are relative to the bottom of the board — that is, the surface of the board with the copper traces and pads. Thus, the numbers 1 through 10 are on the left and the numbers 11 through 20 are on the right. When you flip the board over to insert components from the top of the board, you'll have to mentally reverse the numbers: 1 through 10 is on the right and 11 through 20 is on the left.

TIP

When installing components onto the PCB, use an alligator clip as a temporary clamp to hold the components flush against the board. This will enable you to turn the board upside down so that you can solder the leads to the pads. If you don't clamp the component to the board, the component will fall out when you turn the board upside down, or you'll be tempted to hold the component in place with one finger while you solder the leads. Bad idea: Resistors get really hot when you solder them.

Here are the steps for building the coin-toss circuit on a preprinted PCB.

1. Break the PCB in half.

The preprinted circuit board comes with two identical sections. You need just one of those sections for this project, so you can break the board in half and save the other half for another project. (To break the board, just grab an end in each hand and snap it in two.)

2. Insert the 555 timer IC.

Remember that the dot or notch on the 555 IC marks pin 1. Install the chip so that pin 1 is in strip 4 and pin 8 is in strip 14. Then solder the chip carefully into place. (See Chapter 7 of this minibook for tips on soldering.)

3. Install the jumper wires.

This circuit needs a total of nine jumper wires. Cut the jumper wires from the 22-gauge solid wire and carefully strip the insulation from each end. Use needle-nosed pliers to bend the bare end of each jumper wire down, insert both ends into the appropriate holes, solder the leads to the pads, and then use your wire cutters to snip the excess of the end of each lead.

The following table provides the PCB strip locations for each jumper wire. Use your own judgment to determine which hole in the indicated strip to place the jumper wire in. Whenever possible, use the shortest possible path for each jumper wire.

Jumper Number	From strip	To strip
1	9	10
2	19	20
3	5	16
4	4	10
5	7	19
6	2	6
7	1	5

Jumper Number	From strip	To strip
8	2	12
9	14	19

4. Install the resistors.

There are four resistors to be installed. Use the following table to install each resistor into its correct location. Bend the leads down and insert each resistor into the correct holes, solder the resistor in place, and then snip off the excess wire from the ends of the leads.

Resistor Number	Value	Colors	From strip	To strip
R1	1 kΩ	Brown, black, red	15	20
R2	10 kΩ	Brown, black, orange	11	15
R3	470 Ω	Yellow, purple, brown	13	19
R4	470 Ω	Yellow, purple, brown	3	10

5. Install the capacitor.

Install the capacitor into strips 5 and 10. Push the capacitor all the way in until it's flush with the board. Then solder the leads to the pad and trim off the excess wire.

6. Install the LEDs.

Remember that LEDs are directional and must be installed in the correct direction or they won't work. One lead is shorter than the other to help you tell which lead is which. This short lead is the cathode, and the longer lead is the anode.

The following table shows where to install the LEDs:

LED Color	Cathode (short lead)	Anode (long lead)
Red	12	13
Green	3	2

When you install the LEDs, do *not* push the LED in until it is flush with the circuit board. Instead, push just a little bit of the leads into the holes so that the LED stands up about an inch from the top of the board.

7. Install the jumper wires for the metal contacts.

Cut two, 2-inch lengths of stranded wire and strip about 3/8 inch of insulation from each end. Solder one end of each wire into holes in strips 1 and 11 and leave the other ends free. When the circuit is installed in its final enclosure, you connect the ends of these wires to the metal posts that the user will touch to activate the coin-toss circuit.

TIP

Feeding the stranded wire through the holes in the circuit board can be tricky. First, carefully twist the loose strands until there are no stragglers protruding from the end of the wire. Then, carefully push the wire through the hole. If any of the strands get caught and refuse to go through the hole, pull the wire out and try again.

8. Connect the push button.

Cut a 2-inch length of stranded wire and strip about 3/8 inch of insulation from each end. Solder one end to either terminal on the push button (it doesn't matter which). Push the other end through a hole in strip 10 and solder it in place.

9. Connect the battery snap connector.

Strip off about 3/8 inch of insulation from the end of both leads. Then, solder the black lead to the free terminal on the push button (the terminal you did not use in Step 8) and solder the red lead to a hole in strip 20 on the PCB.

10. Connect the 9 V battery to the snap connector.

The red LED should immediately light up, indicating that the circuit is ready to do its decision-making work.

11. Turn off your soldering iron.

You're done!

Test the circuit by pinching both of the free jumper wires between your fingers. The LEDs should alternately flash until you let go, at which time one or the other will remain lit.

Figure 6-16 shows the completed circuit in operation.

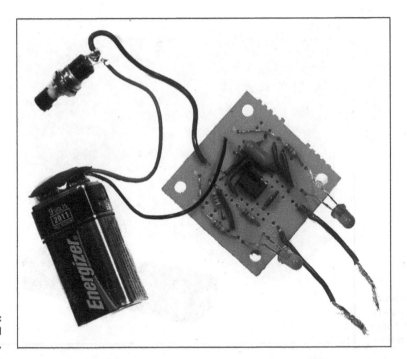

FIGURE 6-16:
The completed
coin-tosser PCB.

Finding an Enclosure for Your Circuit

When your circuit board is finished, the final step to completing your project is to mount it in a nice enclosure such as a plastic, metal, or wooden box. You can purchase plastic or metal boxes specifically designed for electronics projects from most electronic parts suppliers. Most RadioShack stores stock a half dozen or so different sizes in their stores, but you'll find a better assortment of sizes if you shop online. Figure 6-17 shows an assortment of boxes I picked up at my local RadioShack.

If you don't want to spend the money for a bona fide electronic project box, here are a few alternative ways to find the perfect enclosure for your project:

>> **Shop discount department stores for small storage boxes.** You might find one that's just the right size and shape for much less money than an official project box of the same size would cost. (If the storage box is an unattractive color, you can always give it a quick coat of paint.)

>> **In the electrical department of any hardware store, you'll find inexpensive plastic and metal boxes designed for household wiring.** Many of these boxes can be adapted for your electronic projects.

FIGURE 6-17:
Project boxes
come in a
variety of shapes
and sizes.

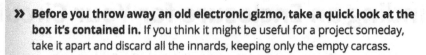

>> **Before you throw away an old electronic gizmo, take a quick look at the box it's contained in.** If you think it might be useful for a project someday, take it apart and discard all the innards, keeping only the empty carcass.

WARNING

Be careful whenever you disassemble any electronic device. Make sure you have first completely removed the power source and watch out for large capacitors that may be holding on to their charge.

>> **If you frequent yard sales, be on the lookout for items that might be useful as containers for your projects.**

Working with a project box

Most project boxes are made of plastic or metal and have a detachable lid that's held on with four screws, one at each corner of the lid. To gain access to the insides of the box, you simply remove the screws to release the lid.

The inside of the box may be completely smooth, or it may contain ridges or mounting studs designed to make it easier to mount components inside the box. If there are no such accoutrements inside the box, you'll have to devise your own method of attaching the various bits and pieces that need to go inside. Here are some tips:

>> **You'll need a good assortment of small drill bits to drill holes through the box to mount your components.** You'll need to drill holes to mount the circuit board as well as such things as battery holders, switches, LEDs, speakers, and whatever else your project may require.

>> **Make a good sketch of your project box and how its parts will be arranged before you start drilling holes.** When you're sure that you have everything laid out the way you want, use a marker to indicate the exact position of the holes you need to drill.

>> **To mount the circuit board, use *standoffs* to provide some empty space between the board and the case.** A *standoff* is a screw that allows you to mount the board so that it is raised above the bottom of the project box. You can purchase standoffs from any electronic parts supplier, but they are surprisingly expensive, often as much as 45 cents each.

TIP

If you have an ample supply of nuts and bolts, you can fashion your own standoffs simply by cutting a short length of plastic tubing — my favorite material is ¼-inch drip irrigation hose — and feeding a long bolt through it.

>> **Consider mounting the circuit board on the back of the lid rather than inside the body of the box.** This sometimes frees up more room within the box for larger items such as batteries or a speaker.

>> **Most switches can be mounted to a box by drilling a hole large enough to allow the neck of the switch to pass through.** The switch comes with a nut that you can tighten over the neck of the screw to secure the screw to the box.

>> **Some components don't have mounting nuts to secure them to the box.** For them, a small amount of epoxy or other glue can help set the component in place.

>> **Use stranded wire for connections within the project box.** Stranded wire holds up better to the handling it occasionally gets when you open the box, for example, to change the batteries.

Mounting the coin-toss circuit in a box

In this section, you finish the coin-toss project by mounting its circuit board in a plastic project box along with a 9 V battery, a power button, and the two metal contacts that the user can touch to toss the coin.

All the parts you need for this project can be purchased at most RadioShack stores — with the exception of the standoffs, which I got at a local hardware store.

In addition to the circuit board assembled earlier in this chapter, you'll need the following materials:

Quantity	Description
1	Project enclosure 5 x 2½ x 2 inches
1	9 V battery holder
8	½-inch 6-32 standoff male-to-female
6	6-32 nuts (to fit standoffs)
4	6-32¼-inch inch bolts (to fit standoffs)

When you've collected your parts, here's the procedure for putting the project together:

1. **Drill the required mounting holes in the lid.**

   You need to drill a total of eight holes in the lid. Four are for mounting the circuit board, two are for the LEDs, and two are for the metal finger contacts.

   Figure 6-18 provides a template you can use. The smaller holes are $\frac{3}{16}$ inch, the two larger holes are ¼ inch. **Note:** The distance between the four holes at the top of the lid must be exactly 1⅜ inches so that they line up with the mounting holes in the circuit board. The measurements for the other holes don't need to be as precise.

2. **Drill a $\frac{5}{16}$-inch hole for the push button near the top of the left side of the box.**

   The exact location isn't that important. I suggest you hold the box in the palm of your left hand to determine the best location to drill, where your thumb will be able to easily reach the button.

3. **Mount the push button in the box.**

   Remove the nut from the neck of the push button, pass the neck through the hole you drilled in the preceding step from the inside of the box, and then thread the nut onto the neck from the outside of the box and tighten it down with pliers.

4. **Use Epoxy or hot glue to glue the 9 V battery holder to the base of the box.**

   Position the holder near the top of the box, adjacent to the hole you drilled for the switch.

5. **Mount the circuit board standoffs.**

   Mount four of the ½-inch standoffs on the underside of the lid by inserting the threaded end of the standoff through the appropriate hole and attaching a 6-32 nut on the other side. See the left side of Figure 6-19 for guidance on how to attach these four standoffs.

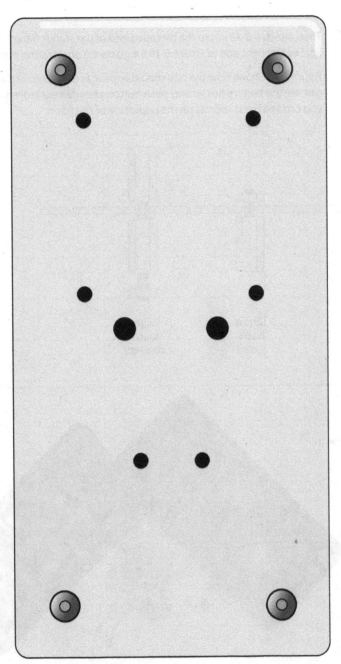

FIGURE 6-18:
Location for
drilling holes
in the lid of the
project box.

6. Mount the finger touch contacts on the lid.

To make the metal contacts that the user can touch to toss the coin, first insert the threaded end of one of the standoffs through the hole from the top side of the lid, and then screw a second standoff into it from the bottom of the lid.

Then attach a 6-32 nut to the threaded end of the standoff that's beneath the lid. See the right side of Figure 6-19 for guidance on installing these standoffs.

Figure 6-20 shows how the box should appear at this point. In this figure, you can see the battery holder and push button already installed in the box, and you can see the standoffs on the underside of the lid.

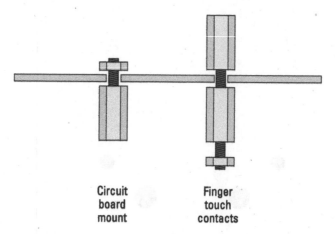

Circuit
board
mount

Finger
touch
contacts

FIGURE 6-19:
Installing the
standoffs.

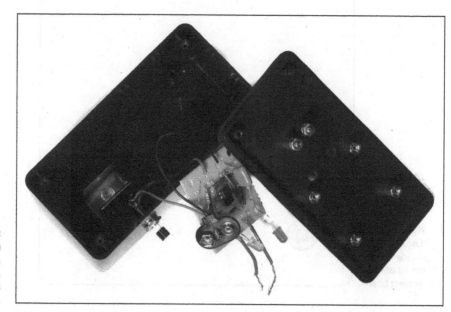

FIGURE 6-20:
The box with
the push button,
battery holder,
and standoffs
installed.

7. **Mount the circuit board to the standoffs.**

To complete this step, you must first bend the LEDs around so that they wrap over the edge of the circuit card and then face straight down. Be very careful when you bend the LEDs so that you don't break the leads or damage any of the solder joints. Figure 6-21 shows what the circuit board looks like when the LEDs are properly turned around.

When the LEDs are ready, position the circuit board over the four standoffs. The LEDs should slide right into the two $\frac{3}{16}$-inch holes you drilled for them in Step 1. If they don't, just nudge them a bit to make them fit. When everything is in place, secure the circuit board to the standoffs using the 6-32 bolts.

FIGURE 6-21:
The circuit board with the LEDs turned around.

8. **Connect the finger contacts.**

Attach the free ends of the two jumper wires that are connected to the circuit board to the two finger contacts. To connect each wire, wrap the stripped end of the wire tightly around the threaded part of the standoff. Then, attach a 6-32 nut to the standoff and tighten it with pliers.

Figure 6-22 shows what the project looks like with the circuit board installed into the lid and the jumpers connected to the finger contacts.

9. **Install the battery.**

Insert the battery into the holder, and then connect the snap adapter to the battery.

10. **Attach the lid to the box.**

Carefully flip the lid over and secure it to the box using the screws that came with the project box.

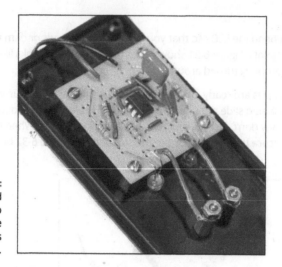

FIGURE 6-22:
The circuit board
attached to
the lid and the
finger contacts
connected.

11. **Turn it on and toss a coin!**

You're finally ready to use the coin-tosser project to help you make decisions. Hold the project box in your left hand and depress the push button with your thumb. Then, touch the index finger of your right hand to the two finger contacts, and watch the LEDs alternate. When you're ready, let go and see whether the red or green light stays lit.

Figure 6-23 shows the finished project.

FIGURE 6-23:
The completed
coin-toss project.

Chapter **7**

The Secrets of Successful Soldering

S oldering is one of the basic skills of building electronic projects. Although you can use solderless breadboards to build test versions of your circuits, sooner or later you'll want to build permanent versions of your circuits. To do that, you need to know how to solder.

In this chapter, you learn the basics of soldering: how soldering works, what tools and equipment you need for soldering, and how to create the perfect solder joint. You also learn how to correct your mistakes when (not if) you make them.

Keep in mind throughout this chapter that soldering is a skill that takes a bit of practice to master. It's not nearly as hard as playing the French horn, but solder-ing does have a learning curve. When you're first getting started, you'll feel like you're all thumbs, and your solder joints may look less than perfect. But stick with it — with a little practice, you'll get the hang of soldering in no time at all.

Understanding How Solder Works

Before we get into the nuts and bolts of making a solder joint, spend a few minutes thinking about what soldering is and what it isn't.

Soldering refers to the process of joining two or more metal objects by heating them, and then applying *solder* to the joint. Solder is a soft metal made from a combination of tin and lead. When the solder melts, which happens at about 700 degrees Fahrenheit, it flows over the metals to be joined. When the solder cools, it locks the metals together in a connection.

TIP

You can get lead-free solder to avoid the dangers of lead poisoning. However, lead-free solder is much more difficult to work with than normal lead-and-tin solder, so I suggest you avoid lead-free solder until you've mastered the art of soldering.

WARNING

Seven hundred degrees Fahrenheit is pretty hot — certainly hot enough to burn your skin instantly on contact — so soldering is an inherently dangerous task. It isn't extremely dangerous, as the amount of solder you actually use and the tools you use to solder are fairly small. So any burns you do receive are likely to be small. But they can be painful, so you should take great care whenever you're soldering.

Soldering is especially useful for electronics because not only does it create a strong physical connection between metals, but it also creates an excellent conductive path for electric current to flow from one conductor to another. This is because the solder itself is an excellent conductor. For example, you can create a reasonably good connection between two wires simply by stripping the insulation off the ends of each wire and twisting them together. However, current can flow through only the areas that are actually physically touching. Even when twisted tightly together, most of the surface area of the two wires won't actually be touching. But when you solder them, the solder flows through and around the twists, filling any gaps while connecting the entire surface area of both wires.

TIP

Soldering isn't the same as braising or welding. *Braising* is similar to soldering, but instead of solder, metals with a higher melting point (usually 850 degrees Fahrenheit or more) are used. Braising forms a stronger bond than soldering. *Welding* is an entirely different process. In welding, the metals to be joined are actually melted. In liquid form, the metals intermix and they cool to form an extremely strong bond.

Procuring What You Need to Solder

Before you can start soldering, you need to acquire some stuff, as described in the following sections.

Buying a soldering iron

A *soldering iron* — also called a *soldering pencil* — is the basic tool for soldering. Figure 7-1 shows a typical soldering iron.

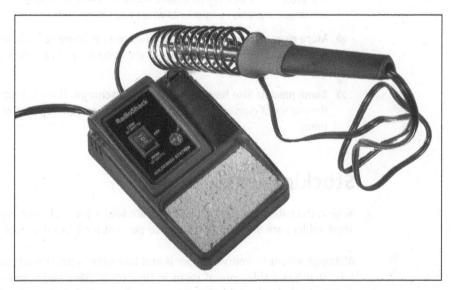

FIGURE 7-1:
A soldering iron.

Here are some things to look for when purchasing a soldering iron:

>> **The wattage rating should be between 20 and 50 watts.** Note that the wattage doesn't control how hot the soldering iron gets. Instead, it controls how fast it heats up and how fast it regains its normal operating temperature after completing each solder joint. (Each time you solder, the tip of the soldering iron cools a bit as it transfers its heat to the wires you're joining and to the solder itself. A higher-wattage soldering iron can maintain a stable temperature longer while you're soldering a connection and can reheat itself faster in between.)

>> **The tip should be replaceable.** When you buy your soldering iron, buy a few extra tips so you'll have replacements handy when you need them.

>> **Although you can buy a soldering iron by itself for under $10, I suggest you spend a few more dollars and buy a *soldering station* that includes an integrated stand.** A good, secure place to rest your soldering iron when not in use is essential. Without a good stand, I guarantee that your workbench will soon be covered with unsightly burn marks. (The soldering iron, shown in Figure 7-1, includes a soldering station.)

>> **A ground three-prong power plug is desirable but not essential.**

The grounding helps prevent static discharges that can damage some sensitive electronic components when you solder them.

>> **More expensive models have built-in temperature control.** Although not necessary, temperature control is a nice feature if you'll be doing a lot of soldering.

>> **Some models also have built-in static discharge.** This can help eliminate the chance of damaging your circuit's components while you're soldering them.

Stocking up on solder

Solder, the soft metal that's used to create solder joints, is an alloy of tin and lead. Most solders are 60 percent tin and 40 percent lead, but that ratio may vary a bit.

Although solder is wound on spools and looks like wire, it's actually a thin hollow tube that has a thin core of rosin in the center. This rosin, called the *flux*, plays a crucial role in the soldering process. It has a slightly lower melting point than the tin/lead alloy, and so it melts just a few moments before the tin/lead mixture melts. The flux prepares the metals to be joined by cleaning and lubricating the surfaces to be joined.

Solder comes in various thicknesses, and you'll need to have several different thicknesses on hand for different types of work. I suggest you start with three spools: 0.062 inch, 0.032 inch, and 0.020 inch. You'll use the 0.032 inch for most work, but the thick stuff (0.062 inch) comes in handy for soldering larger stranded wires — and the fine solder (0.020 inch) is useful for delicate soldering jobs on small components.

WARNING

Solder is about 40 percent lead, and as you probably know, lead poisoning is a very real health hazard. Fortunately, your actual exposure to lead while soldering is pretty small. The smoke that often puffs up when you're soldering is from the rosin flux, not from the lead or tin melting. Even so, it's a good idea to always work in a well-ventilated area, and to wash your hands after soldering to remove any residual lead.

You can purchase lead-free solder, although it's considerably more expensive than regular solder made from lead and tin. However, lead-free solder is more difficult to work with than normal solder and is less reliable. If you're concerned about the long-term effects of working with leaded solder, I suggest you switch to lead-free solder but only after you've become proficient at working with leaded solder.

WARNING

My attorney advised me to rewrite the previous paragraph to say that under no circumstance should you ever use solder that contains lead. He also insisted that I warn you to never plug your soldering iron in, lest it get hot and create the potential for burns, and that you should never ever connect your electronic circuits to actual electricity because it could create a shock hazard. In fact, you shouldn't even read this book because it could damage your eyes or cause a paper cut. Or you might accidentally drop it on your foot.

WARNING

Do *not* buy acid-flux solder, which plumbers use to solder pipes. The acid in this type of solder will destroy your electronics projects.

Other goodies you need

Besides a soldering iron and solder, there are a few additional things you'll need to have on hand for successful soldering. To wit:

>> **Third-hand tool or vise:** It takes at least three hands to solder: one to hold the items you're soldering, one to hold the soldering iron, and one to hold the solder. Unless you actually have three hands, you need to use a third-hand tool, a vise, or some other resourceful device to hold the items you're soldering so you can wield the soldering iron and the solder. (See Chapter 3 of this minibook for photographs of a third-hand tool and a hobby vise.)

>> **A sponge:** Used to clean the tip of the soldering iron.

>> **Alligator clips:** They serve two purposes when soldering. First, you can use an alligator clip as a clamp to hold a component in place while you solder it and, second, as a heat sink to avoid damaging a sensitive component when

soldering the component's leads. (A *heat sink* is simply a piece of metal attached to a heat-sensitive component that helps dissipate heat conducted by the component.)

WARNING

>> **Eye protection:** Always wear eye protection when soldering. Sometimes hot solder pops and flies through the air. Your eye and melted solder aren't a good mix.

>> **Magnifying glass:** Soldering is much easier if you do it through a magnifying glass so that you can get a better look at your work. You can use a tabletop magnifying glass, or a magnifying glass attached to a third-hand tool, or special magnifying goggles.

>> **Desoldering braid and desoldering bulb:** These tools are used to undo soldered joints when necessary to correct mistakes. For more information, see the section "Desoldering" later in this chapter.

Preparing to Solder

Before you start soldering, you should first prepare your soldering iron by clean-ing and tinning it. Follow these steps:

1. **Turn on your soldering iron.**

   It will take about a minute to heat up.

2. **When the iron is hot, clean the tip.**

   The best way to clean your soldering iron is to wipe the tip of the iron on a damp sponge. As you work, you should wipe the iron on the sponge frequently to keep the tip clean.

3. **Tin the soldering iron.**

   Tinning refers to the process of applying a light coat of solder to the tip of the soldering iron. Tinning the tip of your soldering iron helps the solder flow more freely once it heats up. To tin your soldering iron, melt a small amount of solder on the end of the tip. Then, wipe the tip dry with your sponge.

TIP

You should clean the tip of your soldering iron frequently as you work — ideally, immediately after every joint you solder. You should also tin the soldering iron occasionally. Usually, it's sufficient to tin the iron once at the start of each project. But if the solder disappears from the tip of the iron, you should tin it again.

THE TEN SOLDERING COMMANDMENTS

Have you heard of the Ten Soldering Commandments? In truth, they're actually more like guidelines than commandments. But if you heed them, it will go well with your solder joints, your children's solder joints, and your children's children's solder joints. So let it be written; so let it be done.

I. Thou shalt wear eye protection whenever thou solderest, lest thy get molten solder in thine eye.

II. Thou shalt not touch the heated end of thy soldering iron, lest thy burn thyself.

III. Thou shalt not fashion molten solder into false globs.

IV. Thou shalt wash thy hands after thou solderest, to remove vile lead contamination from upon thy hands before thou eatest.

V. Thou shalt provide bright illumination upon thine objects which thou solderest, that thou might see clearly the way unto which the solder may be applied.

VI. Thou shalt not spill thy excess solder upon thy neighbor's pad lest thy create unintended pathways through which current may flow.

VII. Thou shalt not leave thine hot soldering iron unattended.

VIII. Thou shalt not covet thy neighbor's professional-grade temperature-controlled soldering station.

IX. Thou shalt not apply solder directly upon thine soldering iron, but shalt instead apply solder to the objects which thou solderest, that their heat may melteth thy solder.

X. Thou shalt always place thine hot soldering iron in a suitable holder.

Soldering a Solid Solder Joint

The most common form of soldering when creating electronic projects is soldering component leads to copper pads on the back of a printed circuit board. If you can do that, you'll have no trouble with other types of soldering, such as soldering two wires together or soldering a wire to a switch terminal.

TIP

Just for fun, try saying "Soldering a Solid Solder Joint" real fast ten times.

The following steps outline the procedure for soldering a component lead to a PC board:

1. **Inspect the component leads to make sure they're clean.**

   If the component is really old, you may need to touch it up with an emery board to remove any accumulated gunk.

2. **Pass the component leads through the correct holes.**

   Check the circuit diagram carefully to be sure that you have installed the component in the correct location. If the component is polarized (such as a diode or an integrated circuit), verify that the component is oriented correctly. You don't want to solder it in backward.

3. **Secure the component to the circuit board.**

   If the component is near the edge of the board, the easiest way to secure it is with an alligator clip. You can also secure the component with a bit of tape.

4. **Clamp the circuit board in place with your third-hand tool or vise.**

   Orient the board so that the copper-plated side is up. If you're using a magnifying glass, position the board under the glass.

5. **Make sure you have adequate light.**

   If you have a desktop lamp, adjust it now so that it shines directly on the connection to be soldered.

6. **Touch the tip of the soldering iron to both the pad and the lead at the same time.**

   It's important that you touch the tip of the soldering iron to both the copper pad and the wire lead. The idea is to heat them both so that solder will flow and adhere to both.

   The easiest way to achieve the correct contact is to use the tip of the soldering iron to press the lead against the edge of the hole, as shown in Figure 7-2.

7. **Let the lead and the pad heat up for a moment.**

   It should take only a few seconds for the lead and the pad to heat up sufficiently.

8. **Apply the solder.**

   Apply the solder to the lead on the opposite side of the tip of the soldering iron, just above the copper pad. The solder should begin to melt almost immediately.

   Figure 7-3 shows the correct way to apply the solder.

FIGURE 7-2:
Positioning the
soldering iron.

WARNING

Do *not* touch the solder directly to the soldering iron. If you do, the solder will melt immediately, and you may end up with an unstable connection, often called a *cold joint*, where the solder doesn't properly fuse itself to the copper pad or the wire lead.

FIGURE 7-3:
Applying the
solder.

9. **When the solder begins to melt, feed just enough solder to cover the pad.**

As the solder melts, it will flow down the lead and then spread out onto the pad. You want to feed just enough solder to completely cover the pad, but not enough to create a big glob on top of the pad.

TIP

Be stingy when applying solder. It's more common to have too much solder than too little, and it's a lot easier to add a little solder later if you didn't get quite enough coverage than it is to remove solder if you applied too much.

10. **Remove the solder and soldering iron and let the solder cool.**

Be patient — it will take a few seconds for the solder to cool. Don't move anything while the joint is cooling. If you inadvertently move the lead, you'll create an unstable cold joint that will have to be resoldered.

11. **Trim the excess lead by snipping it with wire cutters just above the top of the solder joint.**

Use a small pair of wire cutters so you can trim it close to the joint.

Checking Your Work

After you've completed a solder joint, you should inspect it to make sure the joint is good. Look at it under a magnifying glass, and gently wiggle the component to see if the joint is stable. A good solder joint should be shiny, and it should fill but not overflow the pad, as shown in Figure 7-4.

Nearly all bad solder joints are caused by one of three things: not allowing the wire and pad to heat sufficiently, applying too much solder, or melting the solder with the soldering iron instead of with the wire lead. Here are some indications of a bad solder joint:

» **The pad and lead aren't completely covered with solder, enabling you to see through one side of the hole through which the lead passes.** Either you didn't apply quite enough solder, or the pad wasn't quite hot enough to accept the solder.

» **The lead is loose in the hole or the solder isn't firmly attached to the pad.** One possible reason for this is that you moved the lead before the solder had completely cooled.

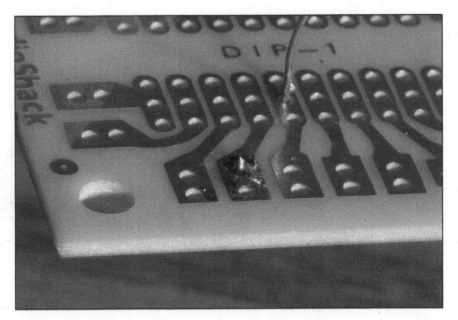

FIGURE 7-4:
A good solder
joint.

>> **The solder isn't shiny.** Shiny solder indicates solder that heated, flowed, and then cooled properly. If the solder gets just barely hot enough to melt, then flows over a wire or pad that isn't heated sufficiently, it will be dull when it cools. (Unfortunately, the new lead-free solder almost always cools dull, so it looks like a bad solder joint even when the joint is good!)

>> **Solder overflows the pad and touches an adjacent pad.** This can happen if you apply too much solder. It can also happen if the pad didn't get hot enough to accept the solder, which can cause the solder to flow off the pad and onto an adjacent pad. If solder spills over from one pad to an adjacent pad, your circuit may not work correctly.

Desoldering

Desoldering refers to the process of undoing a soldered joint. You may have to do this if you discover that a solder joint is less than satisfactory, if a component fails, or if you connect your circuit incorrectly.

To desolder a solder connection, you need a desoldering bulb and a desoldering braid. Figure 7-5 shows these tools.

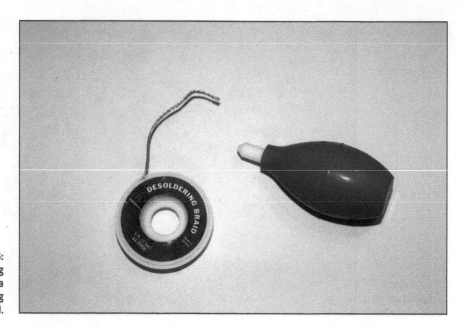

FIGURE 7-5:
A desoldering
bulb and a
desoldering
braid.

Here are the steps for undoing a solder joint:

1. **Apply the hot soldering iron to the joint you need to remove.**

Give it a second for the solder to melt.

2. **Squeeze the desoldering bulb to expel the air it contains, and then touch the tip of the desoldering bulb to the molten solder joint and release the bulb.**

As the bulb expands, it will suck the solder off the joint and into the bulb.

3. **If the desoldering bulb didn't completely free the lead, apply heat again and touch the remaining molten solder with the desoldering braid.**

The desoldering braid is specially designed to draw solder up much like a candlewick draws up wax.

4. **Use needle-nose pliers or tweezers to remove the lead.**

WARNING

Do *not* try to remove the lead with your fingers after you have desoldered the connection. The lead will remain hot for a while after you have desoldered it.

Chapter 8

Measuring Circuits with a Multimeter

Wouldn't it be great if every circuit you ever built worked right the first time you built it? You'd quickly develop a reputation as an electronic genius and in no time at all you'd be the president of Intel.

But in the real world, a circuit doesn't always work right the first time. When it doesn't, you can scratch your head, stare at it, put a dead chicken in a bag and swing it over your head, or you can pull out your test equipment and analyze the circuit to find out what went wrong.

In this chapter, you learn how to use one of the electronic guru's favorite tools: the multimeter. Learn how to use it well. It will be your trusty companion throughout your electronic journeys.

Looking at Multimeters

You know those late-night commercials that try to sell you those amazing kitchen gadgets that are like a combination blender, juicer, food processor, mixer, and snow cone maker all in one? A multimeter is kind of like one of those crazy kitchen gadgets, except that, unlike the crazy kitchen gadgets, a multimeter really *can* do all the things it claims it can do. And, unlike the crazy kitchen gadgets, the things a multimeter can do turn out to be genuinely useful.

Along with a good soldering iron, a good multimeter is the most important item in your toolbox. Learn how to use it, and your electronic exploits will be much more fruitful.

Figure 8-1 shows a simple, inexpensive multimeter. This one can be purchased from almost any hardware store for under $20. If you shop around, however, you can often find a basic multimeter for under $10 and sometimes for as little as $5.

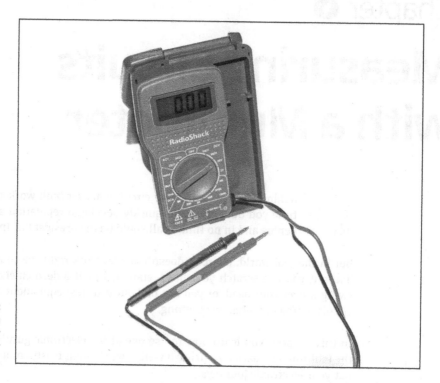

FIGURE 8-1:
You can buy a basic multimeter like this one for under $20.

Of course, you can also spend much more, but if you're just getting started, an inexpensive multimeter is fine. Eventually, you'll want to invest a little more money in a better-quality multimeter.

The multimeter shown in Figure 8-1 is a *digital* multimeter, which displays its values using a digital display that shows the actual numbers for the measurements being taken. The alternative to a digital multimeter is an *analog* multimeter, which shows its readings by moving a needle across a printed scale. To determine the value of a measurement, you simply read the scale behind the needle.

Figure 8-2 shows a typical analog multimeter. This one happens to be one of my favorites. Even though it's old enough to be called vintage, it's still an excellent and accurate meter. One of the benefits of spending a little more to buy quality equipment is that the equipment will give you many years — sometimes decades — of reliable service.

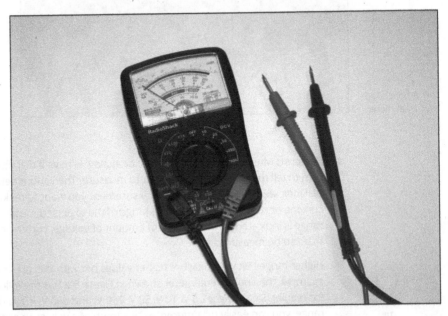

FIGURE 8-2:
An analog
multimeter.

For the most precise measurements, you can pick up a lab-quality bench-top multimeter such as the one shown in Figure 8-3. These meters usually cost $100 or more but offer additional features such as data logging, USB or network connectivity, and a wide array of test leads useful for different circumstances.

The following paragraphs describe the various parts that make up a typical multimeter:

>> **Display or meter:** Indicates the value of the measurement being taken. In a digital multimeter, the display is a number that indicates the amperage (current), voltage, or resistance being measured. In an analog meter, the current, voltage, or resistance is indicated by a needle that moves across a printed scale. To read the value, you look straight down at the needle and read the scale printed behind it.

FIGURE 8-3:
A bench-top
multimeter.

TIP

>> **Selector:** Most multimeters — digital or analog — have a dial that you can turn to tell the meter what you want to measure. The various settings on this dial indicate not only the type of measurement you want to make (voltage, current, or resistance) but also the range of the expected measurements. The range is indicated by the maximum amount of voltage, current, or resistance that can be measured.

Higher ranges let you measure higher values but with less precision. For example, the analog multimeter shown in Figure 8-2 has the following ranges for reading DC voltage: 2.5 V, 10 V, 50 V, 250 V, and 500 V. If you use the 2.5 V range, you can easily tell differences of a tenth of a volt, such as the difference between 1.6 and 1.7 V. But when the range is set to the 500 V range, you'll be lucky to pick out differences of 10 volts.

>> **On/off switch:** Some multimeters don't have an on/off switch. Instead, one of the positions on the selector dial is Off. Other multimeters have a separate on/off switch. If your meter doesn't give you any readings, check to make sure the power switch is turned on or that the battery doesn't need to be replaced.

>> **Test leads:** The *test leads* are a pair of red and black wires with metal probes on their ends. One end of these wires plugs into the meter. You use the other end to connect to the circuits you want to measure. The red lead is positive; the black lead is negative.

What a Multimeter Measures

A *meter* is a device that measures electrical quantities. A multimeter, therefore, is a combination of several different types of meters all in one box. At the minimum, a multimeter combines three distinct types of meters (ammeter, voltmeter, and ohmmeter) into a single device, as described in the following sections.

Ammeter

As you learn in Chapter 2 of this minibook, *current* is the flow of electric charge through a conductor. Current is measured in units called *amperes*. It should come as no surprise, then, that a meter that measures amperage is called an *ammeter*.

Don't ask me why the *p* is dropped to form the word *ammeter*, rather than *ampmeter*. After all, the short form of the word *ampere* is *amp*, not *am*. Go figure.

Very few electronic circuits have currents so strong that they can be measured in actual amperes. So ammeters usually measure current in *milliamperes*, also called a *milliamp* and usually abbreviated *mA*. One mA is one-thousandth of an ampere; in other words, there are 1,000 milliamps in an amp.

TECHNICAL STUFF

The world's first ammeter was invented by a Dutch physicist named Hans Christian Oersted in 1821, when he accidentally left a compass next to a wire that had an electric current flowing through it. Hans noticed that when the current flowed, the needle moved away from its normal northerly orientation and pointed toward the wire. This is because current moving through a wire creates a magnetic field around the wire, and the magnetic field was strong enough to attract the magnetized end of the compass needle.

After fooling around with it a bit, Hans discovered that the more current he ran through the wire, the farther the needle strayed away from north. It didn't take him long to figure out that this discovery could be utilized to measure the amount of current flowing through a circuit. Analog ammeters work by this very same principle even today.

Hans Christian Oersted wasn't the famous writer of children's stories; that was Hans Christian Andersen. In a strange twist of history, though, Hans Christian Oersted was close friends with Hans Christian Andersen. It's entirely possible that they were the founding members of some secret society of Evil Mad-Scientist Children's Book Writers. Perhaps we'll read about them in Dan Brown's next novel.

Voltmeter

In Chapter 2 of this minibook, you learn about a second fundamental quantity of electricity, *voltage*, a term that refers to the difference in electric charge between two points. If those two points are connected to a conductor, a current will flow through the conductor. Thus, voltage is the instigator of current.

The unit of voltage is, naturally, the *volt*, and a device that measures voltage is called a *voltmeter*.

It turns out that, all other things being equal, a change in the amount of voltage between two points results in a corresponding change in current. Thus, if you can keep things equal, you can measure voltage by measuring current, and you already know of a device that can measure current: It's called an ammeter.

The basic difference between an ammeter and a voltmeter is that in an ammeter, you let current run directly through the meter so that you can measure the amount of current. In a voltmeter, the current is first run through a very large resistor and then through the ammeter, and the device makes the necessary calculations as follows.

You haven't learned it yet (unless you're eagerly skipping around the book), but in Book 2, Chapter 1, you find out that there's a direct relationship between voltage, resistance, and current in an electrical circuit. In particular, if you know any two of these quantities, the third one is easy to calculate. In a voltmeter, a large fixed resistance is used, and the ammeter measures the current. Because you know the amount of the fixed resistance and the amount of current, you can easily calculate the amount of voltage across the circuit.

Don't worry; you don't have to do any math to calculate this voltage. The voltmeter does the math for you. In an analog voltmeter, the calculation is built in to the scale that's printed on the meter, so all you have to do is look at the position of the needle on the scale to read the voltage. In a digital voltmeter, the voltage is automatically calculated and displayed digitally.

TIP

For a brief explanation of the relationship between current, voltage, and resistance, see the sidebar, "A sneak peek at Ohm's law."

Ohmmeter

As you know, a *resistor* is a material that resists the flow of current. How much the current is restricted is a function of the amount of resistance in the resistor, which is measured in units called *ohms*. The symbol for ohms is the Greek letter omega (Ω). A device that measures resistance is called an *ohmmeter*.

Like voltage, resistance can also be measured with an ammeter. Remember how I say in the previous section that there's a direct relationship between voltage, resistance, and current in any circuit, and that if you know any two of these quantities you can easily calculate the third? To recap, to measure voltage, a voltmeter provides a fixed resistance, uses an ammeter to measure the current, and then uses the resistance and current to calculate the voltage.

To measure the resistance of a circuit, an ohmmeter provides a fixed amount of voltage across the circuit, uses an ammeter to measure the current that flows through the circuit, and then uses the amount of voltage provided by the meter and the amount of current read by the meter to calculate the resistance.

As with voltage, you don't have to do this calculation; the calculation is automatically made by digital multimeters and is built in to the meter scale for analog multimeters. Thus, all you have to do is read the display or the needle on the meter to determine the resistance.

Other measurements

All multimeters can measure current, voltage, and resistance. Some multimeters can perform other types of measurements as well. For example, some meters can measure the capacitance of capacitors, and some meters can test diodes or transistors. These features are handy, but not essential.

Schematic symbols for meter functions

Ammeter, voltmeter, and ohmmeters are often included in schematic diagrams. When they are, the following symbols are used:

Symbol	Meaning
—(A)—	Ammeter
—(V)—	Voltmeter
—(Ω)—	Ohmmeter

A SNEAK PEEK AT OHM'S LAW

In Book 2, Chapter 2, you take a closer look at resistors and also take an in-depth look at *Ohm's law,* which is one of the most important mathematical relationships in electronics. But without going into all the details of Ohm's law here, I want to at least give you a sneak preview.

Ohm's law describes a fundamental relationship between current, voltage, and resistance in an electrical circuit. Remember from Chapter 2 that voltage is a difference in electrical charge between two points, and that if those two points are connected by a conductor, current will flow through the conductor.

Other than exotic superconductors (which exist only in laboratory experiments), no conductor is perfect. All conductors have a certain amount of resistance that inhibits the flow of current. The greater this resistance, the less the current will flow. The less this resistance is, the more the current will flow.

Ohm's law is a mathematical formula that formalizes the relationship between current, voltage, and resistance. The formula is this:

$$I = \frac{V}{R}$$

In other words, the amount of current running through a circuit is equal to the amount of voltage across the circuit divided by the amount of resistance in the circuit. The amount of current in amperes is represented by the letter *I* (don't ask why; it just is). *V* represents voltage in volts, and *R* represents resistance in ohms.

Using basic math, you can use this equation to calculate the voltage if you know the current and the resistance. Then, the formula becomes:

$$V = IR$$

In other words, voltage is equal to current times resistance.

Similarly, you can calculate resistance if you know the current and the voltage. Then, the formula becomes:

$$R = \frac{V}{I}$$

In other words, resistance equals the voltage divided by the current.

Using Your Multimeter

In the following sections, I show you how to use your multimeter to measure current, voltage, and resistance in a simple circuit. The circuit being measured consists of just three components: a 9 V battery, a light-emitting diode (LED), and a resistor. The schematic for this circuit is shown in Figure 8-4.

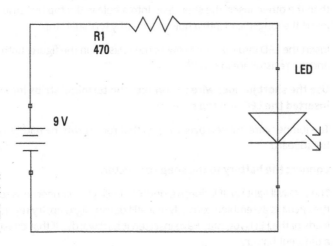

FIGURE 8-4:
A simple circuit with a battery, a resistor, and an LED.

If you want to follow along with the measuring procedures detailed in the following sections, you can build this circuit on a solderless breadboard. You'll need the following parts:

>> Small solderless breadboard

>> 470 Ω, ¼ W resistor

>> Red LED, 5 mm

>> 9 V battery snap connector

>> 9 V battery

>> Short length of jumper wire (1 inch or less)

Figure 8-5 shows the circuit installed on the breadboard. Here are the steps for building this circuit:

1. **Connect the battery snap connector.**

   Insert the black lead in the top bus strip and the red lead in the bottom bus strip. Any hole will do, but it makes sense to connect the battery at the very end of the breadboard.

2. Connect the resistor.

Insert one end of the resistor into any hole in the bottom bus strip. Then, pick a row in the nearby terminal strip and insert the other end into a hole in that terminal strip.

3. Connect the LED.

Notice that the leads of the LED aren't the same length; one lead is shorter than the other. Insert the short lead into a hole in the top bus strip, and then insert the longer lead into a hole in a nearby terminal strip.

TIP

Insert the LED into the same row as the resistor. In the figure, both the LED and the resistor are in row 26.

4. Use the short jumper wire to connect the terminal strips into which you inserted the LED and the resistor.

The jumper wire will hop over the gap that runs down the middle of the breadboard.

5. Connect the battery to the snap connector.

The LED will light up. If it doesn't, double-check your connections to make sure the circuit is assembled correctly. If it still doesn't light up, try reversing the leads of the LED (you may have inserted it backwards). If that doesn't work, try a different battery.

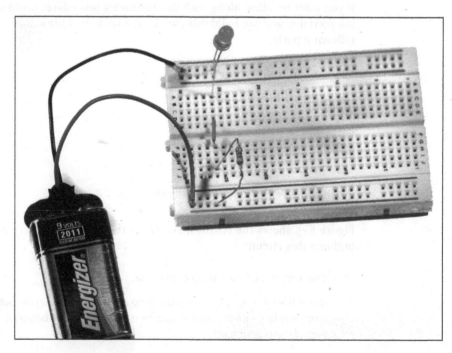

FIGURE 8-5:
The LED circuit assembled on a breadboard.

WARNING

Do *not* connect the LED directly to the battery without a resistor. If you do, the LED will flash brightly, and then it will be dead forever.

Measuring current

Electric current is measured in amperes, but actually in most electronics work, you'll measure current in milliamps, or mA. To measure current, you must connect the two leads of the ammeter in the circuit so that the current flows through the ammeter. In other words, the ammeter must become a part of the circuit itself.

The only way to measure the current flowing through the LED circuit that's shown in Figure 8-4 is to insert your ammeter into the circuit. Figure 8-6 shows one way to do this. Here, the ammeter is inserted into the circuit between the LED and the resistor.

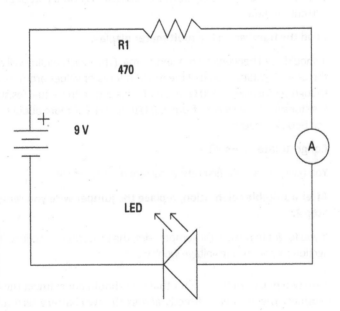

FIGURE 8-6:
Using an ammeter to measure current flow in the LED circuit.

Note that it doesn't matter where in this circuit you insert the ammeter. You'll get the same current reading whether you insert the ammeter between the LED and the resistor, between the resistor and the battery, or between the LED and the battery.

To measure the current in the LED circuit, follow these steps:

1. **Set your multimeter's range selector to a DC milliamp range of at least 20 mA.**

   This circuit uses direct current (DC), so you need to make sure the multimeter is set to a DC current range.

2. **Remove the jumper wire that connects the two terminal strips.**

   The LED should go dark, as removing the jumper wire breaks the circuit.

3. **Touch the black lead from the multimeter to the LED lead that connects to the terminal strip (not the bus strip).**

4. **Touch the red lead from the multimeter to the resistor lead that connects to the terminal strip (not the bus strip).**

   The LED should light up again, as the ammeter is now a part of the circuit, and current can flow.

5. **Read the number on the multimeter display.**

   It should read between 12 mA and 13 mA. (The exact reading will depend on the exact resistance value of the resistor. Resistor values aren't exact, so even though you're using a 470 Ω resistor in this circuit, the actual resistance of the resistor may be anywhere from 420 Ω to 520 Ω. For more about this effect, see Book 2, Chapter 2.)

6. **Congratulate yourself!**

   You have made your first official current measurement.

7. **After a suitable celebration, replace the jumper wire you removed in Step 2.**

   If you forget to replace the jumper wire, the procedure described in the next section for measuring voltage won't work.

WARNING   There are two places in this circuit that you should *not* connect the ammeter. First, don't connect the ammeter directly across the two battery terminals. This effectively shorts out the battery. It will get real hot, real fast. Second, don't connect one lead of the ammeter to the positive battery terminal and the other directly to the LED lead. That will bypass the resistor, which will probably blow out the LED.

If you want to experiment a little more, try measuring the current at other places in the circuit. For example, remove the battery snap connector from the battery, and then reconnect it so that just the negative battery terminal is connected. Then, touch the red meter lead to the positive battery terminal and the black lead to the lead of the resistor that's connected to the bus strip (not the lead that's connected to the terminal strip). This measures the current by inserting the ammeter between the resistor and the battery. You should get the same value that you got when you measured between the LED and the resistor.

You can use a similar method to measure the current between the LED and the negative battery terminal. Again, the result should be the same.

Measuring voltage

Measuring voltage is a little easier than measuring current because to measure voltage, you don't have to insert the meter into the circuit. Instead, all you have to do is touch the leads of the multimeter to any two points in the circuit. When you do, the multimeter displays the voltage that exists between those two points.

For example, Figure 8-7 shows how you can insert a voltmeter into the LED circuit so that you can measure voltage. In this case, the voltage is measured across the battery. It should read in the vicinity of 9.3 V. (9 V batteries generally provide a bit more than a full 9 V unless you've placed a load on the circuit.)

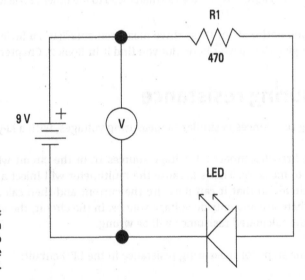

FIGURE 8-7:
Using a voltmeter to measure voltage in the LED circuit.

To measure voltages in the LED circuit, first put the circuit back together (assuming you took it apart to measure currents). Then spin the multimeter dial to a range whose maximum is at least 10 V. Now just touch the leads to different spots in the circuit. To measure the voltage across the entire circuit as shown in Figure 8-7, touch the black lead to the LED lead that's inserted into the negative bus strip, and touch the red lead to the resistor lead that's inserted into the positive bus strip.

TIP

If you connect the leads backward when measuring voltage, you'll read a negative voltage. On a digital voltmeter, the meter will display a negative number. On an analog meter, the needle will move in the wrong direction. (Unless the voltage is large, this won't damage your meter.)

Here's an interesting exercise. Write down the following three voltage measurements:

>> **Across the battery:** Connect the red meter lead to the resistor lead that's inserted into the positive bus strip and the black meter lead to the LED lead that's inserted into the negative bus strip.

>> **Across the resistor:** Connect the red meter lead to the resistor lead that's inserted into the positive bus and the black meter lead to the other resistor lead.

>> **Across the LED:** Connect the black meter lead to the LED lead that's inserted into the negative bus and the red meter lead to the other LED lead.

What do you notice about these three measurements? (It's a little bit of a puzzle, so I won't give the answer here. But you find it in Book 2, Chapter 2.)

Measuring resistance

Measuring resistances is similar to measuring voltages, with a key difference:

REMEMBER

You must first disconnect all voltage sources from the circuit whose resistance you want to measure. That's because the multimeter will inject a known voltage into the circuit so that it can measure the current and then calculate the resistance. If there are any outside voltage sources in the circuit, the voltage won't be fixed, so the calculated resistance will be wrong.

Here are the steps for measuring resistance in the LED circuit:

1. **Remove the battery.**

   Just unplug it from the battery snap connector and set the battery aside.

2. **Turn the meter selector dial to one of the resistance settings.**

TIP

   If you have an idea of what the resistance is, pick the smallest range that's greater than the value you're expecting. Otherwise, pick the largest range available on your meter.

3. **If you're using an analog meter, calibrate it.**

Analog meters must first be calibrated before they can give an accurate resistance measurement. To calibrate an analog meter, touch the two meter leads together. Then, adjust the meter's calibration knob until the meter indicates 0 resistance.

4. **Touch the meter leads to the two points in the circuit for which you wish to measure resistance.**

For example, to measure the resistance of the resistor, touch the meter leads to the two leads of the resistor. The result should be in the vicinity of 470 Ω.

You can learn much more about measuring resistances in Book 2, Chapter 2. But until then, here are a few additional thoughts to tide you over:

>> When you measure the resistance of an individual resistor or of circuits consisting of nothing other than resistors, it doesn't matter what direction the current flows through the resistor. Thus, you can reverse the multimeter leads and you'll still get the same result.

>> Some components such as diodes pass current better in one direction than in the other. In that case, the direction of the current does matter. You can learn more about this effect in Book 2, Chapter 5.

>> Resistors aren't perfect. Thus, a 470 Ω resistor rarely provides exactly 470 Ω of resistance. The usual tolerance for resistors is 5 percent, which means that a 470 Ω resistor should have somewhere between 446.5 Ω and 495.5 Ω of resistance. For most circuits, this amount of imprecision doesn't matter. But in circuits where it does, you can use the ohmmeter function of your multimeter to determine the exact value of a particular resistor. Then, you can adjust the rest of your circuit accordingly. (More on this in Book 3, Chapter 2.)

Chapter 9

Catching Waves with an Oscilloscope

Electronics can generate some gnarly waves, dude. You'll encounter all kinds of totally hairy waves in your electronic circuits. You meet sine waves, which just roll along nice and easy, like corduroy on the horizon. Sawtooth waves are epic: They ride up slow, then drop you way fast. And of course, square waves. They're just, you know, *square*.

Your basic multimeter, which you learn about in the preceding chapter, is essential. You can't do electronics without one. In this chapter, I tell you about another incredibly useful tool, called an *oscilloscope*. Although a voltmeter can give you a simple number that represents voltage, an oscilloscope can draw a picture of voltage. And, as they say, a picture is worth a thousand words — er, numbers.

Why is a picture so much more valuable than a number, when it comes to voltage? Because in all but the simplest circuits, voltage is always in motion; it's always changing, and an oscilloscope is the perfect tool for observing voltage in motion.

As I said, an oscilloscope — often just called a scope — is an incredibly useful tool to have on your workbench. Although an oscilloscope is a bit expensive, you can pick up a good digital scope for a few hundred dollars. You can get by for a while without an oscilloscope, but eventually you'll want to get one.

The purpose of this chapter is really twofold. First, I want to show you how to use an oscilloscope should you manage to get your hands on one. Second, and perhaps more importantly, I want to convince you to start saving your pocket change so that someday you'll be able to afford one. Once you get an oscilloscope on your workbench, you'll wonder how you ever managed without it.

Understanding Oscilloscopes

Figure 9-1 shows a typical, midrange oscilloscope, which can be purchased for about $500. A scope like this one will serve you well for many years.

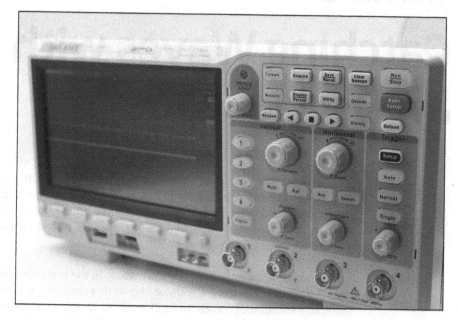

FIGURE 9-1:
A typical
oscilloscope.

The most obvious feature of any oscilloscope is its screen. On older oscilloscopes, the screen is a cathode-ray tube (CRT) similar to an older television or computer monitor. On newer oscilloscopes, the screen is an LCD display like a flat-screen computer monitor.

Whether CRT or LCD, the purpose of the screen is the same: to display a simple graph of an electric signal. This graph, called a *trace*, shows how voltage changes over time. The horizontal axis of this graph, reading from left to right, represents time. The vertical axis, going up and down, represents voltage.

Figure 9-2 shows a typical oscilloscope display showing a very common type of trace known as a *sine wave*. Before I tell you about the sine wave, though, there are a few things to notice about the display:

» Gridlines are printed on the display. On most oscilloscopes, these lines are 1 cm apart, with ten horizontal and eight vertical divisions.

» The vertical and horizontal lines in the middle are thicker than the other lines and include hash marks, usually 2 mm apart. These hash marks help you pinpoint the exact position of the trace between the major intervals.

» Various knobs and dials on the oscilloscope let you set the *scale* at which the graph of the waveform is plotted.

» The vertical divisions represent voltage. On most oscilloscopes, you can set the voltage scale to as little as 0.5 mV (half of one millivolts) and as much as 10 V or more. The oscilloscope usually represents 0 V by the horizontal line in the middle — so lines in the top half of the display represent positive voltage, and lines in the bottom half are negative voltage. Because there are four vertical divisions above the center line and four divisions below it, the display can show voltages between +4 V and –4 V when the voltage scale is set to 1 V. If you set the scale to 2 V, the display can show voltages between +8 V and –8 V.

» The horizontal divisions represent time. The range within which you can set your scope depends on the quality of the scope. For mine, the maximum is 100 seconds and minimum time is 1 ns — that's 1 nanosecond, which is one billionth of a second.

To draw a waveform, the oscilloscope actually draws a single dot that moves across the screen from left to right. Each passage of the dot from the left edge of the screen to the right edge is called a *sweep*. The vertical position of the dot indicates the voltage, and the speed that the dot moves is determined by the time interval, which is sometimes called the *sweep time*. Because there are a total of ten horizontal divisions, the dot sweeps the display once every 2 seconds if the sweep time is set to 0.2 s.

Most waveforms in electronics repeat at much smaller intervals than 2 seconds, so you'll usually want to shorten the time interval. As you work with your oscilloscope, you'll usually need to adjust the sweep time until at least one full cycle of the waveform you're examining can be shown within the screen. (Note also that many scopes have an Auto button, which automatically adjusts the scope to stabilize the waveform.)

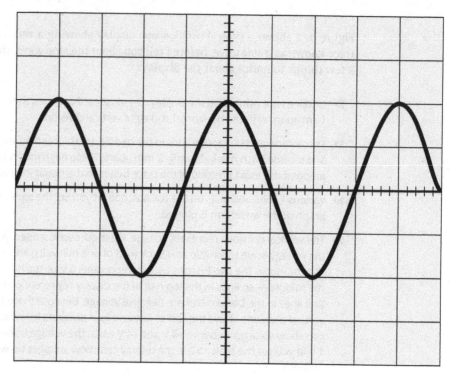

FIGURE 9-2:
An oscilloscope
trace showing a
sine wave.

Examining Waveforms

Waveforms are the characteristic patterns that oscilloscope traces usually take. These patterns indicate how the voltage in the signal changes over time — does it rise and fall slow or fast, is the voltage change steady or irregular, and so on.

There are four basic types of waveforms that you'll run into over and over again as you work with electronic circuits. These four waveforms are shown in Figure 9-3. They are:

>> **Sine wave:** The voltage increases and decreases in a steady curve. If you remember your trigonometry class from high school, you might remember a trig function called *sine*, which has to do with angles measured in right triangles.

Few of us want to go back to high school trigonometry, so that's all I'm going to say about the mathematics behind sine waves. Suffice it to say, however, that sine waves are found everywhere in nature. For example, sine waves can be found in sound waves, light waves, ocean waves — even the bouncing of a slinky is a sine wave.

Sine wave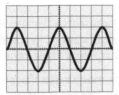

Square wave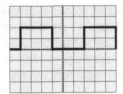

Triangle wave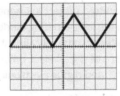

Sawtooth wave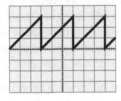

FIGURE 9-3:
Four common
waveforms.

And, most importantly from the standpoint of electronics, the alternating
current voltage that is provided in the public power grid is in the form of a
sine wave. In an alternating current sine wave, voltage increases steadily until
a peak voltage is reached. Then, the voltage begins to decrease until it reaches
zero. At that point, the voltage becomes negative, which causes the current
flow to reverse direction. Once it's negative, the voltage continues to change
until it reaches its peak negative voltage, and then it begins to increase until it
reaches zero again. The voltage then becomes positive, the current reverses,
and the sine wave cycle repeats.

The number of times a sine wave (or any other wave, for that matter) repeats
itself in a given unit of time (typically 1 second) is called its *frequency*.
Frequency is measured in units called *hertz*, abbreviated *Hz*. The alternating
current available from a standard electrical outlet cycles 60 times a second.
Thus, the frequency of household AC is 60 Hz. Most waveforms found in
electronic circuits have a much higher frequency than household alternating
current, typically in the range of several thousand hertz (*kilohertz*, or *kHz*) or
millions of hertz (*megahertz*, or *MHz*).

» **Square wave:** Represents a signal in which a voltage simply turns on, stays on for a while, turns off, stays off for a while, and then repeats. The graph of such a wave shows sharp, right-angle turns, which is why it's called a square wave.

In actual practice, most circuits that attempt to create square waves don't do their job perfectly. As a result, the voltage rarely comes on absolutely instantly, and it rarely shuts off absolutely instantly. Thus, the vertical parts of the square wave in Figure 9-3 aren't actually vertical in the real world. In addition, sometimes the initial voltage overshoots the target voltage by a little bit, so the initial vertical uptake goes a little too high for a very brief moment, and then settles down to the right voltage.

Square waves are found in many electronic circuits. For example, the 555 timer IC used in the coin-toss project in Chapter 6 of this minibook produces square waves that flash the light-emitting diodes on and off, and digital logic circuits (for example, computer circuits) rely almost exclusively on square waves to represent the ones and zeros of digital electronics. (I tell you more about the 555 timer IC in Book 3, Chapter 2, and digital electronics in Book 5.)

» **Triangle wave:** Voltage increases in a straight line until it reaches a peak value, and then it decreases in a straight line. If the voltage reaches zero and then begins to rise again, the triangle wave is a form of direct current. If the voltage crosses zero and goes negative before it begins to rise again, the triangle wave is a form of alternating current.

» **Sawtooth wave:** This one is a hybrid of a triangle wave and a square wave. In most sawtooth waves, the voltage increases in a straight line until it reaches its peak voltage, and then the voltage drops instantly (or as close to instantly as possible) to zero, and immediately repeats.

TECHNICAL STUFF

Sawtooth waves have many interesting applications. One of the most appropriate for the purposes of this chapter is within an oscilloscope that has a CRT display. Here's a very simplified explanation of how the CRT in an oscilloscope works: It shoots a beam of electrons at a specially coated glass surface that glows when electrons hit it and uses electromagnets to steer the beam. Electromagnets above and below the beam steer it vertically; electromagnets to the right and left steer it horizontally.

To create the sweep of the electron beam from left to right, a sawtooth wave is applied to the electromagnets on the left and right of the beam. As the voltage increases, the electromagnet produces an increasingly stronger magnetic field, which pulls the beam toward the right side of display. When the voltage reaches its peak and drops instantly back to zero, the magnetic field collapses, and the electron beam snaps back to the left side of the display.

Changing the sweep rate of the oscilloscope is simply a matter of changing the frequency of the sawtooth wave applied to the horizontal electromagnets in the oscilloscope's CRT.

Calibrating an Oscilloscope

Quick: What were the first words spoken from the surface of the moon?

If you guessed, "That's one small step for a man," you'd be off by a long shot. Neil Armstrong and Buzz Aldrin had been on the moon for several hours by the time Neil said that.

If you guessed, "Houston, Tranquility Base here; the Eagle has landed," you'd be close, but still not quite right.

Contrary to popular belief, the first words spoken from the surface of the moon were spoken by Buzz Aldrin, not Neil Armstrong. Those first words were: "Engine Stop. ACA out of detent. Auto mode control, both auto. Descent engine command override, off. Engine arm, off. 413 is in."

Before Neil Armstrong could make his historic announcement that the Eagle had landed on the moon, Buzz (being the lunar module pilot) had to quickly verify the settings of some key controls within the lunar module to ensure that everything was working well.

In the same way, before you make your historic first waveform measurement, you must first verify the settings of some key controls on your oscilloscope to ensure that everything is working well. The exact steps you need to follow to set up your oscilloscope vary depending on the exact type and model of your scope, so be sure to read the instruction manual that came with your scope. But the general steps should be as follows:

1. **Examine all the controls on your scope and set them to normal positions.**

   For most scopes, all rotating dials should be centered, all push buttons should be out, and all slide switches and paddle switches should be up.

2. **Turn your oscilloscope on.**

   If it's the old-fashioned CRT kind, give it a minute or two to warm up.

3. **Set the VOLTS/DIV control to 1.**

   This sets the scope to display one volt per vertical division. Depending on the signal you're displaying, you may need to increase or decrease this setting, but one volt is a good starting point.

4. **Set the TIME/DIV control to 1 ms.**

   This control determines the time interval represented by each horizontal sweep of the display. Try turning this dial to its slow setting, such as 5 seconds. Then, turn the dial one notch at a time and watch the dot speed up until it becomes a solid line.

5. **Set the Trigger switch to Auto.**

The Auto position enables the oscilloscope to stabilize the trace on a common trigger point in the waveform. If the trigger mode isn't set to Auto, the waveform may drift across the screen, making it difficult to watch.

6. **Connect a probe to the input connector.**

If your scope has more than one input connector, connect the probe to the one labeled *A*.

Oscilloscope probes include a probe point, which you connect to the input signal and a separate ground lead. The ground lead usually has an alligator clip. When testing a circuit, this clip can be connected to any common ground point within the circuit. In some probes, the ground lead is detachable, so that you can remove it when it isn't needed.

7. **Touch the end of the probe to the scope's calibration terminal.**

This terminal provides a sample square wave that you can use to calibrate the scope's display. Some scopes have two calibration terminals, labeled *0.2 V* and *2 V*. If your scope has two terminals, touch the probe to the 2 V terminal.

TIP

For calibrating, it's best to use an alligator clip test probe. If your test probe has a pointy tip instead of an alligator clip, you can usually push the tip through the little hole in the end of the calibration terminal to hold the probe in place.

It isn't necessary to connect the ground lead of your test probe for calibration.

8. **If necessary, adjust the TIME/DIV and VOLTS/DIV controls until the square wave fits nicely within the display.**

For example, see Figure 9-4.

9. **If necessary, adjust the Y-POS control to center the trace vertically.**

10. **If necessary, adjust the X-POS control to center the trace horizontally.**

11. **If necessary, adjust the Intensity and Focus settings to get a clear trace.**

12. **Congratulate yourself!**

You're now ready to begin viewing the waveforms of actual electronic signals.

TIP

Remember that the controls of every oscilloscope make and model are unique. Be sure to read the owner's manual that came with your oscilloscope to see if there are any other setup or calibration procedures you need to follow before feeding real signals into your scope.

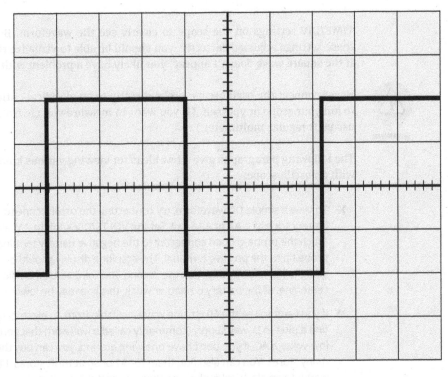

FIGURE 9-4:
An oscilloscope trace showing a square wave.

Also, be aware that each probe you use needs to be calibrated using a procedure called *compensation*. This is usually done by turning a small screw in the probe using a nonmetallic screwdriver until a square-wave signal shown on the scope is perfectly square. If you fail to adjust the compensation properly, you'll get incorrect readings from your scope.

Displaying Signals

The basic procedure for testing a circuit with an oscilloscope is to attach the ground connector of the scope's test lead to a ground point in the circuit, and then touch the tip of the probe to the point in the circuit that you want to test.

For example, if you want to verify that the output from a pin of an integrated circuit is emitting a square wave, touch the oscilloscope probe to the pin and look at the display on the scope. Note that you may need to adjust the VOLTS/DIV and

TIME/DIV settings on the scope to clearly see the waveform. But once you get those settings adjusted correctly, you should be able to visualize the square wave. If the square wave doesn't appear, you likely have a problem with the circuit.

WARNING

Never connect the oscilloscope probe directly to an electrical outlet. You're likely to kill your scope or yourself. (If you want to measure voltage from an outlet, just use your regular multimeter.)

The following paragraphs give a few ideas for viewing various kinds of waveforms with an oscilloscope:

>> To view a simple DC waveform, try connecting the oscilloscope to a 1.5 V battery such as a AA or AAA cell. Set the VOLTS/DIV knob to 2 V, and then touch the probe ground connector to the negative battery terminal and the probe tip to the positive terminal. The resulting display should be a simple straight line midway between the second and third vertical division above the centerline. (If the battery is dead or weak, this line may be lower.)

>> If you want to see the 60 Hz sine wave available from an electrical wall outlet, find a plug-in power supply (commonly called a *wall wart*) that generates low-voltage AC. If you don't have one lying around, you can buy them new at many stores. You can also find them for $1 or so at thrift stores. Plug the wall wart into an electrical outlet, and then connect the oscilloscope probe to the wall wart's low-voltage plug. Adjust the VOLTS/DIV and TIME/DIV settings until you can see the sine wave.

>> If you want to see what an audio waveform looks like, find a short ⅛-inch audio cable that's male on both ends. Plug one end into the headphone jack of any audio device, such as a radio or an iPod. Then, connect the probe's ground lead to the shaft of the plug on the free end of the audio cable and touch the probe tip to the tip of the audio plug, as shown in Figure 9-5. After fiddling with the VOLTS/DIV and TIME/DIV settings, you should see a display of the jumbled waveform that's typical of audio signals.

TIP

>> As you build circuits while working your way through this book, keep your oscilloscope handy. Don't hesitate at any time to pick up the oscilloscope probe and check out the signals that are being generated at various points within your circuit. Connect the probe's ground clip to any ground point in the circuit, and then touch the probe tip to every loose wire and exposed pin that you can to see what's going on inside the circuit.

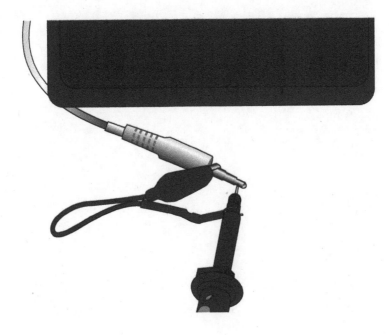

FIGURE 9-5:
Connecting an
oscilloscope
probe to an audio
plug.

2

Working with Basic Electronic Components

Contents at a Glance

IN THIS CHAPTER

» **Examining the basic elements of circuits**

» **Looking at different batteries that you can use to power your circuits**

» **Categorizing the various switches that you can use to control your circuits**

» **Learning the difference between series and parallel circuits**

» **Building a variety of circuits to demonstrate basic circuits**

Chapter **1**

Working with Basic Circuits

ook 1 introduces you to the underlying concepts of electricity and gives you an overview of the tools and skills you need to start working with electronics. Now that you know what electricity is, and you've gone shopping for some basic tools (such as a solderless breadboard, a soldering iron, and a multimeter), it's time to start learning how electronic circuits work.

This chapter explores the basic concepts of a circuit. The circuits I cover in this chapter are very simple, consisting of nothing other than batteries that supply voltage, wires that carry current, and lamps that consume power (because they light up in the process!). I also throw in some switches that let you turn the circuit on and off.

Though simple, the circuits presented here are a great introduction to more complicated circuits. The remaining chapters in this minibook add additional elements that are basic to all circuits, such as resistors, capacitors, coils, diodes, transistors, and integrated circuits. By the time you get through this minibook, you'll be familiar with all the basic components of electronic circuits.

What Is a Circuit?

A *circuit* is a complete course of conductors through which current can travel. Circuits provide a path for current to flow. To be a circuit, this path must start and end at the same point. In other words, a circuit must form a loop.

For example, Figure 1-1 shows a simple circuit that includes two components: a battery and a lamp. The circuit allows current to flow from the battery to the lamp, through the lamp, then back to the battery. Thus, the circuit forms a complete loop.

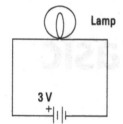

FIGURE 1-1:
A simple circuit.

Of course, circuits can be more complex. However, all circuits can be distilled down to three basic elements:

>> **Voltage source:** A voltage source causes current to flow. In Figure 1-1, the voltage source is the battery.

>> **Load:** The load consumes power; it represents the actual work done by the circuit. Without the load, there's not much point in having a circuit.

  In Figure 1-1, the load is the lamp. In complex circuits, the load is a combination of components, such as resistors, capacitors, transistors, and so on.

>> **Conductive path:** The conductive path provides a route through which current flows. This route begins at the voltage source, travels through the load, and then returns to the voltage source. This path must form a loop from the negative side of the voltage source to the positive side of the voltage source.

In Figure 1-1, the two lines that travel between the battery and the lamp represent the conductive path. The conductive path can be intricate in a complex circuit, but it must still form a loop from the negative side of the voltage source to the positive side.

The following paragraphs describe a few additional interesting points to keep in mind as you ponder the nature of basic circuits:

>> When a circuit is complete and forms a loop that allows current to flow, the circuit is called a *closed circuit*. If any part of the circuit is disconnected or disrupted so that a loop is not formed, current cannot flow. In that case, the circuit is called an *open circuit*.

TIP

Open circuit is an oxymoron. After all, the components must form a complete path to be considered a circuit. If the path is open, it isn't a circuit. Therefore, *open circuit* is most often used to describe a circuit that has become broken, either on purpose (by the use of a switch, which I discuss later in this chapter) or by some error, such as a loose connection or a damaged component.

>> *Short circuit* refers to a circuit that does not have a load. For example, Figure 1-2 shows a short circuit; the lamp is connected to the circuit but a direct connection is present between the battery's negative terminal and its positive terminal, too.

WARNING

Current in a short circuit can flow at dangerously high levels. Short circuits can damage electronic components, cause a battery to explode, or maybe start a fire.

TIP

The short circuit shown in Figure 1-2 illustrates an important point about electrical circuits: It is possible — common, even — for a circuit to have multiple pathways for current to flow. In Figure 1-2, the current can flow through the lamp as well as through the path that connects the two battery terminals directly.

Current flows everywhere it can. If your circuit has two pathways through which current can flow, the current doesn't choose one over the other; it chooses both. However, not all paths are equal, so current doesn't flow equally through all paths. In the circuit shown in Figure 1-2, current will flow much more easily through the short circuit than it will through the lamp. Thus, the lamp will not glow because nearly all the current will bypass the lamp in favor of the easier route through the short circuit. Although a small amount of current will flow through the lamp, the current flowing through the lamp will not be sufficient to cause the lamp to visibly glow.

To determine how much current flows through a given path, you use a mathematical formula that is explained in Chapter 2 of this minibook. Nevertheless, when one of the available paths is a short circuit, you needn't bother with the formula because nearly all the current will flow through the short circuit.

TIP

>> You can imagine electric current flowing through a circuit from the positive side of the voltage source to the negative side because this is usually how you visualize a circuit when you study it. For example, in Figure 1-1, you can think of the current as flowing in a clockwise direction, starting at the positive terminal of the battery, flowing through the path at the left side of the diagram to the lamp at the top of the diagram, through the lamp and then through the path at the right side of the diagram and finally returning to the negative terminal of the battery.

As you see in Book 1, Chapter 2, this way of thinking about current flow is called *conventional current.* In reality, the electric charge — and the electrons — in the circuit flow from the negative side of the voltage source through the circuit to the positive side of the voltage source.

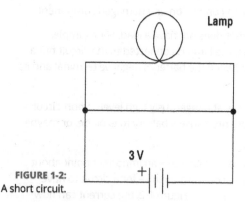

FIGURE 1-2:
A short circuit.

Using Batteries

The easiest way to provide a voltage source for a circuit is to include a battery. There are plenty of other ways to provide voltage, including AC adapters (which you can plug into the wall) and solar cells (which convert sunlight to voltage). However, batteries remain the most practical source of juice for most of the circuits you build in this book.

A *battery* is a device that converts chemical energy into electrical energy in the form of voltage, which in turn can cause current to flow. A battery works by immersing two plates made of different metals into a special chemical solution called an *electrolyte.* The metals react with the electrolyte to produce a flow of charges that accumulate on the negative plate, called the *anode*. The positive plate, called the *cathode*, is sucked dry of charges. As a result, a voltage is formed between the two plates. These plates are connected to external terminals to which you can connect a circuit to cause current to flow.

Figure 1-3 shows a simplified diagram of how a battery works. A bowl filled with the right kind of chemical plus an anode and a cathode made of the right kind of metal gives you a working battery.

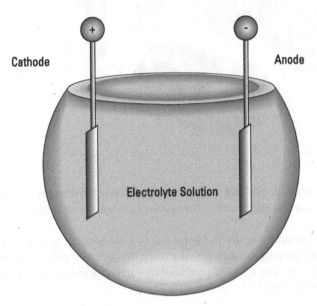

Cathode

Anode

Electrolyte Solution

FIGURE 1-3:
What goes on inside a battery.

TECHNICAL STUFF

Technically, Figure 1-3 shows a *cell*, not a battery. A *battery* is a combination of two or more cells.

Batteries come in many different shapes and sizes, but for the purposes of this book, you need concern yourself only with a few standard types of batteries, all of which are available at any grocery, drug, or department store. Figure 1-4 shows the most common sizes available.

Cylindrical batteries come in four standard sizes: AAA, AA, C, and D. Regardless of the size, these batteries provide 1.5 V each; the only difference between the smaller and larger sizes is that the larger batteries can provide more current. Technically, AAA, AA, C, and D cells are just that — single cells, not batteries. But if it were a crime to call them batteries, our prisons would be more overloaded than they already are.

FIGURE 1-4:
Common batteries.

The cathode, or positive terminal, in a cylindrical battery is the end with the metal bump. The flat metal end is the anode, or negative terminal.

The rectangular battery in Figure 1-4 is a 9 V battery. It truly is a battery because that little rectangular box actually contains six small cells, each about half the size of a AAA cell. The 1.5 volts produced by each of these small cells combine to create a total of 9 volts.

Here are a few other things you should know about batteries:

WARNING

>> Besides AAA, AA, C, D, and 9 V batteries, many other battery sizes are available. Most of those batteries are designed for special applications, such as digital cameras, hearing aids, laptop computers, and so on.

>> All batteries contain chemicals that are toxic to you and to the environment. Treat them with care, and dispose of them properly according to your local laws. Don't just throw them in the trash.

>> You can (and should) use your multimeter to measure the voltage produced by your batteries. Set the multimeter to an appropriate DC voltage range (such as 20 V). Then, touch the red test lead to the positive terminal of the battery and the black test lead to the negative terminal. The multimeter will tell you the voltage difference between the negative and positive terminals. For cylindrical batteries (AAA, AA, C, or D) it should be about 1.5 V. For 9 V batteries, it should be about 9 V.

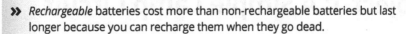

TECHNICAL STUFF

TIP

» *Rechargeable* batteries cost more than non-rechargeable batteries but last longer because you can recharge them when they go dead.

» The technical term for a cell that cannot be recharged is *primary cell.* A rechargeable cell is properly called a *secondary cell.* However, I would be shocked if you walked into any store that sells batteries and asked for secondary cells and the salesperson knew what you were talking about.

» The easiest way to use batteries in an electronic circuit is to use a battery holder, which is a little plastic gadget designed to hold one or more batteries.

» Wonder why they sell AAA, AA, C, and D cells but not A or B? Actually, A cell and B cell batteries do exist. However, those sizes never really caught on in consumer devices, so they aren't readily available at retail stores.

Working with Basic Circuits

Building a Lamp Circuit

Project 1 shows you how to build a simple circuit that uses a battery to light a lamp. Although this circuit is quite simple, it helps illustrate the basic principles I've explained so far. Figure 1-5 shows the circuit assembled.

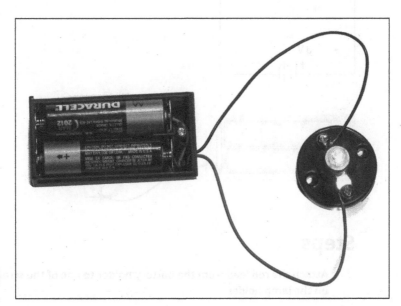

FIGURE 1-5: A simple lamp circuit.

Project 1: A Simple Lamp Circuit

In this project, you build a simple circuit that connects a lamp to a battery. You also use a multimeter to measure the voltage and current in the circuit. To assemble and test the circuit, you need a small Phillips-head screwdriver and a multimeter.

Parts

>> Two AA batteries

>> One 2 x AA battery holder

>> One flashlight lamp holder

>> One 2.33 V flashlight lamp

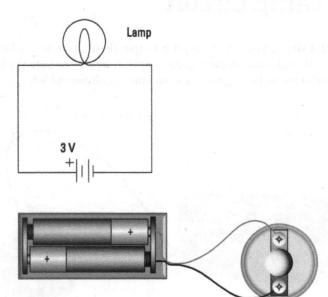

Steps

1. **Attach the red lead from the battery holder to one of the screw terminals on the lamp holder.**

   Loosen the screw terminal a bit to create a gap under the screw. Next, bend the stripped end of the red lead into a hook shape (you can do this easily by wrapping the wire around the tip of the screwdriver). Insert the stripped end of the lead beneath the terminal screw, and then tighten the screw to secure the wire.

2. **Attach the black lead to the other terminal on the lamp holder.**

3. **Insert the lamp into the lamp holder.**

4. **Insert the batteries into the holder.**

   The lamp should light.

5. **Set the multimeter range to the lowest DC voltage setting that will measure at least 3 volts.**

   My meter has a 5 V DC range, so I used that setting.

6. **Touch the red multimeter lead to the lamp holder terminal that the red battery holder lead is attached to, and then touch the black lead to the other terminal.**

   The meter should read about 3 volts.

7. **Disconnect the red battery lead from the lamp holder.**

   This breaks the circuit, so the lamp will go out.

8. **Turn the multimeter dial to your highest DC mA setting.**

   On mine, the highest setting is 1,000 DC mA.

9. **Touch the stripped end of the red battery lead to the tip of the red multimeter lead, and then touch the tip of the black multimeter lead to the unconnected terminal on the lamp holder.**

   The meter should read approximately 250 mA. If the largest DC mA range on your multimeter is less than 250 mA, you may not get an accurate reading. However, you should get an indication that the current exceeds the maximum for the range.

   Additionally, when you test the current, the lamp comes back on because the meter completes the circuit.

Working with Switches

Switches are an important part of most electronic circuits. In the simplest case, most circuits contain an on/off switch to turn the circuit on and off. In addition to the on/off switch, many circuits contain additional switches that control how the circuit works or activate different features of the circuit.

Switches are mechanical devices with two or more *leads* (or *terminals*) that are internally connected to metal contacts which can be opened or closed by the person operating the switch. When the switch is in the On position, the contacts are brought together to complete the circuit so that current can flow. When the contacts are together, the switch is closed. When the contacts are apart, the switch is open and current cannot flow.

The following sections describe two ways to categorize switches: by the method used to operate the switch and by the connections made by the switch.

The many ways to throw the switch

One way to categorize switches is by the movement a person uses to open or close the contacts. Figure 1-6 shows many different switch designs. The most common are

>> **Slide switch:** A *slide switch* has a knob that you can slide back and forth to open or close the contacts.

>> **Toggle switch:** A *toggle switch* has a lever that you flip up or down to open or close the contacts. Common household light switches are examples of toggle switches.

>> **Rotary switch:** A *rotary switch* has a knob that you turn to open and close the contacts. The switch in the base of many tabletop lamps is an example of a rotary switch.

>> **Rocker switch:** A *rocker switch* has a seesaw action. You press one side of the switch down to close the contacts, and press the other side down to open the contacts.

>> **Knife switch:** A *knife switch* is the kind of switch Igor throws in a Frankenstein movie to reanimate the creature. In a knife switch, the contacts are exposed for everyone to see.

>> **Push-button switch:** A *push-button switch* is a switch that has a knob that you push to open or close the contacts. In some push-button switches, you push the switch once to open the contacts and then push again to close the contacts. In other words, each time you push the switch, the contacts alternate between opened and closed.

Other push-button switches are *momentary contact switches,* where contacts change from their default state only when the button is pressed and held down. The two types of momentary contact switches are

- *Normally open (NO):* In a normally open switch, the default state of the contacts is open. When you push the button, the contacts are closed. When you release the button, the contacts open again. Thus, current flows only when you press and hold the button.

- *Normally closed (NC):* In a normally closed switch, the default state of the contacts is closed. Thus, current flows until you press the button. When you press the button, the contacts are opened and current does not flow. When you release the button, the contacts close again and current resumes.

FIGURE 1-6: There are switches for every need.

Making connections with poles and throws

Another way to classify switches is by the connections they make. If you were under the impression that switches simply turn circuits on and off, guess again. Two important factors that determine what types of connections a switch makes are

» **Poles:** A switch *pole* refers to the number of separate circuits that the switch controls. A *single-pole* switch controls just one circuit. A *double-pole* switch controls two separate circuits.

   A double-pole switch is like two separate single-pole switches that are mechanically operated by the same lever, knob, or button.

» **Throw:** The number of *throws* indicates how many different output connections each switch pole can connect its input to. Figure 1-7 shows the two most common types, single-throw and double-throw:

- A *single-throw* switch is a simple on/off switch that connects or disconnects two terminals. When the switch is closed, the two terminals are connected and current flows between them. When the switch is opened, the terminals are not connected, so current does not flow.

- A *double-throw* switch connects an input terminal to one of two output terminals. Thus, a double-pole switch has three terminals. One of the terminals is called the *common terminal*. The other two terminals are often referred to as *A* and *B*. When the switch is in one position, the common terminal is connected to the A terminal, so current flows from the common terminal to the A terminal, but no current flows to the B terminal. When the switch is moved to its other position, the terminal connections are reversed: Current flows from the common terminal to the B terminal, but no current flows through the A terminal.

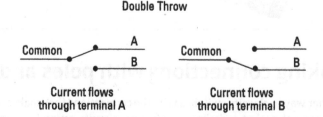

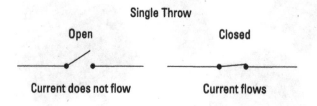

FIGURE 1-7:
Single- and double-throw switches.

Switches vary in both the number of poles and the number of throws. In theory, any number of poles and any number of throws is possible. However, most switches have one or two poles and one or two throws. This leads to four common combinations, as described in the following paragraphs. The symbols used in schematic diagrams for each of these switches are shown in the margins.

>> **SPST (single pole, single throw):** A basic on/off switch that turns a single circuit on or off. An SPST switch has two terminals: one for the input and one for the output.

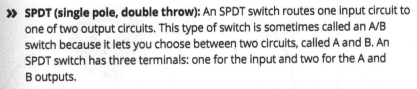

» **SPDT (single pole, double throw):** An SPDT switch routes one input circuit to one of two output circuits. This type of switch is sometimes called an A/B switch because it lets you choose between two circuits, called A and B. An SPDT switch has three terminals: one for the input and two for the A and B outputs.

» **DPST (double pole, single throw):** A DPST switch turns two circuits on or off. A DPST switch has four terminals: two inputs and two outputs.

» **DPDT (double pole, double throw):** A DPDT switch routes two separate circuits, connecting each of two inputs to one of two outputs. A DPDT switch has six terminals: two for the inputs, two for the A outputs, and two for the B outputs.

Here are a few other points to ponder concerning the arrangement of poles and throws:

» Switches with more than two poles or more than two throws are not commonplace, but they do exist. Rotary switches lend themselves especially well to having many throws. For example, the rotary switch in a multimeter typically has 16 or more throws, one for each range of measurement the meter can make.

» A common variation of a double-throw switch is to have a middle position that does not connect to either output. Often called *center open,* this type of switch has three positions but only two throws. For example, an SPDT center open switch can switch one input between either of two outputs, but in its center position, neither output is connected.

TIP

» If you're stocking up on switches just to have them on hand, you're better off buying DPDT switches rather than single-pole or single-throw switches because a DPDT can be used when a circuit calls for a simpler SPST, SPDT, or DPST switch. You can use a DPDT switch when a simpler type is called for because there's no law that says you have to wire all the contacts on the switch. For example, to use a DPDT switch as an SPST switch, you just use one of the poles and one of the throws and leave the other connections unused.

Building a Switched Lamp Circuit

Project 2 presents a simple construction project that lets you explore the use of a simple on/off switch to control a lamp. Figure 1-8 shows the assembled project.

FIGURE 1-8:
The switched
lamp project.

This project and the remaining projects in this chapter use a DPDT knife switch from RadioShack, as shown in Figure 1-9. It's unlikely that you'd use a knife switch in an actual electronic circuit. However, a knife switch like this is a perfect tool for learning the ins and outs of working with switches. For one thing, it is entirely exposed, so you can see how it works. Additionally, because they come on their own base and have screw terminals, connecting them in temporary circuits is simple because you don't have to do any soldering.

As you can see in the figure, the knife switch is a double pole, double throw (DPDT) switch, which means it operates like two SPDT switches that are mechanically linked. I numbered the six terminals on the switch 1X, 1A, 1B, 2X, 2A, and 2B. The 1 and 2 designate which of the two circuits is being switched. The X terminals are the input terminals in the center of the switch, and the A and B terminals are for the two possible outputs. Thus, when the switch is flipped one way, 1X is connected to 1A and 2X is connected to 2A. When the switch is flipped the other way, 1X is connected to 1B and 2X is connected to 2B.

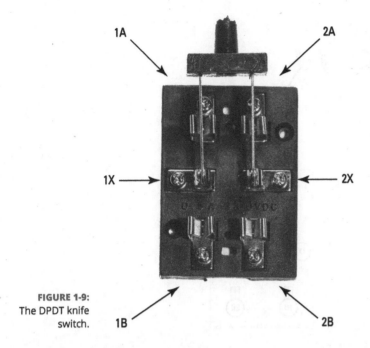

1A

2A

1X ←

→ 2X

FIGURE 1-9:
The DPDT knife
switch.

1B

2B

Project 2: A Lamp Controlled by a Switch

In this project, you build a simple circuit that connects a lamp to a battery and uses a switch to turn the lamp on and off. To assemble and test the circuit, you need a small Phillips-head screwdriver, wire cutters, and a wire stripper.

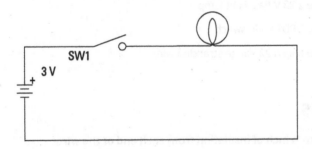

SW1

3 V

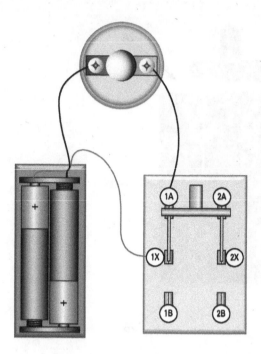

Parts

>> Two AA batteries

>> One battery holder

>> One lamp holder

>> One 2.33 V flashlight lamp

>> One DPDT knife switch

>> One 6-inch 22-gauge stranded wire

Steps

1. **Strip ½ inch of insulation from each end of the wire.**

2. **Open the knife switch.**

   Lift its handle to the upright position so that no connections are made.

3. **Attach the red lead from the battery holder to terminal 1X on the knife switch.**

4. **Attach the black lead to one of the terminals on the lamp holder.**

5. **Use the 6-inch wire to connect terminal 1A of the knife switch to the other terminal of the lamp holder.**

6. **Insert the batteries into the holder.**

7. **Close the knife switch on the A side.**

   The lamp should light up.

If you want to experiment with different variations of this project, try the following:

>> Try moving the switch from the positive side of the circuit to the negative side. In other words, connect the red lead from the battery holder to the lamp and connect the black lead to the 1X terminal on the switch.

   The circuit will function the same. This shows that the location of a switch in a circuit often doesn't matter. If the circuit is broken anywhere, current cannot flow. Thus, whether the switch is before or after the lamp doesn't matter.

>> Cut a second 6-inch piece of wire and strip the insulation from both ends. Then, wire the circuit so that the red battery lead goes to switch terminal 1X, the black lead goes to switch terminal 2X, one of the wires goes from switch terminal 1A to one of the lamp terminals, and the other wire goes from switch terminal 2A to one of the lamp terminals.

   Now you've created the circuit shown in Figure 1-10. In this circuit, the knife switch is used as a DPST (double pole, single throw) switch to interrupt the circuit on both the negative and the positive side of the lamp.

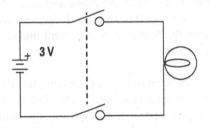

FIGURE 1-10:
Using a DPST
switch to control
a lamp.

Understand Series and Parallel Circuits

Whenever you have circuits that consist of more than one component, those components must be linked together. The two ways to connect components in a circuit are in series and in parallel. Figure 1-11 illustrates how you might use series and parallel circuits to connect two lamps in a single circuit.

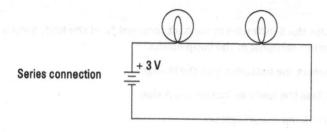

Series connection

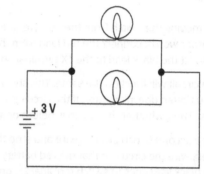

FIGURE 1-11:
Lamps connected
in series and in
parallel.

Parallel connection

In a *series* connection, components are connected end to end, so that current flows first through one, then through the other. As you can see in the first circuit in Figure 1-11, the current goes through one lamp and then the other. The lamps are strung together end to end.

One drawback of series connections is that if one component fails in a way that results in an open circuit, the entire circuit is broken and none of the components will work. For example, if either one of the lamps in the series circuit in Figure 1-11 burns out, neither lamp will work. That's because current must flow through both lamps for the circuit to be complete.

In the *parallel* connection shown in Figure 1-12, each lamp has its own direct connection to the battery. This arrangement avoids the if-one-fails-they-all-fail nature of series connections. In a parallel connection, the components do not depend on each other for their connection to the battery. Thus, if one lamp burns out, the other will continue to burn.

TECHNICAL
STUFF

An interesting thing happens with voltage when components are connected in series: The voltages present at each component are divided up. For example, in a circuit with a 3 V battery and two identical lamps connected in series, each lamp will see only one and a half volts. If you connected three identical lamps in series, each lamp would see only one volt.

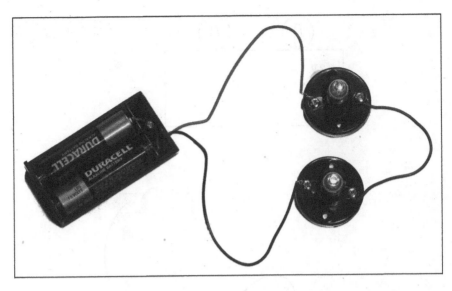

FIGURE 1-12:
Lamps connected
in series.

You can measure the voltage seen by any component in a circuit by setting your multimeter to an appropriate voltage range and then touching the leads to both sides of the component. The voltage you measure there is called the component's *voltage drop*.

Building a Series Lamp Circuit

In Project 3, you build a circuit that connects two lamps in series, a simple circuit. Then, you use your multimeter to measure the voltages at various points in the circuit. The completed project is shown in Figure 1-12.

Project 3: A Series Lamp Circuit

In this project, you connect two lamps in a series circuit. The lamps are powered by a pair of AA batteries. To build this project, you need a small Phillips-head screwdriver, wire cutters, wire strippers, and a multimeter.

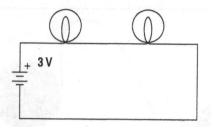

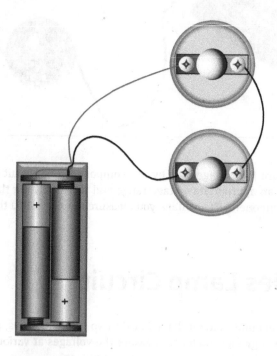

Parts

» Two AA batteries

» One battery holder

» Two lamp holders

» Two 2.33 V flashlight lamps

» One 6-inch 22-gauge stranded wire

Steps

1. Strip ½ inch of insulation from each end of the wire.

2. Attach the red lead from the battery holder to one of the terminals on one of the lamp holders.

3. Attach the black lead to one of the terminals on the other lamp holder.

4. Use the 6-inch wire to connect the unused terminal of the first lamp holder to the unused terminal of the second lamp holder.

5. Insert the batteries into the holder.

   Both lamps light.

TIP

   Notice that the lamps are dim. That's because in a series circuit made with two identical lamps, each of the two lamps sees only half the total voltage.

6. Remove one of the lamps from its holder.

   The other lamp goes out. This is because in a series circuit, a failure in any one component breaks the circuit so none of the other components will work.

7. Replace the lamp you removed in Step 6.

8. Set your multimeter to a DC voltage range that can read at least 3 volts.

9. Touch the leads to the two terminals on the first lamp holder.

   The multimeter should read approximately 1.5 V. (If you're using an analog meter and the needle moves backward, just reverse the leads.)

10. Touch the leads to the two terminals on the other lamp holder.

   Again, the multimeter should read approximately 1.5 V.

11. Touch the red lead of the meter to the terminal that the red lead from the battery is connected to, and touch the black meter lead to the terminal that the black battery lead is connected to.

   This measures the voltage across both lamps combined. The meter will indicate 3 V.

Building a Parallel Lamp Circuit

In Project 4, you build a circuit that connects two lamps in parallel, and you use your multimeter to measure voltages within various points in the circuit. The completed project is shown in Figure 1-13.

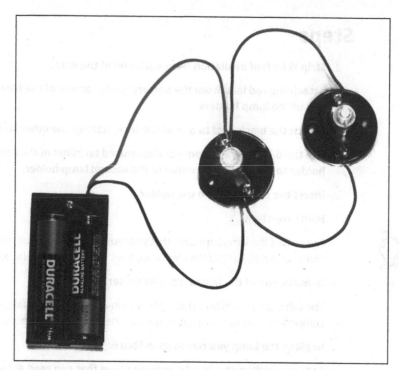

FIGURE 1-13:
Lamps connected
in parallel.

Project 4: A Parallel Lamp Circuit

In this project, you connect two lamps in a series circuit. The lamps are powered by a pair of AA batteries. To build this project, you need a small Phillips-head screwdriver, wire cutters, wire strippers, and a multimeter.

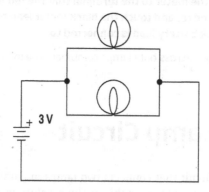

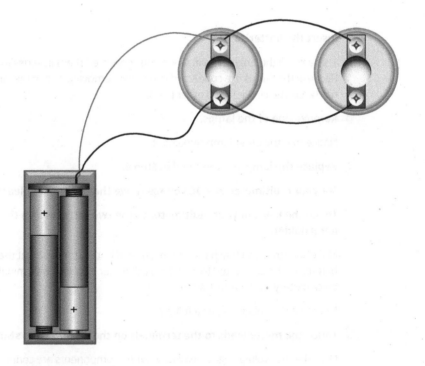

Parts

>> Two AA batteries

>> 1 battery holder (RadioShack 2700408)

>> Two lamp holders (RadioShack 2720357)

>> Two 2.33 V flashlight lamps (RadioShack 2721175)

>> Two 6-inch 22-gauge stranded wires

Steps

1. **Strip ½ inch of insulation from each end of the wires.**

2. **Attach the red lead from the battery holder to one of the terminals on the first lamp holder.**

3. **Attach the black lead to the other terminal on the first lamp holder.**

4. **Use the two wires to connect each of the terminals on the first lamp holder to the corresponding terminal on the second lamp holder.**

This wiring connects the two lamp holders in parallel.

5. **Insert the batteries.**

   The lamps light, brighter than when you connected them in series in Project 3. Because the lamps are connected in parallel, removing one of the lamps does not break the circuit to the other lamp.

6. **Remove one of the lamps.**

   Notice that the other lamp remains lit.

7. **Replace the lamp you removed in Step 6.**

8. **Set your multimeter to a DC voltage range that can read at least 3 volts.**

9. **Touch the leads of your multimeter to the two terminals on the first lamp holder.**

   Make sure that you touch the red meter lead to the terminal that the red battery lead is connected to and the black meter lead to the terminal that the black battery lead is attached to.

   Note that the voltage reads a full 3 V.

10. **Touch the meter leads to the terminals on the second lamp stand.**

    Note that the voltage again reads 3 V. When components are connected in parallel, the voltage is not divided among them. Instead, each component sees the same voltage. That's why the lamps light at full intensity in the parallel circuit.

Using Switches in Series and Parallel

Just as lamps can be connected in series or parallel, switches can also be connected in series or parallel. For example, Figure 1-14 shows two circuits that each use a pair of SPST switches to turn a lamp on or off. In the first circuit, the switches are wired in series. In the second, the switches are wired in parallel.

The interesting thing to note about wiring switches in series is that both switches must be closed in order to complete the circuit. A great example of switches wired in series is in the typical nuclear-war movie, where two people must flip a switch in order to launch the missiles. Switches wired in series means that Denzel Washington *and* Gene Hackman must both agree to launch the missiles.

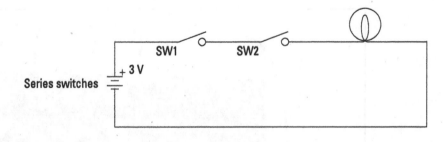

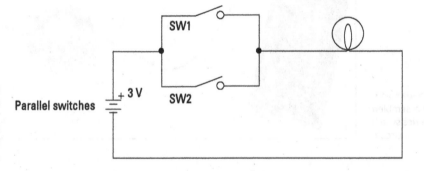

FIGURE 1-14:
Schematic
diagrams for
series and
parallel switch
circuits.

When switches are wired in parallel, closing either switch will complete the circuit. Thus, parallel switches are often used when you want the convenience of
controlling a circuit from two different locations. If the nuclear missile switches
were wired in parallel, either Denzel Washington *or* Gene Hackman could fire the
missiles.

Building a Series Switch Circuit

Project 5 presents a simple project that uses two switches to open or close a circuit that lights a lamp. The switches are wired in series, so both switches must be
closed to light the lamp. Figure 1-15 shows the completed project.

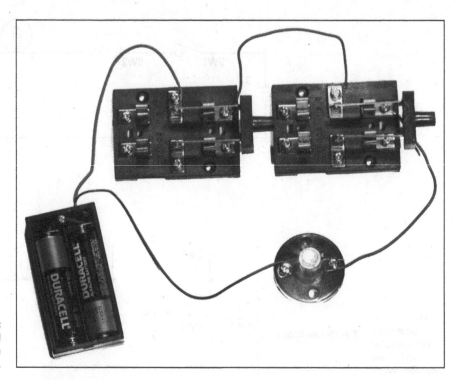

FIGURE 1-15:
The assembled series switch circuit.

Project 5: A Series Switch Circuit

In this project, you build a simple circuit that uses two knife switches to control a single lamp. To complete this project, you need a small Phillips-head screwdriver, wire cutters, and wire strippers.

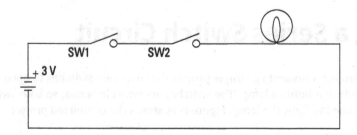

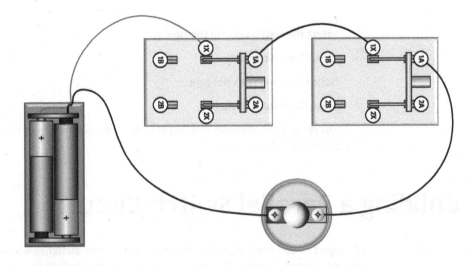

Parts

>> Two AA batteries

>> One battery holder

>> One lamp holder

>> One 2.33 V flashlight lamp

>> Two DPDT knife switches

>> Two 6-inch 22-gauge stranded wires

Steps

1. **Strip ½ inch of insulation from each end of the wires.**

2. **Open both switches.**

   Move the handles to the upright position so the contacts are not connected.

3. **Attach the red lead from the battery holder to terminal 1X of one of the switches.**

4. **Attach the black lead to one of the terminals on the lamp holder.**

5. **Connect one of the 6-inch wires from terminal 1A of the first switch to terminal 1X of the second switch.**

6. **Connect the other 6-inch wire from terminal 1A of the second switch to on the unused terminal of the lamp holder.**

7. **Insert the batteries into the holder.**

8. **Close the first switch.**

   Notice the lamp does not light.

9. **Close the second switch.**

   The lamp lights. With two switches in series, both switches must be closed for the circuit to be complete.

Building a Parallel Switch Circuit

In Project 6, you build a simple circuit that uses two switches wired in parallel to control a lamp. Because the switches are wired in parallel, the lamp will light if either of the switches is closed. Figure 1-16 shows the completed project.

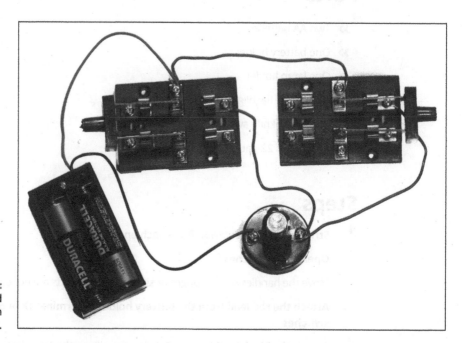

FIGURE 1-16: The assembled parallel switch circuit.

Project 6: A Parallel Switch Circuit

This project is a circuit that uses two switches wired in parallel to control a lamp. To complete this project, you need a small Phillips-head screwdriver, wire cutters, and wire strippers.

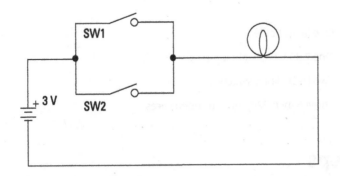

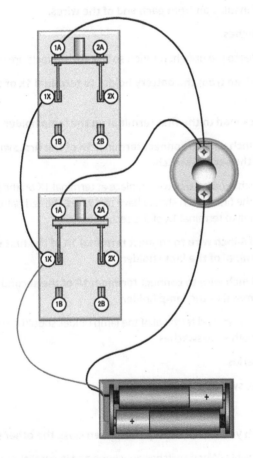

Parts

» Two AA batteries

» One battery holder

>> One lamp holder

>> One 2.33 V flashlight lamp

>> Two DPDT knife switches

>> Three 6-inch 22-gauge stranded wires

Steps

1. **Strip ½ inch of insulation from each end of the wires.**

2. **Open both switches.**

   Move the handles to the upright position so that the contacts are not connected.

3. **Attach the red lead from the battery holder to terminal 1X of the first switch.**

4. **Attach the black lead to the first terminal on the lamp holder.**

5. **Use the first 6-inch wire to connect terminal 1X of the first switch to terminal 1X of the second switch.**

   Be sure to leave the red battery wire in place at terminal 1X on the first switch. Terminal 1X on the first switch should have two wires: the red battery terminal and the wire going to terminal 1X of the second switch.

6. **Use the second 6-inch wire to connect terminal 1A of the first switch to the second terminal of the lamp holder.**

7. **Use the third 6-inch wire to connect terminal 1A of the second switch to the second terminal of the lamp holder.**

   In other words, the second terminal of the lamp holder should be connected to terminal 1A on both knife switches.

8. **Insert the batteries.**

9. **Close one of the switches.**

   The lamp lights.

10. **Open the switch you closed in Step 9 and then close the other switch.**

    Again, the lamp lights. When switches are connected in parallel, current flows through the circuit when either of the switches is closed.

11. **Close both switches.**

    The lamp remains lit because current continues to flow when both switches are closed.

12. Open both switches.

The lamp goes out. With switches wired in parallel, at least one of the switches must be closed in order for current to flow.

Switching between Two Lamps

In Project 7, you build a simple circuit that uses a single pole, double throw (SPDT) to switch a circuit between one of two lamps. In other words, one of two lamps will light depending on the position of the switch. This type of switching is a common requirement in electronic circuits. Figure 1-17 shows the completed circuit.

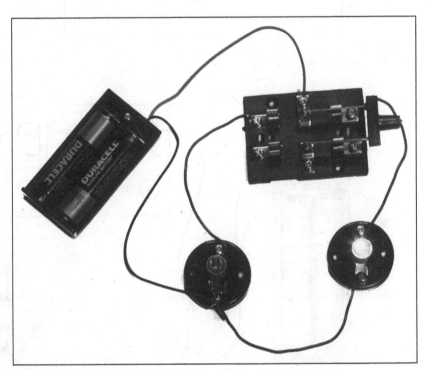

FIGURE 1-17:
The switch controls two lamps.

Project 7: Controlling Two Lamps with One Switch

In this project, you build a circuit that uses a single switch to control two lamps. When the switch is in the first position, the first lamp lights and the second lamp is dark. When the switch is in the second position, the second lamp lights and the first lamp is dark. You need a small Phillips-head screwdriver, wire cutters, and wire strippers to build this project.

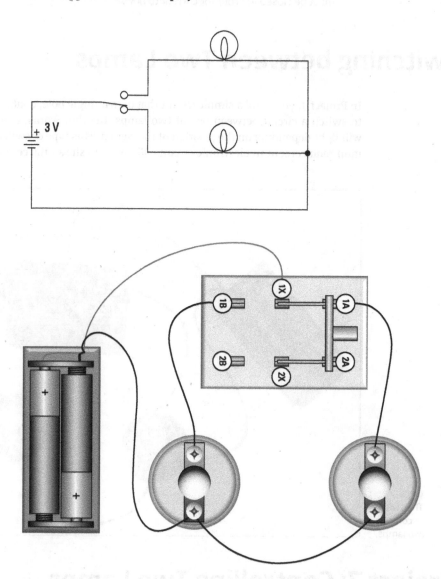

Parts

>> Two AA batteries

>> One battery holder

>> Two lamp holders

>> Two 2.33 V flashlight lamps

>> One DPDT knife switch

>> Three 6-inch 22-gauge stranded wires

Steps

1. **Strip ½ inch of insulation from each end of the wires.**

2. **Open the switch.**

   Lift the handle into the upright position so the contacts are not connected.

3. **Attach the red lead from the battery holder to terminal 1X of the switch.**

4. **Attach the black lead to one of the terminals on the first lamp holder.**

5. **Connect one of the 6-inch wires from terminal 1A of the switch to one terminal of the second lamp holder.**

6. **Connect another 6-inch wire from terminal 1B of the switch to the unused terminal of the first lamp holder.**

7. **Use the third 6-inch wire to connect the unused terminal of the second lamp holder to the terminal of the first lamp holder that the black battery lead is connected to.**

8. **Insert the lamps into the lamp holders.**

9. **Insert the batteries into the holder.**

10. **Flip the switch to the A position.**

    The second lamp lights.

11. **Change the switch to the B position.**

    The first lamp lights.

An interesting variant of the circuit in Project 7 uses both poles of the DPDT knife switch to switch the circuit on both the negative and positive sides of the lamp. For this circuit, you need four 6-inch wires. Connect the six terminals of the DPDT switch as follows:

Terminal	Connect To
1X	Red battery lead
1A	Terminal 1 of second lamp

Terminal	Connect To
1B	Terminal 1 of first lamp
2X	Black battery lead
2A	Terminal 2 of second lamp
2B	Terminal 2 of first lamp

Figure 1-18 shows this circuit assembled.

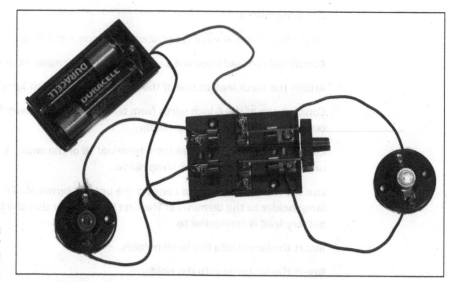

FIGURE 1-18:
Another
way to control
two lamps.

Building a Three-Way Lamp Switch

Many homes and offices have hallways that have a light switch on both ends. You can turn the light on or off by flipping either switch. This kind of switching arrangement is called a *three-way switch*.

Have you ever wondered how these three-way switches work? If you think about it, the switches are puzzling. If the light is on, flipping either switch will turn it off. If the light is off, flipping either switch will turn it on. Say you flip one switch to its On position and the light goes on. Now go to the other switch, flip it to turn the light off, and come back to the first switch. It is still in its On position, but the light is off. To turn the light back on, you can flip the first switch again.

In other words, sometimes the light is on when the switch is up, sometimes it is on when the switch is down. How can this be?

The answer is that both switches are single pole, double throw switches, and they are wired in series such that either both switches must be up or both must be down to complete the circuit. If one switch is up and the other is down, the circuit is open.

TIP

Sometimes electricians install one of the switches upside down or wire the three-way switch backwards just to confuse you. Then, the switches work backwards: If both switches are up or if both switches are down, the circuit is open and the lamp lights only when one switch is up and the other is down. But that's not the normal way to wire a three-way switch.

In Project 8, you build a simple circuit that uses two single pole, double throw switches to show how a three-way light switch works. Figure 1-19 shows the completed project.

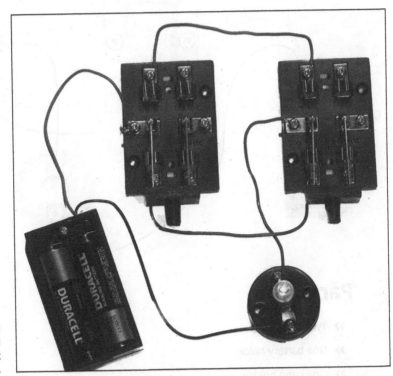

FIGURE 1-19:
The completed three-way light switch circuit.

Project 8: A Three-Way Light Switch

In this project, you create a three-way switch circuit in which a single lamp is controlled by either of two switches. You need a small Phillips-head screwdriver, wire cutters, and wire strippers to complete this project.

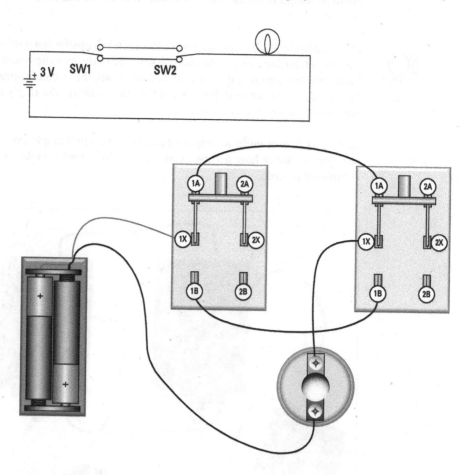

Parts

» Two AA batteries

» One battery holder

» One lamp holder

» One 2.33 V flashlight lamp

>> Two DPDT knife switches

>> Three 6-inch 22-gauge stranded wires

Steps

1. **Strip ½ inch of insulation from each end of the wires.**

2. **Open both switches.**

   Move the handles to their upright positions so the contacts are not connected.

3. **Attach the red lead from the battery holder to terminal 1X of the first switch.**

4. **Attach the black lead to one of the terminals on the lamp holder.**

5. **Connect one of the 6-inch wires from terminal 1A of the first switch to terminal 1A of the second switch.**

6. **Connect another 6-inch wire from terminal 1B of the first switch to terminal 1B of the second switch.**

7. **Connect the last 6-inch wire from terminal 1X of the second switch to the unused terminal of the lamp socket.**

8. **Insert the lamps into the lamp holder.**

9. **Insert the batteries into the holder.**

10. **Flip the switches to see the operation of the three-way switch.**

    The lamp lights only when both switches are in the A position or when both switches are in the B position.

Reversing Polarity

The final project in this chapter — Project 9 — shows you a common trick: using a DPDT switch to reverse the polarity of a circuit. One common use for this trick is powering a DC motor with the circuit. When you reverse the polarity of a DC motor, the motor spins in the opposite direction. Thus, you can use a DPDT switch to control the direction in which a DC motor turns. Figure 1-20 shows the assembled project.

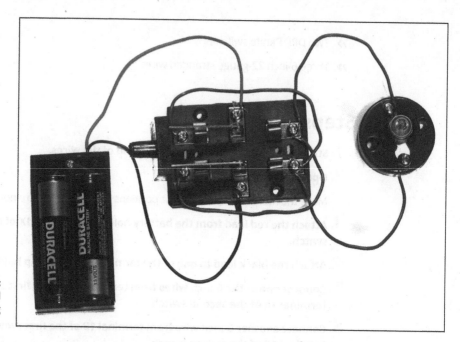

FIGURE 1-20:
The assembled
polarity-reversing
circuit.

Project 9: A Polarity-Reversing Circuit

In this project, you build a circuit that uses a DPDT switch to reverse a circuit's polarity. In other words, flipping the switch from one position to the other reverses the direction in which current flows through the circuit. To build this circuit, you need a small Phillips-head screwdriver, wire cutters, and wire strippers.

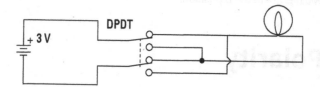

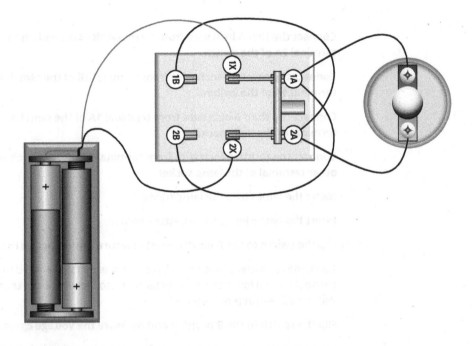

Parts

>> Two AA batteries

>> One battery holder

>> One lamp holder

>> One 2.33 V flashlight lamp

>> One DPDT knife switch

>> Four 6-inch 22-gauge stranded wires

Steps

1. **Strip ½ inch of insulation from each end of the wires.**

2. **Open the switch.**

   Move the handles to their upright positions so the contacts are not connected.

3. **Attach the red lead from the battery holder to terminal 1X of the switch.**

4. **Attach the black lead from the battery to terminal 2X of the switch.**

5. **Connect the first 6-inch wire from terminal 1B of the switch to terminal 2A of the switch.**

6. **Connect the second 6-inch wire from terminal 2B of the switch to terminal 1A of the switch.**

7. **Connect the third 6-inch wire from terminal 1A of the switch to one terminal of the lamp socket.**

8. **Connect the fourth 6-inch wire from terminal 2A of the switch to the other terminal of the lamp socket.**

9. **Insert the lamps into the lamp holder.**

10. **Insert the batteries into the battery holder.**

11. **Flip the switch to the A position and measure the voltage at the lamp.**

   Touch the red meter probe to the lamp terminal that's connected to switch terminal 1A, and touch the black meter probe to the other lamp terminal. You should read approximately +3 V.

12. **Flip the switch to the B position and measure the voltage again.**

   You should read approximately –3 V. (If you're using an analog meter, you have to reverse the probes to read the negative voltage.)

Chapter **2**

Working with Resistors

M y favorite science fiction villains are the Borg from *Star Trek: The Next Generation.* They had a saying: "Resistance is futile."

In the *Star Trek* world, it turned out that resistance against the Borg wasn't futile. Picard and the rest of the crew of the *Enterprise* resisted and eventually triumphed over the Borg.

In the electronics world, resistance is not futile. In fact, resistance can be very useful. Without resistance, electronics would not be possible. Electronics is all about manipulating the flow of current, and one of the most basic ways to manipulate current is to reduce it through resistance. Without resistance, current would flow unregulated and there would be no way to coax it into doing useful work.

In this chapter, you learn what resistance is and how to work with *resistors*, which are little devices that let you intentionally introduce resistance into your circuits. You also learn about a fundamental relationship in the nature of electricity: the relationship between voltage, current, and resistance. This relationship is expressed in a simple mathematical formula called *Ohm's law*. (Not to worry; the math isn't complicated. If you know how to multiply and divide, you can understand Ohm's law.) And finally, you learn the most common ways resistors are used in circuits.

What Is Resistance?

As you already know, a *conductor* is a material that allows current to flow, and an *insulator* is a material that doesn't. Good conductors allow current to flow with abandon, without impediments. Examples of good conductors include the metals copper and aluminum. Carbon is also an excellent conductor. Good insulators, on the other hand, erect solid walls that completely block current. Examples of good insulators include glass, Teflon, and plastic. The key factor that determines whether a material is a conductor or an insulator is how readily its atoms give up electrons to move charge along. Most atoms are very possessive of their electrons and are therefore good insulators. But some atoms don't have a strong hold on their outermost electrons. Those atoms are good conductors.

If a conductor and an insulator are mixed together, the result is a compound that conducts current but not very well. Such a compound has *resistance* — that is, it resists the flow of current. The degree to which the compound resists current depends on the exact mix of elements that make up the compound.

For example, a conducting material such as carbon might be mixed with an insulating material such as ceramic. If the mix is mostly carbon, the overall resistance of the mixture will be low. If the mix is mostly ceramic, the overall resistance will be high.

TECHNICAL STUFF

The truth is, *all* materials have some resistance. Even the best conductors have a small but measurable amount of resistance. The only exceptions are certain materials called *superconductors* that, when chilled to unbelievably low temperatures, conduct with 100 percent efficiency. Unfortunately, you can't buy superconductors at RadioShack, and even if you could, you can't buy a freezer at Sears that is powerful enough to chill the stuff down to absolute zero.

Measuring Resistance

Resistance is measured in units called *ohms*, represented by the Greek letter omega (Ω). The standard definition of *one ohm* is simple: It's the amount of resistance required to allow one ampere of current to flow when one volt of potential is applied to the circuit. In other words, if you connect a one-ohm resistor across the terminals of a one-volt battery, one amp of current will flow through the resistor.

A single ohm (1 Ω) is actually a very small amount of resistance. Resistances in the hundreds, thousands, or even millions of ohms are usually called for in electronic circuits.

In Book 1, Chapter 8, you learn that you can measure the resistance of a circuit using an *ohmmeter*, which is a standard feature found in most multimeters. The procedure is simple: First, you disconnect all voltage sources from the circuit; then, you touch the ohmmeter's two probes to the ends of the circuit and read the resistance (in ohms) on the meter.

Here are a few other points to consider about resistance and ohms:

>> The abbreviations *k* (for *kilo*) and M (for *mega*) are used for thousands and millions of ohms. Thus, a 1,000-ohm resistance is written as 1 kΩ, and a 1,000,000-ohm resistance is written as 1 MΩ.

TECHNICAL STUFF

>> For the purposes of most electronic circuits, you can assume that the resistance value of ordinary wire is zero ohms (0 Ω). In reality, however, only superconductors have a resistance of 0 Ω. Even copper wire has some resistance. Because of that, the resistance of wire is usually measured in terms of ohms per kilometer or per mile. Electronic circuits usually deal with wires that are at most a few inches or feet long, not kilometers or miles.

>> Short circuits also have essentially zero resistance.

TECHNICAL STUFF

>> Just as ordinary wire and short circuits can be considered to have zero resistance, insulators and open circuits can be considered to have infinite resistance, and in reality, there's no such thing as completely infinite resistance. If you connect two wires to the terminals of a battery and hold the wires apart, a voltage difference exists between the ends of those two wires, and a very small current will travel between them — even through the air because air doesn't have infinite resistance. This current is extraordinarily small — too small to even measure — but it's there nonetheless. Electric currents are literally everywhere.

>> The unit *ohm* is named after the famous German physicist Georg Ohm, who was the first to explain the relationship between voltage, current, and resistance in 1827.

>> Actually, the discovery was first made by a British scientist named Henry Cavendish more than 45 years earlier, but Cavendish never published his work. If he had, resistance would be measured in cavens, not ohms.

Looking at Ohm's Law

The term *Ohm's law* refers to one of the fundamental relationships found in electric circuits: that, for a given resistance, current is directly proportional to voltage. In other words, if you increase the voltage through a circuit whose resistance is fixed, the current goes up. If you decrease the voltage, the current goes down.

Ohm's law expresses this relationship as a simple mathematical formula:

$$V = I \times R$$

In this formula, V stands for voltage (in volts), I stands for current (in amperes), and R stands for resistance (in ohms). (You may be wondering why V stands for voltage here but in other equations voltage is sometimes represented by the letter E. Although scientists sometimes argue over whether V or E should be used in various circumstances, for the most part V and E are interchangeable when referring to voltage.)

Here's an example of how to calculate voltage in a circuit with a lamp powered by the two AA cells. Suppose you already know that the resistance of the lamp is 12 Ω, and the current flowing through the lamp is 250 mA, which is the same as 0.25 A. Then, you can calculate the voltage as follows:

Ohm's law is incredibly useful because it lets you calculate an unknown voltage, current, or resistance. In short, if you know two of these three quantities you can calculate the third.

Go back (if you dare) to your high school algebra class and remember that you can rearrange the terms in a simple formula such as Ohm's law to create other equivalent formulas. In particular:

>> If you don't know the voltage, you can calculate it by multiplying the current by the resistance:

$$V = I \times R$$

>> If you don't know the current, you can calculate it by dividing the voltage by the resistance:

$$I = \frac{V}{R}$$

>> If you don't know the resistance, you can calculate it by dividing the voltage by the current:

$$R = \frac{V}{I}$$

To convince yourself that these formulas work, look again at the circuit with a lamp that has 12 Ω of resistance connected to two AA batteries for a total voltage of 3 V. Then you can calculate that the current flowing through the lamp is 0.25 A (or 250 mA), as follows:

$$I = \frac{V}{R} = \frac{3V}{12\Omega} = 0.25A$$

If you know the battery voltage (3 V) and the current (250 mA, which is 0.25 A), you can calculate that the resistance of the lamp is 12 Ω, like this:

$$R = \frac{V}{I} = \frac{3V}{0.25A} = 12\Omega$$

Wasn't going back to high school algebra fun? Next thing you know, you're going to start looking for a prom date.

TIP

The most important thing to remember about Ohm's law is that you must always do the calculations in terms of volts, amperes, and ohms. For example, if you measure the current in milliamps (which you usually will in electronic circuits), you must convert the milliamps to amperes by dividing by 1,000. For example, 250 mA is 0.25 A.

Here are a few other things you should keep in mind concerning Ohm's law:

TECHNICAL STUFF

» Remember how I say in the preceding section that the definition of one ohm is the amount of resistance that allows one ampere of current to flow when one volt of potential is applied to it? This definition is based on Ohm's law. If V is 1 and I is 1, then R must also be 1.

$$R = \frac{V}{I} = \frac{1V}{1A} + 1\Omega$$

» If you wonder why the symbols for voltage and resistance are V and R, which make perfect sense, but the symbol for current is I, which makes no sense, it has to do with history. The unit of measure for current — the ampere — is named after André-Marie Ampère, a French physicist who was one of the pioneers of early electrical science. The French word he used to describe the strength of an electric current was *intensité* – in English, *intensity*. Thus, amperage is a measure of the intensity of the current. Hence the letter I.

» In the interest of international cooperation, the term *volt* is named for the Italian scientist Alessandro Volta, who invented the first electric battery in 1800. (Actually, his full name was Count Alessandro Giuseppe Antonio Anastasio Volta. But that won't be on the test.)

Introducing Resistors

A *resistor* is a component that's designed to provide a specific amount of resistance in a circuit. Because resistance is an essential element of nearly every electronic circuit, you'll use resistors in just about every circuit that you build.

Although resistors come in a variety of sizes and shapes, the most common type of resistor for hobby electronics is the *carbon film resistor,* as shown in Figure 2-1. These small resistors consist of a layer of carbon laid down on an insulating material and contained in a small cylinder, with wire leads attached to both ends. The resistor itself is about ¼ inch long, and the leads are about an inch long, making the entire thing about 2¼ inches long.

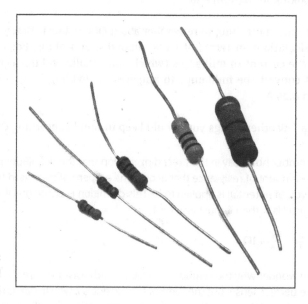

FIGURE 2-1:
Carbon film
resistors.

TIP

Resistors are blind to the polarity in a circuit. Thus, you don't have to worry about installing them backwards. Current can pass equally through a resistor in either direction.

In schematic diagrams, a resistor is represented by a jagged line, like the one shown in the margin. The resistance value is typically written next to the resistor symbol. In addition, an identifier such as R1 or R2 is also sometimes written next to the symbol.

In some schematics, particularly those drawn in Europe, the symbol shown in the margin is used instead of the jagged line.

Resistors are used for many reasons in electronic circuits. The three most popular are

REMEMBER

>> **Limiting current:** By introducing resistance into a circuit, resistors can limit the amount of current that flows through the circuit. In accordance with Ohm's law, if the voltage in a circuit remains the same, the current will decrease if you increase the resistance.

Many electronic components have an appetite for current that must be regulated by resistors. One of the best known are light-emitting diodes (LEDs), which are a special type of diode that emits visible light when current runs through it. Unfortunately, LEDs don't know when to step away from the table when it comes to consuming current. That's because they have very little internal resistance. Unfortunately, LEDs don't have much tolerance for current, so too much current will burn them out. As a result, it's always prudent — essential, in fact — to place a resistor in series with an LED to keep the LED from burning itself up. (You can learn more about LEDs in Chapter 5 of this minibook.)

You can use Ohm's law to your advantage when using current-limiting resistors. For example, if you know what the supply voltage is and you know how much current you need, you can use Ohm's law to determine the right resistor to use for the circuit, as explained in the preceding section.

>> **Dividing voltage:** You can also use resistors to reduce voltage to a level that's appropriate for specific parts of your circuit. For example, suppose your circuit is powered by a 3 V battery but a part of your circuit needs 1.5 V. You could use two resistors of equal value to split this voltage in half, yielding 1.5 V. For more information, see the section "Dividing Voltage," later in this chapter.

>> **Resistor/capacitor networks:** Resistors can be used in combination with capacitors for a variety of interesting purposes. You learn about this use of resistors in Chapter 3 of this minibook.

Reading Resistor Color Codes

You can determine the resistance provided by a resistor by examining the *color codes* that are painted on the resistor. These little stripes of bright colors indicate two important factoids about the resistor: its resistance in ohms and its *tolerance*, which indicates how close to the indicated resistance value the resistor actually is.

Most resistors have four stripes of color. The first three stripes indicate the resistance value, and the fourth stripe indicates the tolerance. Some resistors have five stripes of color, with four representing the resistance value and the last one the tolerance.

TIP

If you're uncertain from which side of the resistor to read the colors, start with the side closest to the color stripe. The first stripe is usually painted very close to the edge of the resistor; the last stripe isn't as close to the edge.

Reading a resistor's value

To read a resistor's color code, see Table 2-1. Here's the procedure for determining the value of a resistor with four stripes:

1. **Orient the resistor so you can read the stripes properly.**

   You should read the stripes from left to right. The first stripe is the one that's closest to one end of the resistor. If this stripe is on the right side of the resistor, turn the resistor around so the first stripe is on the left.

2. **Look up the color of the first stripe to determine the value of the first digit.**

   For example, if the first stripe is yellow, the first digit is 4.

3. **Look up the color of the second stripe to determine the value of the second digit.**

   For example, if the second stripe is violet, the second digit is 7.

4. **Look up the color of the third stripe to determine the multiplier.**

   For example, if the third stripe is brown, the multiplier is 10.

5. **Multiply the two-digit value by the multiplier to determine the resistor's value.**

   For example, 47 times 10 is 470. Thus, a yellow-violet-brown resistor is 470 Ω.

TIP

If a resistor has five stripes, the first three stripes are the value digits, and the fourth stripe is the multiplier. The fifth stripe is the tolerance, as described in the next section.

Here are a few examples that should help you understand how to read resistor codes:

Color Stripes	Digit Values	Multiplier (in Ohms)	Resistor Value
Brown–black–brown	10	10	100 Ω
Brown–black–red	10	100	1 kΩ
Red–red–orange	22	1 k	22 Ω
Red–red–yellow	22	10 k	220 kΩ
Yellow–violet–black	47	1	47 Ω

TABLE 2-1 ## Resistor Color Codes (Resistance Values)

Color	Digit	Multiplier
Black	0	1
Brown	1	10
Red	2	100
Orange	3	1 k
Yellow	4	10 k
Green	5	100 k
Blue	6	1 M
Violet	7	10 M
Gray	8	100 M
White	9	1,000 M
Gold		0.1
Silver		0.01

Understanding resistor tolerance

The value indicated by the stripes painted on a resistor provides an estimate of the actual resistance. The exact resistance varies by a percentage that depends on the *tolerance* factor of the resistor.

For example, a 22 kΩ resistor with a ±5 percent tolerance actually has a value somewhere between 5 percent above and 5 percent below 22 kΩ, which works out to somewhere between 20.9 kΩ and 23.1 kΩ. A 470 Ω resistor with a 10 percent tolerance has an actual value somewhere between 423 Ω and 517 Ω.

NOTICING STANDARD RESISTOR VALUES

In theory, there are 100 different combinations of colors for the first two bands, covering the range of values from 00 through 99. However, in practice there are only a few color combinations that are commonly encountered. These color combinations represent standardized values that allow manufacturers to produce resistors that will be useful in a wide variety of applications.

For example, take the value 47, represented by the color code yellow-violet. 47 happens to be one of the preferred resistor numbers, so you can easily obtain resistors of 4.7 Ω, 47 Ω, 470 Ω, 4.7 kΩ, 47 kΩ, 470 kΩ, and 4.7 MΩ.

But 45 is not one of the preferred values. Thus you won't find 45 Ω or 450 Ω resistors.

Although there are several different systems for standardizing preferred resistor values, the most common system uses 12 different standard values:

First Two Colors	Standard Value
Brown, black	10
Brown, red	12
Brown, green	15
Brown, gray	18
Red, red	22
Red, violet	27
Orange, orange	33
Orange, white	39
Yellow, violet	47
Green, blue	56
Blue, gray	68
Gray, red	82

These values are standardized values that are designed to provide a wide range of resistance values. Although not exact, each value is approximately 1.2 times larger than the previous value.

Why the approximations? Simply because it costs more money to manufacture resistors to very close tolerances, and for most electronic circuits, a 5 percent or 10 percent margin of error is perfectly acceptable. For example, if you're building a circuit to limit the current flowing through a component to 200 mA, it probably won't matter much if the actual current is a little above or below 200 mA. Thus, a 5 percent or 10 percent tolerance resistor is acceptable.

If your application demands higher precision, you can spend a bit more money to buy higher-tolerance resistors. But ±5 percent or ±10 percent tolerance resistors are fine for most work, including all the circuits presented in this book (unless otherwise indicated).

The tolerance of a resistor is indicated in the resistor's last color stripe, as listed in Table 2-2.

TABLE 2-2

Resistor Color Codes (Tolerance Values)

Color	Tolerance
Brown	±1%
Red	±2%
Orange	±3%
Yellow	±4%
Gold	±5%
Silver	±10%
No band	±20%

Understanding Resistor Power Ratings

Resistors are like brakes for electric current. Like the brakes in your car, resistors work by applying the electrical equivalent of friction to flowing current. This friction inhibits the flow of current by absorbing some of the current's energy and dissipating it in the form of heat. Whenever you use a resistor in a circuit, you must make sure that the resistor is capable of handling the heat.

The *power rating* of a resistor indicates how much power a resistor can handle before it becomes too hot and burns up. You may remember from Book 1, Chapter 2 that power is measured in units called *watts*. The more watts a resistor can handle, the larger and more expensive the resistor is.

REMEMBER

Most resistors are designed to handle ⅛ W or ¼ W. You can also find resistors rated for ½ W or 1 W, but they're rarely needed in the types of electronic projects you build in this book. Unless otherwise stated, all the resistors used in this book are rated at ¼ W.

TIP

Unfortunately, you can't tell a resistor's power rating just by looking at it. Unlike resistance and tolerance, there are no color codes for wattage. However, the size of the resistor is a good indicator of its power rating. Power ratings are written on the packaging when you buy new resistors. After you work with them for a while, you'll come to quickly recognize the size difference between resistors of different power ratings.

If you want to be safe, you can calculate the power demands required of a particular resistor in your circuits. First, use Ohm's law to calculate the voltage across the resistor and the current that will pass through the resistor. For example, if a 1,000 Ω resistor will have 3 V across it, you can calculate that 30 mA of current will flow through the resistor by dividing the voltage by the resistance (3 V ÷ 1,000 Ω = 0.03 A, which is 30 mA).

Once you know the voltage and the current, you can calculate the power that will be dissipated by the resistor by using the power formula you learn about in Book 1, Chapter 2:

$$P = I \times V$$

Thus, the power dissipated by the resistor will be just 0.09 W, well under the maximum that can be handled by a ¼ W (0.25 W) resistor. (A ⅛ W resistor should be able to handle this amount of power, too, but it's always better to err on the large side when it comes to power ratings.)

Limiting Current with a Resistor

One of the most common uses for resistors is to limit the current flowing through a component. Some components, such as light-emitting diodes, are very sensitive to current. A few milliamps of current is enough to make an LED glow; a few hundred milliamps is enough to destroy the LED.

To use a resistor to limit current through a component, you simply place the resistor in series with the component. For example, Figure 2-2 shows two ways to place a resistor in series with an LED. As you can see, the version on the left places the resistor on the positive voltage side of the LED (called the *anode*). The version on the right places the resistor on the negative side of the LED (called the *cathode*).

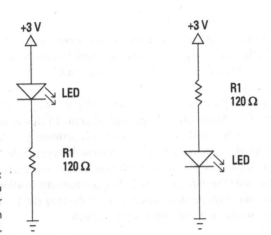

FIGURE 2-2:
Two ways to place a resistor in series with an LED.

TECHNICAL STUFF

You could say that the version on the left places the current-limiting resistor *before* the LED and the version on the right places the resistor *after* the LED. This is consistent with the conventional notion that electric current flows from positive to negative. It doesn't matter whether the resistor is before or after the LED, as long as it's in series with the LED.

Project 10 shows you how to build a simple circuit that demonstrates how a resistor can be used to limit current to an LED, with the resistor placed after the LED, between the LED and the ground connection. Figure 2-3 shows the assembled circuit.

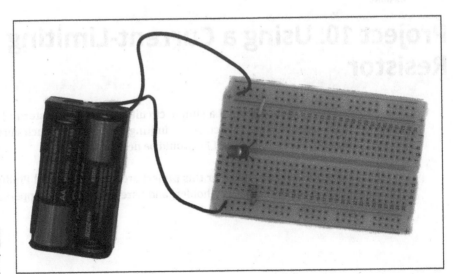

FIGURE 2-3:
The assembled circuit for Project 10.

Before we get into the construction of the circuit, here's a simple question: Why a 120 Ω resistor? Why not a larger or a smaller value? In other words, how do you determine what size resistor to use in a circuit like this?

The answer is simple: Ohm's law, which can easily tell you what size resistor to use, but you must first know the voltage and current. In this case, the voltage is easy to figure out: You know that two AA batteries provide 3 V. To figure out the current, you just need to decide how much current is acceptable for your circuit. The technical specifications of the LED tell you how much current the LED can handle. In the case of a standard 5 mm red LED, the maximum allowable current is 28 mA. To be safe and to make sure that you don't destroy the LED with too much current, round the maximum current down to 25 mA.

To calculate the desired resistance, you divide the voltage (3 V) by the current (0.025 A). The result is 120 Ω.

TIP

You don't need to stress too much about getting the exact right resistor value to limit current for an LED. If the resistance value you need doesn't correspond to one of the standard resistor values, just round up to the next standard value. For example, if you need a 135 Ω resistor, round up and use a 150 Ω resistor. And if you don't happen to have a 150 Ω resistor on hand, a 220 Ω resistor will do.

WARNING

Do *not* connect the LED directly to the battery without a resistor. If you do, the LED will flash brightly, and then it will be dead forever.

Project 10: Using a Current-Limiting Resistor

In this project, you build a simple circuit that uses a resistor to limit the current to an LED. Without this current-limiting resistor, too much current would flow through the LED and the LED would be destroyed.

The only tools you need for this project are perhaps a small Phillips-head screwdriver to open the battery holder, and wire cutters and strippers to cut and strip the jumper wire.

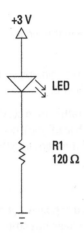

+3 V

LED

R1
120 Ω

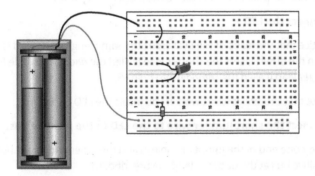

Parts

» Two AA batteries

» One battery holder

» One red LED, 5 mm

» One resistor, 120 Ω (brown, red, brown)

» One solderless breadboard

» One 1-inch jumper wire

Steps

1. **Connect the battery holder.**

   Orient the breadboard so that hole A1 is near the lower left, corresponding to
   the orientation of the diagram. Then, insert the black lead in the negative (–)

bus strip on the bottom side of the breadboard and the red lead in the positive (+) bus strip on the top side of the breadboard.

TIP

Any location on the correct bus will be fine, but I recommend you insert the leads into the holes at the very end of the breadboard.

TIP

The leads on most battery holders are stranded rather than solid. After you strip ¼ inch or so of insulation off the end of each lead, you'll need to twist the stranded wires together tightly so you can insert the leads into the holds on the breadboard.

2. **Connect the resistor.**

Insert one end of the resistor into the breadboard in hole A5, and insert the other lead into any nearby hole in the negative voltage bus strip on the bottom side (the strip that's connected to the black battery wire).

3. **Connect the LED.**

Notice that the leads of the LED aren't the same length; one lead is shorter than the other. Insert the shorter lead (called the *anode*) into hole E5 and the longer lead (called the *cathode*) into hole F5.

Note that the circuit won't work if you insert the LED backward.

4. **Connect the short jumper wire to the LED to the positive bus.**

Insert one end of the jumper in hole J5 and the other end in any hole on the positive bus at the upper side of the breadboard.

5. **Insert the batteries.**

The LED lights up. If it doesn't, double-check your connections to make sure that the circuit is assembled correctly. If it still doesn't light up, try a different battery.

Combining Resistors

Suppose you've designed the perfect circuit, and it calls for a 1,100 Ω resistor in a critical spot. You run down to your local RadioShack and are miffed because you can't find any 1,100 Ω resistors. So you go online and search your favorite online suppliers and are even more miffed to discover that they don't have 1,100 Ω resistors either. You can buy 1 kΩ resistors and 100 Ω resistors, but no one seems to have any 1,100 Ω resistors.

Is it time to give up? Or must you settle for a 1 kΩ resistor and hope it will be close enough?

Certainly not!

All you have to do is use two or more resistors in combination to create the resistance that you need. Such a combination of resistors is sometimes called a *resistor network*. You can freely substitute a resistor network for a single resistor whenever you want.

There are two basic ways to combine resistors: in series (strung end to end) and in parallel (side by side). The following sections explain how you calculate the total resistance of a network of resistors in series and in parallel.

You'll need to put your thinking cap on when you read through the following sections. Ohm's law is simple enough, but the math calculations required to calculate parallel resistors can get a little complex. The math isn't horribly complicated, but it isn't trivial, either.

Combining resistors in series

Calculating the total resistance for two or more resistors strung end to end — that is, in series — is simple: You simply add the resistance values to get the total resistance.

For example, if you need 1,100 ohms of resistance and can't find an 1,100 Ω resistor, you can combine a 1 kΩ resistor and a 100 Ω resistor in series. Adding these two resistances together gives you a total resistance of 1,100 Ω.

You can place more than two resistors in series if you want. You just keep adding up all the resistances to get the total resistance value. For example, if you need 1,800 Ω of resistance, you could use a 1 kΩ resistor and eight 100 Ω resistors in series.

Figure 2-4 shows how serial resistors work. Here, the two circuits have identical resistances. The circuit on the left accomplishes the job with one resistor; the circuit on the right does it with three. Thus, the circuits are equivalent.

REMEMBER

Any time you see two or more resistors in series in a circuit, you can substitute a single resistor whose value is the sum of the individual resistors. Similarly, any time you see a single resistor in a circuit, you can substitute two or more resistors in series as long as their values add up to the desired value.

TIP

The total resistance of resistors in series is always greater than the resistance of any of the individual resistors. That's because each resistor adds its own resistance to the total.

Combining resistors in parallel

You can also combine resistors in parallel to create equivalent resistances. However, calculating the total resistance for resistors in parallel is a bit more complicated than calculating the resistance for resistors in series.

As the circuit shown in Figure 2-5 illustrates, when you combine two resistors in parallel, current can flow through both resistors at the same time. Although each resistor does its job to hold back the current, the total resistance of two resistors in parallel is always less than the resistance of either of the resistors because the current has two pathways through which to go.

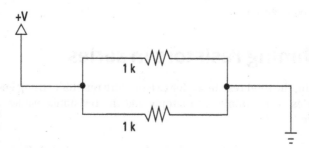

FIGURE 2-5:
Resistors in
parallel.

So how do you calculate the total resistance for resistors in parallel? Very carefully. Here are the rules:

>> First, the simplest case: resistors of equal value in parallel. In this case, you can calculate the total resistance by dividing the value of one of the individual resistors by the number of resistors in parallel. For example, the total resistance of two 1 kΩ resistors in parallel is 500 Ω, and the total resistance of four 1 kΩ resistors is 250 Ω.

Unfortunately, this is the only case that's simple. The math when resistors in parallel have unequal values is more complicated.

>> If only two resistors of different values are involved, the calculation isn't too bad:

$$R_{total} = \frac{R1 \times R2}{R1 + R2}$$

>> In this formula, R1 and R2 are the values of the two resistors.

>> Here's an example, based on a 2 kΩ and a 3 kΩ resistor in parallel:

$$R_{total} = \frac{R1 \times R2}{R1 + R2} = \frac{2,000\Omega \times 3,000\Omega}{2,000\Omega + 3,000\Omega} = 1,200\Omega$$

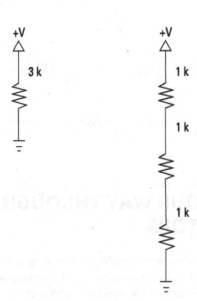

FIGURE 2-4:
Combining resistors in series.

>> For three or more resistors in parallel, the calculation begins to look like rocket science:

$$R_{total} = \cfrac{1}{\cfrac{1}{R1} + \cfrac{1}{R2} + \cfrac{1}{R2} \cdots}$$

>> The dots at the end of the expression indicate that you keep adding up the reciprocals of the resistances for as many resistors as you have.

>> In case you're crazy enough to actually want to do this kind of math, here's an example for three resistors whose values are 2 kΩ, 4 kΩ, and 8 kΩ:

$$R_{total} = \cfrac{1}{\cfrac{1}{2,000\ \Omega} + \cfrac{1}{4,000\ \Omega} + \cfrac{1}{8,000\ \Omega}} = \frac{1}{0.000875\ \Omega} = 1,142.857\ \Omega$$

>> As you can see, the final result is 1,142.857 Ω. That's more precision than you could possibly want, so you can probably safely round it off to 1,142 Ω, or maybe even 1,150 Ω.

Mixing series and parallel resistors

Resistors can be combined to form complex networks in which some of the resistors are in series and others are in parallel. For example, Figure 2-6 shows a network of three 1 kΩ resistors and one 2 kΩ resistor. These resistors are arranged in a mixture of serial and parallel connections.

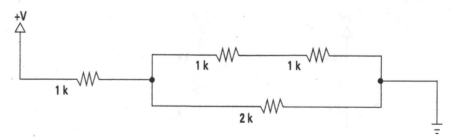

FIGURE 2-6:
Resistors in series
and parallel.

CONDUCTING YOUR WAY THROUGH PARALLEL RESISTORS

The parallel resistance formula makes more sense if you think about it in terms of the opposite of resistance, which is called *conductance*. Resistance is the ability of a conductor to block current; conductance is the ability of a conductor to pass current. Conductance has an inverse relationship with resistance: When you increase resistance, you decrease conductance, and vice versa.

Because the pioneers of electrical theory had a nerdy sense of humor, they named the unit of measure for conductance the *mho*, which is *ohm* spelled backward. The mho is the reciprocal (also known as inverse) of the ohm. To calculate the conductance of any circuit or component (including a single resistor), you just divide the resistance of the circuit or component (in ohms) into 1. Thus, a 100 Ω resistor has $\frac{1}{100}$ mho of conductance.

When circuits are connected in parallel, current has multiple pathways it can travel through. It turns out that the total conductance of a parallel network of resistors is simple to calculate: You just add up the conductances of each individual resistor. For example, suppose you have three resistors in parallel whose conductances are 0.1 mho, 0.02 mho, and 0.005 mho. (These are the conductances of 10 Ω, 50 Ω, and 200 Ω resistors, respectively.) The total conductance of this circuit is 0.125 mho (0.1 + 0.02 + 0.005 = 0.125).

One of the basic rules of doing math with reciprocals is that if one number is the reciprocal of a second number, the second number is also the reciprocal of the first number. Thus, since mhos are the reciprocal of ohms, ohms are the reciprocal of mhos. To convert conductance to resistance, you just divide the conductance into 1. Thus, the resistance equivalent to 0.125 mho is 8 Ω (1 ÷ 0.125 = 8).

It may help you to remember how the parallel resistance formula works when you realize that what you're really doing is converting each individual resistance to conductance, adding them up, and then converting the result back to resistance. In other words, convert the ohms to mhos, add them up, and then convert them back to ohms. That's how — and why — the resistance formula actually works.

The way to calculate the total resistance of a network like this is to divide and conquer. Look for simple series or parallel resistors, calculate their total resistance, and then substitute a single resistor with an equivalent value. For example, you can replace the two 1 kΩ resistors that are in series with a single 2 kΩ resistor. Now, you have two 2 kΩ resistors in parallel. Remembering that the total resistance of two resistors with the same value is half the resistance value, you can replace these two 2 kΩ resistors with a single 1 kΩ resistor. You're now left with two 1 kΩ resistors in series. Thus, the total resistance of this circuit is 2 kΩ.

Deceptively simple, eh?

Combining resistors in series and parallel

Project 11 lets you do a little hands-on work with some simple series and parallel resistor connections so you can see firsthand how the calculations described in the previous three sections actually work in the real world. You'll probably find that due to the individual variations of actual resistors (due to their manufacturing tolerances) the calculated resistances don't always match the resistance of the actual circuits. But in most cases, the variations aren't significant enough to affect the operation of your circuits.

In this project, you assemble five resistors into three different configurations. The first has all five resistors in series. The second has them all in parallel. And the third creates a network of two sets of parallel resistors that are connected in series.

Project 11: Resistors in Series and Parallel

In this project, you experiment with resistors in series and in parallel. You need a multimeter with an ohmmeter function to measure the resistances.

Parts

» One solderless breadboard (RadioShack 2760003)

» Five 1 kΩ ¼ W resistors (brown-black-red)

» One multimeter with an ohmmeter function

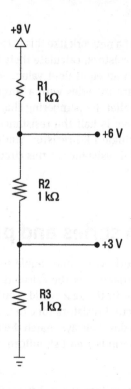

+9 V

R1
1 kΩ

+6 V

R2
1 kΩ

+3 V

R3
1 kΩ

Combining resistors in series and parallel

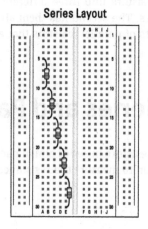

Series Layout

Project 11: Resistors in Series and Parallel

Parts

Parallel Layout

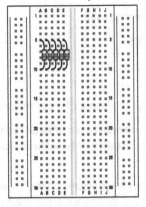

Series/Parallel Layout

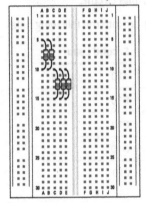

Steps

1. **Set your multimeter to its ohmmeter function with a range large enough to measure at least 5 kΩ of resistance.**

   On my ohmmeter, the closest range is 20 kΩ.

2. **Insert the five 1 kΩ resistors in series.**

   Use the following holes on the breadboard:

Resistor	First Lead	Second Lead
1	A5	A10
2	B10	B15
3	C15	C20

Resistor	First Lead	Second Lead
4	D20	D25
5	E25	E30

Refer to the Series Layout diagram for the layout of these resistors.

3. **Using your multimeter, measure the resistance of each resistor individually, and then measure the total resistance of two, three, four, and five resistors in series.**

To do these measurements, place the meter probes on the leads at the breadboard holes indicated in Table 2-3. Write your actual measurements in the column on the far right.

4. **Rearrange the resistors into a parallel circuit.**

Remove the resistors and reinsert them into the following holes:

Resistor	First Lead	Second Lead
1	A5	A10
2	B5	B10
3	C5	C10
4	D5	D10
5	E5	E10

Refer to the Parallel Layout diagram for the layout of these resistors.

5. **Use your ohmmeter to measure the resistance of the parallel resistor circuit.**

Because there are five 1 kΩ resistors, your measurement should be approximately 200 Ω.

6. **Rearrange the resistors into a series/parallel network.**

Remove resistors 3, 4, and 5 and insert them as follows:

Resistor	First Lead	Second Lead
1	A5	A10
2	B5	B10
3	C10	C15
4	D10	D15
5	E10	E15

Refer to the Series/Parallel Layout diagram for the layout of these resistors.

This configuration has the first two resistors in one parallel circuit and the other three in a second parallel circuit. The two parallel circuits are connected to form a series circuit.

7. **Use your ohmmeter to measure the resistance of two parallel circuits and the entire circuit.**

   Record your measurements in Table 2-4.

   You're done! Unless you're a glutton for punishment. In that case, feel free to experiment with other combinations of series and parallel resistor circuits. Grab some resistors with other values and throw them into the mix. Each time, do the math to predict what the resulting resistance should be. With some practice, you'll get good at calculating resistor networks.

TABLE 2-3   Series Resistance Measurements

Red Lead	Black Lead	Number of Resistors	Expected Measurement	Your Measurement
A5	A10	1	1 kΩ	
B10	B15	1	1 kΩ	
C15	C20	1	1 kΩ	
D20	D25	1	1 kΩ	
E25	E30	1	1 kΩ	
A5	B15	2	2 kΩ	
A5	C20	3	3 kΩ	
A5	D25	4	4 kΩ	
A5	E30	5	5 kΩ	

TABLE 2-4   Series/Parallel Resistance Measurements

Red Lead	Black Lead	Number of Resistors	Expected Measurement	Your Measurement
A5	A10	2	500 Ω	
C10	C15	3	333 Ω	
A5	C15	5	833 Ω	

Dividing Voltage

One interesting and useful property of resistors is that if you connect two resistors together in series, you can tap into the voltage at the point between the two resistors to get a voltage that is a fraction of the total voltage across both resistors. This type of circuit is called a *voltage divider*, and it is a common way to reduce voltage in a circuit. Figure 2-7 shows a typical voltage divider circuit.

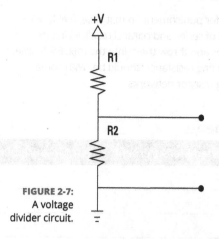

FIGURE 2-7:
A voltage divider circuit.

When the two resistors in the voltage divider are of the same value, the voltage is cut in half. For example, suppose your circuit is powered by a 9 V battery, but your circuit really only needs 4.5 V. You could use a pair of resistors of equal value across the battery leads to provide the necessary 4.5 V.

When the resistors are of different values, you must do a little math to calculate the voltage at the center of the divider. The formula is as follows:

$$V_{out} = \frac{V_{in} \times R2}{R1 + R2}$$

For example, suppose you're using a 9 V battery, but your circuit requires 6 V. In that case, you could create a voltage divider using a 1 kΩ resistor for R1 and a 2 kΩ resistor for R2. Here's the math:

$$V_{out} = \frac{9V \times 2,000 \ \Omega}{1,000 \ \Omega + 2,000 \ \Omega} = \frac{18,000 \ V}{3,000} = 6 \ V$$

As you can see, these resistor values cut the voltage down to 6 V.

TIP

Here's another way to look at the same problem: 6 V is two-thirds of 9 V. So, you need a voltage divider that will give you two-thirds of 9 V. You can use any three resistors of the same value to accomplish this, dividing the voltage with one of the resistors on one side of the divider and the other two resistors on the other side.

Dividing Voltage with Resistors

In Project 12, you build a simple voltage divider circuit on a solderless breadboard to provide either 3 V or 6 V from a 9 V battery. The assembled circuit is shown in Figure 2-8.

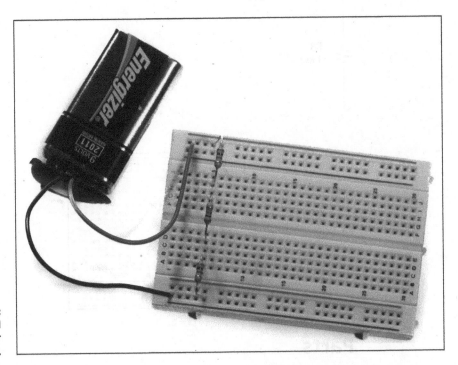

FIGURE 2-8:
The assembled voltage divider circuit.

Project 12: A Voltage Divider Circuit

In this project, you build a simple voltage divider circuit using three 1 kΩ resistors, and you use the ohmmeter function of your multimeter to measure the effect of the divider.

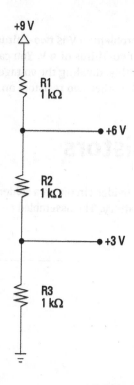

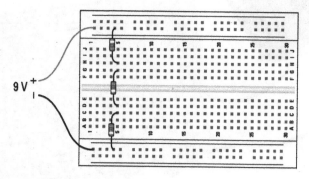

Parts

» One solderless breadboard

» One 9 V battery

» One 9 V battery snap holder

» Three 1 kΩ ¼ W resistors (brown-black-red)

» One multimeter with a voltmeter function

Steps

1. **Connect the battery holder.**

   Insert the black lead into the first hole in the ground bus on the side nearest column A. Then insert the red lead in the first hole in the positive voltage bus on the side nearest column J.

2. **Insert the resistors.**

   Place the resistor leads into the holes indicated in the following table:

Resistor	First Lead	Second Lead
1	Ground bus near row 5	C5
2	D5	G5
3	H5	Positive bus near row 5

3. **Connect the battery.**

4. **Use the voltmeter to observe how the voltage is divided.**

   Set the voltmeter to a range that will measure at least 10 V DC. Then take the measurements indicated in Table 2-5 by touching the meter leads to the resistor leads that are plugged into the holes indicated in the table:

   You're done!

   Be sure to unplug the battery from the circuit. If you leave it plugged in, current will continue to flow through the series resistor circuit, and the battery will soon go dead.

TABLE 2-5 **Series Resistance Measurements**

Black Lead	Red Lead	Expected Measurement	Your Measurement
Ground bus	Positive bus	9 V	
Ground bus	H5	6 V	
Ground bus	C5	3 V	
D5	G5	3 V	

Varying Resistance with a Potentiometer

Many circuits call for a resistance that can be varied by the user. For example, most audio amplifiers include a volume control that lets the user turn the volume up or down, and you can create a simple light dimmer by varying the resistance in series with a lamp.

A variable resistor is properly called a *potentiometer*, or just *pot* for short. A potentiometer is simply a resistor with three terminals. Two of the terminals are permanently fixed on each end of the resistor, but the middle terminal is connected to a wiper that slides in contact with the entire surface of the resistor. Thus, the amount of resistance between this center terminal and either of the two side terminals varies as the wiper moves.

Figure 2-9 shows how a typical potentiometer looks from the outside. The resistive track and slider (properly called the *wiper*) are enclosed within the metal can, and the three terminals are beneath it. The rod that protrudes from the top of the metal can is connected to the wiper so that when the user turns the rod, the wiper moves across the resistor to vary the resistance.

FIGURE 2-9:
A potentiometer.

Figure 2-10 shows how a potentiometer works on the inside. Here, you can see that the resistor is made of a semicircular piece of resistive material such as carbon. The two outer terminals are connected to either end of the resistor. The wiper, to which the third terminal is connected, is mounted so that it can rotate across the resistor. When the wiper moves, the resistance between the center terminal and the other two terminals changes.

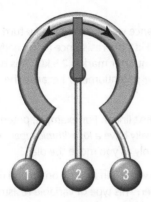

FIGURE 2-10:
How a
potentiometer
works.

The symbol used for a potentiometer in schematic diagrams is shown in the margin. As you can see, the center tap of the resistor is indicated by an arrow that is meant to reflect that the value of the resistance at this terminal varies when the wiper moves.

Potentiometers are rated by their total resistance. The resistance between the center terminal and the two other terminals always adds up to the total resistance rating of the potentiometer. For example, the two resistances split by a 100 Ω potentiometer always add up to 100 Ω. When the dial is exactly in the center, both resistances are 50 Ω. As you move the wiper one way or the other, one resistance increases while the other decreases; but in all cases, the total of the two resistances always adds up to 100 Ω.

Here are a few other rambling thoughts to keep in mind about potentiometers:

>> Potentiometers come in a wide variety of shapes and sizes. With a little hunting around in stores or on the Internet, you should be able to find the perfect potentiometer for every need.

>> Some potentiometers are very small and can be adjusted only by the use of a tiny screwdriver. This type of pot is called a *trim pot*, designed to make occasional fine-tuning adjustments to your circuits.

>> Some potentiometers have switches incorporated into them so that when you turn the knob all the way to one side or pull the knob out, the switch operates to either open or close the circuit.

TIP

>> When the wiper reaches one end of the resistor or the other, the resistance between the center terminal and the terminal on that end is essentially zero. Keep this point in mind when you're designing circuits. To avoid circuit paths with no resistance, it's common to put a small resistor in series with a potentiometer.

>> In some potentiometers, the resistance varies evenly as you turn the dial. For example, if the total resistance is 10 kΩ, the resistance at the halfway mark is 5 kΩ, and the resistance at the one-quarter mark is 2.5 kΩ. This type of potentiometer is called a *linear taper* potentiometer because the resistance change is linear.

Many potentiometers, however, aren't linear. For example, potentiometers designed for audio applications usually have a *logarithmic taper*, which means that the resistance doesn't vary evenly as you move the dial.

>> Some variable resistors have only two terminals: one on an end of the resistor itself, the other attached to the wiper. This type of variable resistor is properly called a *rheostat*, but most people use the term *potentiometer* or *pot* to refer to both two- and three-terminal variable resistors.

Chapter **3**

Working with Capacitors

My favorite science fiction gadget is the device invented by Doc Brown in *Back to the Future*, which enabled him send his friend Marty McFly back to 1955: the flux capacitor. Doc Brown's famous flux capacitor was able to convert 1.21 gigawatts of power into a distortion in the space-time continuum to alter the course of history.

In this chapter, you can examine ordinary capacitors, the little brothers of the fictitious flux capacitor. Although a time-travelling flux capacitor would be pretty useful if such a thing existed, real capacitors do exist, and are incredibly useful. In fact, capacitors are among the most useful of all electronic components. You'll use one or more capacitors in almost every circuit you build.

WARNING

In Book 1, Chapter 4, I warn you about the dangers of working with large capacitors. Before we embark on our Journey of Capacitance, I want to remind you of that warning: Large capacitors can be dangerous because they can hold large charges long after they've been disconnected from power. If you're working on a circuit that has capacitors bigger than a thimble, be careful! You should discharge these capacitors before you work with them. Really large capacitors should be discharged through a high-current resistor. Note that none of the capacitors used in this book is large enough to be dangerous, with the exception of the capacitors used in power supplies, as described in Book 4, Chapter 2.

What Is a Capacitor?

To begin to understand what a capacitor is, consider this question: What makes current flow within a conductor? In Book 1, Chapter 2, you learn that opposite charges attract one another and like charges repel one another. The attraction or repulsion is caused by a force known as the *electromagnetic force*. All charged particles exhibit an electric field that is associated with this force. This field is what draws *electrons* (negative charges) toward *protons* (positive charges), and it's also this field that pushes electrons away from one another.

Within a conductor such as copper, the electric fields produced by individual electrons create a constant movement of electrons within the conductor. However, this movement is completely random. When voltage is applied across the two ends of a conductor, an electric field is created, which causes the movement of electrons to become organized. The electric field pushes electrons through the conductor in an orderly fashion from the negative side of the voltage to the positive side. The result: current.

Electric fields are magical. Unlike current, an electric field doesn't need a conductor to travel. An electric field can reach right across an insulator and push away or tug on charges on the other side of the insulator.

This might seem like magic, but you've encountered it before. Blow up a balloon, rub it fast against your shirt, and then hold it up to your hair. Rubbing the balloon creates a static charge in the balloon. When this charge comes close to your hair, its electric field tugs the charges in your hair, which is magically drawn upward toward the balloon.

A *capacitor* is an electronic component that takes advantage of the apparently magical ability of electric fields to reach out across an insulator. It consists of two flat plates made from a conducting material such as silver or aluminum, separated by a thin insulating material such as Mylar or ceramic, as shown in Figure 3-1. The two conducting plates are connected to terminals so that a voltage can be applied across the plates.

Note that because the two plates are separated by an insulator, a closed circuit *isn't* formed. Nevertheless, current flows — for a moment, anyway.

How can this be? When the voltage from a source such as a battery is connected, the negative side of the battery voltage immediately begins to push negative charges toward one of the plates. Simultaneously, the positive side of the battery voltage begins to pull electrons (negative charges) away from the second plate.

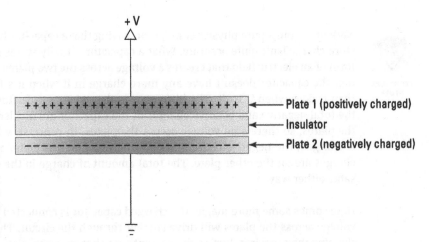

+ V

+++++++++++++++++++++++++++++ ← — Plate 1 (positively charged)
← — Insulator
------------------------------- ← — Plate 2 (negatively charged)

FIGURE 3-1:
A capacitor
creates an electric
field between two
charged plates
separated by an
insulator.

What permits current to flow is the electric field that quickly builds up between the two plates. As the plate on the negative side of the circuit fills with electrons, the electric field created by those electrons begins to push electrons away from the plate on the other side of the insulator, toward the positive side of the battery voltage.

As this current flows, the negative plate of the capacitor builds up an excess of electrons, whereas the positive side develops a corresponding deficiency of electrons. Thus, voltage is developed between the two plates of the capacitor. (Remember, the definition of *voltage* is the difference of charge between two points.)

But there's a catch: This current flows only for a brief time. As the electrons build up on the negative plate and are depleted from the positive plate, the voltage between the two plates increases because the difference in charge between the two plates increases. The voltage continues to increase until the capacitor voltage equals the battery voltage. Once the voltages are the same, current stops flowing through the circuit, and the capacitor is said to be *charged*.

At this point the magic gets even better. Once a capacitor has been charged, you can disconnect the battery from the capacitor, and the voltage will remain in the capacitor. In other words, although the voltage in the capacitor is created by the battery, this voltage isn't dependent on the battery for its continued existence. Disconnect the battery, and the voltage remains across the two plates of the capacitor.

Thus, capacitors have the ability to store charge — an ability known as *capacitance*.

Working with Capacitors

TECHNICAL
STUFF

Note that from a pure physics standpoint, saying that a capacitor has the ability to store charge isn't quite accurate. What a capacitor actually stores is energy in the form of an electric field that creates a voltage across the two plates. Strictly speaking, the capacitor doesn't have any more charge in it when it's fully "charged" than it does when it isn't. The difference is that when a capacitor isn't charged, the total negative and positive charges of the capacitor are divided evenly among the plates, so there's no voltage across the plates. In contrast, when a capacitor is charged, the negative charges are concentrated on one plate, and the positive charges are on the other plate. The total amount of charge in the capacitor is the same either way.

Here comes some more magic: If a charged capacitor is connected to a circuit, the voltage across the plates will drive current through the circuit. This is called *discharging* the capacitor. Just as the current that charges a capacitor lasts only for a short time, the current that results when a capacitor discharges also lasts only for a short time. As the capacitor discharges, the charge difference between the two plates decreases, and the electric field collapses. Once the two plates have reached equilibrium, the voltage across the plates reaches zero, and no current flows.

Here are a few additional things you should know about capacitors before moving on:

» The most common symbol used for capacitors in schematic diagrams is simply two parallel lines separated by a gap, as shown in the margin.

» An alternative symbol uses a straight line and a curved line to represent the plates. The curved line is generally used on the negative side of the circuit.

» Although some capacitors aren't sensitive to polarity, many others are. This sensitivity has to do with the choice of materials used to create the capacitors: With some materials, connecting the voltage in the wrong direction can damage the capacitor. Capacitors that have distinct positive and negative terminals are called *polarized capacitors*. A plus sign is used in the schematic diagram to indicate the polarity, as shown in the margin.

» Sometimes capacitors are called simply *caps*.

TECHNICAL
STUFF

» The insulating material between the two conducting plates is properly called *dielectric*, a term that refers to the ability of the insulating layer to become polarized by the electric field that exists between the two plates when they become charged.

» Long ago, capacitors were called *condensers*. The term *capacitor* came into popular use in the 1920s. My grandfather, who was born in 1910, used the term *condenser* when he introduced me to electronics in the 1970s. (He also treated the word *mile* as plural. So, if I were to have asked him, "How far away are we from Merced?," he'd answer, "About 5 mile.")

Counting Capacitance

Capacitance is the term that refers to the ability of a capacitor to store charge. It's also the measurement used to indicate how much energy a particular capacitor can store. The more capacitance a capacitor has, the more charge it can store.

Capacitance is measured in units called *farads* (abbreviated F). The definition of one farad is deceptively simple. A one-farad capacitor holds a voltage across the plates of exactly one volt when it's charged with exactly one ampere per second of current.

Note that in this definition, the "one ampere per second of current" part is really referring to the amount of charge present in the capacitor. There's no rule that says the current has to flow for a full second. It could be one ampere for one second, or two amperes for half a second, or half an ampere for two seconds. Or it could be 100 mA for 10 seconds or 10 mA for 100 seconds.

One ampere per second corresponds to the standard unit for measuring electric charge, called the *coulomb*. So another way of stating the value of one farad is to say that it's the amount of capacitance that can store one coulomb with a voltage of one volt across the plates.

It turns out that one farad is a huge amount of capacitance, simply because one coulomb is a very large amount of charge. To put it into perspective, the total charge contained in an average lightning bolt is about five coulombs, and you need only five one-farad capacitors to store the charge contained in a lightning strike. (Some lightning strikes are much more powerful, as much as 350 coulombs.)

It's a given that Doc Brown's flux capacitor was in the farad range because Doc charged it with a lightning strike. But the capacitors used in electronics are charged from much more modest sources. *Much* more modest. In fact, the largest capacitors you're likely to use have capacitance that is measured in millionths of a farad, called *microfarads* and abbreviated μF. And the smaller ones are measured in millionths of a microfarad, also called a *picofarad* and abbreviated pF.

Here are a few other things you should know about capacitor measurements:

>> Like resistors, capacitors aren't manufactured to perfection. Instead, most capacitors have a margin of error, also called *tolerance*. In some cases, the margin of error may be as much as 80 percent. Fortunately, that degree of variance rarely has a noticeable effect on most circuits.

TIP

>> The μ in μF isn't an italic letter *u*; it's the Greek letter *mu*, which is a common abbreviation for *micro*.

Working with Capacitors

>> It's common to represent values of 1,000 pF or more in µF rather than pF. For example, 1,000 pF is often written as 0.001 µF, and 22,000 pF is often written as 0.022 µF.

TECHNICAL STUFF

>> Historical note: Like many units of measure in electronics, the farad was named after one of the all-time great pioneers of electricity, an Englishman named Michael Faraday, who did the groundbreaking research into magnetism and its relationship with electric current.

>> As an example of how much Faraday and physicist James Maxwell are respected for their work, consider that Albert Einstein kept portraits of three people on the wall of his study at Princeton. They were Sir Isaac Newton (considered by many to be the greatest physicist of all time), James Maxwell, who developed the theory of electromagnetism (considered by many to be as great an accomplishment as Newton's work with gravity), and Michael Faraday.

Reading Capacitor Values

If there's enough room on the capacitor, most manufacturers print the capacitance directly on the capacitor along with other information such as the working voltage and perhaps the tolerance. However, small capacitors don't have enough room for all that. Many capacitor manufacturers use a shorthand notation to indicate capacitance on small caps.

If you have a capacitor that has nothing other than a three-digit number printed on it, the third digit represents the number of zeros to add to the end of the first two digits. The resulting number is the capacitance in pF. For example, 101 represents 100 pF: the digits 10 followed by one additional zero.

If there are only two digits listed, the number is simply the capacitance in pF. Thus, the digits 22 indicate a 22 pF capacitor.

Table 3-1 lists how some common capacitor values are represented using this notation.

TABLE 3-1

Capacitance Markings

Marking	Capacitance	Capacitance
101	100 pF	0.0001 µF
221	220 pF	0.00022 µF
471	470 pF	0.00047 µF
102	1,000 pF	0.001 µF
222	2,200 pF	0.0022 µF
472	4,700 pF	0.0047 µF
103	10,000 pF	0.01 µF
223	22,000 pF	0.022 µF
473	47,000 pF	0.047 µF
104	100,000 pF	0.1 µF
224	220,000 pF	0.22 µF
474	470,000 pF	0.47 µF
105	1,000,000 pF	1 µF
225	2,200,000 pF	2.2 µF
475	4,700,000 pF	4.7 µF

You may also see a letter printed on the capacitor to indicate the tolerance. You can interpret the tolerance letter according to Table 3-2.

Notice that the tolerances for codes P through Z are a little odd. For codes P and W, the manufacturer promises that the capacitance will be no less than the stated value but may be as much as 100 percent or 200 percent over the stated value. For codes S, X, and Z, the actual capacitance may be as much as 20 percent below the stated value or as much as 50 percent, 40 percent, or 80 percent over the stated value. For example, if the marking is 101P, the actual capacitance is no less than 100 pF but may be as much as 200 pF. If the marking is 101Z, the capacitance is between 80 pF and 180 pF.

TABLE 3-2

Capacitor Tolerance Markings

Letter	Tolerance
A	±0.05pF
B	±0.1pF
C	±0.25pF
D	±0.5pF
E	±0.5%
F	±1%
G	±2%
H	±3%
J	±5%
K	±10%
L	±15%
M	±20%
N	±30%
P	-0%,+100%
S	-20%, +50%
W	-0%, +200%
X	-20%, +40%
Z	-20%, +80%

The Many Sizes and Shapes of Capacitors

Capacitors come in all sorts of shapes and sizes, influenced mostly by three things: the type of material used to create the plates, the type of material used for the dielectric, and the capacitance. Figure 3-2 shows some of the most common varieties of capacitors.

The most common types of capacitors are

>> **Ceramic disk:** The plates are made by coating both sides of a small ceramic or porcelain disk with silver solder. The ceramic or porcelain disk is the dielectric, and the silver solder forms the plates. Leads are soldered to the plates, and the entire thing is dipped in resin.

Ceramic disk capacitors are small and usually have low capacitance values, ranging from 1 pF to a few microfarads. Because they're small, their values are usually printed using the three-digit shorthand notation described earlier in this chapter, in the section "Reading Capacitor Values."

Ceramic disk capacitors aren't polarized, so you don't have to worry about polarity when you use them.

>> **Silver mica:** The dielectric is made from mica, and this capacitor is sometimes referred to simply as a *mica capacitor.* As with ceramic capacitors, the plates in a silver mica capacitor are made from silver. Electrodes are joined to the plates, and then the capacitor is dipped in epoxy.

Silver mica capacitors come in about the same capacitance range as ceramic disk capacitors. However, they can be made to much higher tolerances — as close as 1 percent in some cases. Like ceramic disk capacitors, silver mica capacitors aren't polarized.

Although ceramic disk and mica capacitors are constructed in a similar way, they're easy to tell apart. Ceramic disk capacitors are thin, flat disks and are nearly always a dull, light-brown color. Silver mica capacitors are thicker, bulge at the ends where the leads are attached, and are shiny and sometimes colorful — red, blue, yellow, and green are common colors for silver mica capacitors. (Interestingly enough, I've never seen a silver mica capacitor.)

>> **Film:** The dielectric is made from a thin filmlike sheet of insulating material, and the plates are made from filmlike sheets of metal foil. In some cases, the plates and the dielectric are then tightly rolled together and enclosed in a metal or plastic can. In other cases, the layers are stacked and then dipped in epoxy.

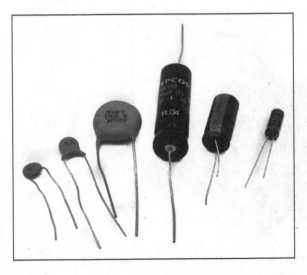

FIGURE 3-2:
Capacitors are made in many different shapes and sizes.

Depending on the materials used, capacitance for film capacitors can be as small as 1,000 pF or as large as 100 μF. Film capacitors aren't polarized.

WARNING

>> **Electrolytic:** One of the plates is made by coating a foil film with a highly conductive, semiliquid solution called *electrolyte*. The other plate is another foil film on which an extremely thin layer of oxide has been deposited; this thin layer serves as the dielectric. The two layers are then rolled up and enclosed in a metal can.

Electrolytic capacitors are polarized, so you must be sure to connect voltage to it in the proper direction. If you apply voltage in the wrong direction, the capacitor may be damaged and might even explode.

You find these two common types of electrolytic capacitors:

- *Aluminum:* Can be quite large, with as much as a tenth of a farad or more (100,000 μF).

- *Tantalum:* Are smaller, ranging up to about 1,000 μF.

>> **Variable:** A capacitor whose capacitance can be adjusted by turning a knob. One common use for a variable capacitor is to tune a radio circuit to a specific frequency.

In the most common type of variable capacitor, air is used as the dielectric, and the plates are made of rigid metal. As shown in Figure 3-3, several pairs of plates are typically used in an intermeshed arrangement. One set of plates is fixed (not moveable), but the other set is attached to a rotating knob. When you turn the knob, you change the amount of surface area on the plates that overlap. This, in turn, changes the capacitance of the device.

The schematic symbol for a variable capacitor is shown in the margin.

FIGURE 3-3:
A variable
capacitor.

Calculating Time Constants for Resistor/Capacitor Networks

As I mention earlier in the "What Is a Capacitor" section, when you put a voltage across a capacitor, it takes a bit of time for the capacitor to fully charge. During this time, current flows through the capacitor. Similarly, when you discharge a capacitor by placing a load across it, it takes a bit of time for the capacitor to fully discharge. Knowing exactly how much time it takes to charge a capacitor is one of the keys to using capacitors correctly in your circuits, and you can get that information by calculating the RC time constant.

Calculating the exact amount of time required to charge or discharge a capacitor requires more math than I'm willing to throw at you in this book. I don't have any qualms about tossing up a little simple algebra, but I draw the line at calculus. Fortunately, you can skip the calculus and use just simple algebra if you're willing to settle for close approximations rather than exactitudes. Given that most resistors and capacitors have a tolerance of plus or minus 5 or 10 percent anyway, using approximate calculations won't have any significant effect on your circuits.

But before I tell you about making these approximate calculations, it's important that you understand the concepts behind the calculus, even if you don't actually do the calculus. So bear with me.

When a capacitor is charging, current flows from a voltage source through the capacitor. In most circuits, a resistor is working in series with the capacitor as well, as shown in Figure 3-4.

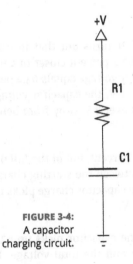

FIGURE 3-4:
A capacitor
charging circuit.

TECHNICAL STUFF

Actually, a circuit always has resistance, even if you don't use a resistor. That's because there's no such thing as a perfect conductor, so even solid wire has some resistance. For the purposes of this discussion, assume that there is a resistor in the circuit, but its resistance might be 0 Ω.

The rate at which the capacitor charges through a resistor is called the *RC time constant* (the *RC* stands for *resistor-capacitor*), which can be calculated simply by multiplying the resistance in ohms by the capacitance in farads. Here's the formula:

$$T = R \times C$$

For example, suppose the resistance is 10 kΩ and the capacitance is 100 μF. Before you do the multiplication, you must first convert the μF to farads. Since one μF is one-millionth of a farad, you can convert μF to farads by dividing the μF by one million. Therefore, 100 μF is equivalent to 0.0001 F. Multiplying 10 kΩ by 0.0001 F gives you a time constant of 1 second (10,000 × 0.0001 = 1).

Note that if you want to increase the RC time constant, you can increase either the resistance or the capacitance, or both. Also note that you can use an infinite number of combinations of resistance and capacitance values to reach a desired RC time constant. For example, all the following combinations of resistance and capacitance yield a time constant of one second:

Resistance	Capacitance	RC Time Constant
1 kΩ	1,000 μF	1 s
10 kΩ	100 μF	1 s
100 kΩ	10 μF	1 s
1 MΩ	1 μF	1 s

So what is the significance of the RC time constant? It turns out that in each interval of the RC time constant, the capacitor moves 63.2 percent closer to a full charge. For example, after the first interval, the capacitor voltage equals 63.2 percent of the battery voltage. So if the battery voltage is 9 V, the capacitor voltage is just under 6 V after the first interval, leaving it just over 3 V away from being fully charged.

In the second time interval, the capacitor picks up 63.2 percent, not of the full 9 V of battery voltage, but 63.2 percent of the difference between the starting charge (just under 6 V) and the battery voltage (9 V). Thus, the capacitor charge picks up just over two additional volts, bringing it up to about 8 V.

This process keeps repeating: In each time interval, the capacitor picks up 63.2 percent of the difference between its starting voltage and the total voltage. In theory, the capacitor will never be fully charged because with the passing of each RC time constant the capacitor picks up only a percentage of the remaining available charge. But within just a few time constants, the capacity becomes very close to fully charged.

The following table gives you a helpful approximation of the percentage of charge that a capacitor reaches after the first five time constants. For all practical purposes, you can consider the capacitor fully charged after five time constants have elapsed.

RC Time Constant Interval	Percentage of Total Charge
1	63.2%
2	86.5%
3	95.0%
4	98.2%
5	99.3%

TECHNICAL STUFF

As I mentioned a moment ago, in theory a capacitor will never become fully charged. But in reality, the capacitor does eventually become fully charged. The difference between theory and reality here is that in mathematics, we can use as many digits beyond the decimal point as we want. In other words, there's no limit to how small a number can be. But there is a limit to how small a charge can be: A charge can't be smaller than a single electron. When enough time constants have elapsed, the difference between the battery voltage and the capacitor charge is a single electron. Once that electron joins the charge, the capacitor is full.

Combining Capacitors

In the preceding chapter, you learn that you can combine resistors in series or parallel networks to create any arbitrary resistance value you need. You can do the same with capacitors. But as you learn in the following sections, the formulas for calculating the total capacitance of a capacitor network are the reverse of the rules you follow to calculate resistor networks. In other words, the formula you use for resistors in series applies to capacitors in parallel, and the formula you use for resistors in parallel applies to capacitors in series. Isn't it funny how science sometimes likes to mess with your mind?

Combining capacitors in parallel

Calculating the total capacitance of two or more capacitors in parallel is simple: Just add up the individual capacitor values to get the total capacitance. For example, if you combine three 100 µF capacitors in parallel, the total capacitance of the circuit is 300 µF.

This rule makes sense if you think about it for a moment. When you connect capacitors in parallel, you're essentially connecting the plates of the individual capacitors. So connecting two identical capacitors in parallel essentially doubles the size of the plates, which effectively doubles the capacitance.

Figure 3-5 shows how parallel capacitors work. Here, the two circuits have identical capacitances. The first circuit accomplishes the job with one capacitor, the second does it with three. Thus, the circuits are equivalent.

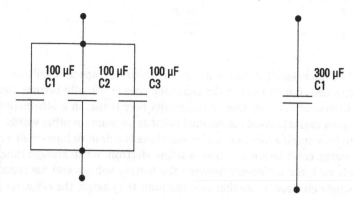

FIGURE 3-5:
Combining capacitors in parallel.

REMEMBER

Whenever you see two or more capacitors in parallel in a circuit, you can substitute a single capacitor whose value is the sum of the individual capacitors. Similarly, any time you see a single capacitor in a circuit, you can substitute two or more capacitors in parallel as long as their values add up to the original value.

TIP

The total capacitance of capacitors in parallel is always greater than the capacitance of any of the individual capacitors. That's because each capacitor adds its own capacitance to the total.

Connecting capacitors in series

You can also combine capacitors in series to create equivalent capacitances, as shown in Figure 3-6. When you do, however, the math is a little more complicated. It turns out that the calculations required for capacitors in series are the same as calculating resistors in parallel.

FIGURE 3-6:
Combing
capacitors
in series.

Here are the rules for calculating capacitances in series:

>> If the capacitors are of equal value, you're in luck. All you must do is divide the value of one of the individual capacitors by the number of capacitors. For example, the total capacitance of two, 100 µF capacitors is 50 µF.

>> If only two capacitors are involved, use this calculation:

$$C_{total} = \frac{C1 \times C2}{C1 + C2}$$

>> In this formula, C1 and C2 are the values of the two capacitors.

>> Here's an example, based on a 220 µF and 470 µF capacitor in series:

$$C_{total} = \frac{C1 \times C2}{C1 + C2} = \frac{220\ \mu F \times 470\ \mu F}{220\ \mu F + 470\ \mu F} = 149.855\ \mu F$$

>> For three or more capacitors in series, the formula is this:

$$C_{total} = \frac{1}{\frac{1}{C1} + \frac{1}{C2} + \frac{1}{C3} \cdots}$$

>> Note that the ellipsis at the end of the expression indicates that you keep adding up the reciprocals of the capacitances for as many capacitors as you have.

>> Here is an example for three capacitors whose values are 100 µF, 220 µF, and 470 µF:

$$C_{total} = \frac{1}{\frac{1}{100\ \mu F} + \frac{1}{220\ \mu F} + \frac{1}{470\ \mu F}} = \frac{1}{0.016673\ \mu F} = 59.977\ \mu F$$

>> As you can see, the final result is 59.9768 µF. Unless your name happens to be Spock, you probably don't care about the answer being so precise, so you can safely round it to an even 60 µF.

Working with
Capacitors

Putting Capacitors to Work

Now that you know all about how capacitors work and you understand about charging, discharging, time constants, and series and parallel capacitors, you're probably wondering what capacitors are actually used for in real-life circuits. It turns out that capacitors are incredibly useful. The following paragraphs describe the most common ways that capacitors are used:

>> **Storing energy:** Once charged, a capacitor has the ability to store a lot of energy and discharge it when needed. One of the most familiar uses of this ability is in a camera flash. Other examples are in devices such as alarm clocks that need to stay powered up even when household power is disrupted for a few moments.

>> **Timing circuits:** Many circuits depend on capacitors along with resistors to provide timing intervals. For example, in Chapter 6 of this minibook you learn how to use a resistor and capacitor together along with a transistor to create a circuit that flashes an LED on and off.

>> **Stabilizing shaky direct current:** Many electronic devices run on direct current but derive their power from household AC. These devices use a power-supply circuit that converts the AC to DC. As you learn in Book 4, Chapter 2, an important part of this circuit is a capacitor that creates steady DC voltage from the constantly changing voltage of an AC circuit.

>> **Blocking DC while passing AC:** Sometimes you need to prevent direct current from flowing while allowing alternating current to pass. A capacitor can do this. The basic trick is to choose a capacitor that will fully charge and discharge within the alternating current cycle. Remember that as the capacitor is charging and discharging, it allows current to flow. So as long as you can keep the capacitor charging and discharging, it will pass the current.

>> **Filtering certain frequencies:** Capacitors are often used in audio or radio circuits to select certain frequencies. For example, a variable capacitor is often used in a radio circuit to tune the circuit to a certain frequency.

Charging and Discharging a Capacitor

One of the most common uses for capacitors is to store a charge that you can discharge when it's needed. Project 13 presents a simple construction project on a solderless breadboard that demonstrates how you can use a capacitor to do this. You connect an LED to a 3 V battery power supply and use a capacitor so that when you disconnect the battery from the circuit, the LED doesn't immediately go out. Instead, it continues to glow for a moment as the capacitor discharges.

Note that this project utilizes a small push button that can be mounted directly to your solderless breadboard. Unlike most of the other components used in this book's projects, this one doesn't have a RadioShack part number. However, such buttons are available inexpensively from many online sources. Use your favorite search engine to search the web for *DIP Push Button*.

Figure 3-7 shows the completed project.

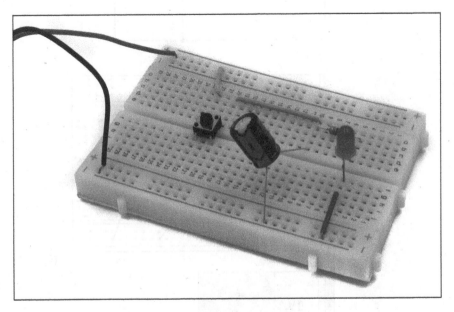

FIGURE 3-7:
The capacitor discharge circuit (Project 13).

Project 13: Charging and Discharging a Capacitor

In this project, you build a circuit that places a capacitor in parallel with an LED. When power is applied to the LED, the capacitor charges. When power is disconnected, the capacitor discharges, which causes the LED to remain lit for a short while after the power has been disconnected.

You need wire cutters and strippers to cut and strip the jumper wires.

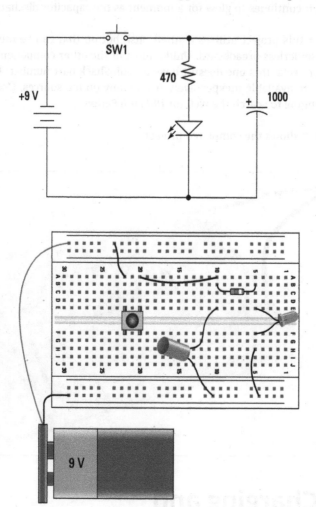

Parts

>> One solderless breadboard

>> One 9 V battery

>> One 9 V battery snap holder

>> One 470 Ω ¼ W resistor

>> One 1,000 μF electrolytic capacitor

>> One red LED, 5 mm

>> One normally open DIP breadboard push button

>> Three short (approximately 1-inch) 22-gauge solid jumper wires

>> One 1-inch 22-gauge solid jumper wire

Steps

Throughout these steps, use the negative (−) bus strip on the bottom side of the board for ground and the positive (+) bus strip on the top of the board as the positive voltage.

1. **Connect the battery holder.**

   Insert the red lead in the last hole of the left-side positive bus strip (located near hole A30) and insert the black lead in the last hole of the left-side negative bus strip (located near hole J30).

2. **Mount the push button on the breadboard.**

   The push button should be inserted in holes E22, E20, F22, and F20.

3. **Connect the resistor.**

   Insert the resistor into the breadboard at holes B5 and B10.

4. **Insert the LED.**

   Insert the long lead into hole E5 and the short lead into hole F5.

   Note that the circuit won't work if you insert the LED backward. The short lead must be on the negative side of the circuit.

5. **Insert the capacitor.**

   The negative terminal of the capacitor, marked by a series of dashes on the capacitor, should go into a hole in the bottom-side negative ground strip near row 10. The other lead should go in hole E10.

6. **Insert the three jumper wires.**

   The first jumper wire should go from hole A10 to hole A20.

   The second jumper wire should go from hole A22 to a nearby hole on the top-side positive bus strip.

   The third jumper wire should go from hole H5 to a nearby hole on the bottom-side negative bus strip.

7. **Attach the 9 V battery to the snap connector.**

8. **Press the push button.**

    The LED lights up. If it doesn't, make sure you've inserted the LED correctly.

    If the LED still doesn't light up, the push button may not be oriented correctly on the breadboard. Remove the button, rotate it 90 degrees, and reinsert it.

9. **After a few seconds, release the button.**

    Watch how the LED goes out. Instead of turning off instantly, it fades over the course of a few seconds. That's because when the battery is disconnected from the circuit, the capacitor discharges through the LED. As the capacitor discharges, the LED goes gradually dim and then goes out altogether.

Here are some additional points you might want to ponder or things you might want to try after you complete Project 13:

TIP

» Pull the LED out of the breadboard, and then push the button. After a few seconds, release the button. Then, use the voltmeter function of your multimeter to measure the voltage across the two leads of the capacitor. Notice that even though the battery is disconnected from the circuit, the capacitor reads 9 V.

» Try varying the size of the capacitor to see what happens. If you use a smaller capacitor, the LED extinguishes more quickly when you remove the power. If you use a larger capacitor, the LED fades more slowly.

» You may not be able to find a capacitor larger than 1,000 µF, but you can increase the capacitance by adding more capacitors in parallel with the 1,000 µF capacitor.

» Try adding a resistor in series with the capacitor. To do that, simply replace the third jumper wire (the one from H5 to the negative bus strip) with a resistor.

Blocking DC while Passing AC

One of the important features of capacitors is their ability to block direct current while allowing alternating current to pass. This works because of the way a capacitor allows current to flow while the plates are charging or discharging, but halts current in its tracks once the capacitor is fully charged. Thus, as long as the source voltage is constantly changing, the capacitor allows current to flow. When the source voltage is steady, the capacitor blocks current from flowing.

In Project 14, you build a simple circuit that demonstrates this principle in action. The circuit places a resistor and a light-emitting diode in series with a capacitor. Then, it uses a double-pole double-throw knife switch so that you can switch the voltage source for the current between a 9 V battery (direct current) and a 9 V AC power adapter (alternating current). The circuit also includes a push button that shorts out the capacitor, causing it to discharge.

Figure 3-8 shows the finished project.

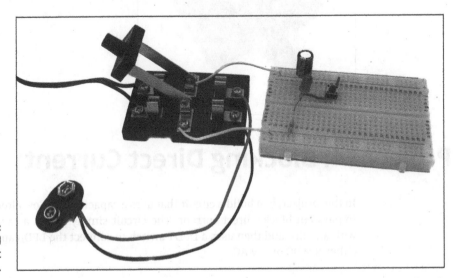

FIGURE 3-8:
The finished
AC/DC circuit
(Project 14).

TIP

The trick to successfully building this circuit is finding a 9 V AC power adapter. If you go looking for one at your local RadioShack or at a department store, you might find one, but it will probably cost $20 or more. If you search the Internet, you can probably find one for more like $10 or $15.

Figure 3-9 shows the one I used for this project; I purchased it at a thrift store for $1. Notice the markings on the back of this power adapter:

    Input: 120 V AC 60 Hz 15 W

    Output: 9 V AC 1,000 mA

These are the specifications you want to look for when you search for an AC adapter to use for this project. In particular, the output voltage must be listed at 9 V AC.

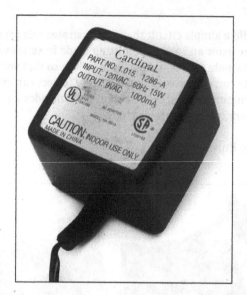

FIGURE 3-9:
A 9 V AC power
adapter.

Project 14: Blocking Direct Current

In this project, you build a circuit that uses a capacitor to allow alternating current to pass but blocks direct current. The circuit simply places a capacitor in series with an LED, and then uses a DPDT switch to connect the LED/capacitor circuit to either 9 V DC or 9 V AC.

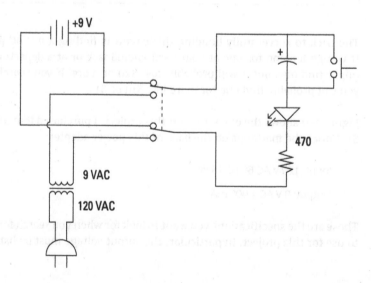

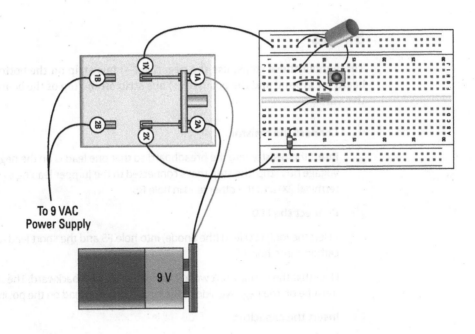

To 9 VAC
Power Supply

9 V

Parts

» One solderless breadboard

» One 9 V battery

» One 9 V battery snap holder

» One 9 V AC power adapter

» One 470 Ω ¼ W resistor (yellow-violet-brown)

» One 1,000 µF electrolytic capacitor

» One red LED, 5 mm

» One DPDT knife switch

» One tactile push button (such as Jameco part number 149948)

» Two 1-inch 22-gauge solid jumper wires

» Two 3-inch 22-gauge solid jumper wires

Steps

Throughout these steps, use the negative (–) bus strip on the bottom side of the board for ground and the positive (+) bus strip on the top of the board as the positive voltage.

1. **Connect the resistor.**

   Insert the resistor into the breadboard so that one lead is on the negative voltage bus strip (the strip that's connected to the jumper leading to switch terminal 2X) and the other lead in hole B5.

2. **Connect the LED.**

   Insert the long LED lead (the anode) into hole F5 and the short lead (the cathode) into hole E5.

   Note that the circuit won't work if you insert the LED backward. The short lead must be on the negative side of the circuit, the long lead on the positive side.

3. **Insert the capacitor.**

   The negative terminal of the capacitor, marked by a series of dashes on the capacitor, should go hole I5. The other lead should go in a hole on the positive voltage bus strip near row 5.

   As with the LED, the circuit won't work if the capacitor is inserted backward. The positive lead of the capacitor must be on the positive voltage bus.

4. **Insert the push button in holes G11, I11, G13, and I13.**

   The push button must be oriented such that when the button is pushed, row 11 connects to row 13. (Although the button used here is a double-pole switch, that doesn't matter because both poles are connected to the same row of the breadboard; holes F through J of each row of the breadboard are connected internally.)

   You can usually tell how the leads on a push button are oriented by looking at the back of the button; there's usually a groove that indicates the path the current will flow when the button is pushed. Orient the button so that this groove runs lengthwise on the breadboard, so that when the button is pushed, row 11 will connect to row 13.

   Note also that you may need to use a pair of small pliers to adjust the button's leads so that they fit nicely in the 0.1-inch spacing of the breadboard holes.

5. **Insert one of the short jumper wires to connect hole J13 to any nearby hole on the positive bus strip.**

6. **Insert the other short jumper wires to connect holes G5 and F11.**

7. Connect the battery snap holder to the knife switch.

Connect the red lead to terminal 1A on the knife switch, and then connect the black lead to terminal 2A.

8. Connect the 9 V AC adapter to the knife switch.

The 9 V AC adapter should be connected to terminals 1B and 2B on the knife switch. The easiest way to connect the power adapter is to cut the plug off the end of the power adapter's cable, and then strip about ⅜-inch of insulation from the ends of the two wires that make up the cable.

9. Connect the knife switch to the breadboard.

Use one of the 3-inch solid jumper wires to connect terminal 1X on the knife switch to the first hole in the positive bus strip on the top of the solderless breadboard. Then use the other 3-inch jumper wire to connect terminal 2X on the knife switch with the first hole in the ground bus strip at the bottom of the breadboard.

10. Move the switch to the vertical position so that neither side of the switch is closed.

11. Insert the 9 V battery.

12. Plug the 9 V AC power adapter into a wall outlet.

The circuit is now done and ready for testing.

13. Close the knife switch toward the A contacts.

This connects the 9 V battery to the circuit. The LED initially turns on, but after a few moments it begins to fade. When it goes completely out, the capacitor is fully charged. When it's charged, the capacitor won't allow direct current to pass, so the LED remains dark.

14. Push the button momentarily.

This shorts out the leads of the capacitor, causing it to discharge. When you release the button, the LED will turn on again but begin to fade again as the capacitor discharges.

15. Flip the switch to the B contacts.

This action disconnects the battery from the circuit and instead connects the AC power adapter. Now the LED comes on and stays on. As long as the AC power source is applied across the circuit, the capacitor allows the current to flow, and the LED lights up.

16. Flip the switch back to the A contacts.

Because the capacitor is now fully charged, the LED goes out immediately.

You're done! Congratulations! You have now witnessed an important aspect of how capacitors work.

IN THIS CHAPTER

» **Examining the mystery of induction**

» **Learning what reactance is**

» **Learning how to combine inductors in series and parallel**

» **Seeing how inductors are used in electronic circuits**

Chapter 4

Working with Inductors

I have a lot of books about electronics on my bookshelf, covering a wide range of topics. Some of them are new; some are a few decades old. My favorite of them all is a called *A Course in Electrical Engineering*, by Chester L. Dawes. It was written in 1920, when electronics was in its infancy. Radio was brand new. Vacuum tubes were just catching on. The transistor wouldn't be invented for another quarter of a century.

You'd think that an electrical engineering book written in this era would be completely obsolete. But it's amazing how much of this book is still accurate. For example, Ohm's law hasn't changed since 1920. Dawes's explanation of Ohm's law is as good as any I've ever read.

What fascinates me most about this old book is the way it starts. Chapter 1 is titled "Magnetism and Magnets," and it begins like this:

> Magnets and magnetism are involved in the operation of practically all electrical apparatus. Therefore an understanding of their underlying principles is essential to a clear conception of the operation of all such apparatus.

This makes perfect sense when you consider the state of electrical technology in those days. We've moved beyond simple magnetism in our ability to exploit the amazing properties of electricity. Yet the basic relationship that exists between electric current and magnetism is still at the heart of many different types of essential circuits.

For example, a power adapter that converts 120 V AC from a wall outlet to a safer level, say 9 V, uses a transformer that relies on magnetism to step down the voltage. An analog multimeter (the kind with a needle that moves with voltage, current, or resistance) relies on magnetism to deflect the needle. And electric motors convert electric current into magnetism, which is then converted into motion.

In this chapter, I turn your attention to a special class of components called *inductors*, which exploit the nature of magnetism and its relationship to electric current. Inductors are sort of like cousins to capacitors, in that they can be used to do similar things and they play by similar rules. For example, an important characteristic of a capacitor is its ability to oppose changes in voltage. Inductors have a similar ability to oppose changes, not in voltage, but rather in current.

TIP

This chapter is a bit of a departure from the other chapters in this minibook, in that it contains no construction projects. The practical applications for inductors are in circuits that you might not be ready to build yet, such as radio tuners or power supplies. You learn the important concepts of how inductors work in this chapter so that later, when it's time to use an inductor in an actual circuit, you'll know what it does. You can always return to this chapter at that time to refresh your memory on how inductors work. In fact, I won't hold it against you if you just skim through this chapter on your first reading through this book, or even skip it altogether for now. You can always come back to this chapter when the need arises.

What Is Magnetism?

When Albert Einstein was 5 years old and sick in bed, his father gave him a compass to play with. Young Albert saw that no matter how he spun the compass around the needle always swung back to point north, and he was amazed. Years later, he wrote that this compass was what launched his lifelong interest in physics. He realized then that there was "something behind things, something deeply hidden."

In the case of the compass, that deeply hidden thing that Einstein marveled at was *magnetism.* We're all familiar with magnetism, though exactly what it is remains mysterious. A *magnetic field* extends through space and either attracts or repels certain materials. Materials that are strongly affected by magnetic fields are called *magnetic.* Materials that create magnetic fields are called *magnets.*

The north and south of magnetism

Magnetic fields and electric fields are distinct things, but they're closely related and have much in common. For example, magnetic fields are polarized in much the same way that electric fields are polarized. Electric fields exist between electric charges of opposite polarity (negative and positive). Magnetic fields exist between opposite magnetic poles, called *north* and *south*.

Just as opposite electric charges attract and like charges repel, opposite magnetic poles attract and like magnetic poles repel. That's why if you take two strong magnets and try to push the two north poles together or the two south poles together, the magnet fights back, and you won't be able to get the poles together. But if you turn one of the magnets around and try to hook up the north pole with the south pole, the magnets attract each other, and you'll have trouble keeping them apart.

A magnetic field has a distinct shape that you can visualize by using a simple experiment that's commonly done in grade school. All you need is a bar magnet, some iron filings, and a piece of paper. Put the paper over the magnet and sprinkle the filings on the paper. The filings magically line up to reveal the shape of the magnetic field, as shown in Figure 4-1. As you can see, the filings align themselves in lines that are aligned from one pole of the magnet to the other.

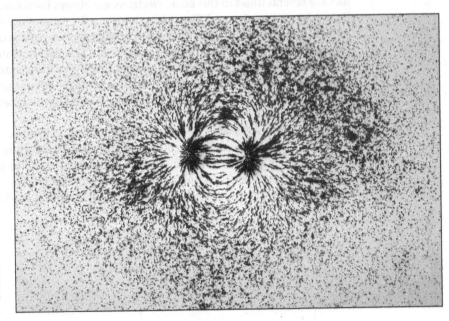

FIGURE 4-1: The shape of a magnetic field revealed by iron filings.

Pondering permanent magnets

A *permanent magnet* is a material that creates its own magnetic field. Some naturally occurring materials, such as lodestone, are inherently magnetic and produce magnetic fields all on their own. But most permanent magnets are made from materials that aren't inherently magnetic but become magnetic when they're exposed to a powerful magnetic field.

You probably have several permanent magnets around your home. In fact, you probably have a few on your refrigerator, holding up your kid's kindergarten picture or your shopping list.

Examining Electromagnets

Permanent magnets create a magnetic field all by themselves. An *electromagnet* relies on the key relationship between electricity and magnetism to create a magnetic field. Specifically, whenever electric current flows, a magnetic field is created. This magnetic field is created by the moving charges. In short, an electron in motion becomes a magnet.

As I say several times in this book, electrons are always in motion, so aren't there little magnets everywhere? The answer is yes, in the same sense that electric current is everywhere. But when the motion of electrons within a material is random, the magnetic fields created by the electrons are oriented randomly, and so they end up just canceling each other out. But when you give the electrons a nudge with voltage, they all start moving in the same direction. This strengthens and organizes the magnetic fields so they can combine to form one large magnetic field.

The magnetic field created by current flowing through a single wire is measurable but small. However, if you coil the wire tightly as shown in Figure 4-2, the magnetic fields are strengthened because of their proximity to one another. For example, you can create a simple electromagnet by wrapping several feet of small, insulated wire around a pencil, a ballpoint pen, or any other rigid tube or cylinder.

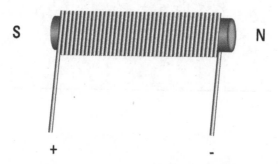

FIGURE 4-2:
An electromagnet.

The strength of the magnetic field of such a coil depends on several factors, the most important being these:

>> **The number of turns in the coil.**

>> **The amount of current flowing through the electromagnet.** Increasing the current increases the strength of the electromagnet exponentially. For example, if you double the current, the electromagnet is four times as strong.

>> **The material used for the core that the electromagnet is wrapped around.** Any coil wrapped around an iron core will be about ten times as strong as the same coil wrapped around an inert core such as air, plastic, or wood.

TIP

The polarity of an electromagnet is easy to determine: The negative side of the coil is the magnet's north pole, while the positive side is the south pole.

Inducing Current

Electromagnets are possible because a moving current creates a magnetic field. Interestingly enough, the reverse is also true: A moving magnetic field creates an electric current. In other words, if you wave a magnet over a wire, you create a current in the wire. This effect is called *electromagnetic induction*, or sometimes just *magnetic induction*.

Remember our old friend Michael Faraday? In Chapter 3 of this minibook, I mention that even though he didn't invent capacitors, his name is the basis of the *farad*, which is the unit of measure we use to measure capacitors. In 1831, Faraday discovered electromagnetic induction. It's one of the most important discoveries in the history of electrical science, as it's at the heart of nearly all forms of electrical power generation. Coal-burning, hydroelectric, and even nuclear power plants all use moving water to spin turbines that are connected to generators. Those generators use the principle of electromagnetic induction to turn the spinning motion of magnetic fields into electric current.

You can increase the strength of the current induced in a wire by coiling the wire into turns so that you can expose a greater length of wire to the magnetic field. Also, it doesn't matter whether it's the magnet or the coil that moves. Either way, a current is induced in the coil if the coil is moving relative to the magnet, provided of course that the coil passes through the magnet's magnetic field.

REMEMBER

Note that whenever current is flowing, there must be voltage. Thus, in addition to causing current to flow through the coil, electromagnetic induction creates a voltage across the coil.

Inductance and the art of resisting change

An *inductor* is a coil that's designed for use in electronic circuits. Figure 4-3 shows several typical inductors. As you can see, Inductors are simple devices, consisting of nothing more than a coil of wire, often wrapped around an iron core.

FIGURE 4-3:
Inductors are simple devices.

Inductors take advantage of an important characteristic of coils called *self-inductance*, also called just *inductance*. The inductor's ability to exploit the idea of self-inductance is a stroke of genius.

Self-inductance is similar to electromagnetic induction as described in the previous section, but with a subtle twist. Whereas *electromagnetic induction* refers to the ability of a coil to generate a current when it moves across a magnetic field, *self-inductance* refers to the ability of a coil to create the very magnetic field that then induces the voltage. In other words, with self-inductance, the coil feeds back upon itself. A voltage applied across the coil causes current to flow, which creates a magnetic field, which in turn creates more voltage.

Inductance happens only when the current running through the coil changes. That's because only a *moving* magnetic field induces voltage in a coil. Whenever the current changes in a coil, the magnetic field created by the current grows or shrinks, depending on whether the current increases or decreases. When the magnetic field grows or shrinks, it's effectively moving, so a voltage is inducted in the coil as a result of this movement. When the current stays steady, no inductance occurs.

Self-inductance is a tricky concept to get your mind around, so don't panic if this doesn't make sense to you the first time through. It took me awhile to get this concept when I first learned about it. Let's go over the idea in more detail, point by point:

>> **When voltage is applied across a coil, the voltage causes current to flow through the coil.** Remember, current always requires voltage, and voltage always results in a current when applied across a conductor.

>> **The current flowing through the coil creates a magnetic field around the coil.** Keep in mind that the coil that creates the magnetic field is itself within the field and can therefore be influenced by it.

>> **If the current flowing through the coil changes, the magnetic field created by the current also changes.** The magnetic field grows or shrinks, depending on whether the current increases or decreases. Either way, the changing magnetic field is in effect moving.

>> **Because the magnetic field is moving, voltage is induced in the coil.** This is an additional voltage, on top of the voltage that's driving the main current through the coil.

>> **The amount of voltage induced by the changing magnetic field depends on the speed in which the current changes.** The faster the current changes, the more the magnetic field moves and therefore the more voltage is induced.

>> **The polarity of the induced voltage depends on whether the current is increasing or decreasing.** This is because the direction of movement in the magnetic field depends on whether the field is growing or shrinking, and the voltage induced by a moving magnetic field depends on which direction the field is moving, according to the following rules:

- When the current increases, the polarity of the induced voltage is opposite to the polarity of the voltage driving the coil. This inducted voltage is often called *back voltage* because it has the opposite polarity as the supply voltage.

- When the current decreases, the resulting self-induced voltage has the same polarity as the supply voltage.

>> **The induced voltage creates a current in the coil that flows either with or against the main coil current, depending on whether the coil current is decreasing or increasing, according to the following rules:**

- When the coil current is increasing, the additional current flows against the main coil current. This has the effect of pushing back against the increasing main current, which effectively slows down the rate at which the current can change.

- When the coil current is decreasing, the additional current flows with the main coil current, thus counteracting the decrease in coil current.

>> **When the coil current stops changing, self-inductance stops.** Thus, when current is steady, an inductor is simply a conductor. (It is also an electromagnet, as the current traveling through it produces a magnetic field.)

Because of self-inductance, an inductor is said to oppose changes in current. If the current increases, an opposite voltage is induced across the coil, which slows the rate at which the current can increase. If the current decreases, a forward voltage is induced across the coil, which slows the rate at which the current decreases. An inductor applies equal opposition to both increases and decreases in current.

It turns out that this ability to oppose changes in current is quite useful in electronic circuits, as I explain later in this chapter, in the section "Putting Inductors to Work."

Here are some additional important details concerning inductors:

>> Inductors can't stop changes in current; they can only slow them down. Measuring how much an inductor can slow down a change in current is the topic of the next section.

>> The magnetic field from one inductor can spill over into a nearby inductor and induce voltage in it, too. To prevent this from happening, many inductors have special shielding around them to keep them magnetically isolated. If you use unshielded inductors in your circuits, be sure to space them as far apart as possible — unless, of course, you *want* them to interact.

Regarding henry

Inductance is only a momentary thing. Exactly how much of a momentary thing depends on the amount of inductance an inductor has. Inductance is measured in units called *henrys* (H).

The definition of one henry is simple: One henry is the amount of inductance necessary to induce one volt when the current in coil changes at a rate of one ampere per second.

As you might guess, one henry is a fairly large inductor. Inductors in the single-digit henry range are often used when dealing with household current (120 V AC at 60 Hz). But for most electronics work, you'll use inductances measured in thousandths of a henry (*millihenrys*, abbreviated *mH*) or in millionths of a henry (*microhenrys*, abbreviated *μH*).

Here are a few additional things to know about inductors and henrys:

>> The plural of *henry* is *henrys*, not *henries*.

>> The letter *L* is often used to represent inductance in formulas. Inductors in schematic diagrams are usually referenced by the letter L. For example, if a circuit calls for three inductors, they will be identified L1, L2, and L3.

   Speaking of schematics, the symbol used to represent inductors in a schematic diagram is shown in the margin.

>> When I first learned about self-inductance and henrys, I hoped that since the henry is a measure of a coil's ability to resist change in current, the henry should be named after someone who was famous for resisting change, such as Professor Henry Higgins from *My Fair Lady*.

   Imagine my disappointment to discover that the henry is named after Joseph Henry. All he ever did was discover the self-inductance and invent the inductor. Well, that plus he was the first Secretary of the Smithsonian Institution. He must have known someone really important to get the henry named after him.

Calculating RL Time Constants

In the preceding chapter, you learn how to calculate the RC time constant for a resistor-capacitor circuit. A similar calculation can be done for inductors, except that instead of calling it the RC time constant, we call it the *RL time constant*. (Remember, *L* is the symbol used to represent inductance.)

The RL time constant indicates the amount of time that it takes to conduct 63.2 percent of the current that results from a voltage applied across an inductor. (Hmm. Where have we seen 63.2 percent before? Right! It's the same percentage used to calculate time constants in resistor-capacitor networks. The value 63.2 percent derives from the calculus equations used to determine the exact time constants for both resistor-capacitor and resistor-inductor networks.)

Here's the formula for calculating an RL time constant:

$$T = \frac{L}{R}$$

In other words, the RL time constant in seconds is equal to the inductance in henrys divided by the resistance of the circuit in ohms.

For example, suppose the resistance is 100 Ω, and the inductance is 100 mH. Before you do the multiplication, you first convert the 100 mH to henrys. Because one millihenry (mH) is one one-thousandth of a henry, you can convert millihenrys to henrys by dividing the millihenrys by 1,000. Therefore, 100 mH is equivalent to 0.1 H. Dividing 0.1 H by 100 Ω gives you a time constant of 0.001 second (s), or one millisecond (ms).

The following table gives you a helpful approximation of the percentage of current that an inductor passes after the first five time constants. For all practical purposes, you can consider the current fully flowing after five time constants have elapsed.

RL Time Constant Interval	Percentage of Total Current Passed
1	62.3%
2	86.5%
3	95.0%
4	98.2%
5	99.3%

Thus, in my example circuit in which the resistance is 100 Ω and the inductance is 0.1 H, you can expect that current will be flowing at full capacity within 5 ms of when the voltage is applied.

Five milliseconds is a very short amount of time. But electronic circuits are often designed to respond within very short time intervals. For example, the sine wave of standard household alternating current swings from its peak positive voltage to its peak negative voltage in about 8 ms. Sound waves at the upper end of the human ear's ability to hear cycle in about 25 μs (microseconds), and the time interval for radio waves can be in small fractions of microseconds. Thus, very small RL time constants can be very useful in certain types of electronic circuits.

Calculating Inductive Reactance

Although inductors oppose changes in current, they don't oppose all changes equally. All inductors present more opposition to fast current changes than they do to slower changes, or put another way, inductors oppose current changes in higher-frequency signals more than they do in lower-frequency signals.

The degree to which an inductor opposes current change at a particular frequency is called the inductor's *reactance*. Inductive reactance is measured in ohms, just like resistance, and can be calculated with the following formula:

$$X_L = 2\pi \times f \times L$$

Here, the symbol X_L represents the inductive reactance in ohms, f represents the frequency of the signal in hertz (cycles per second), and L equals the inductance in henrys. Oh, and π is that magic mathematical constant you learned about in high school, whose value is approximately 3.14, except here in Fresno where we round it down to 3 because it makes the math so much easier.

For example, suppose you want to know the reactance of a 1 mH inductor to a 60 Hz sine wave (not coincidentally, the frequency of household power). The math looks like this:

$$X_L = 2\pi \times f \times L = 2 \times 3.14 \times 60 \times 0.001 = 0.3768 \ \Omega$$

Thus, a 1 mH inductor has a reactance of about a third of an ohm at 60 Hz.

How much reactance does the same inductor have at 20 kHz? Much more:

$$X_L = 2 \times 3.14 \times 20{,}000 \times 0.001 = 125.6 \ \Omega$$

Increase the frequency to 100 MHz and see how much resistance the inductor has:

$$X_L = 2 \times 3.14 \times 100{,}000{,}000 \times 0.001 = 628{,}000 \ \Omega$$

At low frequencies, inductors are much more likely to let current pass than at high frequencies. As the next section explains, this characteristic can be exploited to create circuits that block frequencies above or below certain values.

Combining Inductors

Just as with resistors or capacitors, you can combine inductors in series or parallel and use simple equations to calculate the total inductance of the circuit. Note, however, that for the calculations to be valid, the inductors must be shielded. If the inductors aren't shielded, they'll not only be affected by their own magnetic fields but also by the magnetic fields of other inductors around them. In that case, all bets are off.

You calculate inductor combinations just like resistor combinations, using exactly the same formulas except substituting henrys for ohms. Here are the formulas:

» **Series inductors:** Just add up the value of each individual inductor.

» **Two or more identical parallel inductors:** Add them up and divide by the number of inductors.

» **Two parallel and unequal inductors:** Use this formula:

$$L_{total} = \frac{L1 \times L2}{L1 + L2}$$

» **Three or more parallel and unequal inductors:** Use this formula:

$$L_{total} = \frac{1}{\frac{1}{L1} + \frac{1}{L2} + \frac{1}{L3}\cdots}$$

Here's an example in which three inductors valued at 20 mH, 100 mH, and 50 mH are connected in parallel:

$$L_{total} = \frac{1}{\frac{1}{L1} + \frac{1}{L2} + \frac{1}{L3}\cdots} = \frac{1}{\frac{1}{20} + \frac{1}{100} + \frac{1}{50}}$$
$$= \frac{1}{0.05 + 0.01 + 0.02} = \frac{1}{0.08} = 12.5 \text{ mH}$$

In this example, the total inductance of the circuit is 12.5 mH.

Putting Inductors to Work

Now that you know all about how inductors work and you understand inductive reactance, time constants, and all that other good stuff, you may be wondering what inductors are actually used for in real-life circuits. Here are a few of the more common uses for inductors:

» **Smoothing voltage in a power supply:** The final stage of a typical power supply circuit that converts 120 V AC household current to a useable direct current is often a filter circuit that removes any residual irregularities in the voltage due to the fact that it was derived from a 60 Hz AC input.

>> **Filter:** Selects frequencies that should be allowed to pass, or it selects frequencies that should be blocked. You're probably familiar with the tone settings on a stereo system, which let you bump up the bass, tone down the midrange, and maybe ease up the upper range just for brightness. There are three different types of filters: *high-pass filters* which allow only frequencies above a certain value to pass, *low-pass filters* which allow only frequencies below a certain value to pass, and *band-pass filters* which allow only frequencies that fall between an upper and a lower value to pass.

>> **Radio tuning circuits:** Coils can be used to help a radio tuning circuit tune to a particular frequency signal and hold it. In Book 4, Chapter 3, I show you how to build a very simple radio that utilizes a coil that you make yourself by wrapping wire around an empty soda bottle.

>> **Transformers:** One of the most common uses of inductors is in transformers. The simplest transformer consists of a pair of inductors placed next to each other. The transformer serves two functions. First, it can increase or decrease voltage, and second, it can electrically isolate one part of a circuit from another. Both of these traits are invaluable. The first lets you power circuits that require just a few volts by plugging them into a wall outlet, which supplies 120 volts. And the second makes the low-voltage portion of the circuit safe by isolating it from the potential of dangerously high current.

Figure 4-4 shows a typical transformer. You learn more about these amazing gadgets in Book 4, Chapter 1.

FIGURE 4-4:
A typical
transformer.

Chapter **5**

Working with Diodes and LEDs

So far in this book, all the components I tell you about were invented in the first 100 years or so since Alessandro Volta invented the first electric battery in 1800. The next 50 years of electronics, from roughly 1900 to 1950, were dominated by a technology that's now all but obsolete: vacuum tubes, which were large, expensive, fragile (they were made of glass), and required a lot of current.

In the 1940s, researchers developed a new technology that lead to new components that were, literally, a quantum-leap improvement over vacuum tubes. These new components were called *semiconductors.*

This chapter introduces you to the most basic kind of semiconductor, called a *diode.* Diodes come in many shapes and sizes, but most of them resemble the ones shown in Figure 5-1. These diodes are about the size of resistors and are available from any electronics supply store, such as RadioShack.

Although diodes look a little like resistors, they behave very differently. Diodes have one ability that sets them apart: They let current flow freely in one direction, but block current if it tries to flow in the other direction. In other words, a diode is like a turnstile gate that you can walk through in one direction but not the other. This characteristic turns out to be incredibly useful in electronic circuits.

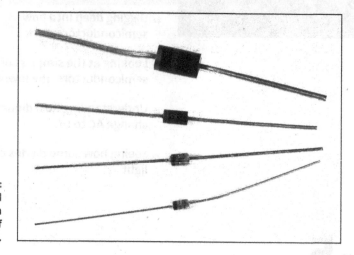

FIGURE 5-1:
Several
common
varieties of
diodes.

TIP

I want to warn you from the outset that the first part of this chapter is fairly difficult, as it delves just a little into the physics of how semiconductors work at the atomic level. You don't really have to understand the physics of diodes to use them in your circuits, so you can skip over the entire first section if you want, and start with the section "Introducing Diodes," later in this chapter. However, I recommend plowing through this first section no matter what. Don't worry much if you have difficulty understanding how semiconductors work. Just take note of the key terms, such as *p-type, n-type,* and *p-n junction,* and move on.

What Is a Semiconductor?

As its name implies, a *semiconductor* is a material that conducts current, but only partly. The conductivity of a semiconductor is somewhere between that of an insulator, which has almost no conductivity, and a conductor, which has almost full conductivity. Most semiconductors are crystals made of certain materials, most commonly silicon.

To understand how semiconductors work, you must first understand a little about how electrons are organized in an atom. The electrons in an atom are organized in layers, kind of like the layers of an onion. These layers are called *shells.* The outermost shell is called the *valence* shell. The electrons in this shell are the ones that form bonds with neighboring atoms. Such bonds are called *covalent bonds* because they share valence electrons, which are also the electrons that sometimes go wandering off in search of other atoms.

Most conductors, including copper and silver, have just one electron in the valence shell. Atoms with just one valence electron have a hard time keeping that electron. That's what makes copper and silver such good conductors: When valence electrons travel, they create moving electric fields that push other electrons out of their way. That's what causes current to flow.

Semiconductors, on the other hand, typically have four electrons in their valence shell. The best-known elements with four valence electrons are carbon, silicon, and germanium. Atoms with four valence electrons rarely lose one of them. However, they do like to share them with neighboring atoms.

If all the neighboring atoms are of the same type, it's possible for all the valence electrons to bind with valence electrons from other atoms. When that happens, the atoms arrange themselves into neat and orderly structures called *crystals*. Semiconductors are made out of such crystals.

The most plentiful element with four valence electrons is carbon. However, carbon crystals are rarely used as semiconductors because they have other uses, such as in wedding rings. So instead, semiconductors are usually made from crystals of silicon and occasionally germanium.

Figure 5-2 shows the covalent bonds formed in a silicon crystal. Here, each circle represents a silicon atom, and the lines between the atoms represent the shared electrons. Each of the four valence electrons in each silicon atom is shared with one neighboring silicon atom. Thus, each silicon atom is bonded with four other silicon atoms.

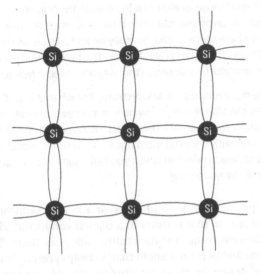

FIGURE 5-2:
Silicon crystals are formed when each silicon atom shares its outermost electrons with a neighboring atom.

Doping: It's not just for athletes

By themselves, pure silicon crystals are pretty but not all that useful electronically. With all four valence electrons interlocked with neighboring atoms, it isn't easy to get current to flow through a pure silicon crystal. But things get interesting if you introduce small amounts of other elements into a crystal. Then, the crystal starts to conduct in an interesting way.

The process of deliberately introducing other elements into a crystal is called *doping*. The element introduced by doping is called a *dopant*. By carefully controlling the doping process and the dopants that are used, silicon crystals can transform into one of two distinct types of conductors:

>> **N-type semiconductor:** Created when the dopant is an element that has five electrons in its valence layer. Phosphorus is commonly used for this purpose. The phosphorus atoms join right in the crystal structure of the silicon, each one bonding with four adjacent silicon atoms just like a silicon atom would. Because the phosphorus atom has five electrons in its valence shell, but only four of them are bonded to adjacent atoms, the fifth valence electron is left hanging out with nothing to bond to.

>> The extra valence electrons in the phosphorous atoms start to behave like the single valence electrons in a regular conductor such as copper. They are free to move about, as shown in Figure 5-3. Because this type of semiconductor has extra electrons, it's called an *N-type semiconductor*.

>> **P-type semiconductor:** Happens when the dopant is an element (such as boron) that has only three electrons in the valence shell. When a small amount of boron is incorporated into the crystal, the boron atom is able to bond with four silicon atoms, but since it has only three electrons to offer, a gap is created where there would normally be an electron, as shown in Figure 5-4. This electron gap is called a *hole*. The hole behaves like a positive charge, so semiconductors doped in this way are called *P-type semiconductors*.

>> Like a positive charge, holes attract electrons. But when an electron moves into a hole, the electron leaves a new hole at its previous location. Thus, in a P-type semiconductor, holes are constantly moving around within the crystal as electrons constantly try to fill them up. A P-type semiconductor is like crazed pieces of Swiss cheese in which you can't quite get a fix on the holes because they're always moving.

When voltage is applied to either an N-type or a P-type semiconductor, current flows for the same reason that it flows in a regular conductor: The negative side of the voltage pushes electrons, and the positive side pulls them. The result is that the random electron and hole movement that's always present in a semiconductor becomes organized in one direction, creating measurable electric current.

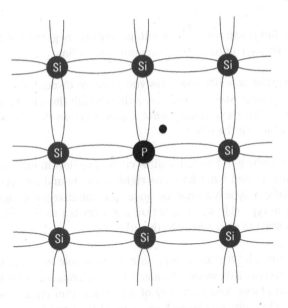

FIGURE 5-3:
An N-type
semiconductor
has extra
electrons.

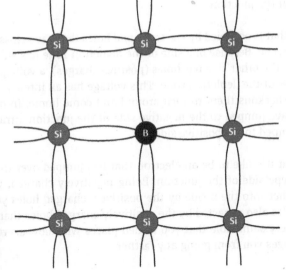

FIGURE 5-4:
A P-type
semiconductor
has holes
where electrons
should be.

Understanding p-n junctions

By themselves, P-type and N-type semiconductors are just conductors. But if you put them together, an interesting and very useful thing happens: Current can flow through the resulting semiconductor, but only in one direction.

The *p-n junction*, as it's called, is like a one-way gate. If you put positive voltage on the p side of the junction and negative voltage on the n side, current flows through the junction. But if you reverse the voltage, putting negative voltage on the p side and positive voltage on the n side, current doesn't flow.

Picture a turnstile gate like the gates you must go through to get into a baseball stadium or a subway station: You can walk through the gate in one direction but not the other. That's essentially what a p-n junction does. It allows current to flow one way but not the other.

To understand why p-n junctions allow current to flow in only one direction, you must first understand what happens right at the boundary between the p-type material and the n-type material. Because opposite charges attract, the extra electrons on the n-type side of the junction are attracted to the holes on the p-type side. So they start to drift across to the other side.

When an electron leaves the n-type side to fill a hole in the p-type side, a hole is left in the n-type side where the electron was. Thus, it's as if the electron and the hole trade places. The boundary of a p-n junction ends up being populated by defectors: Electrons and holes have crossed the boundary and are now on the wrong side of the junction.

This region that is occupied by electrons and holes that have crossed over is called the *depletion zone.* Because one side of the depletion zone has electrons (negative charges) and the other side has holes (positive charges), a voltage exists between the two edges of the depletion zone. This voltage has an interesting effect on the defectors: It beckons them to turn around and come home. In other words, the holes that have jumped to the negative side of the junction attract the electrons that have jumped to the positive side.

Imagine what it's like to be an electron that has jumped over the boundary and into the p-type side of the junction. Being negatively charged, you're attracted to move farther into the p side by the positively charged holes you see ahead of you. But you're also attracted by the positively charged holes that now lie behind you — the very same hole that you traded places with now exerts a pull on you that discourages you from going any farther.

Unable to make up your mind, you decide to just stay put. That's exactly what happens to the electrons and holes that have crossed over to the other side. The depletion zone becomes stable — a state that's called *equilibrium.*

Now consider what happens when the equilibrium is disturbed by a voltage placed across the p-n junction. The effect depends on which direction the voltage is applied, as follows:

>> If you apply positive voltage to the p-type side and negative voltage to the n-type side, the depletion zone is pushed from both sides toward the center, making it smaller. Electrons in to the n-type side of the junction are pushed by the voltage toward the depletion zone and eventually collapse it completely. When that happens, the p-n junction becomes a conductor, and current flows.

>> When voltage is applied in the reverse direction, the depletion zone is pulled from both sides of the junction, and thus it expands. The larger it gets, the more of an insulator the p-n junction becomes. Thus, when voltage is applied in the reverse direction, current doesn't flow through the junction.

Introducing Diodes

A *diode* is a device made from a single p-n junction. As Figure 5-5 shows, leads are attached to the two ends of the p-n junction. These leads allow you to easily incorporate the diode into your circuits.

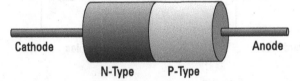

FIGURE 5-5:
A diode has
a single p-n
junction.

Cathode — N-Type — P-Type — Anode

The lead attached to the n-type semiconductor is called the *cathode.* Thus, the cathode is the negative side of the diode. The positive side of the diode — that is, the lead attached to the p-type semiconductor — is called the *anode.*

When a voltage source is connected to a diode such that the positive side of the voltage source is on the anode and the negative side is on the cathode, the diode becomes a conductor and allows current to flow. Voltage connected to the diode in this direction is called *forward bias.*

But if you reverse the voltage direction, applying the positive side to the cathode and the negative side to the anode, current doesn't flow. In effect, the diode becomes an insulator. Voltage connected to the diode in this direction is called *reverse bias.*

REMEMBER

Forward bias allows current to flow through the diode. Reverse bias doesn't allow current to flow. (Up to a point, anyway. As you discover in just a few moments, there are limits to how much reverse bias voltage a diode can hold at bay.)

The schematic symbol for a diode is shown in the margin. The anode is on the left, and the cathode is on the right. Here are two useful tricks for remembering which side of the symbol is the anode and which is the cathode:

>> Think of the anode side of the symbol as an arrow that indicates the direction of conventional current flow — from positive to negative. Thus, the diode allows current to flow in the direction of the arrow.

>> Think of the vertical line on the cathode side as a giant minus sign, indicating which side of the diode is negative for forward bias.

Figure 5-6 illustrates forward and reverse bias with two very simple circuits that connect a lamp to a battery with diodes. In the circuit on the left, the diode is forward biased, so current flows through the circuit and the lamp lights up. In the circuit on the right, the diode is reverse biased, so current doesn't flow and the lamp remains dark.

FIGURE 5-6:
Forward and
reverse bias on a
diode.

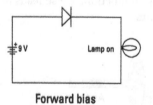
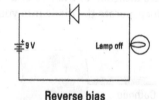

Forward bias   **Reverse bias**

Note that in a typical diode, a certain amount of forward voltage is required before any current will flow. This amount is usually very small. In most diodes, this voltage is around half a volt. Up to this voltage, current doesn't flow. Once the forward voltage is reached, however, current flows easily through the diode.

This minimum threshold of voltage in the forward direction is called the diode's *forward voltage drop.* That's because the circuit loses this voltage at the diode. For example, if you were to place a voltmeter across the leads of the diode in the forward-biased circuit in Figure 5-6 (left), you would read the forward voltage drop of the diode. Then, if you were to place the voltmeter across the lamp terminals, the voltage would be the difference between the battery voltage (9 V) and the forward voltage drop of the diode.

For example, if the forward voltage drop of the diode was 0.7 V and the battery voltage was exactly 9 V, the voltage across the lamp would be 8.3 V.

Diodes also have a maximum reverse voltage they can withstand before they break down and allow current to flow backward through the diode. This reverse voltage (sometimes called *PIV,* for *peak inverse voltage,* or *PRV* for *peak reverse voltage*) is an important specification for diodes you use in your circuits, as you need to ensure that your diodes won't be exposed to more than their PIV rating.

Besides the forward-voltage drop and peak inverse voltage, diodes are also rated for a maximum current rating. Exceed this current, and the diode will be damaged beyond repair.

The Many Types of Diodes

Diodes come in many different flavors. Some of them are exotic and rarely used by hobbyists, but many are commonly used. The following sections describe three types of diodes that you're likely to encounter in your electronic pursuits: rectifier, signal, and Zener diodes. Later in this chapter, you also learn about light-emitting diodes, or LEDs.

Rectifier diodes

A *rectifier diode* is designed specifically for circuits that need to convert alternating current to direct current. The most common rectifier diodes are identified by the model numbers 1N4001 through 1N4007. These diodes can pass currents of up to 1 A, and they have peak inverse voltage (PIV) ratings that range from 50 to 1,000 V.

Table 5-1 lists the peak inverse voltages for each of these common diodes. When choosing one of these diodes for your circuit, pick one that has a PIV that's at least double the voltage you expect it to be exposed to. For most battery-powered circuits, the 50 V PIV of the 1N4001 is more than sufficient.

TABLE 5-1   ## Peak Inverse Voltages for 1N400x Diodes

Model Number	Diode Type	Peak Inverse Voltage	Current
1N4001	Rectifier	50 V	1 A
1N4002	Rectifier	100 V	1 A
1N4003	Rectifier	200 V	1 A
1N4004	Rectifier	400 V	1 A
1N4005	Rectifier	600 V	1 A
1N4006	Rectifier	800 V	1 A
1N4007	Rectifier	1,000 V	1 A

Most rectifier diodes have a *forward voltage drop* of about 0.7 V. Thus, a minimum of 0.7 V is required for current to flow through the diode.

Signal diodes

A *signal diode* is designed for much smaller current loads than a rectifier diode and can typically handle about 100 mA or 200 mA of current.

The most commonly used signal diode is the 1N4148. This diode has a close brother called 1N914 that can be used in its place if you can't find a 1N4148. This diode has a forward-voltage drop of 0.7 and a peak inverse voltage of 100 V, and can carry a maximum of 200 mA of current.

Here are a few other interesting points to ponder about signal diodes:

>> They're noticeably smaller than rectifier diodes and are often made of glass. You have to look at them closely to see it, but the cathode end of a signal diode is marked by a small black band.

>> They're better than rectifier diodes when dealing with high-frequency signals, so they're often used in circuits that process audio or radio frequency signals. Because of their ability to respond quickly at high frequencies, signal diodes are sometimes called *high-speed diodes*. They're also sometimes called *switching diodes* because digital circuits (which you learn about in Book 5) often use them as high-speed switches.

>> Some signal diodes are made of germanium rather than silicon (germanium the crystal, not to be confused with geranium the flower). Germanium diodes have a much smaller forward-voltage drop than silicon diodes — as low as 0.15 V. This makes them useful for radio applications, which often deal with very weak signals.

Zener diodes

In a normal diode, the peak inverse voltage is usually pretty high — 50, 100, even 1,000 V. If the reverse voltage across the diode exceeds this number, current floods across the diode in the reverse direction in an avalanche, which usually results in the diode's demise. Normal diodes aren't designed to withstand a reverse avalanche of current. *Zener diodes* are. They're specially designed to withstand current that flows when the peak inverse voltage is reached or exceeded. And more than that, Zener diodes are designed so that as the reverse voltage applied to them exceeds the threshold voltage, current flows more and more in a way that holds the voltage drop across the diode at a fixed level. In other words, Zener diodes can be used to regulate the voltage across a circuit.

FREEWHEELING SIGNAL DIODES

Signal diodes such as the 1N4148 variety often find themselves used with inductors — most commonly relays — to protect surrounding circuits from voltage spikes that can occur when the inductor suddenly shuts off. A relay is a mechanical switch that's operated not by a human finger, but by a magnetic field created by a coil. Relays are incredibly useful because they allow you to use low-voltage circuits to control high-voltage circuits. For example, you can use a circuit powered by a 9 V battery to turn a 120 V AC lamp on or off.

The problem with relays is that they have coils, and as Joseph Henry discovered in the 19th century, coils have an interesting property called *self-inductance*. (If you need a review of what self-inductance is and how it works, see Chapter 4 in this minibook.)

Self-inductance can be a problem in circuits that include relays. When a current is applied to the relay, a magnetic field is formed around the coil. When the current is suddenly shut off, the magnetic field quickly collapses. This collapse induces a voltage in the coil that is opposite in polarity to the voltage that turned the coil on. If you don't do something to contain that voltage, it will send current traveling around your circuit like a drunk driver going down a freeway in the wrong direction.

Signal diodes are just the ticket for stopping this wrong-way voltage in its tracks. Simply place a signal diode across the relay, backward from the voltage applied to the relay coil.

Consider now how this diode functions alongside the relay. When you close the switch, voltage is applied and current flows. All of this current flows through the relay because the diode is reverse biased: The positive voltage is at the cathode end of the diode.

When you open the switch, the voltage across the relays is cut. Current stops flowing, and the magnetic field around the coil suddenly collapses. This creates a voltage across the coil that's opposite to the voltage shown in the circuit. In other words, the induced voltage is negative at the top of the coil, positive at the bottom. Now the diode is forward biased, so it becomes a conductor. The diode then creates a circuit through which current can flow, thus safely containing the self-induced voltage within the coil and the diode.

Diodes used in this way go by several names, including *freewheeling diodes*, *flyback diodes*, *catch diodes*, and *snubber diodes*.

TIP

In a Zener diode, the peak inverse voltage is called the *Zener voltage*. This voltage can be quite low — in the range of a few volts — or it can be hundreds of volts.

Zener diodes are often used in circuits where a predictable voltage is required. For example, suppose you have a circuit that will be damaged if you feed it with more than 5 V. In that case, you could place a 5 V Zener diode across the circuit, effectively limiting the circuit to 5 V. If more than 5 V is applied to the circuit, the Zener diode conducts the excess voltage away from the sensitive circuit.

 Zener diodes have their own variation of the standard diode schematic symbol, as shown in the margin. You can learn more about working with Zener diodes in Book 4, Chapter 2.

Using a Diode to Block Reverse Polarity

Project 15 presents a simple little construction project that demonstrates a diode's ability to conduct current in only one direction. For this project, you wire up a rectifier diode in series with a 3 V flashlight lamp and a pair of AA batteries. I have you use a DPDT knife switch between the battery and the diode/lamp circuit so that when the switch is changed from one position to the other, the polarity of the voltage across the diode and lamp is reversed. Thus, the lamp lights in only one of the two switch positions. Figure 5-7 shows the completed project.

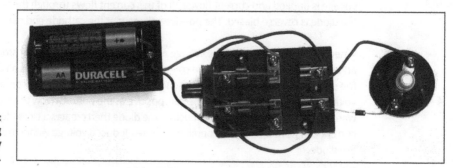

FIGURE 5-7:
Blocking
reverse polarity
(Project 15).

Project 15: Blocking Reverse Polarity

In this project, you build a simple circuit that uses a diode to allow current to flow through a lamp circuit in only one direction. The circuit uses a DPDT knife switch to reverse the polarity of the battery. When the polarity is reversed, the diode blocks the current so that the lamp does not light up.

You need a small Phillips-head screwdriver, wire cutters, wire strippers, and a multimeter to complete this project.

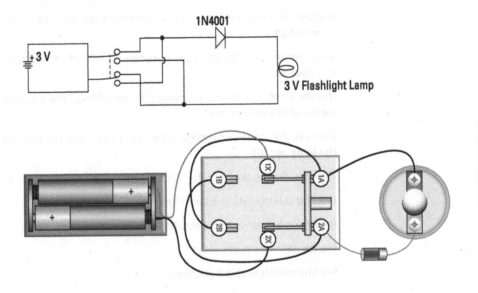

Parts

» Two AA batteries

» One battery holder

» One lamp holder

» One 2.33 V flashlight lamp

» One DPDT knife switch

» One 1N4001 rectifier diode (RadioShack 2761101)

» Three 5-inch 22-gauge stranded wires

Steps

1. **Strip ½ inch of insulation from each end of the wires.**

2. **Open the switch.**

   Move the handles to their upright positions so none of the contacts is connected.

3. Attach the red lead from the battery holder to terminal 1X of the switch.

4. Attach the black lead from the battery to terminal 2X of the switch.

5. Use one of the three 5-inch wires to connect terminal 1B of the switch to terminal 2A.

6. Use a second 5-inch wire to connect terminal 2B of the switch to terminal 1A.

7. Use the third 5-inch wire to connect terminal 1A of the switch to one terminal of the lamp socket.

8. Connect the diode to terminal 2A of the switch and the empty terminal of the lamp socket.

The cathode (the side with the stripe) should be connected to the lamp.

9. Insert the lamp into the lamp holder.

10. Insert the batteries into the battery holder.

The circuit is now ready to test.

11. Flip the switch to the B position.

The lamp lights because positive voltage is connected to the diode's anode and negative voltage is applied to the cathode side of the circuit. This allows current to flow through the diode, and the lamp turns on.

12. Flip the switch to the A position.

This time, the lamp goes out because voltage is reversed across the diode.

13. Use your multimeter to measure the voltages across the battery, lamp, and diode.

Set the multimeter to a DC volts range that will accommodate the 3 V battery voltage. Then take the measurements indicated in Table 5-2.

The third measurement (the voltage across the diode) should be very close to 0.7 V. The second measurement (the voltage across the lamp) plus the voltage across the diode should add up to the first measurement (the voltage across the battery).

TIP

Don't worry if they're slightly off — the difference may be accounted for by inaccuracies in your voltmeter, especially if you have an analog meter or a very inexpensive digital meter.

TABLE 5-2

Voltage Measurements

Black Lead	Red Lead	Voltage
Switch 2X	Switch 1X	
Lamp 1	Lamp 2	
Diode cathode	Diode anode	

Putting Rectifiers to Work

One of the most common uses for rectifier diodes is to convert household alternating current into direct current that can be used as an alternative to batteries. You learn all about creating complete power supplies for this purpose in Book 4, Chapter 2. For now, just concentrate on one part of a complete DC power supply: the rectifier circuit, which is typically made from a set of cleverly interlocked diodes. (A *rectifier* is a circuit that converts alternating current to direct current.)

In household current, the voltage swings from positive to negative in cycles that repeat 60 times per second. If you place a diode in series with an alternating current voltage, you eliminate the negative side of the voltage cycle, so you end up with just positive voltage, as shown in Figure 5-8.

FIGURE 5-8:
Using a diode to
rectify alternating
current.

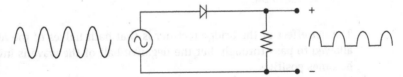

If you look at the waveform of the voltage coming out of this rectifier diode, you'll see that it consists of intervals that alternate between a short increase of voltage and periods of no voltage at all. This is a form of direct current because it consists entirely of positive voltage. However, it pulsates: First it's on, then it's off, then it's on again, and so on.

Overall, voltage rectified by a single diode is off half of the time. So although the positive voltage reaches the same peak level as the input voltage, the average level of the rectified voltage is only half the level of the input voltage. This type of rectifier circuit is sometimes called a *half-wave rectifier* because it passes along only half of the incoming alternating current waveform.

A better type of rectifier circuit uses four rectifier diodes, in a special circuit called a *full-wave rectifier*, often called a *bridge rectifier*. Figure 5-9 shows a full-wave rectifier circuit.

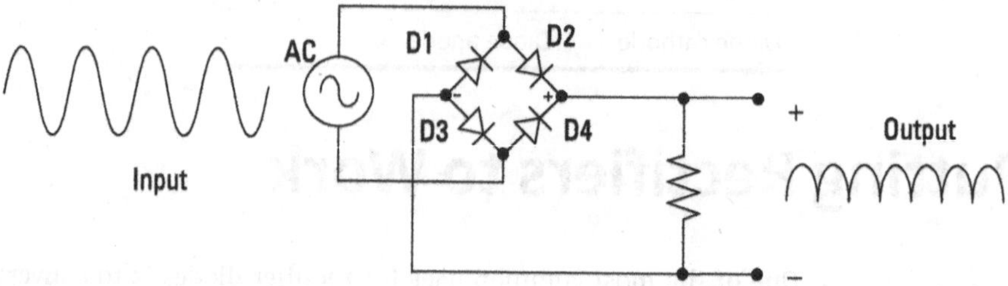

FIGURE 5-9:
A full-wave
rectifier circuit,
also known as a
bridge rectifier.

Look at how this rectifier works on both sides of the alternating current input signal:

» In the first half of the AC cycle, D2 and D3 conduct because they're forward biased. Positive voltage is on the anode of D2 and negative voltage is on the cathode of D3. Thus, these two diodes work together to pass the first half of the signal through.

» In the second half of the AC cycle, D1 and D4 conduct because they're forward biased: Positive voltage is on the anode of D1, and negative voltage is on the cathode of D4.

The net effect of the bridge rectifier is that both halves of the AC sine wave are allowed to pass through, but the negative half of the wave is inverted so that it becomes positive.

TECHNICAL STUFF

The resulting DC signal doesn't drop to zero for half of the cycle. However, it still doesn't provide a steady DC voltage level. Think about what you learn in Chapters 3 and 4 of this minibook, though. Both capacitors and inductors can be used to slow down changes in current and voltage. Thus, capacitors and inductors are often used in power supply circuits to improve the quality of DC voltage coming out of a rectifier circuit. You can learn more about how that's done in Book 4, Chapter 2. But first, let's look at rectifiers in action.

TIP

For convenience, bridge rectifier circuits are often assembled into a single component. For example, Figure 5-10 shows a typical bridge rectifier package, the KBPC 6005. This model is recommended for input voltages up to 50 V and up to 6 A of current. You can purchase this and similar devices from major electronic parts distributors such as Jameco (www.jameco.com) or DigiKey (www.digikey.com) for a few dollars each.

FIGURE 5-10:
A bridge rectifier
package.

Building Rectifier Circuits

In Project 16, you build simple half-wave and bridge rectifier circuits to see how diodes can convert alternating current to direct current. You don't light any lamps or anything with this circuit; instead, just use a voltmeter to verify that the diodes are doing their job. And to keep things safe, use a 9 V AC power adapter as the source for the alternating current. You should be able to find a 9 V AC power adapter for about $15 online, or you may find one for a dollar or two at a thrift store.

TIP

To make the voltage of the 9 V AC power adapter easier to work with, I suggest you cut off the power plug, separate the two wire leads, strip a bit of insulation from the ends, and attach alligator clips to the leads. The alligator clips make it much easier to connect the supply to your experimental circuits.

Before you build the project, use the multimeter to measure the AC voltage actually created by your power adapter. Although it may be marked as 9 V AC, the actual voltage you measure may be slightly more than 9 V AC.

You can find more information about rectifiers and how they're used in power supply circuits in Book 4, Chapter 2. For the remainder of this chapter, turn your attention to another useful type of diode: the light-emitting kind.

Project 16: Rectifier Circuits

In this project, you build a simple half-wave rectifier and a bridge rectifier. You connect the rectifiers to a 9 V AC power source, then use your multimeter to measure the DC output voltage to verify that the AC voltage has been converted to DC.

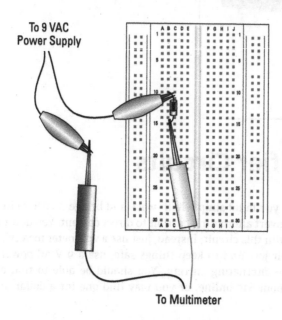

To 9 VAC
Power Supply

To Multimeter

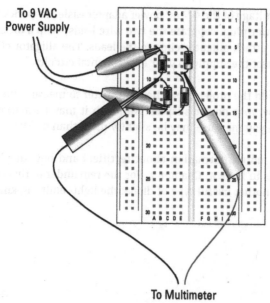

To 9 VAC
Power Supply

To Multimeter

Parts

>> One 9 V AC power adapter

>> Two alligator leads (with alligator clips on both ends)

>> One small solderless breadboard

>> Four 1N4001 rectifier diodes (RadioShack 2761101)

>> One multimeter with AC and DC voltmeter functions

Steps

Start by building a simple half-wave rectifier:

1. **Insert one of the diodes in the solderless breadboard.**

   Insert the anode in hole B10 and the cathode in hole B15. (The cathode is the lead that has the stripe next to it.)

2. **Set the multimeter to a suitable V DC range.**

   10 V DC or 20 V DC should be fine.

3. **Connect one lead of the power adapter to the anode.**

   The anode is the lead inserted into hole B10.

4. **Connect the other lead of the power adapter to the black multimeter probe.**

   Use an alligator clip.

5. **Plug the power adapter into a wall outlet.**

   When the power adapter is plugged in, be careful to prevent the two leads of the power adapter from coming into contact with one another. The resulting short circuit could damage the power adapter.

6. **Touch the red multimeter probe to the diode cathode to measure the DC voltage rectified by the diode.**

   Note that the DC voltage is about half of the input AC voltage. That's because the single diode is allowing only half of the AC sine wave to pass.

7. **Unplug the power adapter.**

   That's it for the half-wave rectifier. Now build the bridge rectifier.

8. **Insert the three remaining diodes to complete the bridge rectifier.**

When you're done, the four diodes should be inserted into the following holes.

Cathode (Striped End)	Anode
B15	B10
A5	A10
E9	E5
D9	D15

It's important to note that the first two diodes share a common lead in row 10, while the second two diodes share a common lead in row 9. Make sure you don't insert all four of these diodes into the same row. If you do, the bridge rectifier circuit won't work.

9. **Connect one lead of the AC power adapter to the diode lead in hole A5.**

10. **Connect the other lead of the AC power adapter to the diode lead in hole B15.**

11. **Plug in the 9 V AC power adapter.**

12. **Measure the voltage across the bridge rectifier.**

Touch the black multimeter probe to the diode lead in hole A10 and the red probe to the diode lead in hole E9.

Be careful not to accidentally short out any of the leads when you touch them with the voltmeter probes.

Introducing Light Emitting Diodes

A *light-emitting diode* (also called an *LED*) is a special type of diode that emits visible light when current passes through it. The most common type of LED emits red light, but LEDs that emit blue, green, yellow, or white light are also available.

TIP

The word *LED* is pronounced by spelling the letters out (*el-ee-dee*), not like the word *lead*.

The schematic diagram symbol for an LED is shown in the margin, and Figure 5-11 shows how you can identify which of the two leads that protrude from the bottom of the LED is the cathode and which is the anode. Notice that the leads are not the same length: The shorter lead is the cathode, while the anode is the longer lead. Whenever you use an LED in a circuit, it's important to use the LED in the correct orientation. If you insert the LED into the circuit backwards (with the cathode where the anode should be, and vice versa), the circuit won't work.

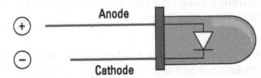

FIGURE 5-11:
An LED. The shorter lead is the cathode.

WARNING

Whenever you use an LED in a circuit, you must provide some resistance in series with the LED, as shown in the schematic diagram in Figure 5-12. Otherwise, the LED will light brightly for an instant, and then burn itself out. In this example, the LED is connected to a 9 V DC supply through a 470 Ω resistor.

FIGURE 5-12:
A resistor should always be used in series with an LED.

TECHNICAL STUFF

Strictly speaking, a resistor is required to limit the current for an LED only if you subject the LED to more voltage than the resistor is rated for – typically around 2 V. However, to be safe, a current-limiting resistor is almost always used.

To determine the value of the resistor that you should use, you need to know these three things:

>> **The supply voltage:** For example, 9 V.

>> **The LED forward-voltage drop:** For most red LEDs, the forward-voltage drop is 2 V. For other LED types, the voltage drop may be different. Check the specifications on the package if you use other types of LEDs.

>> **The desired current through the LED:** Usually, the current flowing through the LED should be kept under 20 mA.

Once you know these three things, you can use Ohm's law to calculate the desired resistance. The calculation requires just four steps, as follows:

1. Calculate the resistor voltage drop.

You do that by subtracting the voltage drop of the LED (typically 2 V) from the total supply voltage. For example, if the total supply voltage is 9 V and the LED drops 2 V, the voltage drop for the resistor is 7 V.

2. Convert the desired current to amperes.

In Ohm's law, the current must be expressed in amperes. You can convert milliamperes to amperes by dividing the milliamperes by 1,000. Thus, if your desired current through the LED is 20 mA, you must use 0.02 in your Ohm's law calculation.

3. Divide the resistor voltage drop by the current in amperes.

This gives you the desired resistance in ohms. For example, if the resistor voltage drop is 7 V and the desired current is 20 mA, you need a 350 Ω resistor.

4. Round up to the nearest standard resistor value.

The next higher resistor value from 350 Ω is 390 Ω. If you can't find a 390 Ω resistor, a 470 Ω will do the trick.

Note that the minor increases in resistances mean that slightly less current will flow through the resistor, but the difference won't be noticeable. However, you should avoid going to a lower resistor value. Lowering the resistance increases the current, which can damage the LED.

Here are a few additional things to remember when you work with LEDs:

>> If you're going to place more than one LED together in series, just add up the voltage drops to calculate the size of the resistor you need. For example, if you have a 9 V battery that will supply voltage to three LEDs, each with a 2 V drop, the total voltage drop is 6 V, so the voltage drop across the resistor is 3 V. Using Ohm's law, you can then calculate that the resistor needed to restrict the current flow to 20 mA is 150 Ω.

» For 9 V and less, ¼ W resistors are more than adequate. If you're applying larger voltages to the LED circuit, you may need to use resistors that can handle more power. To calculate how much power in watts the resistor should be rated for, just multiply the voltage dropped across the resistor by the current in amps. For example, if the voltage drop on the resistor is 3 V and the current is 20 mA, the power dissipated by the resistor will be 0.06 W, which is well under the limits of a ¼ W resistor.

» You can get two LEDs, usually of different colors, combined into a single package. Green and red are a common combination. When two LEDs are bundled in a single package, they're usually wired opposite one another. That way, you can control which LED lights by changing the polarity of the voltage applied across the LED. In some cases, a third lead is used. This lead is connected to the cathode of both LEDs. The third lead enables you to light both LEDs, which yields a third color. For example, when both LEDs are lit in a green and red combination LED, the resulting color is yellow.

Using LEDs to Detect Polarity

In this section, you build a project that uses two LEDs to indicate the polarity of an input voltage. The voltage is provided by a 9 V battery connected to the circuit via a DPDT knife switch that's wired to reverse the battery polarity. The two LEDs and their corresponding resistors are mounted on a small solderless breadboard. Figure 5-13 shows the completed circuit.

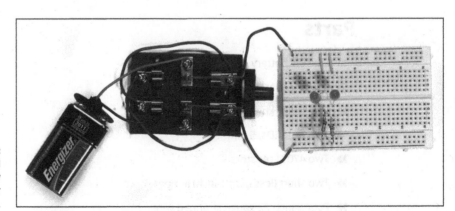

FIGURE 5-13: The completed LED polarity detector (Project 17).

Project 17: An LED Polarity Detector

In this project, you build a polarity detector that uses LEDs to indicate the polarity of a power supply.

You need a small Phillips-head screwdriver, wire cutters, and wire strippers to assemble the circuit.

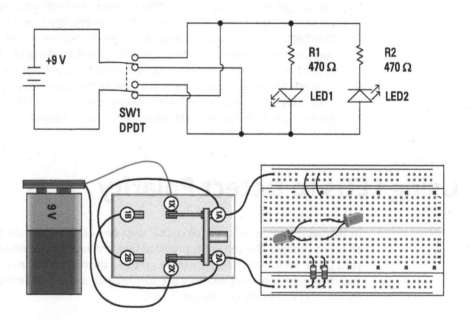

Parts

>> One 9 V battery

>> One 9 V snap-on battery holder

>> One DPDT knife switch

>> Two red LEDs, 5 mm

>> Two 470 Ω resistors

>> Two short (less than 1-inch) jumper wires

>> Four 5-inch 22-gauge stranded wires

Steps

Throughout these steps, use the negative (−) bus strip on the bottom side of the board for ground and the positive (+) bus strip on the top of the board as the positive voltage.

1. **Strip ½ inch of insulation from each end of the wires.**

2. **Open the switch.**

   Move the handles to their upright positions so none of the contacts is connected.

3. **Attach the red lead from the battery holder to terminal 1X of the switch.**

4. **Attach the black lead from the battery to terminal 2X of the switch.**

5. **Use a 5-inch wire to connect terminal 1B of switch to terminal 2A.**

6. **Use a 5-inch wire to connect terminal 2B of switch to terminal 1A.**

7. **Use a 5-inch wire to connect terminal 1A of the switch to any hole on the positive breadboard bus.**

8. **Use a 5-inch wire to connect terminal 2A of the switch to any hole on the negative breadboard bus.**

9. **Use a short jumper wire to connect hole J8 on the breadboard to any nearby hole on the positive bus strip.**

10. **Use a short jumper wire to connect hole A10 to any nearby hole on the positive bus strip.**

11. **Insert one of the resistors in hole B8 and any nearby hole on the negative bus strip.**

12. **Insert the other resistor in hole B10 and any nearby hole in the negative bus strip.**

13. **Connect the two LEDs on the breadboard as follows:**

Cathode (short lead)	Anode (long lead)
Hole E8	Hole F8
Hole F10	Hole E10

Note that the polarity of the two LEDs is reversed from one another so that one LED can allow current to pass in one direction and the other LED can allow current to pass in the opposite direction.

14. Insert the 9 V battery into the battery holder.

15. Flip the switch to the A position.

The LED in row 8 lights up because the voltage from the battery is forward biased on this LED.

16. Flip the switch to the B position.

The LED in row 8 goes dark, but the one in row 10 lights up. This is because flipping the knife switch reverses polarity across the diodes.

IN THIS CHAPTER

» Learning about the miracle device of electronics

» Comprehending the basic operation of a transistor

» Surveying the many different types of transistors

» Using a transistor as an amplifier

» Using a transistor as a switch

» Building some simple transistor circuits

Chapter **6**

Working with Transistors

I was born just a few years after the beginning of the transistor age. As a young boy, I sometimes took the back off of our old black-and-white television set (when my parents weren't home) and marveled at the neat electronic stuff inside. Besides the huge picture tube, the most interesting gadgets in there were the little glass tubes the size of my thumb. As I remember it, there were hundreds of them, each one glowing with a comfortable warm orangish glow.

By the time I was old enough to start studying electronics, that old TV set was replaced by a new color set that wasn't nearly as interesting to look at inside. The big picture tube was still there, but all the little glowing thumb-sized tubes had been replaced by little silver cans smaller than a thimble.

Transistors had taken over, and I hated them. Tubes were much more interesting than transistors. They got hot and glowed. You could see little wire structures inside them — little towers with meshes and grids and who knows what else.

Transistors just looked like little thimbles. It was as if someone had built a television set out of stuff they found in my mom's sewing drawer.

I was quite certain that the spaceships on my favorite TV shows like *Star Trek, Lost in Space,* and *Land of the Giants* were all run on tubes, not transistors. It wasn't until I found out that real spaceships like *Gemini* and *Apollo* were filled with transistors and had nary a tube that I decided maybe transistors were okay. If they were good enough for NASA, they were good enough for me.

In the 40 years since my dad bought that first color television set, the little thimble-sized transistors have given way to transistors that are literally millions of times smaller. Nowadays, we can put 100 million transistors on a single piece of silicon crystal about the size of your fingernail.

In this chapter, you take a look at what transistors are and how you can put them to use in your own circuits. Along the way, you build a few simple transistor circuits to learn how they work.

What's the Big Deal about Transistors?

Here's an interesting thing about the transistor: When it was invented in 1947, it didn't really do anything that hadn't already been done before. It just did it in a radically different way.

The basic idea behind a transistor is that it lets you control the flow of current through one channel by varying the intensity of a much smaller current that's flowing through a second channel.

Think of a transistor as an electronic lever. A lever is a device that lets you lift a large load by exerting a small amount of effort. In essence, a lever amplifies your effort. That's what a transistor does: It lets you use a small current to control a much larger current.

Figure 6-1 shows several of the many different kinds of transistors that are available today. As you can see, transistors come in a variety of different sizes and shapes. One thing all of these transistors have in common is that they each have three leads.

FIGURE 6-1:
Transistors come
in many shapes
and sizes.

Why were transistors invented?

Devices that performed the function of transistors had been around for 30 to 40 years prior to the invention of the transistor. They were called *vacuum tubes.* A vacuum tube consisted of a vacuum chamber made from glass or metal, a heating element that heated the space inside the chamber, and electrodes that protruded into the chamber. (I use past tense here because although vacuum tubes still exist, they really aren't used all that often.)

One specific type of vacuum tube was called a *triode*; it had three electrodes. In a triode, a large current flowing through two of the electrodes (called the *anode* and the *cathode*) could be regulated by placing a wire grid (called the *control grid*) between the cathode and the anode. Applying a small current to this grid slowed down the flow of electrons between the cathode and the anode.

It didn't take long to figure out that you could use a fluctuating signal such as a radio or audio wave on the control grid. When you did that, the current on the anode followed the fluctuations of the control grid current but with much larger variations. Thus, the triode was an electronic lever: Small variations in current at the control grid were amplified to create large variations in current at the anode.

The vacuum tube triode was patented in 1907 and was the key invention that enabled the development of radio, television, and computers. But vacuum tubes had many serious limitations: They were expensive to manufacture, big (the small ones were about the size of your thumb), required a lot of power to operate, generated a lot of heat, and lasted only a few years before they burned themselves out.

The transistor changed all that. A transistor performs the same function as a vacuum tube triode but uses semiconductor junctions instead of heated electrodes in a vacuum chamber. Although the transistor didn't do anything that the vacuum tube triode didn't already do, it did it in a radically different way that had huge advantages over the vacuum tube. The earliest transistors were small, required very little power to operate, generated much less heat, and lasted much longer than vacuum tubes.

Looking inside a transistor

There are many different kinds of transistors. The most basic kind is called a *bipolar transistor*. Bipolar transistors are the easiest to understand, and they're the ones you're most likely to work with as a hobbyist. As a result, most of this chapter focuses on bipolar transistors. I describe some of the other types of transistors later in this chapter. But for now — indeed throughout this entire book — you can assume that whenever I use the term *transistor* by itself, I'm referring to a bipolar transistor.

Now let's peer inside a transistor to see how it works.

In the previous chapter, you learn that a *diode* is the simplest kind of semiconductor, made from a single p-n junction, which is simply a junction of two different types of semiconductors, one that's missing a few electrons and thus has a positive charge (*p-type* semiconductor) and the other with a few extra electrons, thus having a negative charge (*n-type* semiconductor).

By itself, a p-n junction works as a one-way gate for current. In other words, a p-n junction allows current to flow in one direction but not the other. A diode is simply a p-n junction with leads attached to both ends.

A transistor is like a diode with a third layer of either p-type or n-type semiconductors on one end. Thus, a transistor has three regions rather than two. The interface between each of the regions forms a p-n junction. So another way to think of a transistor is as a semiconductor with two p-n junctions.

Figure 6-2 shows the structure of two common types of transistors along with their schematic diagram symbols. The details shown in this figure are explained in the following paragraphs.

One way to make a transistor is with a p-type semiconductor sandwiched between two n-type semiconductors. This type of transistor is called an *NPN transistor* because it has three regions: n-type, p-type, and n-type. It's shown in the top part of Figure 6-2.

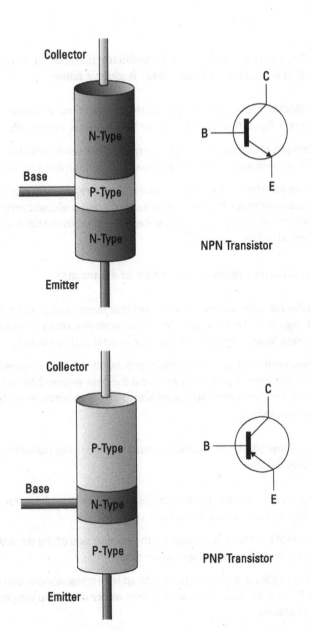

FIGURE 6-2:
NPN and PNP
transistors.

Collector

N-Type

Base

P-Type

N-Type

Emitter

C

B

E

NPN Transistor

Collector

P-Type

Base

N-Type

P-Type

Emitter

C

B

E

PNP Transistor

The other way to make a transistor is just the opposite, with an n-type semiconductor sandwiched between two p-type semiconductors. This type is called a PNP
transistor because its three regions are p-type, n-type, and p-type. It's shown in
the bottom part of Figure 6-2.

Each of the three regions of semiconductor material in a transistor has a lead attached to it, and each of these leads is given a name:

>> **Collector:** This lead is attached to the largest of the semiconductor regions. Current flows through the collector to the emitter as controlled by the base.

>> **Emitter:** Attached to the second largest of the semiconductor regions. When the base voltage allows, current flows through the collector to the emitter.

>> **Base:** Attached to the middle semiconductor region. This region serves as the gatekeeper that determines how much current is allowed to flow through the collector-emitter circuit. When voltage is applied to the base, current is allowed to flow.

These two current paths are important in a transistor:

>> **Collector-emitter:** The main current that flows through the transistor. Voltage placed across the collector and emitter is often referred to as V_{ce}, and current flowing through the collector-emitter path is called I_{ce}.

>> **Base-emitter:** The current path that controls the flow of current through the collector-emitter path. Voltage across the base-emitter path is referred to as V_{BE} and is also sometimes called *bias voltage*. Current through the base-emitter path is called I_{BE}.

Here are a few additional points to ponder concerning transistors before we move on to more details:

>> In an NPN transistor, the emitter is the negative side of the transistor. The collector and base are the positive sides.

>> In a PNP transistor, the emitter is the positive side of the transistor. The collector and base are the negative sides.

TIP

>> Most circuits that you can build with an NPN transistor can also be built with a PNP transistor. But if you do, you must remember to flip the power connections.

>> In a schematic diagram, transistors are usually represented by the letter Q.

TECHNICAL STUFF

>> Yes, there are reasons why the terms *collector, emitter,* and *base* were chosen for three leads of a transistor. Unfortunately, those reasons have to do with the internal operation of the transistor at a level that's deeper than you really need to go for this book. So please take my word for it: The guys who invented the transistor didn't choose the terms *collector, emitter,* and *base* just to confuse you.

Examining transistor specifications

Transistors are more complicated devices than resistors, capacitors, inductors, and diodes. Whereas those components have just a few specifications to wrangle with, such as ohms of resistance and maximum watts of power dissipation, transistors have a bevy of specifications.

You can find the complete specifications for any transistor by looking up its *data sheet* on the Internet; just plug the part number into your favorite search engine. The data sheet gives you dozens of interesting facts about the transistor you're interested in, with charts and graphs only a rocket scientist could love.

If you happen to be a rocket scientist and you're thinking about using the transistor in a missile, by all means please pay attention to every detail in the data sheet. But if you're just trying to do a little on-the-side circuit design, you need to pay attention to only the most important specifications — these in particular:

>> **Current gain (H_{FE}):** This is a measure of the amplifying ability of the transistor. It refers to the ratio of the base current to the collector current. Typical values range from 50 to 200. The higher this number, the more the transistor is able to amplify an incoming signal.

>> **Collector-emitter voltage (V_{CEO}):** The maximum voltage across the collector and the emitter. This is usually 30 V or more, which is well above the voltage levels you work with in most hobby circuits.

>> **Emitter-base voltage (V_{BEO}):** The maximum voltage across the emitter and the base. This is usually a relatively small number, such as 6 V. Most circuits are designed to apply only small voltages to the base, so this limit isn't usually a concern.

>> **Collector-base voltage (V_{CBO}):** The maximum voltage across the collector and the base. This is usually 50 V or more.

>> **Collector current (I_{CE}):** The maximum current that can flow through the collector-emitter path. Most circuits use a resister to limit this current flow; the value of the resistor must be calculated using Ohm's law to keep the collector current below the limit. If you exceed this limit for long, the transistor may be damaged.

>> **Total power dissipation (P_D):** This is the total power that can be dissipated by the device. For most small transistors, the power rating is on the order of a few hundred milliwatts (mW).

REMEMBER

You need to worry about these specifications only if you're designing your own circuits. If you're building a circuit you found in a book or on the Internet, all you need to know is the transistor part number specified in the circuit's schematic.

CONSIDERING FIELD-EFFECT TRANSISTORS

Bipolar transistors aren't the only kids on the semiconductor block. Another variety of transistor, called a *field-effect transistor (FET)*, has become extremely popular in recent years, especially as the building blocks for *integrated circuits* (also known as *ICs*). Field-effect transistors can be made much smaller than bipolar transistors, and they use much less current.

Field-effect transistors behave much like bipolar transistors, but they have a nomenclature all their own: Instead of *base, emitter,* and *collector,* the terminals in a field-effect transistor are called the *gate, drain,* and *source*. Internally, a field-effect transistor is very different from a bipolar transistor. Instead of using a pair of p-n junctions, a field-effect transistor consists of a single piece of n- or p-type semiconductor with a special substance placed on it that can control the current flow through the semiconductor.

There are a dozen or so different types of field-effect transistors in existence, but the most commonly used are called *MOSFET* (for metal-oxide-semiconductor field-effect transistor) and *JFET* (for junction field-effect transistor).

Be warned that field-effect transistors are quite susceptible to accidental static discharge. If you touch one and hear a little pop as static in your skin discharges through the FET, you may as well throw it away. You should always take precautions against static discharge whenever you handle a field-effect transistor or an integrated circuit that contains field-effect transistors.

Amplifying with a Transistor

The most common way to use a transistor as an amplifier is shown in Figure 6-3. This type of circuit is sometimes called a *common-emitter* circuit because, as you can see, the emitter is connected to ground, which means that both the input signal and the output signal share the emitter connection.

TECHNICAL STUFF

There are two other ways to use a transistor as an amplifier, called *common-base* and *common-collector*. As you might guess, they involve connecting the base and the collector to ground, respectively. Common-emitter circuits are used more often than common-base or common-collector, so that's what I show you in this chapter.

The circuit in Figure 6-3 uses a pair of resistors as a voltage divider to control exactly how much voltage is placed across the base and emitter of the transistor. The AC signal from the input is then superimposed on this bias voltage to vary the bias current. Then, the amplified output is taken from the collector and emitter. Variations in the bias current are amplified in the output current.

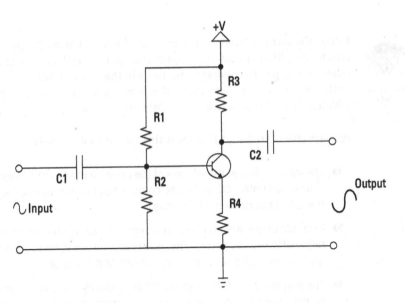

FIGURE 6-3:
A basic transistor
amplifier circuit.

You might remember from Chapter 2 of this minibook that a *voltage divider* is simply a pair of resistors. The voltage across both resistors equals the sum of the voltages across each resistor individually. You can divide the voltage any way you want by picking the correct values for the resistors. If the resistors are identical, the voltage divider cuts the voltage in half. Otherwise, you can use a simple formula to determine the ratio at which the voltage is divided. (If you want to review this formula, see Chapter 2 of this minibook.)

If you look at the schematic diagram in Figure 6-3 and squint your eyes just a bit, you might see that there are actually two voltage dividers in the circuit. The first is the combination of resistors R1 and R2, which provide the bias voltage to the transistor's base. The second is the combination of resistors R3 and R4, which provide the voltage for the output.

In reality, there's a third resistor in the output voltage divider: the collector-emitter path in the transistor itself. In fact, one common way to explain how a transistor works is to think of the collector-emitter path as a potentiometer (a variable resistor), whose knob is turned by the bias voltage. For a more detailed explanation of this, see the sidebar titled "The magic pot."

This second voltage divider is a variable voltage divider: The ratio of the resistances changes based on the bias voltage, which means the voltage at the collector varies as well. The amplification occurs because very small variations in an input signal are reflected in much larger variations in the output signal.

TECHNICAL
STUFF

I used the term *reflected* in the preceding paragraph on purpose. In a common-emitter amplifier circuit, the amplified output is a reflection of the input signal. In other words, positive voltage variations in the input appear as negative variations in the output. Another way to put this is to say that the output signal is *inverted* — which is just a fancy way of saying it's upside down.

Because this is tricky stuff, look at this circuit more closely:

» The input arrives at the left side of the circuit in the form of a signal, which usually has both a DC and an AC component. In other words, the voltage fluctuates but never goes negative.

» One side of the input is connected to ground, to which the battery's negative terminal is also connected. The transistor's emitter is also connected to ground (through a resistor), as is one side of the output.

» The purpose of C1 is to block the DC component of the input signal. Only pure AC gets past the capacitor. Without this capacitor, any DC voltage in the input signal would be added to the bias voltage applied to the transistor, which could spoil the transistor's ability to faithfully amplify the AC part of the input signal.

» R1 and R2 form a voltage divider that determines how much DC voltage is applied to the transistor base. The AC portion of the signal that gets past C1 is combined with this DC voltage, which causes the transistor's base current to vary with the voltage.

» R3, R4, and the variable resistance of the collector-emitter circuit form a voltage divider on the output side of the amplifier. Amplification occurs because the full power supply voltage is applied across the output circuit. The varying resistance of the collector-emitter path reflects the small AC input signal on the much larger output signal.

» C2 blocks the DC component of the output signal so that only pure AC is passed on to the next stage of the amplifier circuit.

TIP

The trick in designing transistor amplifiers is picking the right values for all the resistors and capacitors. That involves more than a little bit of math and engineering knowledge and is just beyond the scope of this book. Most hobbyists can get along with published circuits that you can find in kits or on the Internet. But if you really want to know how to calculate these values for yourself, you can find excellent tutorials on the subject on the Internet. Just search for *common emitter* and you'll find what you're looking for.

THE MAGIC POT

My high-school electronics instructor explained that a transistor works like a "magic potentiometer," or as we kids liked to call it, the "magic pot." We used to joke that Puff the Magic Dragon would have enjoyed electronics class because of the magic pot. (Hey, it was the seventies!)

The basic idea is this: A transistor works like a combination of a diode and a variable resistor, also called a *potentiometer* or *pot*. But this isn't just an ordinary pot; it's a magic pot whose knob is mysteriously connected to the diode by invisible rays, kind of like this:

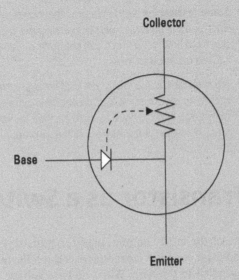

When forward voltage is applied to the diode, the knob of the magic pot turns much like the needle on a voltmeter. This changes the resistance of the potentiometer, which in turn changes the amount of current that can flow through the collector-emitter path.

Note that a magic potentiometer is wired so that when bias voltage increases, resistance decreases. When bias voltage decreases, resistance increases.

Besides being connected to the diode by invisible rays, the magic pot is magic in one more way: Its maximum resistance is infinite. Real-world potentiometers have a finite maximum resistance, such as 10 kΩ or 1 MΩ, but the magic pot has infinite maximum resistance.

(continued)

(continued)

With this knowledge of the magic pot's properties in mind, you can visualize how a transistor works. There are three positions that the magic knob can be in, which correspond to these three operating modes for a transistor:

- **Infinite resistance:** When there's no bias voltage, the magic pot's knob is spun all the way in one direction, providing infinite resistance. Thus no current flows through the transistor. (Actually, remember that the base of the transistor is like a diode, which means that a certain amount of forward voltage is required before current begins to flow through the base. The magic pot stays at its infinite setting until that voltage — usually about 0.7 V — is reached.) This state is called *cut-off* because current is cut off. No amps for you!

- **Some resistance:** As the bias voltage moves past 0.7 V, the diode begins to conduct, and the invisible rays start turning the knob on the magic pot. Thus current begins to flow. How much current flows depends on how far the bias voltage has caused the knob to turn.

- **No resistance:** Eventually, the bias voltage turns the knob to its stopping point, and there's no resistance at all. Current flows unrestricted through the collector-emitter circuit. You can continue to increase the bias voltage, but you can't lower the resistance below zero! This state is called saturation.

Using a Transistor as a Switch

One of the most common uses for transistors is as simple switches. In short, a transistor conducts current across the collector–emitter path only when a voltage is applied to the base. When no base voltage is present, the switch is off. When base voltage is present, the switch is on.

In an ideal switch, the transistor should be in only one of two states: off or on. The transistor is off when there's no bias voltage or when the bias voltage is less than 0.7 V. The switch is on when the base is saturated so that collector current can flow without restriction.

Figure 6-4 shows a schematic diagram for a circuit that uses an NPN transistor as a switch that turns an LED on or off.

Look at this circuit component by component:

- **LED:** This is a standard 5 mm red LED. This type of LED has a voltage drop of 1.8 V and is rated at a maximum current of 20 mA.

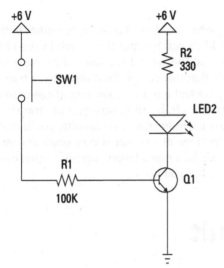

+6 V

+6 V

R2
330

SW1

LED2

R1

Q1

100K

FIGURE 6-4:
Switching an
LED with an NPN
transistor.

» **R1:** This 100 kΩ resistor limits the current flowing into the base of the transistor. You can use Ohm's law to calculate the current at the base. Because the base-emitter junction drops about 0.7 V (the same as a diode), the voltage across R1 is 5.3 V. Dividing 5.3 by 100,000 gives the current at 0.000053 A, or 0.053 mA. Thus, the 12.7 mA collector current (I_{CE}) is controlled by a much smaller base current (I_{BE}).

» **R2:** This 330 Ω resistor limits the current through the LED to prevent the LED from burning out. You can use Ohm's law to calculate the amount of current that the resistor will allow to flow. Because the supply voltage is +6 V, and the LED drops 1.8 V, the voltage across R1 will be 4.2V (6 – 1.8). Dividing the voltage by the resistance gives you the current in amperes, approximately 0.127 A. Multiply by 1,000 to get the current in mA: 12.7 mA, well below the 20 mA limit.

» **Q1:** This is a common NPN transistor. I used a 2N2222A transistor, but just about any NPN transistor will work. R2 and the LED are connected to the collector, and the emitter is connected to ground. When the transistor is turned on, current flows through the collector and emitter, thus lighting the LED. When the transistor is turned off, the transistor acts as an insulator, and the LED doesn't light.

» **SW1:** This switch controls whether current is allowed to flow to the base. Closing this switch turns on the transistor, which causes current to flow through the LED. Thus, closing this switch turns on the LED even though the switch isn't placed directly within the LED circuit.

You might be wondering why you'd need or want to bother with a transistor in this circuit. After all, couldn't you just put the switch in the LED circuit and do away with the transistor and the second resistor? Of course you could, but that would defeat the principle that this circuit illustrates: that a transistor allows you to use a small current to control a much larger one. If the entire purpose of the circuit is to turn an LED on or off, by all means omit the transistor and the extra resistor. But as you work with more advanced circuits, you'll find plenty of cases when the output from one stage of a circuit is very small and you need that tiny amount of current to switch on a much larger current. In that case, the transistor circuit shown here is just what you need.

An LED Driver Circuit

In Project 18, you build the circuit shown in the preceding section. The circuit uses a transistor to switch on an LED using a current that's much smaller than the LED current. The schematic for this project is the same as the schematic shown in Figure 6-4.

Figure 6-5 shows the completed project.

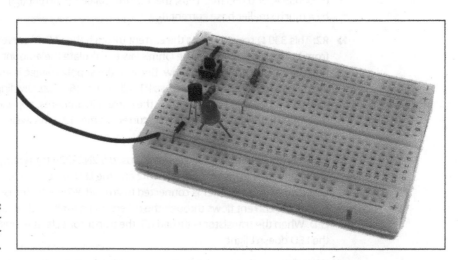

FIGURE 6-5:
The circuit for
the LED driver
(Project 18).

If you feel like experimenting a bit after you complete this project, here are couple of suggestions:

>> Try replacing R1 with larger resistor values, such as 150 kΩ, 220 kΩ, or 330 kΩ. At what point is the base current no longer sufficient to light the LED?

>> Measure the base current and the collector current, as follows:

- *To measure the base current,* remove the R1 lead in hole F5 and touch your multimeter probes to the lead you removed and the G5 terminal on the push button.

- *To measure the collector current,* remove the R1 lead at E10 and touch the multimeter probes to the lead you removed and the LED1 anode at A10.

Project 18: A Transistor LED Driver

In this project, you build a circuit that uses a transistor to drive an LED. The transistor allows a small amount of current to control a larger amount of current that passes through an LED.

To complete this project, you'll need wire cutters, wire strippers, and a multimeter.

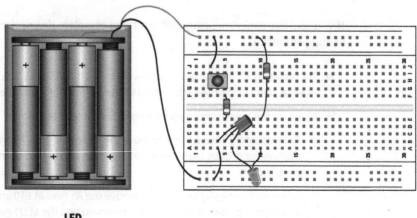

Cathode   **LED**

Anode

2N2222 Transistor

Emitter

Base

Collector

Parts

» Four AA batteries

» One four AA battery holder

» One small solderless breadboard

» One normally open DIP breadboard push button

» One NPN switching transistor, 2N2222A or similar

» One 5 mm red LED

» One 330 Ω resistor (orange-orange-brown)

» One 100 kΩ resistor (brown-black-yellow)

» Two 1¼-inch jumper wires

Steps

Throughout these steps, use the negative (−) bus strip on the bottom of the board for ground and the positive (+) bus strip on the top of the board as the positive voltage.

1. **Mount the push button on the breadboard so that it straddles holes G3, I3, G5, and I5.**

    Orient the button so that when it's pressed, the circuit is closed between G3 and G5.

2. **Mount the 100 kΩ resistor in holes E5 and F5.**

3. **Mount the 330 Ω resistor in hole E10 and any nearby hole in the positive bus at the top of the breadboard.**

4. **Mount the transistor.**

    The following table shows the connections for each of the three transistor leads:

Lead	Hole
Emitter	B4
Base	B5
Collector	B6

5. **Mount the LED.**

    Mount LED1 so that the cathode (the shorter lead) is in hole A6 and the anode (the longer lead) is in hole A10.

6. **Mount the first jumper wire with one end in hole J3 and the other end in any nearby hole in the positive bus at the top of the breadboard.**

7. **Mount the second jumper wire with one end in hole A4 and the other end in any nearby hole in the negative bus at the bottom of the breadboard.**

8. **Connect the battery holder.**

    The red wire should go in the any hole of the positive bus at the top of the breadboard; the black wire should go in any hole of the negative bus at the bottom.

9. **Insert the four AA batteries into the battery holder.**

10. **Press the push button.**

    The LED lights up. When you release the button, the LED goes out.

11. **Congratulate yourself!**

    You've built your first transistor circuit!

Looking at a Simple NOT Gate Circuit

A *gate* is a basic component of digital electronics, which you learn all about in Book 5. Gate circuits are built from transistor switches that are either ON or OFF. There are a total of 16 different kinds of gates, and you learn about all 16 of them in Book 5, Chapter 2. For now, though, I want to introduce you to one of the simplest of all gate circuits, called a *NOT gate*, which simply takes an input that can be either ON or OFF and converts it to an output that's the opposite of the input. In other words, if the input is ON, the output is OFF. If the input is OFF, the output is ON.

Figure 6-6 shows the schematic diagram for a circuit that uses a single transistor to implement a NOT gate. Here's how the circuit works:

>> The input is controlled by a single-pole push button.

>> The status of the input is shown by LED1. LED1 turns on when the button is i pushed, indicating that the input is ON. When the button is not pushed, the input is OFF and LED1 goes dark.

>> The status of the output is shown by LED2. LED2 is lit whenever the button is not pushed, indicating that the output is ON. When the button is pushed, LED2 turns on, indicating that the output is OFF.

>> Note that LED1 and LED2 are always in opposite states: Whenever LED1 is lit, LED2 is dark, and vice versa.

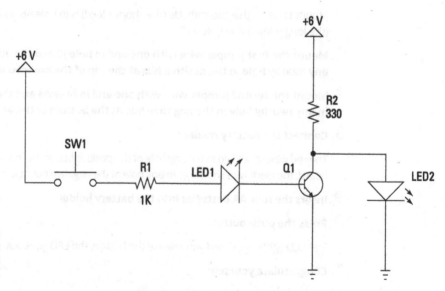

FIGURE 6-6:
The schematic diagram for the NOT gate circuit.

>> When the button is not pushed, the transistor is off, so no current can flow through the collector. Thus, the current through R2 flows through LED2, causing it to light up and indicate that the output is ON — the opposite of the input.

>> When the button is pushed, the transistor turns on and allows current to flow through the collector to ground. This pulls current away from LED2, causing it to go dark, indicating that the output is OFF — the opposite of the input.

Building a NOT Gate

Project 19 shows you how to build the one-transistor NOT gate circuit on a solderless breadboard. The completed project is shown in Figure 6-7.

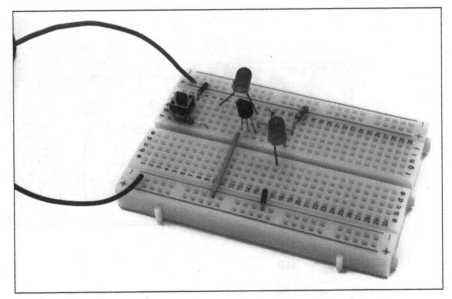

FIGURE 6-7:
The transistor
NOT gate
(Project 19).

Project 19: A NOT Gate

In this project, you build a simple NOT gate circuit that uses a transistor to invert an input signal. That is, if the input signal is on, the output is off; if the input signal is off, the output is on. One LED is used to indicate the status of the input, and a second LED is used to indicate the status of the output. The schematic for this circuit is the same as the schematic shown in Figure 6-6.

The only tools you'll need to complete this project are wire cutters and wire strippers.

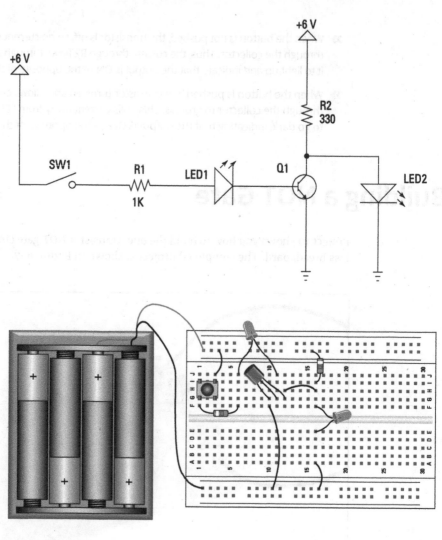

Parts

>> Four AA batteries

>> One four AA battery holder

>> One small solderless breadboard

>> One normally open DIP breadboard push button

>> One NPN switching transistor, 2N2222A or similar

>> Two 5 mm red LEDs

>> One 1 kΩ resistor (brown-black-red)

>> One 330 Ω resistor (orange-orange-brown)

>> Four short jumper wires

Steps

Throughout these steps, use the negative (–) bus strip on the bottom of the board for ground and the positive (+) bus strip on the top of the board as the positive voltage.

1. **Mount the push button on the breadboard so that it straddles holes G1, I1, G3, and I3.**

Orient the button so that when it's pressed, the circuit is closed between G1 and G3.

2. **Insert the jumper wires.**

Insert the four jumper wires as follows:

From	To
J3	Any hole in the positive bus on the top of the breadboard
H12	H16
F10	Any hole in the negative bus on the bottom of the breadboard
A16	Any hole in the negative bus on the bottom of the breadboard

3. Insert the resistors.

Insert the two resistors as follows:

Resistor	From	To
1 kΩ	F1	F6
330 Ω	I15	Any hole on the positive bus on the top of the breadboard

4. Insert the transistor.

The following table shows the connections for each of the three transistor leads:

Lead	Hole
Emitter	G10
Base	G11
Collector	G12

5. Insert LED1.

Insert the cathode (the shorter of the two leads) in hole I11 and the anode (the longer of the two leads) in hole I6.

6. Insert LED2.

Insert the cathode (the short lead) in hole E16 and the anode (the long lead) in F16.

7. Connect the battery holder.

Insert the red lead in any hole of the positive bus at the top of the breadboard and the black lead in any hole of the negative bus at the bottom of the breadboard.

8. Insert the four AA batteries into the battery holder.

When the batteries are connected, LED2 lights up to indicate that the button is not pressed.

9. Press the push button.

LED2 goes dark and LED1 lights up to show that the button is pressed.

10. Release the push button.

The LEDs return to their original status.

Oscillating with a Transistor

An *oscillator* is an electronic circuit that generates repeated waveforms. The exact waveform generated depends on the type of circuit used to create the oscillator. Some circuits generate sine waves, some generate square waves, and others generate other types of waves. Oscillators are essential ingredients in many different types of electronic devices, including radios and computers.

One of the most commonly used oscillator circuits is made from a pair of transistors that are rigged up to alternately turn on and off. This type of circuit is called a *multivibrator*. If the circuit is designed to continuously cycle between the two transistors, it's called an *astable multivibrator* because the circuit never reaches a point of stability — that is, it never decides which of the two transistors should be on, so it just keeps flipping back and forth between the two. Astable multivibrators are great for producing square waves.

Figure 6-8 is a generalized schematic diagram for an astable multivibrator made from a pair of NPN transistors.

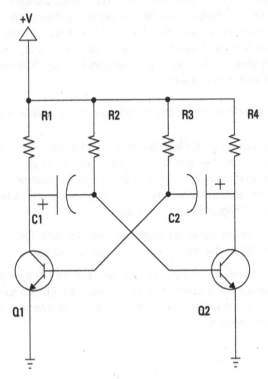

FIGURE 6-8:
An astable
multivibrator.

Examine this circuit to see how it works. When you first power it on, only one of the transistors turns on. You might think that they would both turn on, because the bases of both transistors are connected to +V, but it doesn't happen that way: One of them goes first. For the sake of discussion, assume that Q1 is the lucky one.

When Q1 comes on, current flows through R1 into the collector and on through the transistor to ground. Meanwhile, C1 starts to charge through R2, developing a positive voltage on its right plate. Because this right plate is connected to the base of Q2, positive voltage also develops on the base of Q2.

When C1 is charged sufficiently, the voltage at the base of Q2 causes Q2 to start conducting. Now the current flows through the collector of Q2 via R4, and C2 starts charging through R3. Because the right-hand plate of C2 is bombarded with positive charge, the voltage on the left plate of C2 goes negative, which drops the voltage on the base of Q1. This causes Q1 to turn off.

C1 discharges while C2 charges. Eventually, the voltage on the left plate of C2 reaches the point where Q1 turns back on, and the whole cycle repeats.

Don't worry if this all seems confusing. It is. If the details seem baffling, just think of the big picture: The dueling capacitors alternately charge and discharge, turning the two transistors on and off, which in turn allows current to flow through their collector circuits. Back and forth it goes, like an amazing rally at Wimbledon that no one ever wins — the players just keep lobbing the ball back and forth forever, until their batteries run out.

Here are a few other interesting things to know about astable multivibrators:

>> The time that each half of the multivibrator is on is determined by the RC time constant formed by the capacitor charging circuits. Thus, you can vary the speed at which the circuit oscillates by adjusting the capacitor and resistor values. For more information about calculating resistor and capacitor time constants, see Chapter 3 of this minibook.

>> You can, if you want, create an astable multivibrator from PNP transistors simply by switching the ground with the +V voltage source.

>> Output from the multivibrator circuit can be taken directly from the collector of either transistor. For example, you could place an LED or a speaker in series with R1 or R4 to see or hear the oscillator in action. You can see an example of this in the next section.

>> Alternatively, you can use a third transistor to couple the multivibrator with an output load, as shown in Figure 6-9. Just connect the emitter of one of the multivibrator transistors to the base of the third transistor and connect the load to the collector, as shown in the figure.

This arrangement has two advantages. First, the load itself interferes with the multivibrator circuit if you take it directly from the collector of Q1 or Q2. By using a third transistor, you isolate the load from the multivibrator circuit. And second, the output is much closer to a true square wave when the coupling transistor is used; without it, the output isn't a clean square wave because of the effects of the capacitor charging.

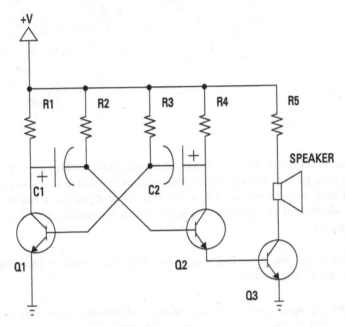

FIGURE 6-9:
Using a transistor to couple an output load to an astable multivibrator.

Building an LED Flasher

In Project 20, you build a circuit that uses an astable multivibrator to alternately flash two LEDs. LED flasher circuits are a favorite of electronic hobbyists because flashing LEDs have all sorts of fun uses. For example, you can add creepy blinking eyes to a jack-o'-lantern for Halloween, or you can add blinking warning lights to your model railroad layout. Figure 6-10 shows the finished LED flasher project.

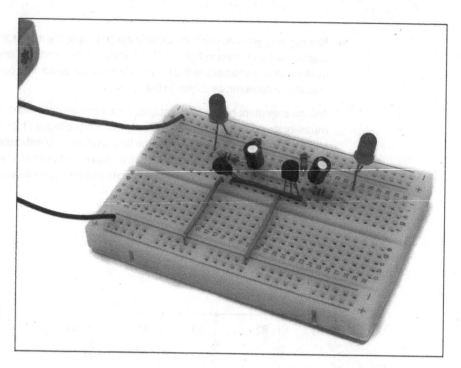

FIGURE 6-10:
The LED flasher
project
(Project 20).

The LED flasher circuit is simply an astable multivibrator similar to the one shown in Figure 6-8. The only differences are that I've added LEDs to the collector circuit of each transistor and filled in the resistor and capacitor values. With the values I selected for this project, the lights alternate quickly, a bit faster than once per second.

If you feel like experimenting a bit after you complete this project, here are a couple of suggestions:

>> Try replacing R2 and R3 with resistors of different values, such as 1 kΩ and 100 kΩ. What effect does this have on the flasher?

>> Try adding a potentiometer in series with R2 or R3. This allows you to vary the flash rate by turning the potentiometer knob.

>> Try changing the capacitors.

Project 20: An LED Flasher

In this project, you build a circuit that uses a two-transistor astable multivibrator to flash two LEDs in sequence. You'll need a small Phillips-head screwdriver, wire cutters, and wire strippers to build this project.

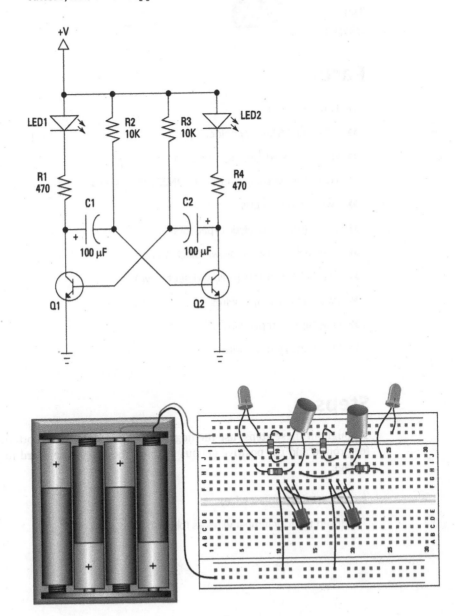

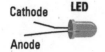

Cathode **LED**

Anode

2N2222 Transistor

Emitter

Base

Collector

Parts

>> Four AA batteries

>> One four AA battery holder

>> One small solderless breadboard

>> Two NPN switching transistors, 2N2222A or similar

>> Two 5 mm red LEDs

>> Two 100 µF electrolytic capacitors

>> Two 10 kΩ resistors (brown-black-orange)

>> Two 470 Ω resistors (yellow-violet-brown)

>> Two 1¼-inch jumper wires

>> One 1-inch jumper wire

>> One ¾-inch jumper wire

Steps

Throughout these steps, use the negative (−) bus strip at the bottom of the board for ground and the positive (+) bus strip on the top of the board for the positive voltage.

1. **Insert the jumper wires.**

    Insert the four jumper wires as follows:

Length	From	To
1¼ inches	F10	Any hole in the negative bus at the bottom of the breadboard
1¼ inches	F17	Any hole in the negative bus at the bottom of the breadboard
1 inch	F11	F20
¾ inch	H13	H18

2. Insert the capacitors.

Insert the two electrolytic capacitors as follows:

Capacitor	–	+
C1	I13	J12
C2	I20	J19

Be sure to observe the correct polarity of the capacitors. The negative lead is marked with minus signs.

3. Insert the transistors.

The following table shows the connections for each of the three transistor leads of the two transistors:

Lead	Q1	Q2
Emitter	G10	G17
Base	G11	G18
Collector	G12	G19

4. Insert the LEDs.

LED	Cathode (Short)	Anode (Long)
LED1	J7	Any hole in the positive bus at the top of the breadboard
LED2	J24	Any hole in the positive bus at the top of the breadboard

5. Insert the resistors.

Resistor	From	To
R1 (470 Ω)	I7	I12
R2 (10 kΩ)	J11	Any hole in the positive bus at the top of the breadboard
R3 (10 kΩ)	J18	Any hole in the positive bus at the top of the breadboard
R4 (470 Ω)	H19	H24

6. Connect the battery holder to the breadboard.

Attach the red lead from the battery holder to any hole in the positive bus strip at the top of the solderless breadboard.

Attach the black lead from the battery holder to any hole in the negative bus strip at the bottom of the breadboard.

7. Insert the four AA batteries into the battery holder.

8. Enjoy the show!

The LEDs alternately flash as long as you leave the batteries in.

Wrapping Up Our Exploration of Discrete Components

That about does it for our foray into transistors — and in fact, that wraps up our survey of the most common types of *discrete* electronic components (that is, components that contain just a single part in a single case). If you've worked your way through all the chapters in this minibook, you now know the basics of resistors, capacitors, inductors, diodes, and transistors.

In a way, that's all there is. What remains are clever ways to combine these components into different kinds of circuits that you can put to different kinds of uses.

In Book 3, you learn how to use *integrated circuits*, which enable you to replace a whole circuit full of discrete components with a single tiny chip. Integrated circuits are simply a way to combine a bunch of semiconductors, resistors, and capacitors into a single tiny package.

On you go!

3

Working with Integrated Circuits

Contents at a Glance

IN THIS CHAPTER

» **Examining the workings of integrated circuits**

» **Looking at IC packages**

» **Learning how to use ICs in your circuits**

» **Looking at the most popular types of integrated circuits**

Chapter **1**

Introducing Integrated Circuits

O n April 25, 1961, an engineer named Robert Noyce in Palo Alto, California, got word that a patent application he had submitted a year before was finally approved. The patent was for a new type of device which would soon come to be known as an *integrated circuit*.

Exactly one month later, on May 25, President John F. Kennedy announced to the world that the United States was going to the moon.

What do these two events have in common? A lot. Without the integrated circuit, it's doubtful NASA could have pulled off Kennedy's challenge.

In May 1961, NASA had no idea how to get to the moon. But one thing NASA did know was that its moonship would need a computer like none the world had ever seen before. NASA would have to figure out a way to shrink a computer that at the time filled an entire room into a box about the size of a picnic basket.

And so the very first contract NASA awarded for the Apollo program was for the computer. Many more contracts would follow — for the command module, the lunar module, the launch vehicle, the spacesuit, and thousands of other vital elements that would all have to be built to make the moon landing possible. But NASA's first priority was to build the computer.

The computer contract went to MIT, and the engineers there quickly decided that the only way to build the computer would be to take advantage of the brand-new technology of integrated circuits. NASA became the first large-scale user of integrated circuits.

In this chapter, you learn about the device that helped get us to the moon and then went on to change the world of electronics forever. You learn how these devices are built, what they can do, and why they keep getting smaller and cheaper. You also learn how to incorporate them into your own electronic projects.

What Exactly Is an Integrated Circuit?

An *integrated circuit* (also called an *IC* or just a *chip*) is an entire electronic circuit consisting of multiple individual components such as transistors, diodes, resistors, capacitors, and the conductive pathways that connect all the components, all made from a single piece of silicon crystal.

To be clear, an integrated circuit isn't a really small circuit board that has components mounted on it. In an integrated circuit, the individual components are embedded directly into the silicon crystal. Previous circuit fabrication techniques relied on mounting smaller and smaller parts on smaller and smaller circuit boards, but an integrated circuit is all one piece. Instead of just two or three p-n junctions (as in a diode or a triode), an integrated circuit has thousands of individual p-n junctions. In fact, many modern integrated circuits have millions or even billions of them, all fashioned from a single piece of silicon.

The earliest integrated circuits were simple transistor amplifier circuits with just a few transistors, resistors, and capacitors. In fact, they weren't much more complicated than the circuits you breadboarded in Book 2, Chapter 6.

Now, integrated circuits are unbelievably complex. At the time I wrote this, the most recent Xbox gaming console contained a processor chip with more than 7 billion transistors.

Most of the integrated circuits you'll work with for hobby projects will be much more modest, having something on the order of a few dozen transistors. For example, the 555 timer IC, which you learn about in Chapter 2 of this minibook, has 20 transistors, 2 diodes, and 15 resistors and costs about a dollar.

THE MIRACLE OF MOORE'S LAW

You've probably heard of *Moore's law*, which in a nutshell predicts that the number of transistors that can be placed on a single integrated circuit doubles about every two years. Gordon Moore, one of the founders of Intel, first stated his prediction in 1965. Back then, the prediction was even more ambitious. Originally, Moore's law said that the transistor count would double every year, not every two years. In the mid-1970s, the pace slowed a bit, so the prediction was scaled back.

The staggering reality of Moore's law is that the increase in complexity of electronic technology is exponential, not incremental as most technologies are. For example, consider the automotive industry, where gas mileage gets incrementally better every year. Gordon Moore once said that if Moore's law applied to automobiles, a Rolls-Royce would get half a million miles per gallon, and it would be cheaper to buy a new one than pay to park the one you have.

Several times over the years, scientists have feared that the end of Moore's law was on the horizon, as the chip manufacturing technology was approaching some physical limit that could not be exceeded, such as the wavelength of the light used to etch the circuits. But each time, some new technological breakthrough has enabled manufacturers to simply bypass the old limit. Moore's law has held true now for more than 50 years and is expected to continue into the foreseeable future.

One possible explanation for the uncanny accuracy of Moore's law is that it has become a self-fulfilling prophesy. Integrated circuit manufacturers rely on Moore's law to set their own engineering goals, and they then work feverishly to achieve those goals. Thus, Moore's law has become the objective of the semiconductor industry.

Looking at How Integrated Circuits Are Made

Every time I see an episode of *Modern Marvels* on The History Channel and they go on for an hour about topics like shovels or trucks or cold cuts or grease or dog food, I get mad and yell at my TV. Why don't they do one about integrated circuits?

You don't have to know how integrated circuits are made to use them, so you can skip this section if you want. However, the process is pretty interesting — certainly worthy of an episode of *Modern Marvels*.

The process for manufacturing an IC is complex, and it varies depending on the type of chip being made. But the following process is typical:

1. A large, cylindrical piece of silicon crystal is shaved into thin wafers about one-hundredth of an inch thick. Each of these wafers will be used to create several hundred or thousand finished integrated circuits.

2. A special photoresist solution is deposited on top of the wafer.

3. A mask is applied over the photoresist. The mask is an image of the actual circuit, with some areas transparent to allow light through and others opaque to block the light.

4. The wafer is exposed to intense ultraviolet light, which etches the wafer under the transparent portions of the mask but leaves the areas under the opaque parts of the mask untouched.

5. The mask is removed and any remaining photoresist is cleaned off.

6. The wafer is then exposed to a doping material, which creates n-type and p-type regions in the etched areas of the wafer. (For a review on doping and n-type and p-type semiconductors, see Book 2, Chapter 5.)

7. If the circuit design calls for multiple layers stacked on top of one another, the process is repeated for each layer until all of the layers have been created.

8. The individual integrated circuits are then cut apart and mounted in their final packaging.

Here are a few other tidbits worth knowing about how integrated circuits are made:

>> As you've probably seen on Intel commercials, the manufacturing process for integrated circuits is done in a clean room, where workers wear special suits and masks. This is necessary because at the scale of the integrated circuits, even the smallest speck of dust is enormous.

>> Each integrated circuit goes through a variety of complicated quality tests after the circuit is finished. The manufacturing process is by no means perfect, so many are discarded.

Integrated Circuit Packages

Integrated circuits come in a variety of different package types, but nearly all the ICs you'll work with in hobby electronics come in a type of package called *dual inline package*, or *DIP*. Figure 1-1 shows several ICs in DIP packages. (One of the ICs is upside down so that you can see its underbelly.)

FIGURE 1-1:
Integrated circuits
in common DIP
packaging.

TIP

Yes, I know that the phrase "DIP package" is redundant because the *P* in *DIP* already stands for *package*, but the phrase "DIP package" is commonly used. Some justify this usage by claiming that the *P* actually stands for *pin*, so there's no redundancy. But that's just easy rationalization. Like it or not, *DIP* stands for *dual inline package*, and the phrase "DIP package" is commonly used and considered correct. Get used to it.

The phrase "DIP chip" is also sometimes used to describe ICs in DIP packages. It has a nice ring to it and sounds like something you would serve at a Super Bowl party.

A DIP package consists of a rectangular plastic or resin case that encloses the IC itself, with two rows of pins on the long sides of the rectangle. The pins on each side jut out a bit from the case, then turn straight down. This arrangement makes the package look like an insect. (In fact, one common way of wiring circuits that use DIP chips is to glue them to a board upside down and solder wires directly to the pins; this technique is called *dead-bug wiring*.)

The pins on each side of a DIP package are spaced exactly 0.1 inch apart, and the two rows of pins are usually spaced 0.3 inch apart, though some larger DIP packages may have wider spacing. In any event, the standard tenth-of-an-inch spacing is perfect for use with solderless breadboards, which have holes spaced at 0.1-inch intervals. In fact, the gap that runs down the center of a solderless breadboard happens to be 0.3 inch, which makes it easy to mount DIP chips such that they straddle the gap, as shown in Figure 1-2.

Introducing Integrated
Circuits

FIGURE 1-2:
Solderless
breadboards are
designed with DIP
chips in mind.

Each pin in a DIP package is numbered. Looking down on the package from above, you'll see an orientation mark, usually a notch, groove, or dot. Orient the package so that this mark is on the top and pin 1 is immediately to the left of the mark. The pins are numbered counterclockwise, working down the left side and then back up the right side until you get to the last pin, which is immediately to the right of the orientation mark.

In case my explanation of DIP pin numbering doesn't make sense, have a look at Figure 1-3.

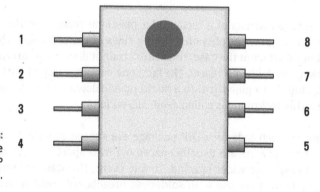

FIGURE 1-3:
Identifying the
pins on a DIP
package.

The DIP package in Figure 1-3 is for an eight-pin DIP. Larger DIPs have more pins, but the numbering scheme is always the same: Pin 1 is to the left of the orientation mark, and the remaining pins are numbered counterclockwise from pin 1.

Using ICs in Schematic Diagrams

In a schematic diagram, an IC is usually represented simply as a rectangle with circuit connections placed conveniently around the rectangle without regard for the physical positioning of the pins. Each pin connection is labeled, as shown in Figure 1-4.

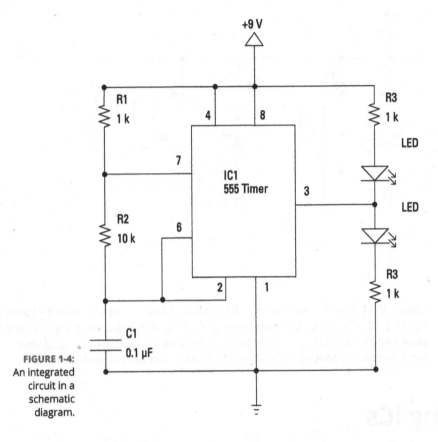

FIGURE 1-4:
An integrated circuit in a schematic diagram.

Notice that the pins in this schematic diagram aren't in the same order as they are in the actual DIP package. Instead, the pins are arranged in the order that best simplifies the schematic diagram. Thus, when you build this circuit, you have to adjust the wiring layout to accommodate the pin arrangement of the DIP package.

Notice also that not all the pins on an Integrated Circuit are always used. Unused pins are usually omitted from the schematic diagram. For example, pin 5 is not used in the circuit shown in Figure 1-4, so it is omitted from the schematic.

Some integrated circuits contain two or more independent circuits that share a common power supply, sort of like conjoined twins. For example, the 556 dual timer chip contains two complete 555 timer circuits within a single 14-pin package. When chips like this are used in a circuit, the schematic diagram may show them separately. For example, Figure 1-5 shows a circuit that calls for a 556 dual timer chip, but each section of the timer is listed separately in the schematic.

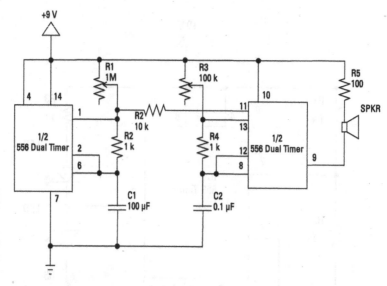

FIGURE 1-5:
Independent
sections within a
single IC are often
shown separately
in a schematic
diagram.

Please don't worry about the details of this circuit (or of the circuit shown in Figure 1-4). My intent here isn't to explain how these circuits work but only to show you how the integrated circuits are depicted in the schematic diagrams. You can learn more about how these circuits work in the next chapter.

Powering ICs

In most DIP integrated circuits, two of the pins are used to provide power to the circuit. One of these is designated for positive voltage, typically identified with the symbol V_{CC}. The other is the ground pin. For example, the 555 timer chip (which you learn about in Chapter 2 of this minibook) requires a positive supply voltage between 4.5–15 V at pin 8, and pin 1 is connected to ground.

In the case of ICs that contain two or more separate circuits, the circuits usually share a common power supply. Thus, even though a 556 dual timer chip contains two separate 555 timer circuits, the chip has just one positive voltage pin and one ground pin.

Also, you should be aware that some integrated circuits call for separate positive and negative supply voltages, not just a positive and a ground connection. You can create a power supply like that using the circuit shown in Figure 1-6.

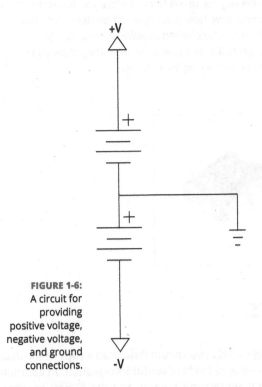

FIGURE 1-6:
A circuit for providing positive voltage, negative voltage, and ground connections.

Avoiding Static and Heat Damage

When you build a circuit board that contains one or more integrated circuits, be careful that you don't damage the IC when you build your circuit. In particular, you should watch out for these two possible problems:

» **Static discharge:** Many integrated circuits can be damaged by static electricity discharged through your fingers when you handle the chips. As a result, make sure that you discharge yourself by touching a grounded metal surface before handling an integrated circuit. You may also want to use an antistatic wristband when handling ICs.

» **Heat damage:** Some integrated circuits are sensitive to heat, so you should take precautions whenever you solder an IC to a circuit board. If possible, attach an alligator clip or other type of heat sink to the pin to help dissipate some of the heat away from the IC itself.

TIP

You can avoid soldering integrated circuits altogether by using DIP sockets, as shown in Figure 1-7. In this figure, the 8-pin IC on the right is matched with an 8-pin DIP socket on the left. When you use a socket like the one in the figure, you solder the pins on the socket to your circuit board before you insert the IC into the socket. That way, any excessive heat is dissipated harmlessly into the empty socket. When the socket is safely soldered in place, you can safely insert the IC into the socket. Sockets are also handy because they allow you to replace a damaged chip without unsoldering the old chip.

FIGURE 1-7: DIP sockets let you avoid soldering delicate integrated circuits.

Reading IC Data Sheets

Before you work with a specific type of IC, you should download a copy of the data sheet for the IC. An IC datasheet contains loads of useful information. In addition to basic information such as the manufacturer's name and the IC part number, you'll find information such as

>> A description of what the circuit does.

>> Detailed pinout descriptions that tell you the purpose of each pin.

>> A diagram of the internal circuitry of the chip. For simple circuits, you may get the entire detailed diagram. For more complicated chips, you'll get a conceptual diagram instead of a detailed schematic.

>> Detailed electrical specifications, such as maximum voltage you can feed the circuit via the V_{CC} pin or the maximum current loads for output pins.

>> Operating conditions such as maximum and minimum temperatures.

>> Charts and graphs that illustrate the circuit's behavior for different operating conditions.

>> Formulas for calculating operating characteristics of the circuit. For example, if the operation of the circuit depends on an external RC (resistor/capacitor) circuit, you'll get formulas for calculating how these external components will affect the operation of the circuit.

>> Sample circuit diagrams.

>> Mechanical descriptions including dimensions.

IC datasheets are available from many sources on the Internet. Use a search engine such as Google to search for the IC part number and the word *datasheet*. For example, to find a datasheet for a 555 timer chip, search for *555 datasheet*.

Popular Integrated Circuits

Literally thousands of different types of integrated circuits are available. Most of these were designed for very specific applications. However, many integrated circuits have been designed for general-purpose use and so are used in a wide variety of circuits.

In the sections that follow, I briefly describe some of the most popular general-purpose integrated circuits. All these ICs have been around for decades, but their multipurpose design, wide availability, and low cost have given them enduring popularity.

555 Timer

The 555 timer chip was invented in 1971 but remains today one of the most popular integrated circuits in use. By some estimates, more than a billion of them are made and sold every year.

As its name implies, the 555 is a timer circuit. The timing interval is controlled by an external resistor/capacitor (RC) network. In other words, by carefully choosing the values for the resistors and capacitors, you can vary the timing duration.

The 555 can be configured in several different ways. In one configuration (called *monostable*), it works like an egg timer: You set it, and then it goes off after a certain period of time has elapsed. In a different configuration (called *astable*), it works like a metronome, triggering pulses at regular intervals.

Besides the basic 555 chip, which comes in an 8-pin DIP package, you can also get a 556 dual timer, which contains two independent 555 timers in a single 14-pin

DIP package. Because many common circuits call for two 555 timers working together, the 556 package is very popular.

You can learn more about the 555 and 556 timers in the next chapter.

741 and LM324 Op-Amp

An *op-amp* is a special type of amplifier circuit that has many applications throughout electronics. Although there are many different types of op-amp circuits, the 741 and LM324 are the most common.

The 741 is a single op-amp circuit in an eight-pin DIP package. It was first introduced in 1968 and is still one of the most widely used integrated circuits ever made. The 741 is one of those ICs that require both positive and negative voltage, as described earlier in the section, "Powering ICs."

The LM324 was introduced in 1972. It consists of four separate op-amp circuits in a single 14-pin DIP package. Unlike the 741, the LM324 doesn't require separate negative and positive voltage supplies.

You can learn more about op-amps in Chapter 3 of this minibook.

78xx Voltage Regulator

The 78xx is a family of simple voltage regulator integrated circuits. A *voltage regulator* is a circuit that accepts an input voltage that can vary within a certain range and produces an output voltage that is a constant value, regardless of fluctuations in the input voltage.

The *xx* in 78xx represents the actual voltage regulated by the chip. For example, a 7805 produces a 5 V output. The input voltage must be at least a couple of volts over the output voltage and can be as high as 35 V.

74xx Logic Family

One of the primary uses for integrated circuits is for digital electronics, and the 74xx is one of the oldest and still most widely used families of digital integrated circuits. The 74xx family includes a wide variety of chips that provide basic building blocks for digital circuits. Thus, you won't find complete microprocessors in the 74xx family. But you will find circuits such as logic gates, flip-flops, counters, buffers, and so on.

Chapter **2**

The Fabulous 555 Timer Chip

The 555 timer chip, developed in 1970, is probably the most popular integrated circuit ever made. By some estimates, more than a billion of them are manufactured every year. Its popularity is well deserved. The 555 is a single-chip version of a commonly used circuit called a *multivibrator*, which is useful in a wide variety of electronic circuits. You can use the 555 chip for basic timing functions, such as turning a light on for a certain length of time, or you can use it to create a warning light that flashes on and off. You can use it to produce musical notes of a particular frequency, or you can use it to control positioning of a servo device. (You learn about servos in Book 6, Chapter 5.) The list goes on and on.

In this chapter, you learn how to use this versatile chip in a variety of circuits. First, I explain how the 555 works and what each of its pins do. Then, you see and build a variety of common 555 circuits.

Looking at How the 555 Works

The 555 can be considered a hybrid of an analog and digital circuit. The output produced by a 555 is purely digital: It's either off (0 V) or it's on (with positive voltage of at least 2.5 V). The timing mechanism within the 555 determines how long the output is on and how long it's off.

The analog part of the circuit lies in how you control how long the output signal is on and how long it's off. You do that by creating an RC network using a resistor and a capacitor. The values you choose for the resistor and capacitor determine the timing interval. I give you all the information you need to choose the correct values for the resistor and capacitor later in this chapter, but if you feel you need a refresher on how RC networks work, see Book 2, Chapter 3.

You can also control whether the 555 does its timing job just once, like an egg timer, or whether it cycles the output on and off repeatedly, like a metronome that keeps ticking over and over again. You do that by connecting the pins of the 555 chip in various ways.

Figure 2-1 shows the arrangement of the eight pins in a standard 555 IC. As you can see, the 555 comes in an 8-pin DIP package.

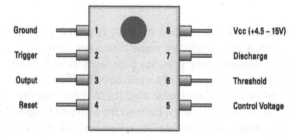

FIGURE 2-1:
Pinout diagram for a 555 timer IC.

The following paragraphs describe the function of each of the eight pins (not in numerical order):

>> **Ground:** Pin 1 is connected to ground.

>> **V_{cc}:** Pin 8 is connected to the positive supply voltage. This voltage must be at least 4.5 V and no greater than 15 V. It's common to run 555 circuits using four AA or AAA batteries, providing 6 V, or a single 9 V battery.

>> **Output:** Pin 3 is the output pin. The output is either low, which is very close to 0 V, or high, which is close to the supply voltage that's placed on pin 8. (On some models of the 555, the output voltage may be as much as 2 V below the supply voltage.) The exact shape of the output — that is, how long it's high and how long it's low, depends on the connections to the remaining five pins.

For more information about using the output from pin 3, see the section "Using the 555 Timer Output" later in this chapter.

>> **Trigger:** Pin 2 is the *trigger*, which works like a starter's pistol to start the 555 timer running. The trigger is an *active low* trigger, which means that the timer starts when voltage on pin 2 drops to below one-third of the supply voltage. When the 555 is triggered via pin 2, the output on pin 3 goes high.

For example, if you want to start the timer by pushing a button, you would connect pin 2 to the supply voltage. Then, to trigger the timer, you simply interrupt this supply voltage. There are several ways to do that, but the most common is to connect a normally open push button between pin 2 and ground. Then, the button is pushed, and the supply voltage is short-circuited to ground (through a current-limiting resistor); so the voltage at pin 2 drops to zero, and the timer is triggered. You see an example of this type of triggering later in this chapter, in the section "Using the 555 in Monostable (One-Shot) Mode."

>> **Discharge:** Pin 7 is called the *discharge*. This pin is used to discharge an external capacitor that works in conjunction with a resistor to control the timing interval. In most circuits, pin 7 is connected to the supply voltage through a resistor and to ground through a capacitor.

>> **Threshold:** Pin 6 is called the *threshold*. The purpose of this pin is to monitor the voltage across the capacitor that's discharged by pin 7. When this voltage reaches two thirds of the supply voltage (Vcc), the timing cycle ends, and the output on pin 3 goes low.

>> **Control:** Pin 5 is the *control* pin. In most 555 circuits, this pin is simply connected to ground, usually through a small 0.01 µF capacitor. (The purpose of the capacitor is to level out any fluctuations in the supply voltage that might affect the operation of the timer.)

In some circuits, a resistor is used between the control pin and Vcc to apply a small voltage to pin 5. This voltage alters the threshold voltage, which in turn changes the timing interval. Most circuits don't use this capability, however. In this chapter, all the 555 circuits simply connect pin 5 to ground through a 0.01 µF capacitor.

>> **Reset:** Pin 4 is the reset pin, which can be used to restart the 555's timing operation. Like the trigger input, reset is an active low input. Thus, pin 4 must be connected to the supply voltage for the 555 timer to operate. If pin 4 is momentarily grounded, the 555 timer's operation is interrupted and won't start again until it's triggered again via pin 2.

TIP

When used in a schematic diagram, the pins of a 555 timer chip are almost always shown in the arrangement depicted in Figure 2-2. As you can see, this arrangement follows the customary order of elements in a schematic diagram: Supply voltage is at the top, ground is at the bottom, inputs are at the left, and outputs are at the right. This arrangement makes it easy to determine the operation of the circuit from the schematic diagram.

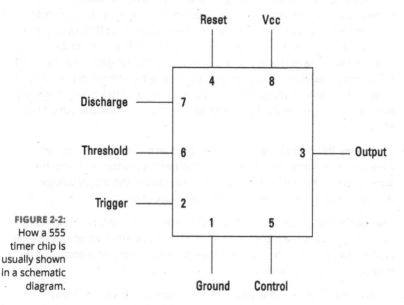

FIGURE 2-2: How a 555 timer chip is usually shown in a schematic diagram.

Understanding 555 Modes

There are three basic ways — called *modes* — that you can wire up a 555 timer IC:

>> **Monostable mode:** Works like an egg timer. When you start it, the timer turns on the output, waits for the time interval to elapse, and then turns the output off and stops.

>> **Astable mode:** Works like a metronome: It keeps running until you turn it off.

>> **Bistable mode:** Isn't really a timer mode. Instead, it uses the trigger input to alternately turn the output on and off. This type of circuit is often called a *flip-flop* and is very commonly used in digital electronics.

The following sections explain how each of these three operating modes work.

Using the 555 in Monostable (One-Shot) Mode

Monostable mode lets you use the 555 timer chip as a single-event timer. This mode is called *monostable* because when wired this way, the 555 has just one stable mode, with the output at pin 3 off. When the 555 is sent a trigger pulse, this stable state is temporarily interrupted for an interval that's determined by the value of a resistor and a capacitor. During this interval, the output at pin 3 goes high, but once the time interval has passed, the 555 returns to its stable state, with pin 3 going low.

Monostable mode is sometimes called *one-shot mode*, which seems a little more descriptive to me. *One-shot mode* conveys the idea that when triggered, the 555 gives one and only one output pulse. When the time interval is reached, the output pulse stops, and the circuit goes quiet until another trigger pulse is detected. Each trigger pulse results in a single output pulse.

Looking at a typical 555 monostable circuit

Figure 2-3 shows the typical wiring for a 555 timer used in monostable mode.

To understand how this circuit works, first look at the way the 10 kΩ resistor and the switch are wired to pin 2, the trigger input. The switch is a normally open push button. When the button isn't depressed, the 10 kΩ resistor provides a voltage input to pin 2, which keeps the trigger input high. With the trigger input high, the output voltage at pin 3 is near zero.

When the push-button switch is depressed, the supply voltage is short-circuited to ground. This causes the voltage at pin 2 to drop to zero, and the timer is triggered. Once the timer is triggered, the output voltage at pin 3 goes high and the timing interval begins.

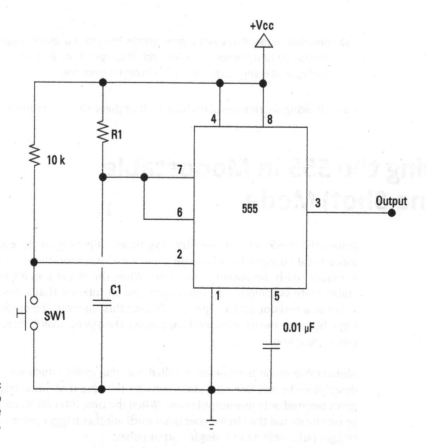

+Vcc

4   8

R1

10 k

7

555   3   Output

6

2

C1

1   5

SW1

0.01 μF

FIGURE 2-3:
A 555 timer chip
in monostable
mode.

Looking at the resistor-capacitor circuit in a monostable timer

Now that you understand how the trigger circuit works, look at how the RC circuit (R1 and C1) works. The resistor and capacitor work together to determine how long the output will remain high. In a nutshell, once the circuit is triggered, C1 begins to charge.

Pins 6 and 7 — the threshold and discharge pins — are tied together in a monostable 555 circuit. Pin 6 watches the voltage across the capacitor. As the capacitor charges, this voltage increases. When the capacitor voltage reaches two-thirds of the Vcc supply voltage, the timing cycle ends, and the output at pin 3 goes low.

The discharge pin (pin 7) charges and discharges the capacitor. To understand how pin 7 works, it may be helpful to visualize the internal workings of pin 7 using the model shown in Figure 2-4. Here, pin 7 is connected to a switch that's controlled by the status of the output at pin 3. When the output is high, the switch is open; when the output is low, the switch is closed. When the switch is closed, a small 10 kΩ resistor within the 555 connects pin 7 to ground. (This isn't really a switch or a 10 kΩ resistor inside the 555, but this model may help you understand how pin 7 works.)

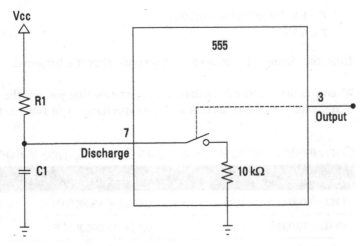

FIGURE 2-4:
The imaginary switch inside the 555 that controls whether pin 7 charges or discharges the capacitor.

When the output on pin 3 is low, the imaginary switch inside the 555 is closed, and pin 7 is connected to ground through the 10 kΩ resistor. This allows the voltage on C1 to discharge through the 555.

But when the output on pin 3 goes high, the imaginary switch inside the 555 is opened. This forces the current flowing through R1 to go through C1, which in turn causes the capacitor to charge at a rate that depends on the values of R1 and the capacitor.

While the capacitor is charging, pin 6 monitors the voltage that builds up across the capacitor. Once this voltage reaches two-thirds of the supply voltage, pin 6 signals the 555 that the timing interval is ended, and the output goes low. This, in turn, closes the imaginary switch inside the 555, which allows the capacitor to discharge.

Calculating the time interval for a monostable circuit

The time interval for a 555 monostable circuit is a measure of how long the output stays high when it's triggered. To calculate the time interval, just use this formula:

$$T = 1.1 \times R \times C$$

Here, T is the time interval in seconds, R is the resistance of R1 in ohms, and C is the capacitance of C1 in farads.

For example, suppose R1 is 500 kΩ, and C1 is 10 μF. Then, you would calculate the time interval like this:

$$T = 1.1 \times 500,000 \ \Omega \ \times \ 0.00001 \ F$$
$$T = 5.5 \ s$$

Thus, the circuit will stay on for 5.5 seconds after it's triggered.

When you make this calculation, it's imperative that you use the correct number of zeros for both the resistance and the capacitance. Use Table 2-1 as a guide:

TABLE 2-1   ## Conversion of Resistance and Capacitance Values

Resistance	Capacitance
1 kΩ = 1,000 Ω	0.01 μF = 0.00000001 F
10 kΩ = 10,000 Ω	0.1 μF = 0.0000001 F
100 kΩ = 100,000 Ω	1 μF = 0.000001 F
1 MΩ = 1,000,000 Ω	10 μF = 0.00001 F
10 MΩ = 10,000,000 Ω	100 μF = 0.0001 F
100 MΩ = 100,000,000 Ω	1,000 μF = 0.001 F

TIP

If you don't want to do these calculations yourself, you can find 555 timer calculators on the web. Just go to Google or any other search engine and search *555 timer calculator*.

Using the 555 in Astable (Oscillator) Mode

Another common way to use a 555 timer is in *astable mode*. The term *astable* simply means that the 555 has no stable state: Just as it gets settled into one state (say, the output at pin 3 high), it switches to the opposite state (output low). Then it switches back to the first state, and so on, ad infinitum. This mode is also called *oscillator mode*, because it uses the 555 as an oscillator, which creates a square wave signal.

Looking at a typical astable circuit

Figure 2-5 shows the basic circuit for a 555 in astable mode.

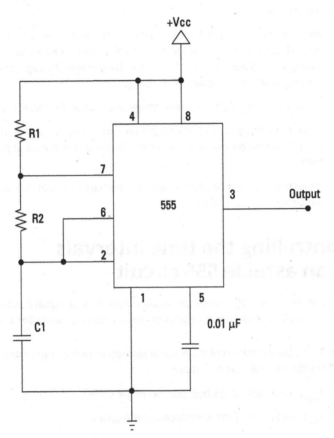

FIGURE 2-5:
A 555 timer chip
in astable mode.

To understand how this circuit works, first notice that the trigger pin (pin 2) is connected directly to C1. In the monostable circuit, the timer was triggered by a switch that short-circuited the voltage applied to pin 2. In the astable circuit, the timer is triggered when the capacitor discharges — once the voltage across the capacitor drops to one-third of the supply voltage, pin 2 triggers the timer to start another cycle.

Here I examine how this timing cycle works, step by step, starting with the output at pin 3 in the high condition:

1. With the output high, the discharge pin (pin 7) is open, forcing current through resistors R2 and C1. This causes the capacitor to charge at a rate that depends on the combined value of R1 and R2 and the value of C1.

2. As the capacitor charges, the voltage at pins 2 and 6 increases.

3. When the voltage at pin 6 (the threshold pin) reaches two-thirds of the supply voltage, the threshold circuitry within the 555 causes the output voltage at pin 3 to go low.

4. When the output at pin 3 goes low, the discharge pin (pin 7) is connected to ground within the 555. This allows C1 to discharge. This discharge occurs through R2, so the value of R2 as well as the value of the capacitor determines the rate at which the capacitor discharges.

5. As the capacitor discharges, the voltage at pins 2 and 6 decreases.

6. When the voltage at pin 2 (the trigger pin) drops to one third of the supply voltage, the trigger circuitry inside the 555 causes the output at pin 3 to go high.

7. When the output at pin 3 goes high, the discharge pin (pin 7) is opened, and the cycle starts over again.

Controlling the time intervals in an astable 555 circuit

The output of a 555 circuit in astable mode is a square wave, as shown in Figure 2-6. There are three important time measurements for a square wave:

>> **T:** The total duration of the wave measured from the start of one high pulse to the start of the next high pulse

>> **T_{high}:** The length of the high portion of the cycle

>> **T_{low}:** The length of the low portion of the cycle

Naturally, the total time T is the sum of T_{high} and T_{low}.

The values of these time constants depend on the values for the two resistors (R1 and R2) and the C1.

Here are the formulas for calculating each of these time constants in seconds:

$$T = 0.7 \times (R1 + 2R2) \times C1$$
$$T_{high} = 0.7 \times (R1 + R2) \times C1$$
$$T_{low} = 0.7 \times R2 \times C1$$

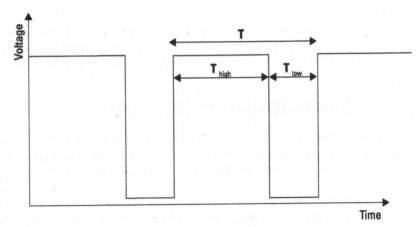

FIGURE 2-6:
Timing the output
wave created by
an astable 555
timer circuit.

There's an interesting fact buried in these calculations that you need to be aware of: C1 charges through both R1 and R2, but it discharges only through R2. That's why you must add the two resistor values for the T_{high} calculation, but you use only R2 for the T_{low} calculation. It's also why you must double R2 but not R1 for the total time (T) calculation.

Now, plug in some real numbers to see how the equations work out. Suppose both resistors are 100 kΩ and the capacitor is 10 μF. Then, the total length of the cycle is calculated like this:

$$T = 0.7 \times (100,000 + 2 \times 100,000) \times 0.00001$$
$$T = 2.1 \text{ seconds}$$

$$T_{high} = 0.7 \times (100,000 + 100,000) \times 0.00001$$
$$T_{high} = 1.4 \text{ seconds}$$

$$T_{low} = 0.7 \times 100,000 \times 0.00001$$
$$T_{low} = 0.7 \text{ seconds}$$

Thus, the total cycle time will be 2.1 seconds, with the output high for 1.4 s and low for 0.7 s.

If you want, you can also calculate the frequency of the output signal by dividing the total cycle time into 1. So, for the above calculations, the frequency is 0.47619 Hz.

If you use smaller resistor and capacitor values, you'll get shorter pulses and higher output frequencies. For example, if you use 1 kΩ resistors and a 0.1 μF capacitor, the output signal will be 48 kHz, and each cycle will last just a few millionths of a second.

You may have also noticed that if the two resistors are the same value, the signal will be high for twice as long as it's off. By using different resistor values, you can vary the difference between the high and low portions of the signal.

Calculating the duty cycle

The *duty cycle* in a 555 circuit is the percentage of time that the output is high for each cycle of the square wave. For example, if the total cycle time is 1 second and the output is high for the first 0.4 second of each cycle, the duty cycle is 40%.

With an astable circuit such as the one shown in Figure 2-5, the duty cycle must always be greater than 50%. In other words, the duration for which the output is high must always be more than the duration during which the output is low.

The explanation for this is pretty simple: For the duty cycle to be 50%, the capacitor would have to charge and discharge through the same resistance. The only way to accomplish that would be to omit R1 altogether, so that the capacitor charged and discharged through R2 only. But the problem with that is that you would end up connecting pin 7 directly to Vcc. With no resistance between pin 7 and the voltage source, the current flowing through pin 7 would exceed the maximum that can be handled by the circuitry inside the 555, and the chip would be damaged.

There's a clever way around this limitation: Place a diode across R2, as shown in Figure 2-7. This diode bypasses R2 when the capacitor is charged. That way, the capacitor charges through R1 and discharges through R2.

When a diode is used in this way, you have complete control over the duration of both the charge and discharge time. If R1 and R2 have the same value, the capacitor takes the same amount of time to charge as it does to discharge, so the duty cycle will be 50%. If R2 is smaller than R1, the duty cycle is less than 50% because the capacitor discharges faster than it charges.

If you use the diode as depicted in Figure 2-7, you must adjust the formulas for calculating the time intervals as follows:

$$T = 0.7 \times (R1 + R2) \times C1$$
$$T_{high} = 0.7 \times R1 \times C1$$
$$T_{low} = 0.7 \times R2 \times C1$$

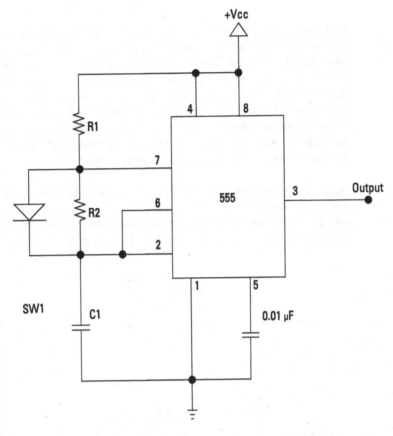

FIGURE 2-7:
Using a diode to separately control the high and low part of the output signal.

Using the 555 in Bistable (Flip-Flop) Mode

A *flip-flop* is a circuit that alternates between two output states. In a flip-flop, a short pulse on the trigger causes the output to go high and stay high, even after the trigger pulse ends. The output stays high until a reset pulse is received, at which time the output goes low.

This type of circuit is called *bistable* because the circuit has two stable states: high and low. The circuit stays low until it's triggered. Then, it stays high until it's reset. This type of circuit is used extensively in computers and other digital circuits.

TIP

For computer applications, the 555 is a poor choice for use as a flip-flop. That's because its output doesn't change fast enough in response to trigger or reset pulses in computer circuits that are driven by high-speed clock pulses. For computer applications, better flip-flop chips are readily available, as you learn in Book 5, Chapter 4.

That being said, the 555 is often used in bistable mode for noncomputer applications where high-speed response isn't necessary. For example, imagine a simple robot that drives itself forward until it bumps into something in front of it, and then drives backward until it bumps into something behind it. The robot would be equipped with contact switches on the front and rear connected to the trigger and reset inputs of a 555 in bistable mode. The robot's drive motor would be connected to the output such that when the output is low, the motor runs forward, and when the output is high, the motor runs backward. Then, the bistable 555 would cause the robot to drive back and forth between two obstacles.

Figure 2-8 shows the schematic for a 555 used in bistable mode. As you can see, this circuit doesn't require a capacitor. That's because in bistable mode, the 555 isn't used as a timer. The highs and lows of the output signal are controlled by the trigger and reset inputs, not by the charging and discharging of a capacitor.

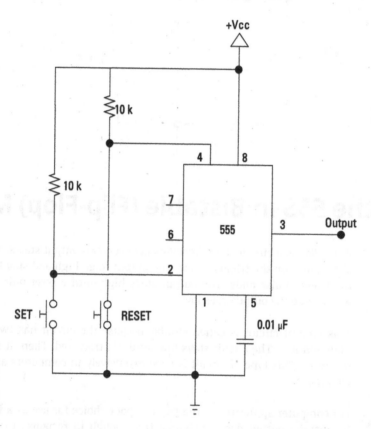

FIGURE 2-8:
The schematic for a 555 timer circuit in bistable mode.

Both the trigger (pin 2) and the reset (pin 4) inputs are connected to Vcc through a 10 kΩ resistor. When the set switch is depressed, pin 2 is shorted to ground. This causes the voltage to bypass pin 2, resulting in a momentary low pulse, which triggers the 555. Once triggered, the output pin goes high.

In astable or monostable mode, the output pin would remain high until the voltage at the threshold pin (pin 6) reaches two-thirds of the supply voltage. However, because pin 6 isn't connected to anything in this circuit, no voltage is ever present on pin 6. Thus, the threshold is never reached, so the output remains high indefinitely until the 555 is reset by a low pulse on the reset pin (pin 4).

The reset input (pin 4) is connected to Vcc in the same manner as the trigger input. As a result, when the reset switch is pressed, pin 4 is short circuited to ground, creating a low pulse, which resets the 555 and brings the output back to low.

Using the 555 Timer Output

The output pin (pin 3) of a 555 can be in one of two states: high or low. In the high state, the voltage at the pin is close to the supply voltage. The low state is 0 V.

There are two ways you can connect output components to the output pin. Figure 2-9 illustrates these two configurations, using an LED as the output device. A resistor is also included in the circuit to limit the current flow. Without the resistor, current will flow through the circuit unimpeded, which will quickly burn out the LED and probably ruin the 555 as well.

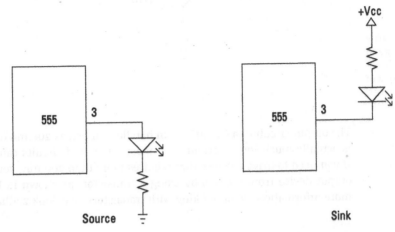

FIGURE 2-9:
Sourcing and
sinking output
current for a 555
timer.

In the circuit on the left, current flows through the LED circuit when the output is high. The current flows from the output pin through the LED and resistor to ground. This output configuration is called *sourcing* because the 555 is the source of the current that drives the output.

In the circuit on the right, current flows through the LED circuit when the output is low. The current flows from the Vcc supply, through the LED and resistor, and into the 555 where it's internally routed to ground through pin 1. This output configuration is called *sinking* because the current is sent into the 555.

Whether you source or sink your output circuit depends on whether you want your output circuit to turn on when the output is high or low.

Figure 2-10 shows that you can combine both sourcing and sinking in a single circuit. Here, two LEDs are connected to the output pin. One is sourced; the other is sunk. In this circuit, the LEDs alternately flash as the output switches from high to low. LED1 lights when the output is low, LED 2 when the output is high.

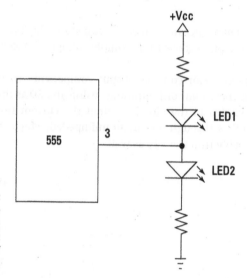

FIGURE 2-10:
You can combine
sourcing and
sinking current
for the output of
a 555 timer.

The output circuit of a 555 timer can handle as much as 200 mA of current, which is actually much more current than most integrated circuits can source or sink. If you need to drive a device that requires more than 200 mA, you can isolate the output device from the 555 by using a transistor, as shown in Figure 2-11. For more information about working with transistors, see Book 2, Chapter 6.

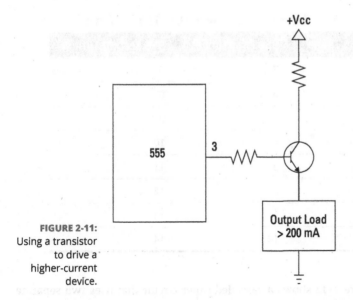

+Vcc

555   3

Output Load
> 200 mA

FIGURE 2-11:
Using a transistor
to drive a
higher-current
device.

Doubling Up with the 556 Dual Timer

If one 555 timer chip is good, two are even better! In fact, it turns out that there are many uses for two (or more) 555 timers in a single circuit — useful enough that you can get two 555 timers in a single chip, called the 556 dual-timer chip.

The 556 dual-timer chip comes in a 14-pin DIP package. The two 555 timers share a common supply and ground pin. The remaining 12 pins are allocated to the inputs and outputs of the individual 555 timers. Table 2-2 lists the pin connections for each of the 555 timers in a 556 dual-timer chip. As an added bonus (no charge!), I also list the pinouts for a standard 555 timer chip in Table 2-2.

One common way to use a 556 dual timer is to connect both 555 circuits in mono-stable (one-shot) mode, with the output pin from the first 555 timer connected to the trigger pin of the second 555 timer. Then, when the output of the first timer goes low, the second timer is triggered. You can connect as many 555 timers as you want in this way, with each timer's output connected to the next timer's trigger so that the timers work in sequence, one after the other.

TABLE 2-2

Pinouts for the 555 Timer and 556 Dual-Timer Chips

Function	555 Timer	556 First Timer	556 Second Timer
Ground	1	7	7
Trigger	2	6	8
Output	3	5	9
Reset	4	4	10
Control	5	3	11
Threshold	6	2	12
Discharge	7	1	13
Vcc	8	14	14

For example, Figure 2-12 shows a cascaded timer circuit that uses two separate 555 timer chips. In this circuit, both of the 555 timer chips are configured in monostable mode, much like the circuit in Figure 2-3. The time interval for the first 555 is controlled by R1 and C1. For the second 555, the interval is controlled by R2 and C2. You can choose whatever values you want for these components to achieve whatever time intervals suit your fancy.

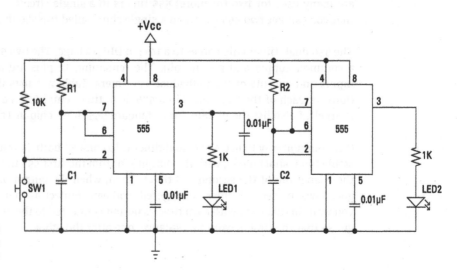

FIGURE 2-12:
555 timers can be
cascaded.

The first 555 chip is triggered when SW1 is depressed, taking pin 2 to ground. This takes the output on pin 3 high, which lights LED1. Notice, however, that pin 3 of the first 555 is connected through a small capacitor to the trigger input of the second 555. As soon as the time interval expires on the first 555, its output goes low, which turns off LED1 and at the same time triggers the second 555, which in turn lights up LED2. LED2 stays lit until C2 charges, and then it goes out. The circuit then waits to be triggered again by a press of the switch.

Figure 2-13 shows how this same circuit can be implemented using a single 556 dual-timer chip. This schematic is nearly identical to the schematic shown in Figure 2-13 but with a few important differences:

>> The two 555 timer circuits are designated as *556 (1)* and *556 (2)* to indicate that these timer circuits are part of a single 556 dual-timer chip.

>> The pin numbers indicate the pin assignments for the two timer circuits of a 556 instead of the pin assignments for a 555.

>> The second timer circuit doesn't show a supply or ground connection. That's because the two timer circuits share a common supply and ground connection, which is shown connected to the first timer.

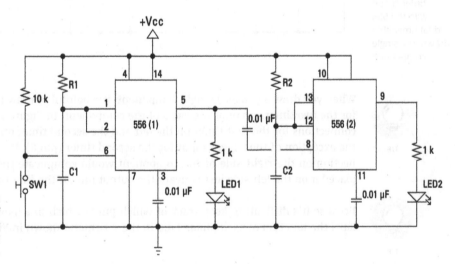

FIGURE 2-13:
The two halves of a 556 dual-timer circuit can be cascaded.

Although it's convenient to show the two halves of a 556 dual timer as separate components in a schematic diagram, you can show the 556 as a single component if you want. Figure 2-14 shows how the schematic for the cascaded timer circuit can be drawn using a single component for the 556 dual timer. This circuit is nearly identical to the circuit shown in Figure 2-13; the only difference is the way the schematic depicts the two sections of the 556 dual-timer chip.

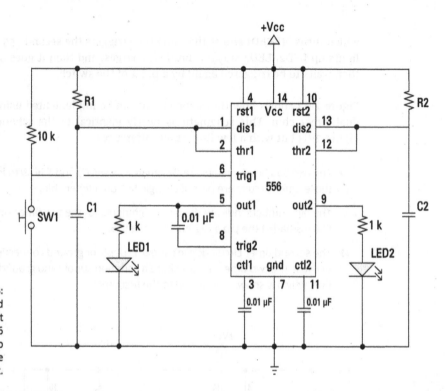

FIGURE 2-14:
The cascaded
timer circuit
with the 556
dual-timer chip
drawn as a single
component.

TIP

When you draw a 556 as a single component, it's helpful to draw the connections for the two timers on opposite sides of the component. In Figure 2-14, I drew the connections for the first timer on the left and the second timer on the right, with the exception of the trigger input for the second timer (pin 8). Drawing that connection on the right side of the component would complicate the diagram, so I placed it on the left side just beneath the output pin for the first timer.

TIP

Because it's difficult to keep track of which pin is which in a 556, it's helpful to label the function of each pin as I did in the schematic shown in Figure 2-14.

Making a One-Shot Timer

In this section, you build a circuit that uses a 555 timer chip in monostable mode. When a trigger switch is pressed, an LED lights and stays lit for approximately 5 seconds. Then, the LED goes dark until the button is pressed again.

Project 21 provides all the information you need to assemble this circuit. The circuit is based on the schematic for the monostable circuit that was shown in

Figure 2-3. The only differences are that an LED is added to the output pin (pin 3) and the resistor, and capacitor values are included in the capacitor charging circuit.

When you are finished, the circuit will look like Figure 2-15. To test the circuit, push the push button. The LED should light, stay lit for just over 5 seconds, and then go back off. It should light again only when you press the push button again.

FIGURE 2-15:
The finished one-shot timer project (Project 21).

If the circuit doesn't work, here are a few things to check:

>> Make sure that the battery is good. (Use a voltmeter to test it.)

>> Carefully double-check all the jumper wires and other components to make sure they're connected properly.

>> Make sure the LED isn't inserted backwards. As a test, pull it out and insert it with the leads reversed.

>> Make sure the electrolytic capacitor is inserted with the negative end on ground side of the circuit.

>> Make sure that the solder connections to the push button are solid.

Project 21: A One-Shot 555 Timer Circuit

In this project, you build a circuit that uses a 555 timer chip in monostable mode to create a one-shot timer. The circuit includes a push button and an LED. When you press the push button, the LED will turn on and stay on for about 5 seconds. Then the LED will turn off and stay off until you press the push button again.

The only tools you'll need to complete this project are wire cutters and wire strippers.

Parts

>> One 9 V battery

>> One 9 V battery snap holder

>> One small solderless breadboard

>> One normally open push-button switch

>> One 555 timer chip

>> One 5 mm red LED

>> One 10 Ω resistor (brown-black-orange)

>> One 470 Ω resistor (yellow-violet-brown)

>> One 470 kΩ resistor (yellow-violet-yellow)

>> One 10 µF electrolytic capacitor

>> One 0.01 µF ceramic disk capacitor

>> One normally open DIP breadboard push button

>> Eight jumper wires (various lengths)

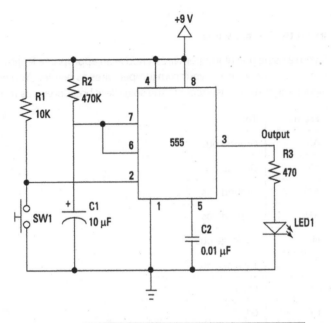

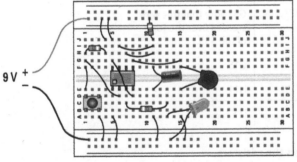

Steps

Throughout these steps, use the negative (−) bus strip at the bottom of the board for ground and the positive (+) bus strip on the top of the board as the positive voltage.

1. **Insert the 555 timer chip.**

   Insert the chip so that it straddles the gap in the center of the solderless breadboard, with pin 1 in hole E5 and pin 8 in hole F5.

2. Insert the jumper wires.

If you're using precut jumper wires, choose an appropriate length for each segment. Otherwise, cut your own jumper wires as needed. You need a total of eight jumper wires, inserted into the solderless breadboard as follows:

From	To
A3	Ground bus
A5	Ground bus
A11	Ground bus
A15	Ground bus
J4	+ voltage bus
J5	+ voltage bus
E1	F1
D6	G1
D8	G5 (Note that this wire crosses over the top of the 555.)
G8	G15
H7	H11
I6	I11

3. Insert the resistors.

Insert the three resistors as follows:

Resistor	From	To
10 kΩ	H1	H4
470 kΩ	J11	+ voltage bus
470 Ω	C7	C13

4. **Insert the capacitors.**

   Insert the two capacitors as follows:

Capacitor	From	To
10 µF	E11	F11
0.01 µF	E15	F15

   Make sure that the negative lead of the electrolytic capacitor is in hole E11.

5. **Insert the LED.**

   The anode (the long lead) should be inserted in hole A13, and the cathode (the short lead) in the ground bus.

6. **Insert the push button.**

   Insert the push button in holes B1, D1, B3, and D3. (Ensure that the switch is oriented such that it closes from B1/D1 to B3/D3.)

7. **Connect the battery.**

   Plug the 9 V battery into the battery snap connector, and then connect the red lead to the positive voltage bus and the black lead to the ground bus.

   You're finished! You can now press the button to turn on the LED. The LED will go dark again after about 5 seconds.

Making an LED Flasher

In this section, you build a circuit that uses a 555 timer chip to alternately flash two LEDs on and off. For this circuit, the 555 is configured in 555 in astable mode, and the resistor values are chosen so that they cause the high and low timings to be very close to one another, about one-tenth of a second each.

Project 22 shows you how to build the LED flasher circuit. The schematic for this project is similar to the one that was shown in Figure 2-5, but it adds a pair of LEDs to the output circuit. LED1 lights when the output is low, and LED2 lights when the output is high. Note that this circuit uses both sinking and sourcing of the output current.

Figure 2-16 shows the completed project.

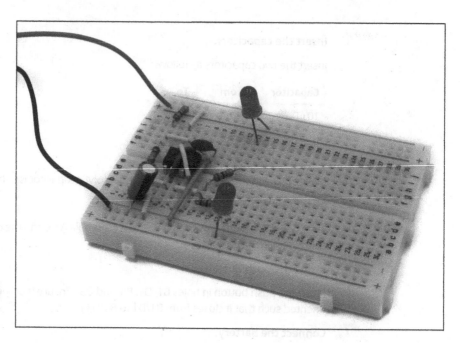

FIGURE 2-16:
The finished
LED flasher
(Project 22).

If the circuit doesn't work, here are a few things to check:

>> Check the battery voltage.

>> Double-check all the jumper wires and resistors to make sure they're inserted in the proper holes.

>> Make sure the diodes and the electrolytic capacitor are inserted correctly. For the LEDs, the anodes must be on the positive side of the circuit and the cathodes on the negative side. For the capacitor, the negative side must be in the ground bus.

TIP

After you finish playing with the completed project, you may want to leave it assembled. Later in this chapter, in Project 23, you'll add a second 555 timer to this circuit to include additional functionality.

Project 22: An LED Flasher

In this project, you build a circuit that alternately flashes a pair of LEDs. The circuit uses a 555 timer configured in astable mode to flash the LEDs. The resistor and capacitor values are chosen so that the duty cycle is close to 50 percent and each LED stays on for about one-tenth of a second.

The only tools you'll need to complete this project are wire cutters and wire strippers.

Parts

>> One 9 V battery

>> One 9 V battery snap holder

>> One small solderless breadboard

>> One 555 timer chip

>> Two 5 mm red LEDs

>> One 10 kΩ resistor (brown-black-orange)

>> Three 1 kΩ resistors (brown-black-red)

>> One 22 µF electrolytic capacitor

>> One 0.01 µF ceramic disk capacitor

>> Eight jumper wires (various lengths)

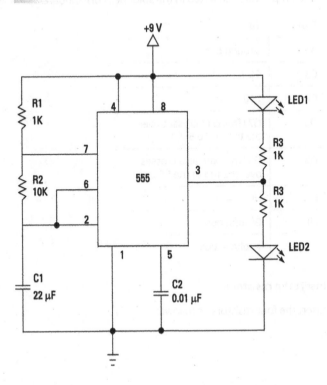

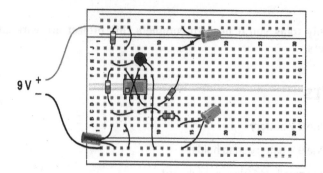

9 V

Steps

1. **Insert the 555 timer chip.**

Insert the chip so that it straddles the gap in the center of the solderless breadboard, with pin 1 in hole E5 and pin 8 in hole F5.

2. **Insert the jumper wires.**

If you're using precut jumper wires, choose an appropriate length for each segment. Otherwise, cut your own jumper wires as needed. You need a total of eight jumper wires, inserted in the solderless breadboard as follows:

From	To
A5	Ground bus
C3	C6
C7	C10
D6	G7 (This wire crosses over the top of the 555.)
D8	G5 (This wire also crosses over the top of the 555.)
H3	H6
F9	Ground bus
J5	+ voltage bus

3. **Insert the resistors.**

Insert the four resistors as follows:

Resistor	From	To
10 kΩ	D3	G3
1 kΩ	J3	+ voltage bus
1 kΩ	B10	B13
1 kΩ	D10	F13

4. **Insert the capacitors.**

Insert the two capacitors as follows:

Capacitor	From	To
0.01 µF	G8	G9
22 µF	A3	Ground bus

Make sure that the negative lead of the electrolytic capacitor is in the ground bus.

5. **Insert the LEDs.**

Insert the two LEDs as follows:

LED	Cathode (Short Lead)	Anode (Long Lead)
LED1	J13	+ voltage bus
LED2	Ground bus	A13

6. **Connect the battery.**

Plug the 9 V battery into the battery snap connector, and then connect the red lead to the positive voltage bus and the black lead to the ground bus. The LEDs should begin flashing immediately when the battery is connected.

You're finished!

After you finish this project, leave it assembled if you intend to build Project 23.

Using a Set/Reset Switch

In this section, you modify the circuit that you built in Project 22 so that the circuit is controlled by two push buttons that function as a set/reset switch. When you connect the power to this circuit, LED1 turns on and stays on. When you press the set push button, the two LEDs start flashing alternately and continue to flash until you press the reset push button.

The circuit for this project uses two 555 timer chips. The first is configured in bistable mode with two push buttons acting as set and reset switches. The second is configured in astable mode, almost identical to the 555 timer that was used in Project 22. The difference is that instead of connecting the astable 555 Timer chip's supply voltage pin (pin 8) directly to the battery, it is connected to the output of the first 555. Thus, the first 555 controls the power to the second 555, so the second 555 flashes the LEDs only when the output of the first 555 is high.

Project 23 shows you how to build this circuit, and the completed project is pictured in Figure 2-17.

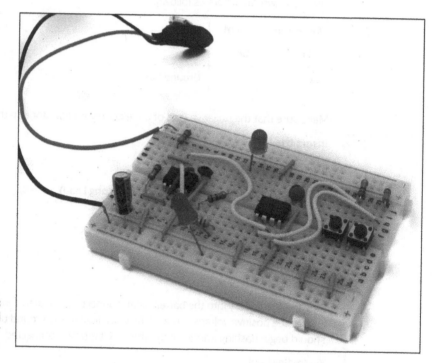

FIGURE 2-17:
The completed circuit for Project 23.

Project 23: An LED Flasher with a Set/Reset Switch

In this project, you expand the circuit you built in Project 22 to add a set/reset switch that controls the flashing of the LEDs. You should build Project 22 before you attempt this project.

The only tools you'll need to complete this project are wire cutters and wire strippers.

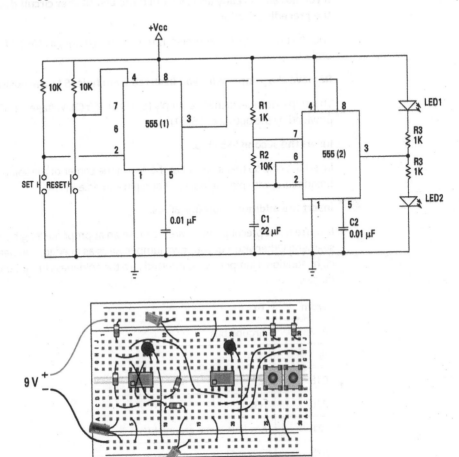

Parts

You will need all the parts from Project 22, plus the following:

>> One 555 timer chip (in addition to the one used in Project 22)

>> Two 10 kΩ resistors (brown-black-orange)

>> One 0.01 µF ceramic disk capacitor

>> One normally open DIP push button

>> Eight jumper wires (various lengths)

Steps

1. **If you haven't already done so, build the LED flasher circuit described in the preceding section.**

   The 555 timer chip in that project is the one designated as *555 (2)* for this project.

2. **Remove the jumper wire you inserted from hole J5 to the positive bus.**

   This step disconnects the Vcc supply (pin 8) of 555 (2). Voltage for 555 (2) is provided by the output of 555 (1).

3. **Insert the second 555 chip.**

   Insert the chip so that it straddles the gap in the center of the solderless breadboard, with pin 1 in hole E17 and pin 8 in hole F17.

4. **Insert the additional jumper wires.**

   If you're using precut jumper wires, choose an appropriate length for each segment. Otherwise, cut your own jumper wires as needed. You need a total of eight additional jumper wires, inserted into the solderless breadboard as follows:

From	To
A17	Ground bus
I5	D19
C18	I27
B20	H30
F21	Any hole in the ground bus
A25	Ground bus
A28	Ground bus

5. **Insert the resistors.**

   Insert the two resistors as follows:

Resistor	From	To
10 kΩ	J27	+ voltage bus
10 kΩ	J30	+ voltage bus

6. **Insert the capacitor.**

The 0.01 µF capacitor should be inserted in holes G20 and G21.

7. **Insert the push buttons.**

Insert the two push-button switches so that they straddle the gap, as follows:

Button	Location
Set	E25, F25, E27, F27
Reset	E28, F28, E30, F30

8. **Connect the battery.**

Plug the 9 V battery into the battery snap connector, and then connect the red lead to the positive voltage bus and the black lead to the ground bus. LED1 should immediately light up. (If it doesn't, double-check all your connections and make sure that the battery isn't dead.)

You're done!

9. **Press the set button to start the LEDs flashing. Let them flash for a while, and then press the reset button to make the flashing stop.**

Making a Beeper

In this section, you use two 555 timer chips to build an audible beeper, with both timers configured in astable mode. One timer generates an audible square-wave tone that is sent to a speaker so the tone can be heard. The other timer generates a much slower frequency that is connected to the reset of the first timer to turn the tone on and off, which creates the beeping effect.

Project 24 shows how to build this circuit. Before you start, have a look for a moment at the project's schematic diagram. For the first timer chip — designated 555 (1) in the schematic — the RC network uses resistors of 1 kΩ and 470 kΩ along with a 1 µF capacitor to produce the 1.5Hz output. The second timer — 555 (2) — uses 4.7 kΩ and a 15 kµ resistors and a 0.01 µF capacitor to create the output tone. The output of the first timer is sent to the reset of the second timer, and the output of the second timer is sent through a 22 µF capacitor to an 8Ω speaker.

Note: You could easily build this circuit using a single 556 dual-timer chip. You'd have to adjust the pin designations on the schematic accordingly.

TIP

The speaker used in this circuit can be any 8Ω speaker. If you have an old unamplified computer speaker, you can use it. I recommend you solder 2- to 3-inch lengths of 20-gauge solid wire to the speaker terminals so you can easily connect the speaker to the breadboard.

Figure 2-18 shows the completed project.

FIGURE 2-18:
The completed
beeper project
(Project 24).

Here are some extra-credit assignments for this project, in case you want to experiment with the circuit a bit:

>> Try building the circuit with a single 556 chip instead of two 555 chips.

>> Replace the 4.7 kΩ R3 with a 1 kΩ resistor, and then add a 1 MΩ potentiometer in series with the resistor. As you turn the potentiometer, the tone changes.

>> Add a 1 MΩ potentiometer in series with R1. Then, as you turn the potentiometer, the beeping rate changes.

Project 24: An Audible Beeper

In this project, you use a pair of 555 timer ICs to build a circuit that produces an audible beep on a small speaker. Both of the 555 timer ICs are configured in astable mode. The first 555 provides the beeping interval, which is about one beep every one-tenth of a second. The second 555 generates an audible frequency that's fed to the speaker to produce the output tone.

To build this project, you'll need wire cutters and wire strippers.

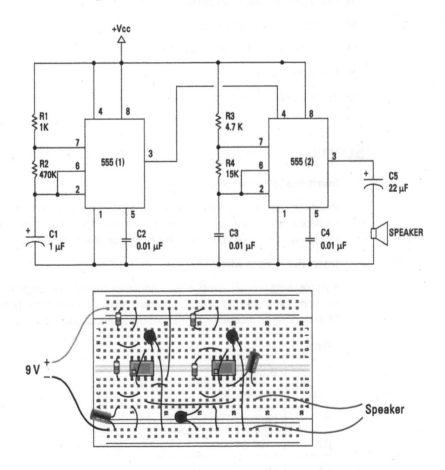

Parts

» One 9 V battery

» One 9 V battery snap holder (RadioShack 2700325)

» One small solderless breadboard (RadioShack 2760003)

» Two 555 timer chips (RadioShack 2761718)

» One 1 kΩ resistor (brown-black-red)

» One 470 kΩ resistor (yellow-violet-yellow)

» One 4.7 kΩ resistor (yellow-violet-red)

» One 15 kΩ resistor (brown-green-orange)

» One 1 μF electrolytic capacitor

» One 22 μF electrolytic capacitor

» Three 0.01 μF ceramic disk capacitors

» One 8-ohm speaker

» 15 jumper wires (various lengths)

Steps

1. **Insert the 555 timer chips.**

   Insert the chips so that they straddle the gap in the center of the solderless breadboard. For the first chip, insert pin 1 in hole E5 and pin 8 in hole F5. For the second chip, insert pin 1 in hole E17 and pin 8 in hole F17.

2. **Insert the jumper wires.**

   If you're using precut jumper wires, choose an appropriate length for each segment. Otherwise, cut your own jumper wires as needed. You need a total of 15 jumper wires, inserted into the solderless breadboard as follows:

From	To
A5	Ground bus
A17	Ground bus
J5	Positive (+) voltage bus
J17	+ voltage bus
C3	C6
H3	H6
C15	C18
G15	G18

From	To
F9	Ground bus
E10	Positive (+) voltage bus
D8	D10
D6	G7 (This wire crosses over the top of the first 555 chip.)
D18	G19 (This wire crosses over the top of the second 555 chip.)
F21	Ground bus

3. **Insert the resistors.**

Insert the four resistors as follows:

Resistor	From	To
R1 – 1 kΩ	J3	Positive (+) voltage bus
R2 – 470 kΩ	E3	F3
R3 – 4.7 kΩ	J15	Positive (+) voltage bus
R4 – 15 kΩ	E15	F15

4. **Insert the capacitors.**

Insert the five capacitors as follows:

Designation	Capacitor	From	To
C1	1 µF Electrolytic	A3	Ground bus
C2	0.01 µF Ceramic Disk	G8	G9
C3	0.01 µF Ceramic Disk	A15	Ground bus
C4	0.01 µF Ceramic Disk	G20	G21
C5	22 µF Electrolytic	D19	D23

Note that for C1, the negative lead should be inserted into the ground bus. For C5, the negative lead should be in D23.

5. **Connect the speaker.**

   One lead from the speaker should be inserted in hole B23. The other can be inserted in any hole in the ground bus.

   Note that speakers aren't sensitive to polarity, so it doesn't matter which lead goes in the ground bus and which goes in B23.

6. **Connect the battery.**

   Plug the 9 V battery into the battery snap connector, and then connect the red lead to the positive voltage bus and the black lead to the ground bus. You should immediately hear the speaker beeping.

   You're finished!

7. **Remove the battery when you can no longer stand the beeping.**

IN THIS CHAPTER

» **Getting familiar with op-amps**

» **Exploring how feedback circuits work with op-amps**

» **Looking at summing amplifiers and comparators**

» **Exploring several popular op-amp packages**

Chapter **3**

Working with Op-Amps

D id you ever play Operation, the game in which you had to use electrified tweezers to remove plastic body parts from little holes in a body? The edges of the holes were metal conductors, so if you touched the edge of the hole with the tweezers while trying to remove the plastic piece inside, a buzzer would sound, and the patient's nose (which was a red light bulb) would light up.

I was never very good at it, because it required nerves of steel and precise control of the tweezers. The slightest jiggle of the tweezers was amplified into a flashing light, a loud buzzer, and the mocking laughter of my brother, who was a much better Operation surgeon than I was.

An operational amplifier (op-amp for short) is kind of like the game Operation. Well, actually, it isn't really, except for the part about the slightest variations in the input (your hand holding the tweezers) being amplified into a huge variation in the output (the flashing red nose, the jarring buzzer, and the ridicule of your brother).

Op-amps are among the most common types of integrated circuits — probably second in popularity only to the 555 timer chip, described in the previous chapter. In this chapter, you learn what an op-amp is and how to build several useful circuits with it. So put on your scrubs, and let's start the operation!

Looking at Operational Amplifiers

An *op-amp* is a super-sensitive amplifier circuit that's designed to amplify the difference of two input voltages. Thus, an op-amp has two inputs and one output. The output voltage is often tens or even hundreds of thousands of times greater than the difference in the input voltages. Thus, a very small difference in the two input voltages — perhaps a few hundredths or even a few thousandths of a volt — can result in a large output voltage.

Although an op-amp is a type of integrated circuit, op-amps were invented long before integrated circuits. Op-amp circuits are a natural for integrated circuits, however, so it wasn't long after the introduction of the first integrated circuits that IC versions of op-amps became available. Today, op-amps are among the most popular types of integrated circuits.

TECHNICAL STUFF

The name *operational amplifier* may be a bit confusing. Originally, the op-amp circuit was created for use as an amplifier in telephone distribution systems. Later, computer engineers discovered that the circuit could easily be adapted to do mathematical operations such as addition, subtraction, multiplication, and division. It was around that time that the term *operational amplifier* was coined, because the circuits are amplifiers that can perform (mathematical) operations. (For more information, see the nearby sidebar "How the op-amp came to be.")

Internally, the simplest op-amps consist of several dozen transistors, and more complicated varieties have many more. In this chapter, I completely ignore the internal circuitry of an op-amp and treat it as what it is: a handy device you can use without understanding how it works. You can thank the engineers who, many decades ago, worked out all the internal details that make an op-amp do its magic.

The standard schematic symbol for an op-amp is a triangle, as shown in Figure 3-1. As you can see, the two inputs are on the left side of the triangle, the output is on the right, and the power supply connections for the circuit are on the top and bottom of the triangle.

Many types of op-amp chips are manufactured today, but all have the five connections shown in Figure 3-1. The following paragraphs describe the function of each of these connections:

>> **+V and −V:** The power supply for an op-amp is provided via two pins usually labeled +V and −V. (These pins may be labeled V_{s+} and V_{s-} instead, but their functions are the same.) Most op-amps require both a positive and a negative voltage power supply, with voltages usually ranging from ±6V to ±18V. This type of power supply is called a *split supply*. The ± indicates that both positive voltage and negative voltage are required. ±6V, for example, means that both +6V and −6V are required.

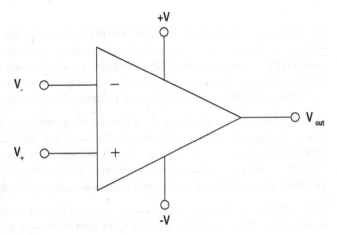

FIGURE 3-1:
Schematic symbol
for an op-amp.

You can easily build a split supply by using two batteries connected end to end, as shown in Figure 3-2. Here, two 9 V batteries are connected to create a ±9V supply. Note that the +9V and –9V are measured relative to ground, which is accessed between the two batteries.

Note that some op-amps don't require split-voltage power supplies. Op-amps that use single power supplies have a ground terminal instead of a –V terminal.

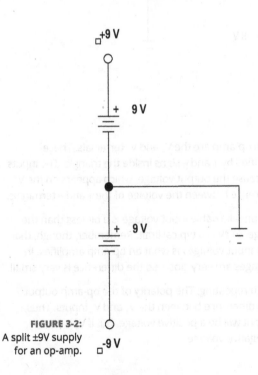

FIGURE 3-2:
A split ±9V supply
for an op-amp.

» **V_{out}:** The output of the op-amp is taken from the V_{out} terminal. The voltage at the output terminal can be positive or negative, depending on the voltage difference between the two input terminals. The maximum voltage is usually a few volts less than the supply voltage at the +V and –V terminals. Thus, if the power supply for an op-amp is ±9V, the maximum output will be around ±7V or ±8V.

Most op-amps can handle only a small amount of current through the output terminal — usually, in the neighborhood of 25 mA or less. As Figure 3-3 shows, the output is passed through an external resistance, designated R_L. The other end of this resistance is connected to ground. Thus, the output current that flows through the op-amp must eventually end up at ground.

Note that the load resistance isn't necessarily in the form of a simple resistor; it can be any other circuit that provides some load resistance, such as the base-emitter circuit of a transistor or even input of another op-amp.

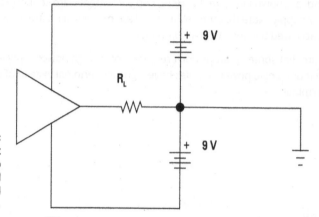

FIGURE 3-3:
The output from an op-amp passes to ground through a load resistance.

» **V₊ and V_–:** The two inputs of an op-amp are the V_+ and V_- terminals. These terminals are sometimes identified by + and – signs inside the triangle. The inputs are called *differential inputs* because the output voltage, which appears on the V_{out} terminal, depends on the difference between the voltage of the + and – terminals.

For most op-amps, the maximum allowable input voltage is a bit less than the maximum power supply voltage. ±12V is a typical limit. Remember, though, that the *difference* between the two input voltages is what an op-amp amplifies. In many cases, the two input voltages are very close, so the difference is very small.

I've said it already, but it's worth repeating: The polarity of the op-amp output depends on the polarity of the difference between the V_+ and V_- inputs. Thus, if V_+ is greater than V_-, the output will be a positive voltage, but if V_+ is less than V_-, the output will be a negative voltage.

In many op-amp circuits, one input is connected to ground. If the V_+ input is grounded, the output polarity is always the opposite of the polarity of the input voltage on the V_- terminal. In other words, negative voltage on V_- will give positive voltage on V_{out}, and positive voltage on V_- will give negative voltage on V_{out}. For this reason, the V_- input is often called the *inverting input* because its polarity is inverted in the output.

If, on the other hand, the V_- input is connected to ground, the polarity of the output is the same as the polarity of the input voltage applied to V_+. Thus, if V_+ is positive, V_{out} will be positive; if V_+ is negative, V_{out} will be negative. For this reason, the V_+ input is called the *noninverting input* because its polarity is the same in the output — that is, the V_- input voltage is *not* inverted.

HOW THE OP-AMP CAME TO BE

The modern operational amplifier dates way back to the early 1930s, when Bell Telephone was just starting to run telephone cables throughout the country. In the early days of the telephone, engineers had a difficult problem with phone lines that ran more than a few thousand feet. Long phone lines needed amplifiers to give their signals a boost, but the amplifiers available at the time were very fidgety — too sensitive to weather (temperature and humidity) and unable to work consistently over the range of voltages used in early phone lines.

A Bell engineer named Harry Black was working on the amplifier problem in 1934 when an idea struck him as he was riding the ferry home from work. This idea was a stroke of genius that seems obvious decades later. Instead of trying to design an amplifier that would have the exact amplitude gain needed for the job, Black's idea was to use an amplifier that had far more gain than was needed — thousands of times more gain, in fact — and then feed some of the output back into the input through a resistor. This feedback circuit would reduce the overall gain of the amplifier based on the amount of resistance in the circuit.

The circuit didn't gain the name *operational amplifier* until the computer age began a decade or so later and computer researchers figured out how to use the amplifier's unique characteristics to perform basic mathematical operations like addition, subtraction, multiplication, and division on the input voltages. Eventually, digital computers replaced the analog computers built from op-amps. Op-amps still play an important role in computers today, however, primarily to provide an interface with real-world input measuring devices such as voltage sensors and moisture detectors.

The original op-amp circuits were built with vacuum tubes. They were large, required several hundred volts to operate, and generated substantial heat. When transistors replaced vacuum tubes in the 1950s, op-amps became smaller, and when integrated circuits were invented in the 1960s, op-amps were among the first chips to be designed.

Understanding Open Loop-Amplifiers

As its name suggests, one of the most basic uses of an op-amp is as an amplifier. If you connect an input source to one of the input terminals and ground the other input terminal, an amplified version of the input signal appears on the out terminal.

An important concept in op-amp circuits is *voltage gain*, which simply represents the amount by which the difference between the two input voltages is multiplied to produce the output voltage. If the input voltage difference is 2 V and the output voltage is 12 V, for example, the voltage gain of the amplifier is 6.

If you simply apply an input signal to the V_ terminal of an op-amp as shown in Figure 3-4, the circuit is called an *open loop-amplifier*. The reason it's called this will become more apparent when you get to the next section. For now, just realize that this type of circuit goes by the name *open loop*.

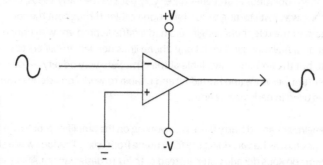

FIGURE 3-4:
A op-amp
configured
as an open
loop-amplifier.

In the open loop op-amp circuit, the V_+ input is connected to ground, and an input signal is placed on the V_ input. In this arrangement, the voltage to be amplified is the same as the voltage of the V_ input. Although Figure 3-4 shows the input as alternating current, the open loop op-amp circuit works for direct current as well.

The voltage gain in an open loop op-amp circuit is extraordinarily high — on the order of tens or even hundreds of thousands. Suppose that you're using an op-amp whose open loop voltage gain is 200,000 and the power supply is ±9 V. In that case, an input voltage of +0.000025 V will result in an output voltage of +5V. An input voltage of +0.00004 V will give you an output voltage of 8 V.

REMEMBER

The output voltage can never exceed the power supply voltage. In fact, the maximum output voltage usually is about 1 V less than the power supply voltage. So if you're using a pair of 9 V batteries to provide a ±9 V power supply, the maximum output voltage is ±8 V. As a result, the most that an open loop op-amp circuit with an open loop gain of 200,000 can reliably amplify is 0.00004 V. If the input voltage difference is any larger than 0.00004 V, the op-amp is said to be *saturated*, and the output voltage will go to the maximum.

I guarantee that no matter how much money you invested in a top-quality volt-meter, it isn't sensitive enough to measure voltages that small. Physicists at Cal Tech may be able to measure voltages that size, but for all practical purposes, 0.00004 V is the same as 0 V.

As a result, one of the basic features of an open loop op-amp circuit is that if the input voltage difference is anything other than zero, the op-amp will be saturated, and the output voltage will be the same at its maximum. So if the maximum output voltage is ±8 V, the output will be one of only three voltages: +8V, 0V, or –8V.

Open loop op-amp circuits may not sound very useful, but actually, they have many useful applications. You see one example in "Using an Op-amp as a Voltage Comparator," later in this chapter.

THE IDEAL OP-AMP

If you read about op-amps on the web or in an electronics book, you'll undoubtedly come across the term *ideal op-amp*. An *ideal op-amp* is a hypothetical op-amp with certain characteristics that real op-amps strive to achieve. Real op-amps come very close to the ideal op-amp, but no op-amp in existence actually achieves the perfection of an ideal op-amp. So you see, in many ways, op-amps are almost human.

Depending on which list you read, an ideal op-amp has anywhere between two and seven characteristics, the most important of which are

- **Infinite open loop gain:** Several times in this chapter, I mention that the open loop gain in an op-amp is very large — on the order of tens or even hundreds of thousands. In an ideal op-amp, the open loop gain is infinite, which means that *any* voltage differential on the two input terminals will result in an infinite voltage on the output. In real op-amps, the output voltage is limited by the power supply voltage. Because the output voltage can't be infinite, the gain can't be infinite either.

- **Infinite input impedance:** *Impedance* represents a circuit's opposition to current flow, whether the current is alternating or direct. In an ideal op-amp, the imped-ance of the two input terminals is infinite, which means that no current enters the op-amp from the inputs. The inputs are able to see and react to the voltage, but that voltage is unable to push any current into the op-amp. What that means in practice is that the op-amp has no effect on the input voltage. In an actual op-amp, a small amount of current — usually, a few milliamps or less — does leak into the op-amp's input circuits.

(continued)

(continued)

- **Zero output impedance:** In an ideal op-amp, the output circuitry has zero internal impedance, which means that the voltage provided from the output is the same regardless of the amount of load placed on it by the circuit to which the output is connected. In reality, most op-amps have an output impedance of a few ohms, which means that the actual voltage provided by the output terminal will vary a small amount depending on the load connected to the output.

- **Zero offset voltage:** The *offset voltage* is the amount of voltage at the output terminal when the two inputs are exactly the same. If you connect both inputs to ground, for example, there should be exactly 0 V at the output. In reality, real world op-amps have a very small voltage on the output even when both inputs are grounded, connected to each other, or not connected to anything at all. For most op-amps, this offset voltage is just a few millivolts.

- **Infinite bandwidth:** The term *bandwidth* refers to the range of alternating current frequencies within which an op-amp can accurately amplify. In an ideal op-amp, the frequency of the input signal has no effect on how the op-amp behaves. In real-world op-amps, the op-amp doesn't perform well above a certain frequency — typically, a few megahertz (millions of cycles per second).

The characteristics are often summed up with the following two "golden rules" of op-amps:

1. **The output attempts to do whatever is necessary to make the voltage difference between the inputs zero.**

   This rule, which applies only to closed loop-amplifier circuits, means that the feedback sent from the output to the input causes the two input voltages to become the same.

2. **The input draws no current.**

   This rule means that the input terminals look at the voltage placed across them but don't allow any current to flow into the op-amp.

Although no actual op-amp is able to live up to the standards of the ideal op-amp, most come pretty close. Close enough, in fact, that you can safely design an op-amp circuit as if the op-amps were ideal. In particular, the two golden rules apply: The feedback will equalize the input voltages, and the op-amp draws no current from the input.

Looking at Closed Loop-Amplifiers

As I explain in the preceding section, open loop op-amp circuits aren't very useful as amplifiers because they're so easily saturated. To make an op-amp useful as an amplifier, you must use it in a *feedback circuit*, which reduces the gain to a more manageable amount so that input voltages that are usable (and even measurable!) can be amplified reliably.

I'm sure that you're already familiar with the concept of feedback. You've probably sat in an auditorium listening to someone talk into a public-address system when suddenly a piercing screech came out of the speakers. That screech was feedback. In the case of the public-address system, the microphone picked up some of the output from the speakers and sent it back through the amplifier again. The result was an annoying high-pitched squeal.

Not all feedback is bad, though. In an op-amp amplifier circuit, feedback is used to reduce the enormous open loop-amplification gain to a more manageable gain, such as 10. To do this, the output signal is fed back into the input via the V_+ terminal. A resistor is used to reduce the voltage that is fed back to the input. This type of circuit is called a *closed loop-amplifier* because a closed circuit path exists between the output and the input. (Now you understand why an op-amp circuit without the feedback loop is called an *open loop-amplifier*.)

The most common op-amp configuration is called an *inverting amplifier* because the voltage of the output is opposite the voltage of the input. Figure 3-5 shows a basic inverting amplifier circuit.

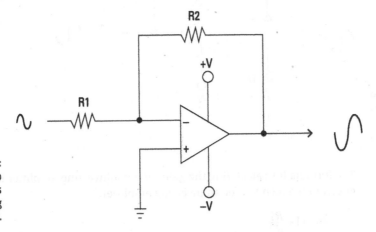

FIGURE 3-5:
An op-amp
configured as
an inverting
amplifier.

In an inverting amplifier circuit, the input signal works its way through a resistor on its way to the V_ input, and the output is looped back into the V_ input through a second resistor. In Figure 3-5, these resistors are designated R1 and R2. You can easily calculate the overall voltage gain of the circuit by using this formula:

$$A_{CL} = -\frac{R2}{R1}$$

Here, the gain is designated A_{CL} (CL stands for *closed loop*).

If R1 is 1 kΩ and R2 is 10 kΩ, the voltage gain of the circuit will be –10. Then if the input voltage is +0.5 V, the output voltage will be –5 V (0.5 × –10).

Note that the negative sign is required because Figure 3-5 is an inverting amplifier circuit, so positive inputs give negative outputs, and vice versa.

A closed loop-amplifier can also be designed as a *noninverting amplifier* in which the output voltage is not reversed. To do that, you simply reverse the inputs, as shown in Figure 3-6. Instead of connecting the input voltage to V_ through a resistor and grounding V_+, you ground V_ through a resistor and connect the input voltage to V_+. The feedback circuit is the same; the output is connected to the V_ input through resistor R1.

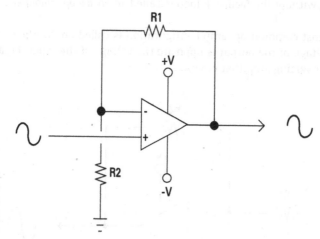

FIGURE 3-6: An op-amp configured as a noninverting amplifier.

The formula for calculating the gain for a noninverting amplifier is a little different from the formula for an inverting amplifier:

$$A_{CL} = 1 + \frac{R1}{R2}$$

If R1 is 10 kΩ and R2 is 1 kΩ, the gain is 11. Thus, an input voltage of +0.5 V will result in an output voltage of +5.5 V.

DELVING DEEPER INTO FEEDBACK

If you're interested in understanding how the feedback circuit works, remember the rule that I describe in the section "Understanding Open Loop-Amplifiers": If the input voltage is anything other than zero, the op-amp will be saturated, and the output voltage will be the maximum allowed. The purpose of the feedback circuit is to return some of the output voltage to the inverting input, which results in the input-voltage differences being driven toward zero. As the voltage approaches zero, the op-amp's gain begins to drop to a useful range.

Suppose that the input voltage difference in an inverting op-amp circuit is +0.5 V. This difference results in the op-amp's becoming saturated, so –8 V appears at the output. A portion of the negative voltage that depends on the voltage divider created by R1 and R2 is returned to the V_ input, which has the effect of reducing the input voltage. That results in a smaller voltage difference but not small enough to prevent the op-amp from still being saturated.

Remember, however, that the feedback loop is just that: a loop. As more and more of the saturated output voltage gets fed back through the loop, the input voltage difference gets closer to zero. When it gets oh-so-close to zero, the output voltage drops to a range between zero and the maximum voltage.

The best feature of the closed loop-amplifier circuit is that the two resistors, which are outside the op-amp IC, give you precise control of the amount of gain that the circuit will ultimately have. All you have to do to get any gain you want (within the limits of the op-amp) is choose the right resistor values.

Using an Op-Amp as a Unity Gain Amplifier

A *unity gain amplifier* is an amplifier circuit that doesn't amplify. In other words, it has a gain of 1. The output voltage in a unity gain amplifier is the same as the input voltage.

You may think that such a circuit would be worthless. After all, isn't a simple piece of wire a unity gain circuit? Sure, but a unity gain amplifier provides one important benefit: It doesn't take any current from the input source. (Remember, that's one of the Golden Rules of the ideal op-amp.) Therefore, it completely isolates the input side of the circuit from the output side of the circuit. Op-amps are often used as unity gain amplifiers to isolate stages of a circuit from one another.

Unity gain amplifiers come in two types: voltage followers and voltage inverters. A *follower* is a circuit in which the output is exactly the same voltage as the input. An *inverter* is a circuit in which the output is the same voltage level as the input but with the opposite polarity.

If you think about it for a moment, you might be able to come up with the circuit for unity gain followers and inverters on your own. The formula for calculating the gain of both an inverting amplifier and a noninverting amplifier requires you to divide R1 by R2, so all you have to do is choose resistor values that will result in a gain of 1.

The following sections explain how to create unity followers and unity inverters.

Configuring a unity follower

A unity gain follower is simply a noninverting amplifier with a gain of 1. Recall that the formula for calculating the value of a noninverting amplifier is this:

$$A_{CL} = 1 + \frac{R2}{R1}$$

To create a unity gain follower, you just omit R2 and connect the output directly to the inverting input, as shown in Figure 3-7. Because R2 is zero, the value of R1 doesn't matter, because zero divided by anything equals zero. So, R1 is usually omitted as well, and the V_ input isn't connected to ground.

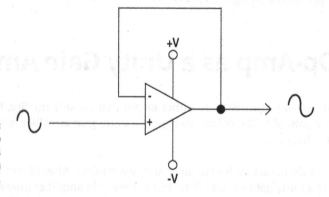

FIGURE 3-7:
An op-amp
configured as
a unity gain
follower.

Configuring a unity inverter

The formula for calculating gain for an inverting amplifier is this:

$$A_{CL} = -\frac{R2}{R1}$$

In this case, all you have to do is use identical values for R1 and R2 to make the amplifier gain equal to 1. Figure 3-8 shows a unity gain inverter circuit using 1 kΩ resistors.

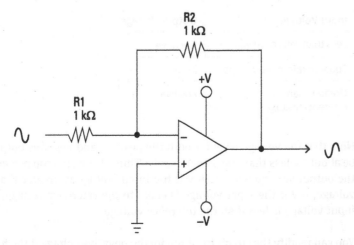

FIGURE 3-8:
An op-amp
configured as
a unity gain
inverter.

Using an Op-Amp as a Voltage Comparator

A *voltage comparator* is a circuit that compares two input voltages and lets you know which of the two is greater. Suppose that you have a photocell that generates 0.5 V when it's exposed to full sunlight, and you want to use this photocell as a sensor to determine when it's daylight. You can use a voltage comparator to compare the voltage from the photocell with a 0.5 V reference voltage to determine whether or not the sun is shining.

It's easy to create a voltage comparator from an op-amp, because the polarity of the op-amp's output circuit depends on the polarity of the difference between the two input voltages. Figure 3-9 shows the basic circuit for an op-amp voltage comparator.

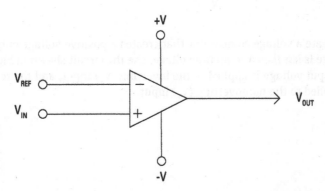

FIGURE 3-9:
An op-amp
configured
as a voltage
comparator.

In the voltage-comparator circuit, first a reference voltage is applied to the inverting input (V_); then the voltage to be compared with the reference voltage is applied to the noninverting input. The output voltage depends on the value of the input voltage relative to the reference voltage, as follows:

Input Voltage	Output Voltage
Less than reference voltage	Negative
Equal to reference voltage	Zero
Greater than reference voltage	Positive

Note that the voltage level for both the positive and negative output voltages will be about 1 V less than the power supply. Thus, if the op-amp power supply is ±9 V, the output voltage will be +8 V if the input voltage is greater than the reference voltage, 0 V if the input voltage is equal to the reference voltage, and –8 V if the input voltage is less than the reference voltage.

You can modify the circuit to eliminate the negative voltage if the input is less than the reference by sending the output through a diode, as shown in Figure 3-10. In this circuit, a positive voltage appears at the output if the input voltage is greater than the reference voltage; otherwise, no output voltage exists.

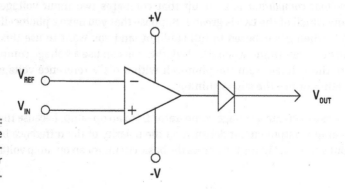

FIGURE 3-10: Using a diode in a voltage-comparator circuit.

To create a voltage comparator that creates a positive voltage output if the input voltage is *less than* a reference voltage, use the circuit shown in Figure 3-11. Here, the input voltage is applied to the inverting (V_) input, and the reference voltage is applied to the noninverting (V_+) input.

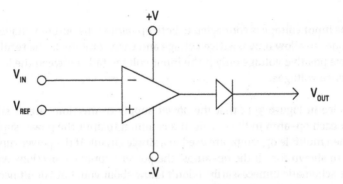

FIGURE 3-11:
A voltage
comparator that
tests for a voltage
that's less than a
reference voltage.

The final voltage-comparator circuit you should know about is the *window comparator*, which lets you know whether the input voltage falls within a given range. A window comparator requires three inputs: a low reference voltage, a high reference voltage, and an input voltage. The output of the window comparator will be a positive voltage only if the input voltage is greater than the low reference voltage and less than the high reference voltage. If the input voltage is less than the low reference voltage, the output will be zero. Similarly, if the input voltage is greater than the high reference voltage, the output will also be zero.

You need two op-amps to create a window comparator, as shown in Figure 3-12. As you can see in the figure, one op-amp is configured to produce positive output voltage only if the input is greater than the low reference voltage ($V_{REF(LOW)}$). The other op-amp is configured to produce positive output voltage only if the input is less than the high reference voltage ($V_{REF(HIGH)}$).

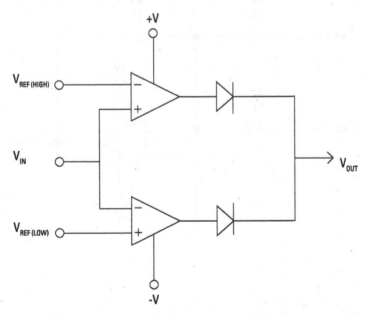

FIGURE 3-12:
Two op-amps
can be used to
create a window
comparator.

The input voltage is connected to both op-amps; the output voltage is sent through diodes to allow only positive voltage and then combined. The resulting output will have positive voltage only if the input voltage falls between the low and high reference voltages.

TIP

Notice in Figure 3-12 that the power supply connections aren't shown separately for each op-amp in the circuit. It's common to omit the power supply connections when multiple op-amps are used in a single circuit. If the power supply connections were shown for all the op-amps, the power supply connections would complicate the schematic unnecessarily. I don't know about you, but I don't need any unnecessary complications in my life. I have enough necessary complications as it is.

Adding Voltages

An op-amp can be used to add or subtract two or more voltages. A circuit that adds voltages is called a *summing amplifier*. A summing amplifier has two inputs and an output voltage that is the sum of the two input voltages but with the opposite polarity. If one of the inputs is +1.5 V and the other is +1.0 V, for example, the output voltage will be –2.5 V.

Figure 3-13 shows a basic circuit for a summing amplifier. For the summing amplifier to work, resistors R1, R2, and R3 should all be the same value.

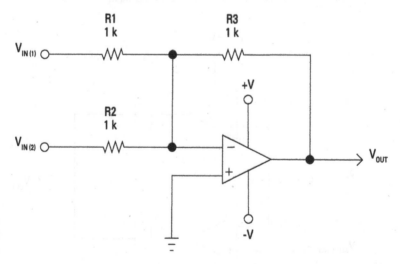

FIGURE 3-13:
A basic summing amplifier circuit.

If all the resistors in a summing amplifier are the same, the output voltage will be the sum of the input voltages. This is the usual way to configure a summing amplifier, though you can vary the resistor values if you want.

If the resistors have different values, each of the input voltages is weighted according to the value of the resistor on its input circuit, which has the effect of multiplying each input voltage by a certain value before the voltages are summed. The exact value by which each input is multiplied depends on the mix of resistors you use.

If R1 is 1 kΩ and R2 is 10 kΩ, for example, the input voltage applied through the 1 kΩ resistor will be multiplied by 10 before being added to the voltage applied through the 10 kΩ resistor. Thus, if the input at R1 is +1 V, and the input at R2 is +2 V, the output voltage will be –12 V. (For this formula to work, R3 must also be 10 kΩ.)

The actual formula for calculating the output voltage based on the input voltages and the resistor values is this:

$$V_{out} = -R3\left(\frac{V1}{R1} + \frac{V2}{R2}\right)$$

I'll leave it to you to work out the math for various combinations of resistor values and input voltages. Here, though, are a few examples that should give you an idea of how the circuit will behave when R1 is 1 kΩ and both R2 and R3 are 10 kΩ:

$V_{In\,(1)}$	$V_{In\,(2)}$	V_{out}
+1 V	+1 V	–11 V
+1 V	+5 V	–15 V
0 V	+5 V	–5 V
+2 V	–5 V	–15 V
–1 V	–5 V	+15 V

One drawback of the summing amplifier is that it inverts the polarity of the input, but you can easily feed the output of a summing amplifier into the input of a unity gain inverter, as shown in Figure 3-14. Here, the second op-amp inverts the polarity of the output from the summing amplifier, which has the effect of returning the output voltage polarity to the polarity of the original inputs. (For more information about the voltage-inverter portion of this circuit, refer to "Using an Op-Amp as a Unity Gain Amplifier," earlier in this chapter.)

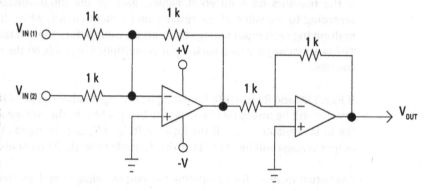

FIGURE 3-14:
A summing amplifier can be combined with a voltage inverter to preserve the input polarity.

One common use for a summing amplifier circuit is as an audio mixer. When this type of circuit is used as an audio mixer, each input is connected to a microphone. The summing amplifier combines all the microphone inputs by adding the voltages from each microphone, and the resulting output is sent on to another amplifier stage.

The resistors in each input circuit are often potentiometers, which allows you to vary the signal level from each input source. When you increase the resistance on one of the input circuits, less of that input is represented in the output mix — especially useful if one of your singers is a bit off key.

A summing amplifier circuit can be extended with additional inputs. Figure 3-15 shows a circuit with four inputs that uses potentiometers to control the level of each input. You can add as many inputs as you want, but you need to ensure that the total combined voltage from all inputs doesn't exceed the power supply voltage (minus a volt or two).

WARNING

Note that op-amps have a maximum output voltage rating. If you're working with output voltages above ±15 V, you should check the maximum output voltage rating of your op-amp to make sure that the op-amp can handle it.

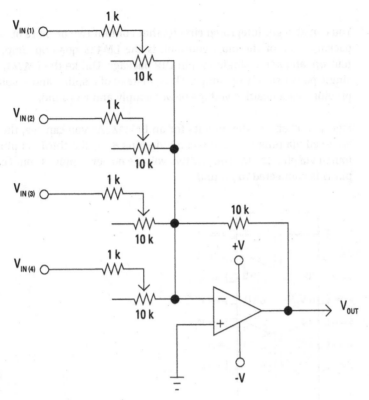

FIGURE 3-15:
A simple audio
mixer with four
inputs.

Working with Op-Amp ICs

So far, all the examples in this chapter have assumed that you're using a generic op-amp in your circuits. When you get around to building an actual op-amp circuit, of course, you'll need to use a real op-amp. Fortunately, op-amp integrated circuits are plentiful, and nearly all stores that sell electronic components sell several types of inexpensive op-amp ICs.

The most popular op-amp IC is the LM741, which comes in a standard eight-pin DIP package. Figure 3-16 shows the pin connections for an LM741 op-amp.

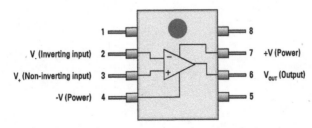

FIGURE 3-16:
Pinouts for the
LM741 op-amp.

You can also get integrated circuits that contain two or more op-amps in a single package. One of the most common is the LM324 quad op-amp, which contains four op-amps in a single 14-pin DIP package. Unlike the LM741, the LM324 uses single power supply op-amps. Thus, instead of a split + and − voltage supply, you provide just a positive voltage power supply and a ground.

Figure 3-17 shows the pinouts for an LM324. As you can see, the first op-amp is accessed via pins 1−3, the second via pins 5−7, the third via pins 8−10, and the fourth via pins 12−14. The positive voltage power supply is connected to pin 4, and pin 11 is connected to ground.

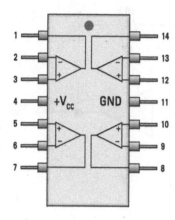

FIGURE 3-17:
Pinouts for the
LM324 quad
op-amp.

4

Beyond Direct Current

Contents at a Glance

Chapter **1**

Getting into Alternating Current

A few years ago, I saw one of my favorite musicals, *Les Misérables*. It has nothing to do with electricity, of course, but at the end of the first act, as the daring gang of revolutionary college students join the rest of the characters singing "One Day More!," they do a really cool marching step in which they march one pace toward the audience, then one pace back, one pace forward, one pace back, and so on. It creates a convincing effect of a mob on the move, even though the mob isn't going anywhere.

That's how alternating current works. So far in this book, I've focused mostly on working with direct current, in which the electric current flows in one direction and one direction only. If the revolutionaries in *Les Misérables* were demonstrating direct current, they would march straight off the stage, over the orchestra pit, and through the audience, out the back doors of the auditorium.

In *alternating current*, the current flows in both directions — forward and backward — much as the revolutionaries march one step at a time forward and backward.

Alternating current is of vital importance in electronics for one simple reason: The electric current you can access by plugging a circuit into a wall outlet happens to be alternating current. Thus, if you want to free your circuits from the tyranny of batteries, which eventually die, you'll need to learn how to make your circuits work from an alternating current power supply.

In this chapter, you take a look at the nature of alternating current and how it can deliver reliable voltage to your home or place of business. You can also look at three fundamental alternating current devices: alternators, which generate alternating current from a source of motion such as a steam turbine or windmill; motors, which turn alternating current into motion; and transformers, which can transfer alternating current from one circuit to another without any physical connection between the circuits. And finally, you learn about the basics of working with alternating current safely.

Incidentally, if you want to see the famous *Les Misérables* march step in action, go to https://youtu.be/x-1B0PL9TJs. You'll see the marching about 3 minutes into the video.

What Is Alternating Current?

As you know, electric current that flows continuously in a single direction is called a *direct current*, or *DC*. In a direct current circuit, current is caused by electrons that all line up and move in one direction. Within a wire carrying direct current, electrons hop from atom to atom while moving in a single direction. Thus, a given electron that starts its trek at one end of the wire will eventually end up at the other end of the wire.

In *alternating current*, the electrons don't move in only one direction. Instead, they hop from atom to atom in one direction for a while, and then turn around and hop from atom to atom in the opposite direction. Every so often, the electrons change direction. In alternating current, the electrons don't move steadily forward. Instead, they just move back and forth.

When the electrons in alternating current switch direction, the direction of current and the voltage of the circuit reverses itself. In public power distribution systems in the United States (including household current), the voltage reverses itself 60 times per second. In some countries, the voltage reverses itself 50 times per second.

The rate at which alternating current reverses direction is called its *frequency*, expressed in hertz. Thus, standard household current in the United States is 60 Hz.

It's important to realize that the voltage in an alternating current circuit doesn't instantly reverse polarity. Instead, the voltage steadily increases from zero until it reaches a maximum voltage, which is called the *peak voltage*. Then, the voltage begins to decrease again back to zero. The voltage then reverses polarity and drops below zero, again heading for the peak voltage but negative polarity. When it reaches the peak negative voltage, it begins climbing back again until it gets to zero. Then the cycle repeats.

The swinging change of voltage is important because of the basic relationship between magnetic fields and electric currents. When a conductor (such as a wire) moves through a magnetic field, the magnetic field induces a current in the wire. But if the conductor is stationary relative to the magnetic field, no current is induced.

Physical movement is not necessary to create this effect. If the conductor stays in a fixed position but the intensity of the magnetic field increases or decreases (that is, if the magnetic field expands or contracts), a current is induced in the conductor the same as if the magnetic field were fixed and the conductor was physically moving across the field.

Because the voltage in an alternating current is always either increasing or decreasing as the polarity swings from positive to negative and back again, the magnetic field that surrounds the current is always either expanding or collapsing. So, if you place a conductor within this expanding and collapsing magnetic field, current will be induced in the conductor. And that current will alternate in concert with the movement of the magnetic field.

It seems like magic! With alternating current, it is possible for current in one wire to induce current in an adjacent wire, even though there is no physical contact between the wires.

The bottom line is this: Alternating current can be used to create a changing magnetic field, and changing magnetic fields can be used to create alternating current.

This relationship between alternating current and magnetic fields makes three important devices possible:

» **Alternator:** A device that generates alternating current from a source of rotating motion, such as a turbine powered by flowing water or steam or a windmill. Alternators work by using the rotating motion to spin a magnet that's placed within a coil of wire. As the magnet rotates, its magnetic field moves, which induces an alternating current in the coiled wire. (Coils of wires are used instead of straight wires simply because coiling up the wire allows a greater length of wire to be exposed to the changing magnetic field.)

» **Motor:** The opposite of an alternator. It converts alternating current to rotating motion. In its simplest form, a motor is simply an alternator that's connected backward. A magnet is mounted on a shaft that can rotate; the magnet is placed within the turns of a coil of wire. When alternating current is applied to the coil, the rising and falling magnetic field created by the current causes the magnet to spin, which turns the shaft.

» **Transformer:** Consists of two coils of wire placed within close proximity. If an alternating current is placed on one of the coils, the collapsing and expanding magnetic field will induce an alternating current in the other coil.

Measuring Alternating Current

With direct current, it's easy to determine the voltage that's present between two points: You simply measure the voltage with a voltmeter.

With alternating current, however, measuring the voltage isn't so simple. That's because the voltage in an alternating current circuit is constantly changing. So, for example, when we say that the voltage at a wall receptacle is 120 VAC, what does that really mean?

There are actually three ways you can measure voltage in an AC circuit. They are illustrated in Figure 1-1. The three ways are

» **Peak voltage:** A measurement of the largest voltage present between 0 V and the highest point on the AC cycle. It's the maximum voltage that the AC voltage attains.

» **Peak-to-peak voltage:** The difference between the highest and lowest peaks of the AC voltage. In most AC voltages, the peak-to-peak voltage is double the peak voltage.

>> **RMS voltage:** The average voltage of the circuit; also called the *mean voltage*. *RMS* stands for *root mean square*, but that's important only if you're studying for an exam or something. RMS voltage is far and away the most common way to specify the voltage of an AC circuit. For example, when we say that the voltage at a household electrical outlet is 120 VAC, what we really mean is that the RMS voltage is 120 V.

If the AC voltage follows a true sine wave, the RMS voltage is equal to 0.707 times the peak voltage. Or to turn it around, the peak voltage is equal to about 1.4 times the RMS voltage. Thus, the actual peak voltage at a household electrical outlet is about 168 V.

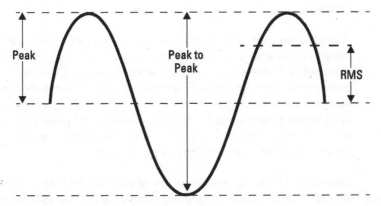

FIGURE 1-1:
Three ways to measure alternating current.

TECHNICAL STUFF

The true RMS voltage is a bit tricky to calculate since it involves some fairly complicated math. RMS is calculated by sampling the actual voltage in very small time increments. Then, the sample voltages are squared, the squares of the voltages are added up, and the average of all the squared values is calculated. Finally, the square root of the average is calculated. This is the actual RMS value.

For a true sine wave, the preceding calculation turns out to be very close to multiplying the peak voltage by 0.707. For AC voltages that aren't true sine waves, however, the actual RMS value can be different than the "multiply by 0.707" shortcut would indicate.

Nearly all AC voltmeters report the RMS voltage, but only more expensive AC voltmeters calculate the actual RMS by sampling the input voltage and doing the sum-of-the-squares thing. Inexpensive voltmeters simply measure the peak voltage and multiply it by 0.707. Fortunately, this is close enough for most purposes.

THE CURRENT WARS

Alternating current is the worldwide standard for power distribution. This wasn't always the case, however. Back in the days when electricity was first being put to practical use, direct current was the normal way to distribute electricity. The biggest champion of direct current was none other than Thomas Edison himself, the Great American Inventor who is credited with inventing just about everything from the light bulb to the phonograph to the motion picture.

In 1880, Edison patented a system of electrical power distribution based on direct current and opened the first public electric utility company in 1882, providing electricity to 59 customers in New York. By 1890, he had more than 100 power plants operating nationwide.

Thomas Edison's biggest rival was a fellow named George Westinghouse, who advocated the use of alternating current for power distribution and promoted a system developed by the brilliant but eccentric inventor Nicola Tesla. Westinghouse promoted the benefits of alternating current over direct current — primarily the benefit that alternating current could transmit power efficiently over much larger distances than direct current. Edison's direct current system required that power plants be located within a few miles of customers. But Tesla's alternating current system could deliver power hundreds of miles away from the power plants.

Edison responded to the negative publicity of direct current the way any true-blooded American marketer would: by launching a smear campaign. In 1887, a man was accidentally killed when he touched bare power lines. Edison had one of his employees develop a method of *intentionally* killing people with electricity. Hence, the electric chair was invented.

Of course, the electric chair used alternating current to electrocute its victims. Edison launched a nationwide publicity campaign to convince the public that alternating current was so dangerous that it was used in prisons to kill condemned murderers. He even went so far as to conduct public executions of stray dogs and, in one case, an elephant. The message was clear: You don't want this dangerous stuff in your house.

Fortunately, the smear campaign didn't work. The benefits of alternating current eventually won out. The turning point came when the alternating current generators at Niagara Falls began operating in 1896, delivering power 20 miles away to Buffalo, NY. By the early part of the twentieth century, nearly all power distribution worldwide was done with alternating current.

Direct current distribution lasted much longer than you might think, however. Con Edison — one of the largest electric companies in the world and the direct descendant of Edison's original electric company — did not convert its last few holdout customers over to alternating current until 2007.

Understanding Alternators

One good way to get your mind around how alternating current works is to look at the device that's most often used to generate it: the *alternator*. An alternator is a device that converts rotary motion, usually from a turbine driven by water, steam, or a windmill, into electric current. By its very nature, an alternator creates alternating current.

Figure 1-2 shows a simplified diagram of how an alternator works. Essentially, a large magnet is placed within a set of stationary wire coils. The magnet is mounted on a rotating shaft that's connected to a turbine or windmill. Thus, when water or steam flows through the turbine or when wind turns the windmill, the magnet rotates.

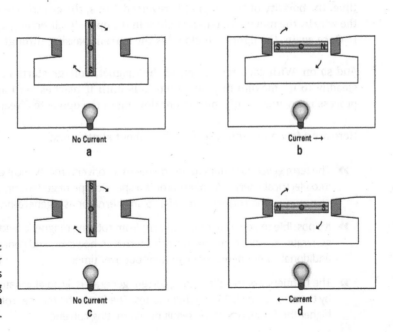

FIGURE 1-2:
An alternator
generates
alternating
current from a
rotating magnet.

As the magnet rotates, its magnetic field moves across the coils of wire. Because of the phenomenon of electromagnetic induction, the moving magnetic field induces an electric current within the wire coils. The strength and direction of this electric current depends on the position and direction of the rotating magnet.

In Figure 1-2, you can see how the current is induced in the wire at four different positions of the magnet's rotation. In part A, the magnet is at its farthest point away from the coils and oriented in the same direction as the coils. At this moment, the magnetic field doesn't induce any electric current at all. Thus, the light bulb is dark.

But as the magnet begins to rotate clockwise, the magnet comes closer to the coils, thus exposing more of its magnetic field to the coils. The moving magnetic field induces a current that gets stronger as the magnet continues to rotate closer to the coils. This causes the light bulb to glow. Soon, the magnet reaches its closest point to the coils, as shown in part B. At this point, the current and the voltage are at their maximum, and the light bulb glows at its brightest.

As the magnet continues to rotate counterclockwise, it now begins to move away from the coil. The moving electric field continues to induce current in the coil, but the current (and the voltage) decreases as the magnet retreats farther away from the coils. When the magnet reaches its farthest point from the coils, shown in part C, the current stops and the light bulb goes dark.

As the magnet continues to rotate, it now gets closer again to the coils. But this time, the polarity of the magnet is reversed. Thus, the electric current induced in the wire by the moving magnetic field is in the opposite direction, as shown in part D. Once again, the light bulb glows as the current passing through it increases.

And so on. With each revolution of the magnet, voltage starts at zero and rises steadily to its maximum point, then falls until it reaches zero again. Then the process is reversed, with the current flowing in the opposite direction.

Here are a few other interesting tidbits about alternators:

>> The term *generator* refers to any device that converts mechanical energy into electrical energy. An alternator is a specific type of generator, so it is common — and correct — to refer to an alternator as a generator.

>> It's possible to generate direct current from rotating magnetic fields. A DC generator is more complex than an alternator, however, and contains additional components that can wear out over time.

>> The frequency of the alternating current generated by an alternator is dictated by the rate at which the magnet rotates. The faster the magnet rotates, the higher the frequency of the resulting alternating current.

>> If you place two sets of coils spaced evenly around the magnet, each forming its own complete circuit, alternating current will be induced in each set of coils. However, the polarities of the two voltages will be mirror images of one another. In other words, when the voltage is positive in one of the circuits, it will be negative in the other. The relationship between the polarity of the circuits is called a *phase*, and a power-generating system with two circuits arranged in this way is called a *two-phase system*. The two circuits are said to be 180° out of phase with one another.

>> If three sets of coils are used, the system is called a *three-phase system*. In a three-phase system, the three circuits are 120° out of phase. Most public power-generation systems are three-phase systems because that results in the most efficient generation of power from the rotating magnetic fields.

Understanding Motors

An electric *motor* converts electrical energy in the form of electric current to rotating mechanical energy. The simplest type of electric motor is essentially the same thing as an alternator. The difference is that instead of using some other mechanical force such as water or steam to turn the magnet — which in turn induces electric current in the coils — electric current is applied to the coils, which in turn causes the magnet to rotate.

I would show you a diagram depicting the operation of a motor, but it would look pretty much like the alternator diagram shown in Figure 1-2. The only difference would be that the light bulb would be replaced by an AC power source. The same force that causes an electric current to be induced in a coil when the coil passes through a moving magnetic field causes a moving magnetic field to be created when a current is passed through a coil. The moving magnetic field, in turn, causes the magnet to rotate. This rotation is transferred to the shaft to which the magnet is attached.

As with alternators, it's also possible to create motors that work with DC circuits instead of AC. As with alternators, DC motors are more complicated than AC motors. In a DC motor, the polarity of the coils must be reversed every half-revolution of the magnet to keep the magnet moving in complete rotations. Usually, metal brushes are used to do this. In an AC motor, the brushes aren't necessary because the alternating current reverses polarity on its own.

Understanding Transformers

In Book 2, Chapter 4, you learn about the basic principles of magnetism and inductance. A transformer is a device that exploits these two principles:

>> A changing current passing through a wire creates a moving magnetic field around the wire.

>> A changing current will be induced in a wire that's exposed to a moving magnetic field.

A transformer combines these two principles by placing two coils of wire in close proximity to one another, as shown in Figure 1-3. When a source of AC is connected to one of the coils, that coil creates a magnetic field that expands and collapses in concert with the changing voltage of the AC. In other words, as the voltage increases across the coil, the coil creates an expanding magnetic field. When the voltage reaches its peak and begins to decrease, the magnetic field created around the coil begins to collapse.

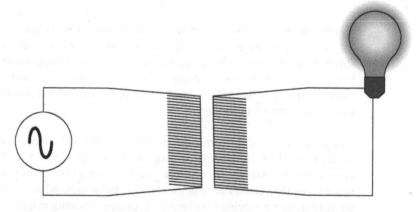

FIGURE 1-3:
A transformer
uses magnetic
induction to pass
current from one
circuit to another.

The second coil is located within the magnetic field created by the first coil. As the magnetic field expands, it induces current in the second coil. The voltage across the second coil increases as long as the magnetic field expands. When the magnetic field begins to collapse, the voltage across the second coil begins to decrease.

Thus, the current induced in the second coil mirrors the current that is passed through the first coil. A small amount of energy is lost in the process, but if the transformer is well constructed, the strength of the current induced in the second coil is very close to the strength of the current passed through the first coil.

The first coil in a transformer — the one that's connected to the AC voltage — is called the *primary coil*. The second coil — the one in which an AC voltage is induced — is called the *secondary coil*. All transformers have both a primary and a secondary coil.

TECHNICAL STUFF

Strictly speaking, not *all* transformers have both a primary and a secondary coil. There is a special type of transformer called an *autotransformer* that has just one coil with multiple taps. In an autotransformer, the single coil performs the functions of both the primary and the secondary coil in an ordinary transformer. I don't cover autotransformers in this book.

One of the most useful characteristics of a transformer is that the voltage induced in the secondary coil is equal to the voltage applied to the primary coil times the ratio of the number of turns in the primary coil and the secondary coil.

In the simplest case, where both the primary and secondary coils have the same number of turns, the voltage induced in the secondary coil will be the same as the voltage applied to the primary coil. But what if the primary coil has more turns than the secondary coil? In that case, the voltage induced in the secondary coil will be less than the voltage applied to the primary.

How much less depends on the ratio of the turns in the primary and secondary coils. If the secondary coil has half as many turns as the primary coil, the voltage induced in the secondary coil will be half the voltage applied to the primary coil. For example, if you apply 110 VAC to the primary coil, 55 VAC will be induced in the secondary coil.

Likewise, if the secondary coil has more turns than the primary coil, the induced voltage will be more than the voltage applied to the primary coil. For example, suppose the primary coil has 1,000 turns and the secondary coil has 2,000 turns. Then, if you apply 110 V to the primary coil, 220 V will be induced in the secondary coil.

A transformer whose primary coil has more turns than its secondary coil is called a *step-down transformer* because it reduces voltage — that is, the voltage at the secondary coil is less than the voltage at the primary coil. Similarly, a transformer that has more turns in the secondary than in the primary is called a *step-up transformer* because it increases voltage.

Although the voltage increases in a step-up transformer, the current is reduced proportionately. For example, if the primary coil has half as many turns as the secondary coil, the voltage induced in the secondary coil will be twice the voltage that's applied to the primary coil, but the current that flows through the secondary coil will be half the current flowing through the primary coil.

Similarly, when the voltage decreases in a step-down transformer, the current increases proportionately. Thus, if the voltage is cut in half, the current doubles. (There is a small amount of loss due to the various inefficiencies, but those losses aren't usually significant.)

This makes perfect sense when you think about it. After all, a transformer can't just conjure up power out of thin air. If it could, we'd have solved the planet's energy problems long ago. But there is no such thing as free energy.

Remember the basic formula for calculating electric power:

$$P = V \times I$$

In other words, power equals voltage times current. A transformer transfers power from the primary coil to the secondary coil. Since the power must stay the same, if the voltage increases, the current must decrease. Likewise, if the voltage decreases, the current must increase.

Transformers are the main reason we use alternating current instead of direct current in large power distribution systems. That's because when you send large amounts of power over a long distance, it's much more efficient to send the power in the form of high voltage and low current. That's why overhead power transmission lines often carry voltages as high as 400,000 VAC. Such high voltages allow the electrical power to be transmitted using much smaller wires than would be required if the same amount of power were transmitted at 120 VAC.

Power distribution systems use large step-up transformers to step voltages generated at power plants up to thousands or hundreds of thousands of volts. Then, as the power gets closer to its final destination (such as your house), a series of step-down transformers drops the voltage down to more manageable levels, until the voltage is dropped to its final level (120 VAC) before it enters your house.

REMEMBER

Transformers work only with alternating current. That's because it's the *change* of the magnetic field created by the primary coil that induces voltage in the secondary coil. To create a changing magnetic field, the voltage applied to the primary coil must be constantly changing. Because DC is a steady, fixed voltage, it creates a fixed magnetic field that won't induce voltage in the secondary coil.

Working with Line Voltage

Line voltage refers to the voltage that's available in standard residential or commercial wall outlets. In the United States, this voltage is almost always in the neighborhood of 120 VAC, though it's commonly referred to as 110 VAC, 115 VAC, or 117 VAC. In other parts of the world, the voltage may be lower or higher.

TIP

In Europe, line voltage is often referred to as *mains voltage* or just *mains*.

Unlike the voltage available from household batteries, line voltage is dangerous. In fact, if you're not careful, line voltage can kill you. Thus, you need to take precautions whenever you build a circuit that works with line voltage. In this chapter, you learn how to use line voltage safely so that neither you nor anyone else gets hurt.

Using Line Voltage in Your Projects

So far, none of the projects presented in this book have involved the use of line voltage. But many — if not most — real-world projects do require that you use line voltage.

The most common reason for using line voltage in a project is to eliminate the need for batteries. Batteries are a convenient source of power for your circuits, but they wear out. For many circuits, you want to provide a power source that will last indefinitely. If you use batteries, they'll eventually lose their charge and have to be replaced. If you use line voltage, you can plug the project in and not have to worry about changing batteries.

Of course, most electronic components require direct current rather than alternating current and at much lower voltage levels than the levels that line voltage supplies. Thus, for your project to use line voltage as its source of power, you need to provide the project with a *power supply* that converts the 120 VAC line voltage to something more useful, such as 5 VDC.

There are at least two ways to accomplish this:

» **By using a power adapter:** The easiest way is to use an external *power adapter,* often called a *wall wart* or a *power brick.* Figure 1-4 shows a typical external power adapter. You can purchase power adapters from just about any store that has a consumer electronics department. Just get one that provides the right level of DC voltage and use it instead of batteries.

» **By building your own power supply:** The alternative to purchasing a power adapter is to build your own power supply circuit. This circuit must accomplish two things. First, it must step the voltage down from 120 VAC to whatever voltage your circuit requires, and second, it must convert the AC voltage to DC voltage. You learn how to design and build a power supply circuit in the next chapter.

The second common reason for using line voltage in a project is if the project needs to control some external device that runs on line voltage, such as a flood lamp or a pump. In that case, your project needs to be able to turn the line voltage on and off.

The most common way to turn a line voltage device on and off from an electronic circuit is to use a device called a *relay,* which is basically an electronic switch that uses a low-current input to control a high-current output. For example, a relay can let you use a 12 VDC circuit to control a separate line voltage circuit. You learn how to use a relay for this purpose later in this chapter, in the section "Using Relays to Control Line-Voltage Circuits."

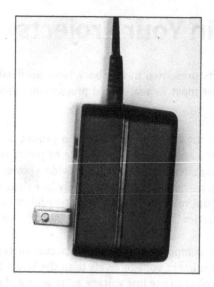

FIGURE 1-4:
An external
power adapter.

Being safe with line voltage

Whenever you build an electronics project that uses line voltage, you must take extra precautions to ensure your safety and the safety of anyone who may come in contact with your project. Line voltage is potentially deadly, so these precautions are absolutely mandatory.

WARNING

Many people are under the mistaken belief that line voltage is not sufficient to cause serious injury or death. That simply isn't true; 120 VAC is more than enough voltage to kill, given the right conditions. In fact, you should treat any voltage above 50 V as potentially lethal.

When you work with line voltage, be sure to take the following precautions:

» *Never* work on the circuit when the power plug is plugged in.

» *Never* leave exposed line-voltage connections anywhere that you or anyone who comes into contact with your project might accidentally touch. *All* line-voltage connections must be completely insulated or contained within an insulated project box or a grounded metal project box.

» *Always* enclose projects that use line voltage in a sealed project box so that stray hands can't accidentally come in contact with bare wires or other components.

» *Always* use grounded power cords if your project is contained in a metal box, and *always* connect the metal box itself to the power cord's ground lead.

>> *Always* use the correct gauge of wire for the amount of current your circuit will be carrying. For more information, see the section, "Wires and Connectors for Working with Line Voltage."

>> *Always* ensure that all line-voltage connections are tight and secure. If you're using stranded wire instead of solid wire, always check for stray strands at your connections.

>> *Always* provide some form of strain-relief for wires that carry line voltage. The most common way to do this is to pass the wire through a grommet-protected hole in the project box and tie the wire into a knot inside the box. The knot will prevent the cord from sliding through the hole in the case and possibly pulling loose.

>> *Always* incorporate a fuse in the primary side of your line-voltage circuit. The fuse will automatically detect when too much current is flowing and immediately break the circuit.

>> *Never* use a fuse that's rated for more than the maximum current your circuit is designed to bear. For example, if you're using a relay that can switch 5 A of current, use a fuse rated for 5 A or less. (For more on fuses, see the section, "Using Fuses to Protect Line-Voltage Circuits," later in this chapter.)

>> *Never* arrange the wiring for your project in a way that causes wires to move or rub against one another. The rubbing will eventually wear off the insulation and create a shock hazard.

>> *Always* be aware of heat sinks that may be hot.

>> *Never* use a three-prong to two-prong adapter to plug a three-prong power connector into a two-prong extension cord. This disables the safety provided by proper grounding and can result in a potentially fatal electric shock if a short circuit occurs.

Understanding hot, neutral, and ground

Before you start working with line voltage in your circuits, you need to understand a few details about how most residential and commercial buildings are wired. The following description applies only to the United States; if you're in a different country, you'll need to determine the standards for your country's wiring.

Standard line voltage wiring in the United States is done with plastic-sheathed cables, which usually have three conductors, as shown in Figure 1-5. This type of cable is technically called *NMB cable*, but most electricians refer to it using its most popular brand name, Romex.

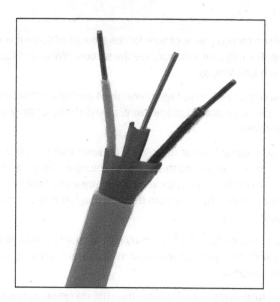

FIGURE 1-5:
NMB cabling.

Two of the conductors in NMB cable are covered with plastic insulation (one white, the other black). The third conductor is bare copper. These conductors are designated as follows:

>> **Hot:** The black wire is the *hot wire*, which provides a 120 VAC current source.

>> **Neutral:** The white wire is called the *neutral wire*. It provides the return path for the current provided by the hot wire. The neutral wire is connected to an earth ground.

>> **Ground:** The bare wire is called the *ground wire*. Like the neutral wire, the ground wire is also connected to an earth ground. However, the neutral and ground wires serve two distinct purposes. The neutral wire forms a part of the live circuit along with the hot wire. In contrast, the ground wire is connected to any metal parts in an appliance such as a microwave oven or coffee pot. This is a safety feature, in case the hot or neutral wires somehow come in contact with metal parts. Connecting the metal parts to earth ground eliminates the shock hazard in the event of a short circuit.

Note that some circuits require a fourth conductor. When a fourth conductor is used, it's covered with red insulation and is also a hot wire.

The three wires in a standard NMB cable are connected to the three prongs of a standard electrical outlet (properly called a *receptacle*), as shown in Figure 1-6. As you can see, the neutral and hot wires are connected to the two vertical prongs at the top of the receptacle (neutral on the left, hot on the right), and the ground wire is connected to the round prong at the bottom of the receptacle.

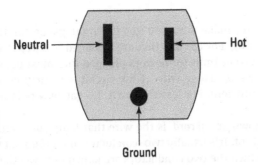

FIGURE 1-6:
A standard
electrical
receptacle.

Neutral

Hot

Ground

You can plug a two-prong or three-prong plug into a standard three-prong receptacle. Two-prong plugs are designed for appliances that don't require grounding. Most nongrounded appliances are *double-insulated*, which means that there are two layers of insulation between any live wires and any metal parts within the appliance. The first layer is the insulation on the wire itself; the second is usually in the form of a plastic case that isolates the live wiring from other metal parts.

Three-prong plugs are for appliances that require the ground connection for safety. Most appliances that use a metal chassis require a separate ground connection.

There is only one way to insert a three-prong plug into a three-prong receptacle. But regular two-prong plugs, which lack the ground prong, can be connected with either prong on the hot side. To prevent that from happening, the receptacles are *polarized*, which means that the neutral prong is wider than the hot prong. Thus, there's only one way to plug a polarized plug into a polarized receptacle. That way, you can always keep track of which wire is hot and which is neutral.

TIP

You should always place switches or fuses on the hot wire rather than on the neutral wire. That way, if the switch is open or the fuse blows, the current in the hot wire will be prevented from proceeding beyond the switch or fuse into your circuit. This minimizes any risk of shock that might occur if a wire comes loose within your project.

Wires and Connectors for Working with Line Voltage

When working with line voltage, you must always use wire that's designed specifically to handle line-voltage currents. Depending on your needs, you may choose to use solid or stranded wire. Stranded wire is usually easier to work with because it's more flexible.

When choosing wire, make sure you get the right gauge for the current your circuit will be carrying. For circuits that are designed to carry 15 A or less (which is the maximum current limit for devices plugged into most household electrical outlets), you can use 14-gauge wire. If the circuit will carry no more than 13 A, 16-gauge wire is sufficient. For less than 10 A, 18-gauge wire is sufficient.

Lamp wire, also known as *zip cord*, is the wire that lamp cords and indoor extension cords are made of. It's usually two-conductor, nongrounded 16- or 18-guage stranded wire in which the two conductors are joined in a way that lets you easily peel them apart. Lamp wire is the easiest wire to use for short connections within your project. You can buy lamp wire in bulk at most hardware stores, or you can buy an inexpensive indoor extension cord at a discount dollar store and cut the ends off.

Make sure that all connections you make with wires carrying line voltage are secure. The easiest way to connect the wires is to use *wire nuts*, illustrated in Figure 1-7. To use wire nuts, strip off 3/8 inch or so of insulation from the two wires to be connected and loosely twist the two ends together. Then, slip the wire nut over the twisted end and tighten the nut onto the connection by pushing down and twisting. When the wire nut is as tight as you can get it, check to make sure that none of the stripped wire extends below the base of the wire nut. For good measure, you can wrap a short strip of black electrical tape around the connection.

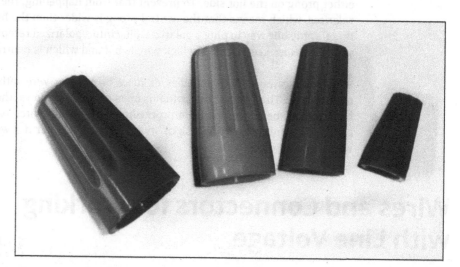

FIGURE 1-7:
Wire nuts.

Another way to make connections is to use a *barrier strip*, as shown in Figure 1-8. Barrier strips come in various sizes and shapes. The small one pictured in the figure can make four connections; the larger one can make eight. To use a barrier strip, simply strip away a short length of insulation from the end of the wires you want to connect and secure them under the terminal screws. If you're using stranded wire, make sure that all the strands are held by the screw. Loose strands can cause short circuits.

Using Fuses to Protect Line-Voltage Circuits

A *fuse* is an inexpensive device that can carry only a certain amount of current. If the current exceeds the rated level, the fuse melts (blows), thus breaking the circuit and preventing the excessive current from flowing. Fuses are an essential component of any electrical system that uses line voltage and has the possibility of short-circuiting or overheating and causing a fire.

The most common type of fuse is the *cartridge fuse*, which consists of a cylindrical body that's usually made of glass, plastic, or ceramic, with two metal ends. The metal ends are the two terminals of the fuse. Inside the body is a thin wire conductor that's designed to melt if the current exceeds the rated threshold. As long as the current stays below the maximum level, the conductor passes the current from one metal end to the other. But when the current exceeds the rated maximum, the conductor melts, and the circuit is broken.

Figure 1-9 shows an *AGC fuse,* which is a small fuse made of glass, 1¼ inches in length and ¼ inch in diameter. This particular fuse is rated at 2 A, but you can get AGC fuses in larger ratings, up to 15 A. (*AGC* stands for *Automotive Glass Cartridge,* but that won't be on the test.)

FIGURE 1-9:
A 2 A, 1¼-x-
¼-inch AGC fuse.

Fuses should always be connected to the hot wire and should be placed before any other component in the circuit. In most projects, the fuse should be the first thing the hot wire connects to after it enters your project enclosure. Figure 1-10 shows how a fuse is depicted in a schematic diagram. Here, the fuse is placed on the hot wire before the lamp.

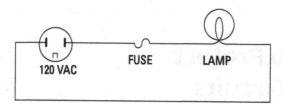

FIGURE 1-10:
A fuse in a
schematic
diagram.

If you plan on using a fuse in your circuit, you'll need to purchase a fuse holder to hold the fuse. For AGC cylindrical fuses, there are two distinct types of fuse hold-ers to choose from. If the fuse will be mounted inside your project's enclosure, you can use a chassis-type fuse holder. If you want the fuse to be accessible from outside the project's enclosure, you should choose a panel-mount holder instead. Figure 1-11 shows both types of holders.

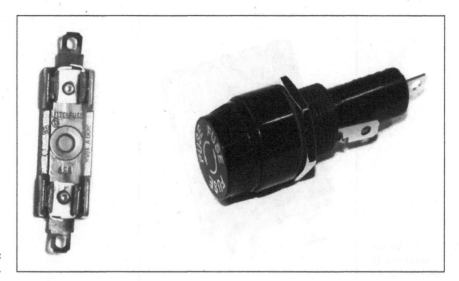

FIGURE 1-11:
Fuse holders.

Using Relays to Control Line-Voltage Circuits

In many projects, you need to turn line-voltage circuits on and off using circuits that use low-voltage DC power supplies. For example, suppose you want to flash a 120 VAC flood lamp on and off at regular intervals. You could build a circuit to provide the necessary timing using a 555 timer IC, as described in Book 3, Chapter 2, but the 555 timer IC requires just a small DC power supply, in the range of 5 to 15 V. And the output current can't exceed 200 mA, not nearly enough to light a flood lamp.

Relays to the rescue!

A *relay* is an electromechanical device that uses an electromagnet to open or close a switch. The circuit that powers the electromagnet's coil is completely separate from the circuit that is switched on or off by the relay's switch, so it's possible to use a relay whose coil requires just a few volts to turn a line voltage circuit on or off.

Figure 1-12 shows a typical relay. For this relay, the coil requires just 12 VDC to operate and pulls just 75 mA, well under the current limit that can be sourced by the 555 timer IC's output pin. But the switch part of this relay can handle up to 10 A of current at 120 VAC, more than enough to illuminate a flood lamp.

FIGURE 1-12:
A relay is a
switch that is
controlled by an
electromagnet.

The switch part of a relay is available in different configurations just like manual switches. The most common switch configuration is double pole, double throw (DPDT), which means that the relay actually controls two separate switches that operate together, and that each switch has both normally open and normally closed contacts.

Figure 1-13 shows a schematic diagram for a simple circuit that uses a 9 VDC circuit with a hand-held push button to turn a 120 VAC lamp on and off. The relay in the circuit has a coil rated for 9 VDC and a switch rating of 10 A at 117 VAC. Thus, only 9 VDC passes through the hand-held push button. If the person holding the switch decides to take it apart, they won't be exposed to dangerous voltage.

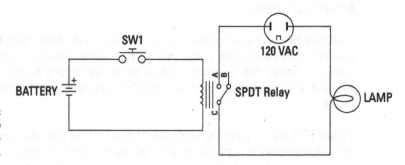

FIGURE 1-13:
Using a relay to
switch a line-
voltage circuit.

Figure 1-14 shows a more complicated circuit, in which a 555 timer IC controls a flood lamp via a relay. Here, one end of the relay coil is connected to the 555 timer IC's output pin (pin 3), and the other end is connected to ground. When the 555's output switches on, the relay closes, and the flood lamp circuit is completed.

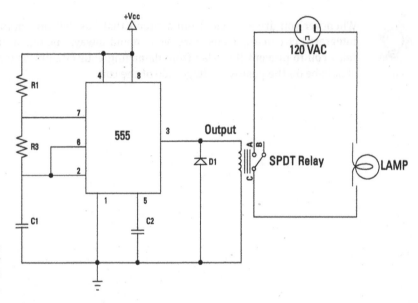

Getting into Alternating Current

FIGURE 1-14:
Driving a relay
from a 555
timer IC.

Note the diode that's placed across the relay coil in this circuit. This diode is required to protect the 555 timer IC from any back-current that might be created within the relay's coil when the coil is energized. Because of electromagnetic induction, relay coils are prone to this problem.

When the coil is energized, it creates a magnetic field that causes the relay's switch contacts to move. However, this magnetic field has a subtle side effect. In the instant that the voltage on the coil goes from zero to the Vss supply voltage, the magnetic field surrounding the coil expands from nothing to its maximum strength. During this expansion, the magnetic field is moving relative to the coil itself. Because of the principal of induction, this moving magnetic field induces a current in the coil itself, in the opposite direction as the current that is energizing the coil.

Depending on the circumstances, this back current can be powerful — powerful enough to overwhelm the output current coming from the 555 timer IC and possibly powerful enough to send current into the output pin, which can damage or destroy the 555 chip. D1 prevents this from happening by providing the equivalent of a short circuit across the coil for current flowing back toward the output pin.

TIP

Whenever you drive a relay from a circuit that has delicate components such as integrated circuits or transistors, you should always include a diode across the relay coil to prevent the relay from damaging your circuits. The diode's cathode should be on the positive voltage side of the coil.

IN THIS CHAPTER

» **Looking at how power supplies work**

» **Stepping down the voltage**

» **Converting AC to DC**

» **Filtering wavy DC**

» **Making the voltage level more reliable through regulation**

Chapter **2**

Building Power Supplies

With very few exceptions, every electronic circuit requires a power supply of some sort. Although some projects run off of solar power or more exotic power sources such as wind turbines, fuel cells, or nuclear reactors, most of the projects you build will get their power from one of two sources: batteries or an electrical outlet.

So far in this book, I assume that all circuits get their power from batteries. In this chapter, you look at how you can get power from an electrical outlet instead. Electrical outlets have a compelling advantage over batteries: Unless there's a power outage, electrical outlets don't go dead like batteries do. However, electrical outlets have an equally compelling disadvantage over batteries: Unless you use really long extension cords, you can't take a project powered from an electrical outlet very far from the outlet.

Most electronic circuits require a relatively low DC voltage, typically in the range of 3 V to 12 V. It's easy to get that range of voltage out of batteries. Since each battery provides about 1.5 V, you just team up two or more batteries to get the right voltage. For example, if your circuit needs 6 V, you can use four batteries connected in series.

Powering a project from an electrical outlet is a little more challenging. First, the 120 V provided at the electrical outlet is much more than most circuits require, so you have to step the voltage down to a more appropriate level. Second, electronic circuits usually require direct current — and the wall outlet provides alternating current — so you have to convert the AC to DC. And third, circuits that run directly on 120 VAC are inherently more dangerous than circuits that run on lower voltages because of the shock danger that accompanies higher voltages.

The circuit that converts 120 VAC to direct current at a lower voltage is called a *power supply.* In this chapter, you learn the basics of creating your own power supplies so you can power your projects from a wall outlet instead of from batteries.

Using a Power Adapter

Before I show you how to build your own power supply circuit, I want to let you know that you can probably purchase a preassembled *power adapter* that will provide the voltage you need for just a few dollars more than you can build the circuit yourself. A power adapter, also called a *wall wart*, is a self-contained power supply circuit that plugs into a wall outlet and provides a specified level of AC or DC voltage as its output. As long as the power adapter supplies the correct voltage, you can use the power adapter instead of batteries in just about any circuit.

When you purchase a power adapter, you need to check the specifications to make sure you're purchasing the correct adapter. The specifications are usually printed on the adapter itself. Look for the following important specifications:

>> **AC or DC:** Not all power adapters supply direct current; some are made to power low-voltage AC devices. So make sure that you get an adapter that provides direct-current output.

>> **Voltage level:** Next, check the output voltage. Some power adapters have a switch that lets you choose from among several output voltages. If you use such an adapter, make sure that you set the switch to the correct output level for your circuit.

>> **Current capacity:** Most power adapters will have a maximum current rating expressed in milliamps. Smaller adapters can handle a few hundred milliamps, whereas larger adapters may be able to handle an ampere or more. Make sure that the adapter you use can handle the current requirements of your project. (Although some power adapters can handle several amps, few can handle more than that.)

>> **Polarity:** Most power adapters use a *barrel connector* to plug the power adapter into the circuit. In nearly all modern power adapters, the center connection of the barrel connector is positive, and the outer connection is

negative. However, some power adapters are wired just the opposite, with negative in the center and positive on the outside. The polarity of the connector should be printed on the adapter along with the voltage and current specifications.

>> **Connector size:** Unfortunately, there are far too many different sizes and styles of connectors used for power adapters. Once you've purchased a power adapter, you can go to a local electronics store such as RadioShack and purchase a jack that is compatible with the connector on the power adapter. Then, you can use the jack to connect the power adapter to your circuit. (Note that some power adapters have different interchangeable plugs in different sizes.)

TIP

Using a power adapter instead of building your own power supply can make your project safer to build and use. That's because the part of your project that is potentially dangerous — the part that works directly with 120 VAC line voltage — is fully contained inside the preassembled power adapter.

However, as you'll soon discover, you get what you pay for when it comes to power supplies. Inexpensive wall warts convert AC to DC and step down the voltage, but most do not provide power that is very clean (that is, a pure level of DC) or stable (that is, with a predictable voltage). Thus, even if you use a wall wart to power your project, you may still need to add circuitry that will improve the quality of the DC supplied by the wall wart.

Understanding What a Power Supply Does

If you want to add your own power supply circuit to a project to convert 120 VAC line voltage to a DC voltage that's suitable for your circuit, you'll have to design a power supply circuit that provides at least three distinct functions:

>> **Voltage transformation:** Reduces the 120 VAC line voltage to the voltage your circuit needs.

>> **Rectification:** Converts the reduced AC voltage to DC voltage. Note that the DC voltage produced by a rectifier circuit is technically direct current, but it isn't steady direct current. Instead, a rectifier produces pulsating direct current in which the voltage fluctuates in sync with the 60 Hz alternating current that's fed into it from the transformation stage.

>> **Filtering:** Smoothes out the ripples in the DC voltage produced by the rectification stage.

Transforming Voltage

You already know that a *transformer* is a device that uses the principal of electro-magnetic induction to transfer voltage and current from one circuit to another. The transformer uses a *primary coil* that's connected to line voltage and a *secondary coil* that provides the output voltage.

In most power supplies, the transformer reduces the voltage. The amount of the voltage reduction depends on the ratio of the number of turns in the primary coil versus the number of turns in the secondary coil. For example, if the secondary coil has half as many turns as the primary coil, the primary coil voltage will be cut in half at the secondary coil. In other words, if 120 VAC is applied to the primary coil, 60 VAC will be available at the secondary coil.

Common secondary voltages for transformers used in low-voltage power supplies range from 6 V to 24 VAC. Note that because some voltage will be lost in the recti-fier and filtering stages, you'll want to choose a secondary coil voltage that's a few volts higher than the final DC voltage your circuit actually needs. (Note, however, that the actual DC voltage level used for most circuits isn't all that critical. So if you're designing a power supply for a circuit that calls for 6 VDC and you use a transformer that provides 6 VAC in its secondary coil, the output from the power supply after it's rectified to DC voltage will be closer to 5 VDC. Most likely, 5 VDC will be close enough, and the circuit will work just fine.

Note that many transformers have more than one *tap* in the secondary coil. A tap is simply a wire connected somewhere in the middle of a coil, effectively dividing a single coil into two smaller coils. Multiple taps let you access several different voltages in the secondary coil. The most common arrangement is a *center-tapped* transformer, which provides two voltages, as shown in Figure 2-1.

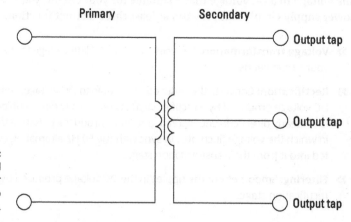

Primary   Secondary

Output tap

Output tap

FIGURE 2-1:
A center-tapped transformer provides two output voltages.

Output tap

In a center-tapped transformer, the voltage measured across the two outer taps is double the voltage measured from the center tap to either one of the two outer taps. Thus, if the voltage across the two outer taps is 24 VAC, the voltage across the center tap and either of the outer taps is 12 VAC.

TECHNICAL STUFF

It's important to note that when a transformer reduces voltage, it increases current. Thus, if a transformer cuts the voltage in half, the current will double. As a result, the overall power in the system (defined as the voltage multiplied by the current) remains the same.

If the current didn't increase as the voltage decreased, the transformer would violate a basic law of physics — the one about conservation of energy, which says that energy can't just disappear. That's a good thing. You don't want to be violating the laws of physics unless you know what you're doing or you're in a science fiction movie, in which case you can violate the laws of physics at will.

REMEMBER

A transformer is strictly an alternating current device. That means:

>> Transformers work only when alternating current is applied to the primary coil. If you apply direct current to the primary coil, no voltage will appear across the secondary coil. (Actually, there will be a brief spike of voltage across the secondary coil the moment voltage is applied to the primary coil, but in most circuits this fleeting voltage is insignificant.)

>> A step-down transformer reduces the voltage from the primary to the secondary coils but *doesn't* convert alternating current to direct current. The voltage at the secondary coil is always AC.

>> A transformer isolates the circuit attached from the secondary coil from the circuit connected to the primary coil. Thus, you can use a transformer to isolate your project from line voltage.

Turning AC into DC

The task of turning alternating current into direct current is called *rectification*, and the circuit that does the job is called a *rectifier*. The most common way to convert alternating current into direct current is to use one or more *diodes*, those handy electronic components that allow current to pass in one direction but not the other. Diodes are covered in detail in Book 2, Chapter 5. You may want to briefly review that chapter before reading further if the concept doesn't sound familiar.

Although a rectifier converts alternating current to direct current, the resulting direct current isn't a steady voltage. It would be more accurate to refer to it as "pulsating DC." Although the pulsating DC current always moves in the same direction, the voltage level has a distinct ripple to it, rising and falling a bit in sync with the waveform of the AC voltage that's fed into the rectifier. For many DC circuits, a significant amount of ripple in the power supply can cause the circuit to malfunction. Therefore, additional filtering is required to "flatten" the pulsating DC that comes from a rectifier to eliminate the ripple. (For more on filtering, see the section, "Filtering Rectified Current," later in this chapter.)

There are three distinct types of rectifier circuits you can build: half-wave, full-wave, and bridge. The following sections describe each of these three rectifier types.

Half-wave rectifier

The simplest type of rectifier is made from a single diode, as shown in Figure 2-2. This type of rectifier is called a *half-wave rectifier* because it passes just half of the AC input voltage to the output. When the AC voltage is positive on the cathode side of the diode, the diode allows the current to pass through to the output. But when the AC current reverses direction and becomes negative on the cathode side of the diode, the diode blocks the current so that no voltage appears at the output.

Half-wave rectifiers are simple enough to build but aren't very efficient. That's because the entire negative cycle of the AC input is blocked by a half-wave rectifier. As a result, output voltage is zero half of the time. This causes the average voltage at the output to be half of the input voltage.

TECHNICAL STUFF

Note the resistor marked R_L in Figure 2-2. This resistor isn't actually a part of the rectifier circuit. Instead, it represents the resistance imposed by the load that will ultimately be placed on the circuit when the power supply is put to use.

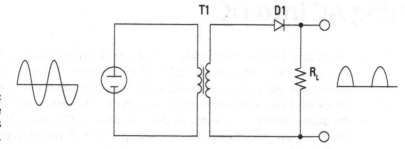

FIGURE 2-2:
A half-wave rectifier uses just one diode.

Full-wave rectifier

A *full-wave rectifier* uses two diodes, which enables it to pass both the positive and the negative side of the alternating current input. The diodes are connected to the transformer, as shown in Figure 2-3.

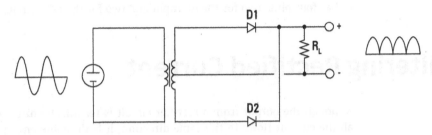

FIGURE 2-3:
A full-wave
rectifier uses
two diodes.

Notice that the full-wave rectifier requires that you use a center-tapped transformer. The diodes are connected to the two outer taps, and the center tap is used as a common ground for the rectified DC voltage. The full-wave rectifier converts both halves of the AC sine wave to positive-voltage direct current. The result is DC voltage that pulses at twice the frequency of the input AC voltage. In other words, assuming the input is 60 Hz household current, the output will be DC pulsing at 120 Hz.

Bridge rectifier

The problem with a full-wave rectifier is that it requires a center-tapped transformer, so it produces DC that's just half of the total output voltage of the transformer. A *bridge rectifier*, shown in Figure 2-4, overcomes this limitation by using four diodes instead of two. The diodes are arranged in a diamond pattern so that, on each half-phase of the AC sine wave, two of the diodes pass the current to the positive and negative sides of the output, and the other two diodes block current. A bridge rectifier doesn't require a center-tapped transformer.

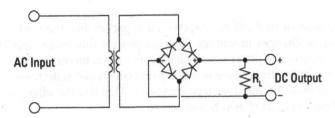

FIGURE 2-4:
A bridge rectifier
uses four diodes.

The output from a bridge rectifier is pulsed DC, just like the output from a full-wave rectifier. However, the full voltage of the transformer's secondary coil is used.

You can construct a bridge rectifier using four diodes, or you can use a bridge rectifier IC that contains the four diodes in the correct arrangement. A bridge rectifier IC has four pins: two for the AC input and two for the DC output.

Filtering Rectified Current

Although the output from a rectifier circuit is technically direct current because all the current flows in the same direction, it isn't stable enough for most purposes. Even full-wave and bridge rectifiers produce direct current that pulses in rhythm with the 60 Hz AC sine wave that originates with the 120 VAC current that's applied to the transformer. And that pulsing current isn't suitable for most electronic circuits.

That's where filtering comes in. The filtering stage of a power supply circuit smoothes out the ripples in the rectified DC to produce a smooth direct current that's suitable for even the most sensitive of circuits.

Filtering is usually accomplished by introducing a capacitor into the power supply circuit, as shown in Figure 2-5. Here, the capacitor is simply placed across the DC output.

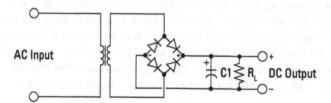

FIGURE 2-5: A capacitor can be used to filter the output from the rectifier.

As you learn in Book 2, Chapter 3, a capacitor has the useful characteristic of resisting changes in voltage. It accomplishes this magic feat by building up a charge across its plates when the input voltage is increasing. When the input voltage decreases, the voltage across the capacitor's plates decreases as well, but more slowly than the input voltage decreases. This has the effect of leveling out the voltage ripple, as shown in Figure 2-6.

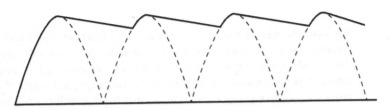

FIGURE 2-6:
A filter circuit
smooths the
output voltage.

The difference between the minimum DC voltage and the maximum DC voltage in the filtering stage is called the *voltage ripple*, or just *ripple*, which is usually measured as a percentage of the average voltage. For example, a 10% ripple in a 5 V power supply means that the actual output voltage varies by 0.5 V.

The filter capacitor must usually be large to provide an acceptable level of filtering. For a typical 5 V power supply, a 2,200 μF electrolytic capacitor will do the job. The bigger the capacitor, the lower the resulting ripple voltage.

TIP

Don't forget to watch the polarity on electrolytic capacitors. The positive side of the capacitor must be connected to the positive voltage output from the rectifier, and the negative side must be connected to ground.

One way to improve the filter circuit is to use two capacitors in combination with a resistor, as shown in Figure 2-7. In this circuit, the first capacitor acts like the capacitor in Figure 2-6, eliminating a large portion of the ripple voltage. The resistor and second capacitor work as an RC network that eliminates the ripple voltage even further.

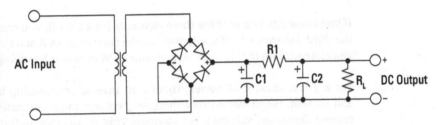

FIGURE 2-7:
Two capacitors
and a resistor
cut ripple voltage
but also reduce
the DC output
voltage.

The advantages of this circuit are that the resulting DC has a smaller ripple voltage and the capacitors can be smaller. The disadvantage is that the resistor drops the DC output voltage. How much depends on the amount of current drawn by the load. For example, if you use a 100 Ω resistor and the load draws 100 mA, the resistor will drop 10 V (100 × 0.1). Thus, to provide a final output of 5 V, the rectifier circuit must supply 15 V because of the 10 V drop introduced by the resistor.

Building Power Supplies

You can also use an inductor in a filter circuit, as shown in Figure 2-8. Unlike a resistor-capacitor filter, an inductor-capacitor filter doesn't significantly reduce the DC output voltage. Although inductor-capacitor filter circuits create the smallest ripple voltage, inductors in the range needed (typically 10 henrys) are large and relatively expensive. Thus, most filter circuits use a single capacitor or a pair of capacitors coupled with a resistor.

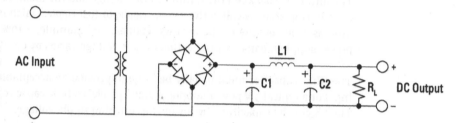

FIGURE 2-8:
An inductor can be used in a filter circuit to minimize DC voltage loss.

Regulating Voltage

The purpose of a power supply is to provide power for an electronic circuit. There's a basic formula for calculating the amount of power circuit uses:

$$P = I \times V$$

Power, measured in watts, is equal to current measured in amperes times voltage measured in volts.

If you know any two of these three elements for a circuit, you can easily calculate the third. For example, if you know that the current is 0.5 A and the voltage is 10 V, you can calculate that the circuit consumes 5 W of power by multiplying 0.5 by 10.

For a given amount of power, there's an inverse relationship between voltage and current. Whenever current increases, voltage must decrease, and whenever current decreases, voltage must increase. This simple fact, unfortunately, has an adverse effect on power supply circuits. When you connect a voltmeter to the output terminals of a power supply, the meter itself draws an almost insignificant amount of current, so the meter reads very close to the voltage you expect to obtain from the power supply.

However, if you connect a circuit that draws significant current from the power supply, the voltage from the power supply will drop in proportion to the current. Depending on the nature of the circuit you're connecting to the power supply, this voltage drop may or may not be a bad thing. Some circuits designed for 12 VDC will work fine if only given 9 VDC. But other circuits are sensitive to the input voltage, so the power supply needs to work harder to make sure it delivers the desired voltage.

To maintain a steady voltage level regardless of the amount of current drawn from a power supply, the power supply can incorporate a *voltage regulator* circuit. The voltage regulator monitors the current drawn by the load and increases or decreases the voltage accordingly to keep the voltage level constant.

TIP

A power supply that incorporates a voltage regulator is called a *regulated power supply*.

You can, if you want, design your own voltage regulator circuit using a couple of transistors, some resistors, and a Zener diode. However, it's far too easy to buy one of the many available integrated circuit voltage regulators. Voltage regulator ICs are inexpensive (under $2) and, with just three pins to connect, easy to incorporate into your circuits.

The most popular type of voltage regulator IC is the *78XX* series, sometimes called the *LM78XX* series. These voltage regulators combine 17 transistors, three Zener diodes, and a handful of resistors into one handy package with three pins and a heat sink that helps dissipate the excess power consumed by the regulator as it compensates for increases or decreases in current draw to keep the voltage at a constant level.

The last two digits of the 78XX ID number indicate the output voltage regulated by the IC. The most popular models are:

Model	Voltage
7805	5
7806	6
7809	9
7810	10
7812	12
7815	15
7818	18
7824	24

Of these, the most common are the 7805 (5 V) and 7812 (12 V).

To use a 78XX voltage regulator, you just insert it in series on the positive side of the power supply circuit and connect the ground lead to the negative side, as shown in Figure 2-9. As this figure shows, it's also a good idea to place a small capacitor (typically 0.1 μF) after the regulator.

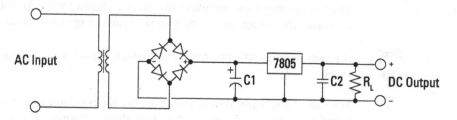

FIGURE 2-9:
Using a 78XX
voltage regulator.

You must supply a voltage regulator with about 3 V more than the regulated output voltage. Thus, for a 7805 regulator, you should give it at least 8 V. The maximum input voltage for a 7805 is 30 V. Remember that the diodes in a bridge rectifier will drop about 3 V from the transformer output, so you'll need a transformer whose secondary delivers at least 11 V to produce 5 V of regulated output.

Eleven-volt transformers are rare, but 12 V transformers are readily available. Thus, a 5 V regulated power supply typically starts with a 12 VAC transformer that delivers 12 V to the bridge rectifier, which converts the AC to DC and drops the voltage down to about 9 V and then delivers the voltage to the filter circuit, which smoothes out the ripples and passes the voltage on to the 7805 voltage regulator, which holds the output voltage at 5 V.

Another popular voltage regulator IC is the LM317, which is an adjustable voltage regulator. An LM317 regulator works much like a 78XX regulator, except that instead of connecting the middle lead directly to ground, you connect it to a voltage divider built from a pair of resistors, as shown in Figure 2-10. The value of the resistors determines the regulated voltage. In Figure 2-10, I used a potentiometer so that the user can vary the output voltage by adjusting the potentiometer.

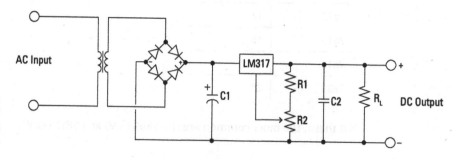

FIGURE 2-10:
Using an LM317
adjustable
voltage regulator.

Chapter **3**

Understanding Radio

t's pet peeve time! One of my pet peeves is the phrase, "The Golden Age of Radio," which implies that the height of radio's popularity was in the 1930s and '40s and that radio has declined in popularity since then.

In reality, radio — the technology if not the audio programming — has never been more popular than it is right now. In the 1930s and '40s, most people knew radio for just one thing: broadcasting audio signals. Today, audio broadcast over radio is as commonplace as ever, but the list of other types of information being broadcast by radio technology has skyrocketed.

First, there was broadcast television, which is nothing more than the combination of audio and video broadcast over radio. Then, there were cellphones, which use radio to extend the world's telephone networks to places that phone cables can't reach. Then there was wireless networking, which replaced bulky computer network cables with data transmitted over radio. And now there are cellular data plans, which transmit Internet data over radio. And there are many other popular uses for radio technology, including radar, GPS navigation systems, and wireless Bluetooth devices.

In this chapter, you learn some of the basic concepts of radio, including what it is, how it works, and how it was discovered. Along the way, you learn a few interesting — and in some cases, sad — stories about the early pioneers of radio. And by the time you're done, you'll learn how to build a crystal radio — the simplest of all radio circuits.

Throughout this chapter, I occasionally violate my own pet peeve and refer to radio as if it's only for the broadcast of sound. Whenever I do, just keep in mind that radio is also used for broadcasting video and digital data as well as other types of information.

Pet peeves are interesting things, aren't they? I think peeves do make interesting pets. They're not as cute as puppies or kittens, but if you get a pet peeve when it's young, it can bring you a lifetime of pleasure. Be careful, however. Like dogs that bark late at night or cats that roam in other people's yards, a pet peeve may annoy your friends and neighbors.

I have always wondered why cities and counties require that you register pet dogs and cats but don't require that you register pet peeves. Just think of the revenue local governments are missing out on! Even a fee as low as $10 per peeve could raise enough money to fill in the potholes on all those roads paved with good intentions, and a portion of the money could go to care for homeless peeves.

Understanding Radio Waves

Most people think of radio as wireless broadcast of sound, most often music and speech. But the term *radio* is actually much broader than that; the broadcast of sound is actually just one application of the extremely useful electrical phenomenon that is called radio.

Radio takes advantage of one of the most interesting of all electrical phenomena: *electromagnetic radiation* (often abbreviated *EMR*), which is a type of energy that travels in waves at the speed of light. EMR travels freely through the air and even in the vacuum of space.

EMR waves can oscillate at any imaginable frequency. The rate of the oscillation is measured in cycles per second, also known as *hertz* (abbreviated Hz). The term *hertz* here does *not* refer to the car rental company. Instead, it honors the great German physicist Heinrich Hertz, who was the first person to build a device that could create and detect radio waves.

Radio is simply a specific range of frequencies of EMR waves. The low end of this range is just a few cycles per second, and the upper end is about 300 billion cycles per second (also known as *gigahertz*, abbreviated *GHz*). That's a pretty big range, but EMR waves with much higher frequencies exist as well and are in fact commonplace. EMR waves with frequencies higher than radio waves go by various names, including infrared, ultraviolet, X-rays, gamma rays, and — most importantly — visible light.

That's right; what we call light is exactly the same thing as what we call radio but at higher frequencies. The frequency of visible light is measured in billions of hertz, also called *terahertz* and abbreviated *THz*. The low end of visible light (red) is around 405 THz and the upper end (violet) is around 790 THz.

So here's an interesting thought to ponder: Radio stations broadcast on a specific frequency. For example in San Francisco, there's a popular radio station called KNBR, which has been broadcasting on the frequency 680 kHz since 1922. There are plenty of other radio stations in the area, but only KNBR broadcasts at 680 kHz.

The term *channel* is often used to refer to a radio station broadcasting at a particular frequency. For example, if someone asks me what radio channel I like to listen to when I'm in San Francisco, I would tell him KNBR.

I might listen to KNBR on my portable radio, which happens to be made of purple plastic. *Purple* is the color we perceive when we see light with frequency right around 680 THz. There are many other colors, but only the color purple is at 680 THz. So in a way, color is the same thing as channel. If EMR waves are vibrating at 680 kHz, they are KNBR radio. If those same EMR waves are vibrating a million times faster, at 680 THz, they are the color purple.

TECHNICAL STUFF

An important concept that's related to frequency is the idea of wavelength. The term *wavelength* refers to the distance between the crests of each cycle of EMR at a particular frequency. Because EMR waves travel at the speed of light, you can calculate the wavelength of a given frequency by dividing the distance that light travels in a single second by the number of cycles per second.

Light is pretty fast: It scoots along at 186,282 miles per second. Thus, the wavelength of an EMR wave oscillating at 100 kHz is about 1.86 miles: 186,282 divided by 100,000.

The higher the frequency, the shorter the wavelength. The wavelength of most AM broadcast radio stations is a few hundred feet. The wavelength of visible light is a very small fraction of an inch.

TECHNICAL STUFF

Sound waves are not a type of EMR. Sound waves are created when particles of matter bump against each other. Thus, you must have matter — such as air or water — to transmit sound waves. Radio waves don't require particles of matter to travel. In fact, radio waves travel best in the vacuum of space, where there's no matter to get in the way.

WHO REALLY INVENTED RADIO?

The history of radio technology is plagued by controversy over the question of who actually invented the thing. The answer most often given is Italian inventor Guglielmo Marconi, but many others made important discoveries that give them good claim to the title.

Here's a rundown on the contenders for the title of The Father of Radio:

- **Marconi:** He was the first person to demonstrate radio successfully and exploit it commercially. In 1901, Marconi sent a message via radio across the Atlantic from England to Canada, though the message was faint, consisted of nothing but the letter *S*, and reception of the message wasn't independently confirmed. Nevertheless, Marconi's accomplishment was astonishing, and he made many important contributions to the technology and business of radio.

- **Tesla:** In 1943, the U.S. Supreme Court ruled that many of Marconi's important radio patents were invalid because Nikola Tesla had already described the devices covered by Marconi's patents. Tesla was a brilliant engineer who is best known for being the champion of alternating current over direct current for power distribution. He publicly demonstrated wireless communication devices as early as 1893. Tesla believed that wireless technology would be used not only for communication, but for power distribution as well.

- **Lodge:** In England, Sir Oliver Lodge was building wireless telegraph systems in the mid-1890s.

- **Popov:** In Russia, Alexander Stepanovich Popov was demonstrating wireless telegraph transmissions around the same time as Lodge.

- **Bose:** In India, Sir Jagadish Chandra Bose was also demonstrating wireless telegraph transmissions in the early 1890s. Whether these demonstrations occurred before, after, or at the same time as other demonstrations by Lodge, Popov, and others is under dispute.

- **Many others:** The list of names of others who did important research and made important discoveries in the last decades of the nineteenth century is long: Heinrich Hertz, Edouard Branly, Roberto de Moura, Ernest Rutherford, Ferdinand Braun, Julio Baviera, and Reginald Fessenden are just a few of the many individuals who made important contributions.

So it seems that no one person has a clear-cut claim to being the first to invent radio. Work was being done all around the world and discoveries were being made, it seems, every day.

It may be that the best answer is that no one "invented" radio. Radio is a natural phenomenon. It was *discovered*, not *invented*.

What was invented were ways to exploit the phenomenon of radio by building devices that could generate radio waves and modulate them to add information, as well as devices that could receive radio waves and extract the information that was added.

Transmitting and Receiving Radio

There are many natural sources of radio waves. But in the later part of the nineteenth century, scientists figured out how to generate radio waves using electric currents. In a nutshell, if you pass an alternating current into a length of wire, radio waves at the same frequency as the alternating current are generated.

Two components are required for radio communication: a *transmitter* and a *receiver*. The transmitter generates radio waves and the receiver detects them. The following sections describe the basic operation of radio transmitters and receivers.

Understanding radio transmitters

A radio transmitter consists of several elements that work together to generate radio waves that contain useful information such as audio, video, or digital data. These components are shown in Figure 3-1 and described here:

>> **Power supply:** Provides the necessary electrical power to operate the transmitter.

>> **Oscillator:** Creates alternating current at the frequency on which the transmitter will transmit. The oscillator usually generates a sine wave, which is referred to as a *carrier wave*.

>> **Modulator:** Adds useful information to the carrier wave. There are two main ways to add this information. The first, called amplitude modulation or AM, makes slight increases or decreases to the intensity of the carrier wave. The second, called frequency modulation or FM, makes slight increases or decreases to the frequency of the carrier wave. For more information about AM and FM, see the sections "Understanding AM Radio" and "Understanding FM Radio," later in this chapter.

(Actually, there is a third method of adding information to a radio signal: by simply turning the signal on and off in a pattern that represents the information. For example, radio signals can send Morse code in this way.)

» **Amplifier:** Amplifies the modulated carrier wave to increase its power. The more powerful the amplifier, the more powerful the broadcast.

» **Antenna:** Converts the amplified signal to radio waves.

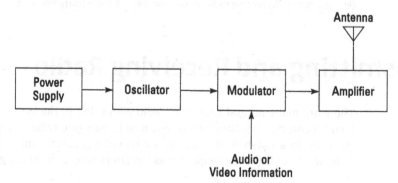

FIGURE 3-1:
The basic components of a radio transmitter.

Understanding radio receivers

A radio receiver is the opposite of a radio transmitter. It uses an antenna to capture radio waves, processes those waves to extract only those waves that are vibrating at the desired frequency, extracts the audio signals that were added to those waves, amplifies the audio signals, and finally plays them on a speaker. Figure 3-2 shows these components, and the following paragraphs explain how each works:

» **Antenna:** Captures the radio waves. Typically, the antenna is simply a length of wire. When this wire is exposed to radio waves, the waves induce a very small alternating current in the antenna.

» **RF amplifier:** A sensitive amplifier that amplifies the very weak radio frequency (RF) signal from the antenna so that the signal can be processed by the tuner.

» **Tuner:** A circuit that can extract signals of a particular frequency from a mix of signals of different frequencies. On its own, the antenna captures radio waves of all frequencies and sends them to the RF amplifier, which dutifully amplifies them all. Unless you want to listen to every radio channel at the same time, you need a circuit that can pick out just the signals for the channel you want to hear. That's the role of the tuner.

The tuner usually employs the combination of an inductor (for example, a coil) and a capacitor to form a circuit that resonates at a particular frequency. This frequency, called the *resonant frequency,* is determined by the values chosen for the coil and the capacitor. This type of circuit tends to block any AC signals at a frequency above or below the resonant frequency.

You can adjust the resonant frequency by varying the amount of inductance in the coil or the capacitance of the capacitor. In simple radio receiver circuits, such as the one you learn about at the end of this chapter, the tuning is adjusted by varying the number of turns of wire in the coil. More sophisticated tuners use a variable capacitor (also called a *tuning capacitor*) to vary the frequency.

» **Detector:** Responsible for separating the audio information from the carrier wave. For AM signals, this can be done with a diode that just rectifies the alternating current signal. What's left after the diode has its way with the alternating current signal is a direct current signal that can be fed to an audio amplifier circuit. For FM signals, the detector circuit is a little more complicated.

» **Audio amplifier:** This component's job is to amplify the weak signal that comes from the detector so that it can be heard. This can be done using a simple transistor amplifier circuit, as described in Book 2, Chapter 6. You can also use an op-amp IC, as described in Book 3, Chapter 3.

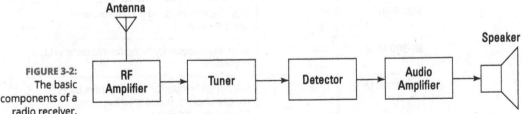

FIGURE 3-2:
The basic components of a radio receiver.

Of course, there are many variations on this basic radio receiver design. Many receivers include additional filtering and tuning circuits to better lock on to the intended frequency — or to produce better-quality audio output — and exclude other signals. Still, these basic elements are found in most receiver circuits.

LOOKING AT THE RADIO SPECTRUM

The term *spectrum* simply means a range of frequencies. Radio is generally considered to be frequencies between 3 Hz to 300 GHz. That broad range of frequencies is carved up into smaller pieces that are used for specific types of radio, as described in the following table:

Frequency	Abbreviation	Description
3–30 Hz	ELF	Extremely low frequency, used for communications with submarines.
30–300 Hz	SLF	Super low frequency, also used for communications with submarines.
300 Hz–3 kHz	ULF	Ultra-low frequency, used for underground communications within mines.
3–30 kHz	VLF	Very low frequency, also for submarine communications and a few other unusual applications.
30–300 kHz	LF	Low frequency, used for navigation, RFID, and a few other applications.
300–3,000 kHz	MF	Medium frequency, used for AM radio.
3–30 MHz	HF	High frequency, used for shortwave and CB radio.
30–300 MHz	VHF	Very high frequency, used for FM radio and television.
300–3,000 MHz	UHF	Ultra-high frequency, used for television, mobile phones, wireless networking, Bluetooth, and so on.
3–30 GHz	SHF	Super high frequency, used for high-speed wireless networking, radar, and communication satellites.
30–300 GHz	EHF	Extreme high frequency, used for microwave communications.
300–3,000 GHz	THF	Tremendously high frequency (no, I didn't make that up), used for exotic applications that border on science fiction.
Above 3,000 GHz		You're off the edge of the spectrum map, mate. Here there be dragons.

Understanding AM Radio

The original method of encoding sound information on radio waves is called *amplitude modulation*, or *AM*. It was developed in the first few decades of the twentieth century. AM is a relatively simple way to add audio information to a carrier wave so that sounds can be transmitted.

One of the simplest forms of AM modulators simply runs the power supply for an oscillator circuit through an audio transformer that is coupled to a microphone or other sound source. Figure 3-3 shows this arrangement.

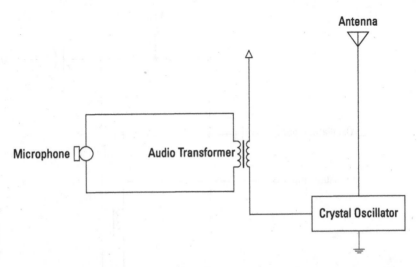

FIGURE 3-3:
The basic AM modulator circuit.

The circuit in Figure 3-3 uses a 1 MHz *crystal oscillator*, which is often used to generate the clock frequencies for microprocessor circuits. 1 MHz is perfect for a simple AM transmitter circuit because 1 MHz falls right in the middle of the band that's used for AM radio transmissions.

Although you can't buy a crystal oscillator at your local RadioShack, you can get it on the Internet. Use Google to search for *1 MHz crystal oscillator*, and you should find several online sources that will sell you one for under $2.

The crystal oscillator is contained in a metal can that has three pins. One pin is for ground, the second pin is the supply voltage (typically 9 VDC), and the third is the oscillator output.

By running the Vss supply through the secondary coil of a transformer with the primary coil connected to an audio input source such as a microphone, the actual voltage supplied to the oscillator will fluctuate based on the variations in the input signal. Because crystal oscillators are very stable, these voltage variations won't affect the frequency generated by the oscillator, but they will affect the voltage of the oscillator output. Thus, the audio input signal will be reflected as voltage changes in the oscillator's output signal.

A better AM modulation circuit uses a transistor, as shown in Figure 3-4. In this circuit, the carrier-wave generated by an oscillator that isn't shown in the circuit is applied to the base of a transistor. Then, the audio input is applied to the transistor's emitter through a transformer. The AM signal is taken from the transistor's collector.

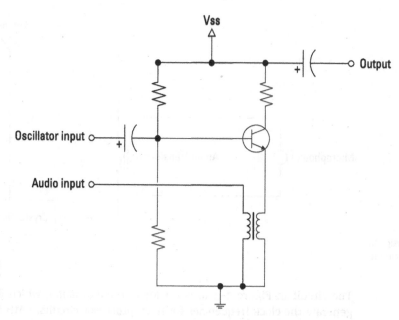

FIGURE 3-4:
Using a transistor for amplitude modulation.

So how does this circuit work? The transistor amplifies the input from the oscillator through the emitter-collector circuit. However, as the audio input varies, it induces a small current in the secondary coil of the transformer. This, in turn, affects the amount of current that flows through the collector-emitter circuit. In this way, the intensity of the output varies with the audio input.

Figure 3-5 shows how a carrier wave is combined with an audio signal to produce an AM radio waveform. As you can see, the carrier wave is a constant frequency and amplitude. In other words, each cycle of the sine wave is of the same intensity. The current of the audio wave varies, however. When the two are combined by the modulator circuit, the result is a signal with a steady frequency, but the intensity of each cycle of the sine wave varies depending on the intensity of the audio signal.

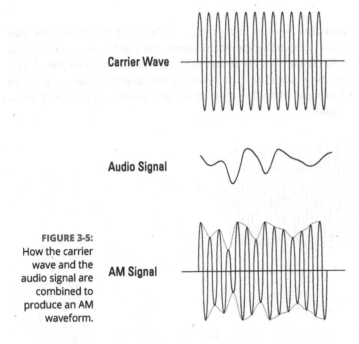

Carrier Wave

Audio Signal

FIGURE 3-5:
How the carrier
wave and the
audio signal are
combined to
produce an AM
waveform.

AM Signal

Understanding FM Radio

AM radio is relatively simple. However, it has several weaknesses. The main draw-back of AM radio is that it's difficult, if not impossible, for an AM radio receiver to distinguish between a signal broadcast by a radio transmitter and spurious signals at the same frequency generated by other sources. The most obvious example of this is lightning. When lightning strikes, it generates a brief but powerful burst of electromagnetic radiation with a very large spectrum of frequencies. The noise generated by a lightning strike includes just about the entire range of frequencies used by AM radio. If you're listening to an AM radio station when the lightning strikes, the sudden burst of radio energy on the frequency you're listening to will be interpreted as sound. Thus, when lightning strikes, you can hear it on the radio.

Signals that interfere with an intentional broadcast are called *static*, and static is the main drawback of AM radio. To counteract static, a better method of superim-posing information on a radio wave, called *frequency modulation* or *FM*, was devel-oped in 1933. (See the sidebar titled "The tragic genius behind FM radio" for the fascinating and sad story about the inventor of FM radio.)

In frequency modulation, the intensity of the carrier wave isn't varied. Instead, the exact frequency of the carrier wave is varied in sync with the audio signal. When the audio signal is higher, the frequency of the broadcast signal goes up a little. When the audio signal is lower, the frequency goes down a bit.

Figure 3-6 shows how this appears in a graph. At the top of the figure, you can see the carrier wave that clocks the specific frequency of the broadcast station. In the middle, you can see the audio signal that is to be superimposed on the carrier wave. And at the bottom, you can see the resulting modulated signal. As you can see, the frequency decreases when the input signal gets lower and increases when the input signal is higher.

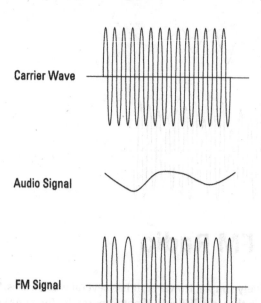

Carrier Wave

Audio Signal

FM Signal

FIGURE 3-6: How the carrier wave and the audio signal are combined to produce an FM waveform.

Note that the frequency variations in a frequency-modulated signal are all within a small proportion of the carrier-wave frequency. Typically, the frequency stays within 100 kHz of the base frequency.

FM radio stations broadcast at frequencies in the range of 88 to 108 MHz, but in the U.S., the base frequency for each station always ends in 0.1, 0.3, 0.5, 0.7, or 0.9. That's why FM radio stations have frequencies such as 89.3 or 107.5, but never 92.0 or 98.6.

Assigning base frequencies in increments of 0.2 MHz gives each station 100 kHz of room on either side of the center frequency for its frequency modulation. Thus, a station broadcasting at 103.1 actually sends signals whose frequencies range from 103.0 to 103.2. Most stations limit the variation from the base frequency to ±75 kHz to help prevent adjacent stations from interfering with one another.

FM modulators usually use a type of electronic component called a *varactor*, which is a type of diode that has an unusual characteristic: It has capacitance like a capacitor, and its capacitance decreases when voltage is applied across the diode. In essence, a varactor is a voltage-controlled variable capacitor. The schematic symbol for a varactor, shown in the margin, looks like a cross between a diode and a capacitor.

Varactors can be used in oscillator circuits to create an oscillator that vibrates faster when voltage increases. This ability makes it ideal for an FM radio modulator. As the voltage of the audio input increases, the capacitance of the varactor decreases and thus the frequency of the oscillator increases. When the voltage decreases, the capacitance of the varactor increases as the oscillator's frequency decreases. Figure 3-7 shows a sample of an FM modulator circuit that uses a varactor.

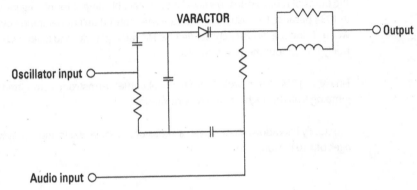

FIGURE 3-7:
An FM modulator circuit that uses a varactor.

THE TRAGIC GENIUS BEHIND FM RADIO

One of the great inventors in the history of radio was a brilliant engineer named Edwin H. Armstrong. Born in 1890, he was fascinated with electrical technology from a very young age. At the age of 14, he started experimenting with wireless radio circuits, building an antenna in his family's backyard that was more than 100 feet tall.

He made his first major contribution to radio technology while he was a junior at Columbia University in 1912. His invention was a circuit that amplified incoming radio signals by feeding them back though the amplifier tube in what came to be called a *regenerative circuit*. It was an important early breakthrough in radio technology that for the first time allowed radio to be heard through a speaker rather than with headphones.

(continued)

(continued)

During World War I, Armstrong invented another type of radio receiver, which he called the *superheterodyne circuit*. The basic principal of the superheterodyne circuit is that a radio signal broadcasting at a high frequency — say 1,500 kHz — can be combined with a nearby frequency from an oscillator — say, 1,560 kHz — in such a way that the original signal could also be detected at 60 kHz — the difference between the original signal's frequency (1,560 kHz) and the oscillator's frequency (1,500 kHz). The superheterodyne circuit may be one of the most important electronic circuits ever invented. It's still used in nearly all radio receivers to this day.

His third great invention came in 1933, when he created a method for transmitting radio signals that wasn't subject to interference from atmospheric disturbances like lightning. His new system was called *frequency modulation*. We know it as FM radio.

Armstrong patented his inventions, but his patents were challenged or ignored by the titans of radio. He lost his lawsuit to protect his patent for his regenerative circuit in 1934 because the justices of the Supreme Court didn't understand how the circuit worked, and the industry challenged his FM radio patents and used his technology freely throughout the 1940s and 1950s.

Finally, in 1954, ill and broke from his legal battles, Armstrong committed suicide by jumping from his high-rise apartment window.

Eventually his widow, Marion, won a series of patent lawsuits and was awarded damages of $10 million.

Building a Crystal Radio

In this section, you learn how to build one of the simplest of all useful electronic circuits: a *crystal radio*, which is a radio receiver that can receive AM radio broadcasts. It's not a particularly good radio receiver. Only one person at a time can listen to it because it uses an earphone or crystal earpiece instead of a speaker. And it's not very sensitive; you'll be lucky if you can receive two or three different stations even if dozens of AM radio stations broadcast in your area.

But what makes the crystal radio unique is that, unlike every other electronic circuit described in this book, a crystal radio has no obvious source of power, no batteries or other power supply. The only source of power used by a crystal radio is the power present in the radio waves themselves.

Crystal radios have been around since the very beginning of radio broadcasting. In the 1920s, it was common for people to build their own crystal radio receivers.

Newspapers and magazines published articles telling readers how to construct crystal radios using mostly household items such as scraps of wood and metal and empty oatmeal boxes, plus a few specialty items including a crystal and a telephone headset.

The total cost for a 1920s crystal radio was around $10. That sounds cheap, but adjusted for inflation, that's more like $125 today. Fortunately, the total cost for a crystal radio today is still around $10.

TIP

If you prefer, you can buy a kit to build your own crystal radio. Amazon sells a nice kit for about $15, and you can find crystal radio kits at many local hobby or school-supply stores. Although you can probably round up the parts separately for less than the cost of a kit, a few of the parts are relatively hard to find outside of a kit. (One option you may want to consider is to buy a kit to get these few hard-to-get parts but build the radio using the instructions found in this chapter.)

Looking at a simple crystal radio circuit

Figure 3-8 shows a basic crystal radio receiver circuit. As you can see, this circuit consists of just a few basic components: an antenna and a ground connection, a coil, a variable capacitor, a diode, and an earphone.

The antenna, of course, captures the radio waves travelling through the air and converts them into alternating current. In order for current to flow, a complete circuit is required. The ground connection is what completes the circuit, allowing current to flow.

The combination of the coil and the capacitor form the tuning circuit. The inductance of the coil combines with the capacitance of the variable capacitor to create a circuit that resonates at a particular frequency, allowing that frequency to pass but blocking other frequencies. In a basic crystal radio such as the one shown in Figure 3-8, the tuning circuit isn't very selective. As a result, you'll probably hear several stations at once. However, it is possible to build more selective tuning circuits that can home in on individual stations.

The diode forms the detector part of the circuit. It simply converts the alternating current signal that comes from the antenna and tuning circuit to direct current. This direct current is extremely small, but it is enough to drive a sensitive piezo-electric earphone, which converts the current to sound.

So that, in a nutshell, is how a crystal radio works. The rest of this chapter shows you how to build a crystal radio of your own.

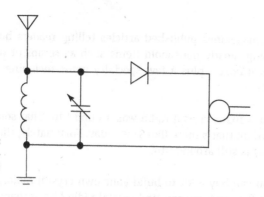

FIGURE 3-8:
Schematic diagram for a crystal radio.

Figure 3-9 shows the finished crystal radio that you can build in this project. Note that there are many ways to build a crystal radio, so the instructions that follow in this chapter are by no means definitive. Use your imagination when looking for materials to build your radio.

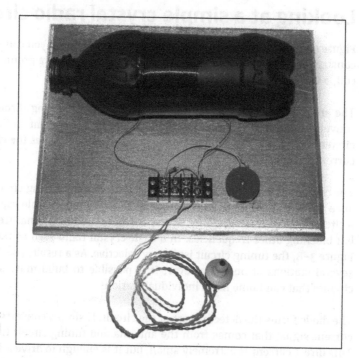

FIGURE 3-9:
A finished crystal radio.

Gathering your parts

You'll need a handful of parts to build your crystal radio. The following is a recommended list:

» At least 50 feet of **antenna wire.** You can use almost any wire for the antenna. I prefer 18-gauge solid hook-up wire, which you can buy at RadioShack.

» A few feet of **hook-up wire** to connect the radio to a ground connection.

» At least 50 feet of 30-gauge, enamel-coated **magnet wire.** You can buy it online at electronic parts suppliers or on Amazon. Just search for "magnet wire."

» Something to wrap the coil on. I used an empty soda bottle.

» A **variable capacitor,** also called a *tuning capacitor.* If you happen to have an old radio lying around — even one that doesn't work — you can harvest the tuning capacitor out of it. Or you can purchase one online — just search your favorite online electronic parts store (or Amazon) for "tuning capacitor."

Note that the variable capacitor is an optional component. If you can't find one, you can still build your crystal set without one; you just won't be able to tune out competing stations.

» A **germanium diode.** Again, you'll probably need to purchase this component online. Just use your favorite search engine to search for *1N34A,* and you'll find several suppliers.

» A **piezoelectric earphone.** Regular earphones like the kind used with an iPod or cellphone won't work. Search for *piezoelectric earphone* and you'll find several suppliers that sell them for about $3.

» A **board** to mount the radio on. About 6 x 9 inches should be sufficient.

» Something to make your electrical connections. I like to use a four-pole barrier strip, but you can improvise.

TIP If you have an old radio that doesn't work lying around, feel free to open it up and harvest its parts. In particular, look for the tuning capacitor. You'll be able to spot it easily because it will be connected to the radio's tuning knob.

TIP The germanium diode, tuning capacitor, and piezoelectric earphone are the three parts that are a bit difficult to find. You may want to purchase a crystal radio kit from a hobby or school-supply store and harvest those three parts from the kit.

Building the coil

When you look at a crystal radio, the first thing you're likely to notice is the large coil. The coil usually consists of 100 turns or more of small-gauge magnet wire wrapped around a non-conductive tube anywhere from 1 to 5 inches in diameter. The coil is an essential part of the radio's tuning circuit.

Many different types of materials can be used to wrap the coil around. Here are a few ideas:

>> An empty 16-ounce soda bottle. That's what I used for the radio built in this chapter. To make the coil look better, you may first want to spray-paint the bottle with black paint. See Figure 3-10. (Don't use metallic paint!)

>> An empty toilet paper roll.

>> An empty oatmeal container.

>> An empty bottle of contact lens fluid or another similarly sized plastic bottle.

>> A 6-inch length of PVC sprinkler pipe 2 or 3 inches in diameter.

>> A 6-inch length of a wooden closet rod.

>> A 6-inch length of a cardboard mailing tube.

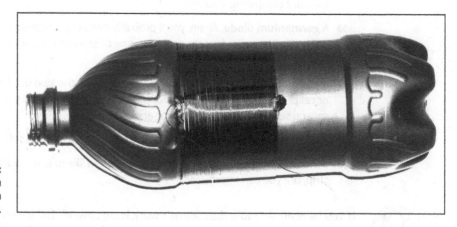

FIGURE 3-10: A coil wound on an empty soda bottle.

In short, any sturdy cylindrical object that is made of an insulating material can be used as the core of your coil. As long as it's cylindrical and not made of metal, you can use it.

As for the choice of wire to use for the coil, the most common is magnet wire, which is coated with thin enamel insulation rather than encased in plastic insulation. Wire wrapped with plastic insulation will work, but enamel insulation is thinner and thus allows the turns to be spaced closer together.

The number of turns in the coil and the diameter of the cylinder you wrap the coil around will determine how much wire you need. You'll want to wrap at least 100 turns. To determine how much wire you'll need for each turn, multiply the diameter of the cylinder by 3.14. Then, multiply the result by the number of turns, and divide by 12 to determine how many feet of wire you'll need.

For example, suppose you're wrapping the coil around a 2-inch cylinder and you want to wrap 100 turns. In this case, each turn will require 6.28 inches of wire (2 inches × 3.14). So you'll need just over 52 feet of wire (6.28 inches × 100 ÷ 12 inches). Allowing a foot or so of extra wire at each end of the coil to connect the coil to the radio circuit, you'll need about 54 feet of wire for the coil.

For the simplest type of crystal radio, the exact number of turns doesn't affect the operation of the radio significantly. So in the example here, if you have a 50-foot spool of wire, you can just wind the coil a few turns short of 100, and the radio will work just as well.

The easiest way to wind a coil is to place the tube you're winding the coil around on a screwdriver blade or other long narrow object so that the tube will spin freely. That way, you can turn the tube and slowly feed wire from its spool onto the tube. This keeps the wire from becoming twisted as you wind the coil. If you have a vice on your workbench, you can clamp the screwdriver horizontally in the vice, and then slide the tube onto the screwdriver so that the tube will spin freely.

To start the coil, you must first attach one end of the magnet wire to one end of the tube. The easiest way to do that is with a dab of hot glue. If you prefer, you can punch a hole through the tube and feed the wire through. Either way, be sure to leave 6 inches or more of wire free. This will give you plenty of wire to connect the coil to the circuit when you finish winding the coil.

Once you've attached one end of the coil, turn the tube slowly while feeding wire from the spool onto the tube. Each half turn or so, use your fingers to carefully scoot the wire you just fed up against the turns you've already wound. The goal is for each turn of wire to be adjacent to the previous turn with no gaps between the turns.

You'll find that you need to keep a bit of tension on the wire as you feed it onto the tube in order to keep the windings nice and tight. If you slip and let go of the tension, several turns may unravel, and you'll have to untangle them to restore the coil's tightness.

TIP

It helps if you wrap the coil in sections of about ten turns each. When you finish each section, dab a little hot glue on it to hold it in place.

When you reach the end of the tube (or run out of wire), use a little hot glue to secure the last turn of the coil, or cut a slit in the tube and slide the wire through it. Be sure to leave about 6 inches of free wire after the last turn.

When you're done, the coil should have a nice, tight appearance with no major gaps between the turns, and you should have about 6 inches of free wire on each end of the coil.

Assembling the circuit

Once you have your coil, the next step is to assemble the various parts of the radio on a base. I recommend you use a piece of wood about 6 × 9 inches. To make your radio look good, consider painting or staining the wood before you assemble the circuit.

Here's a list of the parts you'll need to assemble the circuit:

>> The **coil** you assembled according to the instructions in the previous section of this chapter

>> A four-position **barrier strip**

>> A **germanium diode** (1N34A or similar)

>> A **tuning capacitor** (optional)

>> One length of **hook-up wire**, approximately 1½ inches long

>> Two lengths of **hook-up wire**, approximately 3 inches long

You'll need the following tools to build the crystal radio circuit:

>> **A hot glue gun and some glue sticks**

>> **A Phillips-head screwdriver**

>> **Wire cutters**

>> **Wire strippers**

>> **Soldering iron and some solder**

Figure 3-11 shows the layout for the assembled crystal radio circuit.

In the instructions here, I refer to the individual terminals of the barrier strip by numbering them from left to right, 1 through 4. Terminal connectors in the top row are given the letter A, and those in the bottom row are given the letter B. Thus, the terminal at the top left of the barrier strip is terminal 1A, and the terminal at the bottom right is terminal 4B.

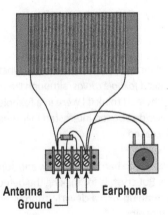

FIGURE 3-11:
Layout for the
crystal radio
circuit.

Antenna ——
Ground ——   —— Earphone

To assemble the crystal radio circuit, follow these steps:

1. **Glue the barrier strip, tuning capacitor, and coil to the board.**

   Use Figure 3-11 to judge the placement of each of these parts. Be sure to give the glue enough time to cool and harden before you continue.

2. **Connect the diode between terminals 1A and 3A.**

   Interestingly enough, the direction in which you connect the diode doesn't matter in a crystal radio circuit.

3. Strip about ⅜ inch of insulation from both ends of all three lengths of hook-up wire.

4. Connect one end of the 1½-inch length of hook-up wire to terminal 2A on the barrier strip, and then connect the other end to terminal 4A.

5. **Use some sandpaper to gently scrape the enamel insulation off the ends of the coil wire.**

6. **Connect the two wires from the coil to terminals 1A and 4A.**

7. **Solder one end of one of the 3-inch hook-up wires to the center lead of the capacitor and one end of the other 3-inch wire to either one of the other leads.**

   It doesn't matter which of the two outside leads you use.

8. **Connect the free ends of the wires you soldered in Step 7 to terminals 1A and 1D of the barrier strip.**

   You're done!

When the radio circuit is assembled, look it over to make sure all the pieces are connected, as shown in Figure 3-11.

Understanding Radio

FOXHOLE RADIOS

In World War II, GIs would often build their own crystal radios from whatever materials they could scrounge up. These were called *foxhole radios*, although I'm not certain how many of them were actually built in foxholes. I think if I were in a foxhole and the enemy was launching mortar shells at me, I wouldn't be too interested in listening to my crystal radio. Nevertheless, that's what they're called.

Wire for making antennas and coils wasn't too hard to come by, and super-sensitive headphones weren't hard to scrounge. But crystals for making the detector part of the circuit were another story. So the GIs came up with a cleverly improvised solution: They used razor blades, pencil lead, and safety pins.

To build the detector for a foxhole radio, the razor blade must be made of blue steel. Most modern blades aren't, but you can fix that by placing the blade in a metal vice and blasting it with a propane torch until it glows red hot. Let it cool before you handle it!

Note that it also doesn't hurt if the razor blade is a bit rusty; the oxide in the rust actually helps.

To build the detector, first glue the razor blade to a piece of wood. Sharpen the pencil, then cut it short (½ inch is long enough). Bend open the safety pin to about 90 degrees and jam the pointed end of the pin into the lead at the end of the pencil that you cut off. Then nail or screw the flat end of the safety pin to the board, positioned so that the tip of the pencil sits on the razor blade.

Connect one wire to the razor blade and the other to the safety pin and wire it into your circuit right where the germanium diode would go. Then, hook your radio up to the antenna and ground, put the earphone in your ear, and drag the pencil tip around to different parts of the razor blade until you hear a signal.

This type of detector is very finicky, so you might have to try different angles and positions, and you might have to try different razor blades or pencils. But once you get it to work, you'll be delighted that you were able to make a radio out of an old razor blade, a safety pin, and a pencil.

Stringing up an antenna

A good, long antenna is vital to the successful operation of a crystal radio. In general, the longer the antenna, the better. If possible, try to make your antenna at least 50 feet. Longer is better.

You can make your antenna from just about any type of wire, insulated or not. You should be able to find suitable wire at your local RadioShack or at most hardware stores.

The best configuration for a crystal radio antenna is to run the wire horizontally between two supports as high off the ground as you can get them, as shown in Figure 3-12. For your antenna, you probably won't find actual poles as shown in the figure. However if you look around, you should be able to find two suitable points to which you can connect the ends of your antenna. Fence posts, trees, basketball hoops, flagpoles, swing sets, or almost any other tall structure will do the trick.

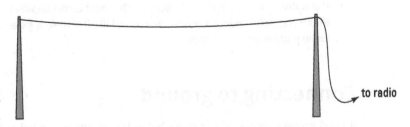

FIGURE 3-12:
Stringing your
antenna.

to radio

Notice that one end of the antenna wire must run to the ground to a convenient place where you can connect it to your crystal radio. You'll need to run this wire to the location at which you intend to operate your radio.

TIP

Wood isn't a great insulator, and most metals, of course, are excellent conductors. It's vital that your antenna is well insulated from the ground. Thus, you must be careful about how you support the ends of the antenna wire to make sure that you don't inadvertently ground the antenna.

If you use insulated wire for the antenna, you can secure the ends to wood by using ½-inch eye screws available from any hardware store. Screw the eye screw into the wood, and then simply tie the end of the antenna wire to it.

If the wire is uninsulated, you'll need to support it with something that doesn't conduct electricity. I suggest browsing the sprinkler parts department of your local hardware store to find a PVC pipe fitting. You can screw this fitting into wood or use duct tape or zip ties to secure it to metal, then loop your antenna wire through the fitting and tie it off.

WARNING

A crystal radio is a relatively safe electronics project, but there are a few dangers associated with the antenna. Here are some things to be careful of:

>> Don't string up your antenna in a thunderstorm! Lightning loves wires, and you don't want to tempt Mother Nature by providing her with a convenient path to discharge her fury through.

>> Likewise, don't operate your crystal radio in a thunderstorm!

>> Do not under any circumstances run your antenna wire anywhere near a power cable or other utility line. That's a sure way to become a statistic.

>> Be very careful if you must climb a ladder to string your antenna. I don't have the statistics to back it up, but I bet that the most common way to seriously injure yourself building a crystal radio is to fall off a 10-foot extension ladder while putting up your antenna.

Connecting to ground

A good ground connection is every bit as important as a good antenna. The best way to create a good ground connection is to use a metal cold water pipe. Assuming you placed your antenna outdoors, you may be lucky enough to find an outdoor water faucet near the end of the antenna. Then, you can connect one end of a length of hookup wire to the water pipe and the other end to your crystal radio. (Note that this won't work if the home uses plastic pipe; it will only work with metal pipe.)

If you can't find a water pipe, get a length of metal rebar and pound it into the ground. The deeper you go, the better the ground connection.

The easiest way to connect a wire to a water pipe (or a piece of rebar) is to use a pipe clamp, which you can find in the plumbing section of any hardware store. Use some coarse sandpaper to sand the pipe where you attach the clamp to improve the electrical connection, especially if the pipe has been painted or varnished. Strip an inch or two of insulation from the end of your ground wire and wrap it around the clamp, then slip the clamp around the water pipe and tighten it down, as shown in Figure 3-13.

FIGURE 3-13:
A good ground
connection.

Using the crystal radio

Once your crystal radio circuit is built, your antenna is up, and your ground wire is connected, it's time to put your crystal radio to the test. Follow these steps:

1. **Connect the two leads of the piezoelectric earphone to terminals 3B and 4B of the barrier strip.**

2. **Connect the antenna lead to terminal 1B of the barrier strip.**

3. **Connect the ground lead to terminal 2B of the barrier strip.**

4. **Put the earphone in your ear.**

   You'll probably immediately hear a radio station.

5. **Turn the knob on the tuning capacitor to hear other stations.**

   The tuner on this crystal radio circuit isn't very sensitive, so you'll probably be able to distinguish just two or three different stations.

   Note that if you didn't add a tuning capacitor, you won't be able to tune to a specific station at all. Instead, you'll probably hear several stations at once. (Even with a tuning capacitor, you may still hear several stations at once. As I said, the tuning circuit for a simple crystal radio like this isn't very discriminating.)

Chapter **4**

Working with Infrared

In this chapter, you learn how to work with circuits that detect the invisible light that's commonly called *infrared*. Infrared light has all sorts of applications for wireless communication, the most common of which is the remote control for your television. Other uses for infrared include night-vision goggles and cameras, and temperature detection.

Have fun!

Introducing Infrared Light

Infrared light is light whose frequency is just below the range of visible red light. Specifically, infrared is light whose frequency falls between 1 THz to 400 THz (one THz is one trillion cycles per second). The infrared spectrum falls right between microwaves and visible light.

In Chapter 1 of this minibook, you learn that there's an inverse relationship between *frequency* and *wavelength*. In other words, the lower the frequency, the longer the wavelength. If you describe infrared in terms of its wavelength rather than its frequency, infrared waves are longer than the waves of visible light but shorter than microwaves. The wavelength of infrared light is between 0.75 to 300 micrometers, which is a millionth of a meter. Thus, at the very bottom edge of the infrared spectrum, the infrared waves are about one-third of a millimeter long. At the upper end, the waves are about one thousandth of a millimeter long. If the waves get any shorter than that, they become visible light.

Figure 4-1 shows the entire spectrum of electromagnetic radiation, so you can see where infrared falls within the grand scheme of things radiation-wise.

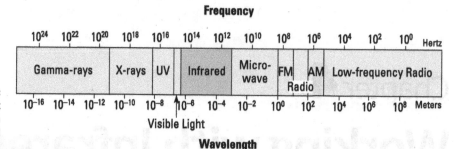

FIGURE 4-1:
Infrared light falls between visible light and microwaves.

Infrared light isn't visible to human eyes. I guess Mother Nature decided that we didn't need to be looking at things in infrared. It's too bad because more than half of all the light energy emitted from our sun is in the form of infrared light. If our eyes could see infrared light as well as the visible light we can see, a sunny day would seem twice as bright.

TECHNICAL STUFF

Sometimes infrared light is confused with heat. That's because we can't see infrared light waves, but we can feel them in the form of heat. In other words, infrared light waves heat surfaces that absorb them. Visible light does this too. That's why it's cooler in the shade than it is in the sun. Because heat is an effect of infrared light, infrared light can be used as a heat source. But infrared light and heat aren't the same thing.

Infrared light is often used to detect objects that we can't see in visible light. One common application of this is night vision. According to a principal of physics called *Planck's law*, all matter emits electromagnetic radiation if its temperature is above absolute zero. Some of that radiation is in the infrared spectrum, so devices that can detect infrared light can literally see in the dark.

To enhance the effect, some night-vision devices actually illuminate an area with infrared light. Because the human eye can't see the infrared light, the area illuminated still appears dark to us, but to a detector sensitive to infrared light, the area is lit up and fully visible.

Another common application of infrared light is for wireless communications across short distances. The best known infrared devices are television remote controls. The remote control unit contains a bright infrared light source, and the television itself includes an infrared detector. When you point the remote control at the television and push a button, the remote control turns on the infrared light source and encodes a message on it. The receiver picks up this signal, decodes the message, and does whatever the message directs it to do — turns up the volume, changes the channel, and so on.

Like visible light, infrared light can be blocked by solid objects and it can bounce off of reflective objects. That's why the remote won't work if your spouse is standing between you and the television. But it's also why you can get around your spouse by pointing the remote at a window. The infrared waves bounce off the glass and, if the angle is right, arrive at the television.

TECHNICAL STUFF

The first wireless remote control was developed by Zenith in 1955. It used ordinary visible light, could turn the TV on or off, and could change channels. It had one nasty defect: You had to position your television in the room so that light from an outside source (such as the setting sun shining through a window) didn't hit the light sensor. Otherwise, the TV might shut itself off right in the middle of the evening news when the sun reached just the right angle and hit the sensor.

Remotes today use complicated encoding schemes to avoid such random misfirings. You're probably familiar with the procedure you must go through when programming a remote control to work with a particular television. This programming is necessary because there's no widely accepted standard for how the codes on a remote control should work, so each manufacturer uses its own encoding scheme.

Detecting Infrared Light

There are several ways to detect infrared light in an electronic circuit, but the most common is with a device called a *phototransistor*. You can buy a phototransistor for a few bucks at RadioShack or any other store that stocks electronic components.

To understand how a phototransistor works, first review how a transistor works. A transistor has three terminals, known as the *base*, *collector*, and *emitter*. Within the transistor, there's a path between the collector and emitter. How well this path conducts depends on whether voltage is applied across the base and the emitter. If voltage is applied, the collector-emitter path conducts well. If there's no voltage on the base, the collector-emitter path doesn't conduct.

In a phototransistor, the base isn't a separate terminal that's connected to a voltage source in your circuit. Instead, the base is exposed to light. When infrared light hits the base, the energy in the light is converted to voltage, and the emitter-collector path conducts.

Thus, infrared light hitting the base has the same effect as voltage on the base of a traditional transistor: The infrared light turns the transistor on. The brighter the infrared light, the better the emitter-collector path conducts.

Figure 4-2 shows a simple circuit that uses an IR phototransistor to detect infrared light. When infrared light is present, the collector-emitter circuit conducts, and the LED lights up. Thus, the LED lights when the phototransistor is exposed to infrared light.

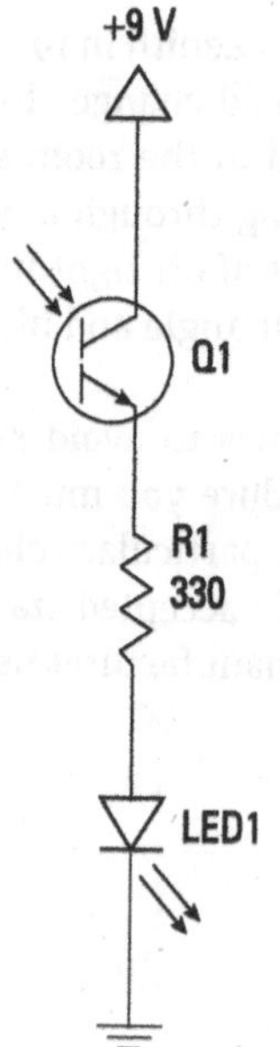

+9 V

Q1

R1
330

LED1

FIGURE 4-2:
A simple infrared
detector circuit.

Project 25 shows you how to build this circuit on a solderless breadboard, and Figure 4-3 shows the assembled circuit.

Once you have assembled this circuit, try exposing the phototransistor to different light sources to see whether they emit infrared light. One sure source of infrared is a TV remote control. Point the remote at the phototransistor and press any button on the remote. You should see the LED flash on and off quickly as it responds to the infrared signals being sent by the remote.

Another interesting source of infrared is an open flame. Be very careful, of course; I don't want you burning down your house just to see if the flames produce infrared light. If you have a small gas lighter, light it up and hold it near the phototransistor.

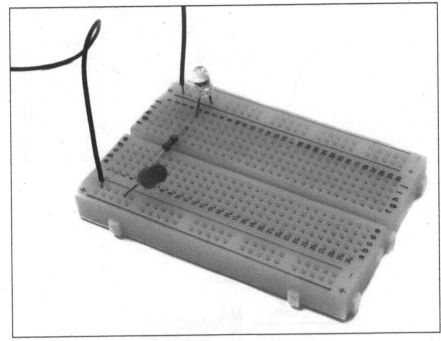

FIGURE 4-3:
The assembled
infrared
detector circuit
(Project 25).

Project 25: A Simple IR Detector

In this project, you build a simple infrared light detector using a phototransistor.
When the phototransistor is exposed to infrared light, the LED lights up.

+9 V

Q1

R1
330

LED1

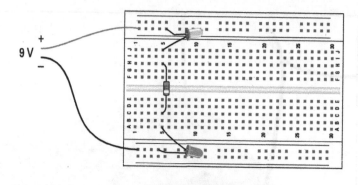

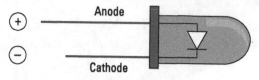

Anode

Cathode

Phototransistor

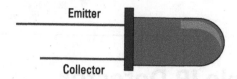

Emitter

Collector

Parts

» One 9 V battery snap holder

» One 9 V battery

» One IR phototransistor

» One red LED

» One 330 Ω ¼ W resistor (orange-orange-brown)

Steps

Throughout these steps, use the negative (–) bus strip on the left side of the board for ground and the positive (+) bus strip on the right side of the board as the positive voltage.

1. **Insert the phototransistor.**

   Insert the phototransistor as follows:

Collector (Long)	Emitter (Short)
Positive bus	J5

2. **Insert the resistor.**

   Insert the resistor as follows:

From	To
C5	H5

3. **Insert the LED.**

   Insert the LED as follows:

Cathode (Short)	Anode (Long)
Ground bus	A5

4. **Connect the battery.**

   Insert the red lead to the positive bus and the black lead to the negative bus.

5. **Expose the phototransistor to an infrared light source.**

   Try a variety of sources, including a TV remote control, a flame (be careful!), and sunlight. Also, try other sources that don't emit infrared, such as an LED flashlight.

Creating Infrared Light

The easiest way to create infrared light is by using a special light-emitting diode (LED) that operates in the infrared spectrum. Infrared LEDs (often called *IR LEDs*) are readily available at RadioShack or any other store that sells electronic parts.

IR LEDs are similar to regular LEDs, except that you can't see the light they emit. The LED itself is usually a dark purple or blue color. Like other LEDs, the cathode lead is shorter than the anode lead.

As with any LED, you must use a resistor in series with an IR LED to prevent excess current from burning out the LED. To calculate the size of the resistor, you need to know three things:

>> **The supply voltage:** For example, 9 V.

>> **The LED forward-voltage drop:** For most infrared LEDs, the forward-voltage drop is 1.3 V.

>> **The desired current through the LED:** Usually, the current flowing through the IR LED should be kept under 50 mA. However, IR LEDs are typically rated for more current than regular LEDs. The ones I buy from RadioShack can handle up to 150 mA.

With these three facts in hand, you can calculate the correct resistor size by using Ohm's law:

1. **Calculate the resistor voltage drop.**

   To do that, subtract the voltage drop of the IR LED (typically 1.3 V) from the total supply voltage. For example, if the total supply voltage is 9 V and the LED drops 1.3 V, the voltage drop for the resistor is 7.7 V.

2. **Convert the desired current to amperes.**

   In Ohm's law, the current must be expressed in amperes. You can convert milliamperes to amperes by dividing the milliamperes by 1,000. Thus, if your desired current through the IR LED is 50 mA, you must use 0.05 A in your Ohm's law calculation.

3. **Divide the resistor voltage drop by the current in amperes.**

   This gives you the desired resistance in ohms. For example, if the resistor voltage drop is 7.6 V and the desired current is 50 mA, you need a 152 Ω resistor.

4. **Choose a standard resistor size that's close to the calculated resistance.**

    A 150 Ω resistor is close enough for a 9 V circuit. If you don't have a 150 Ω resistor, a 220 Ω or 300 Ω resistor will do the job.

Once you've chosen the correct resistor size, just wire it in series with the IR LED, as shown in the schematic in Figure 4-4.

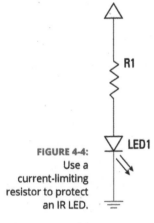

FIGURE 4-4:
Use a
current-limiting
resistor to protect
an IR LED.

Building a Proximity Detector

The combination of an IR LED and a photodiode is often used as a *proximity detector*, a gadget that detects when an object is nearby. There are two ways to build a proximity detector. One is to mount the IR LED and the phototransistor so that they face each other. Then, the infrared light from the IR LED is detected by the phototransistor. If an object comes between the IR LED and the phototransistor, the light is blocked, and the phototransistor turns off.

The other way to build a proximity detector is to mount the IR LED and the IR photodiode next to each other facing the same direction. When an object comes near the IR LED, some infrared light will bounce off the object and be detected by the phototransistor.

You learn how to build two types of proximity detector circuits in the next two sections of this chapter.

Building a Common-Emitter Proximity Detector

A schematic for a simple proximity circuit is shown in Figure 4-5. Although it isn't shown in the schematic, this circuit assumes that IR LED and Q1 are oriented so that Q1 can detect the infrared light emitted by IR LED, either indirectly (for a proximity detector) or directly (for an interrupter).

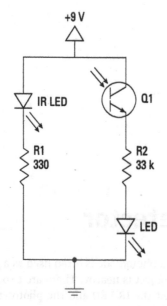

FIGURE 4-5:
A common-emitter proximity detector circuit.

This circuit is called a *common-emitter* circuit because the phototransistor's emitter is common between the phototransistor side of the circuit and the output side of the circuit that's connected to the IR LED. In a common-emitter circuit, the output voltage is on when infrared light is detected by the phototransistor. Thus, the red LED lights up when the path between the IR LED and phototransistor isn't obstructed. If you block the path between the IR LED and phototransistor, the red LED goes dark.

Project 26 shows how to build this circuit configured as an interrupter, and Figure 4-6 shows the finished project. When you connect this circuit to power, the red LED will come on. If you pass an object such as a piece of paper between the IR LED and the phototransistor, the red LED will go off.

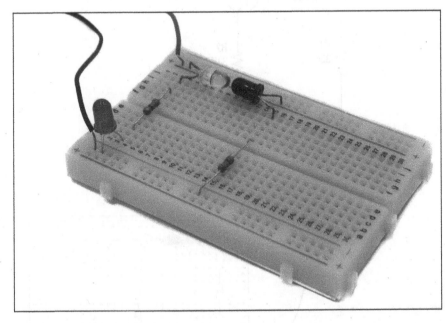

FIGURE 4-6: Using an IR LED and a phototransistor as a proximity detector (Project 26).

The output in this circuit is simply a red LED. However, you could just as easily connect the output to other circuit components. For example, the output could drive a mechanical relay if you want to use the proximity detector to turn on a floodlight or other 120 VAC device, or you could connect the output to a digital logic circuit, as described in Book 5, Chapter 3.

Project 26: A Common-Emitter Proximity Detector

In this project, you build a common-emitter proximity detector that lights a red LED whenever the path between an IR LED and a phototransistor is clear. If anything blocks the path, the red LED goes dark.

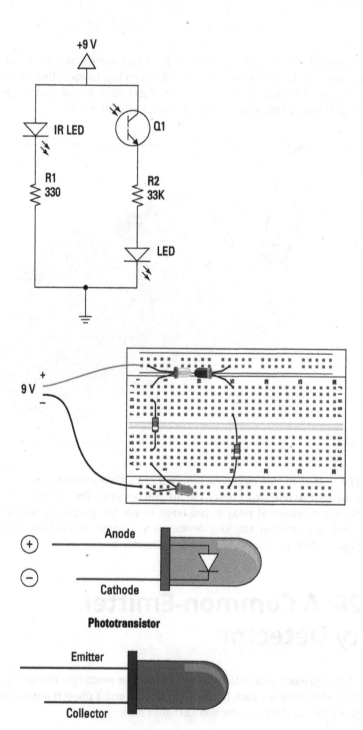

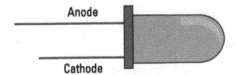

Infrared LED

Anode

Cathode

Parts

» One 9 V battery snap holder

» One 9 V battery

» One IR phototransistor

» One IR LED

» One red LED

» One 33 kΩ ¼ W resistor (orange-orange-orange)

» One 330 Ω ¼ W resistor (orange-orange-brown)

Steps

Throughout these steps, use the negative (–) bus strip on the left side of the board for ground and the positive (+) bus strip on the right side of the board as the positive voltage.

1. **Insert the phototransistor.**

   Insert the phototransistor as follows:

Lead	Hole
Collector (long lead)	Positive bus
Emitter (short lead)	J3

2. **Insert the LEDs.**

   Insert the LEDs as follows:

LED	Cathode (Short)	Anode (Long)
Red	Ground bus	A3
IR	J15	Positive bus

3. Insert the resistors.

Insert the resistors as follows:

Resistor	From	To
R1 (330 Ω)	F15	Ground bus
R2 (33 kΩ)	C3	H3

4. Connect the battery.

Attach the red lead to the positive bus and the black lead to the negative bus.

The red LED will illuminate when the power is connected.

5. Use a piece of paper or other flat object to block the light path between the IR LED and the phototransistor.

The red LED will go dark when you interrupt the light path between the IR LED and the phototransistor.

Building a Common-Collector Proximity Detector

Figure 4-7 shows a circuit that uses a *common-collector* circuit, in which the collector is the common point between the phototransistor circuit and the LED output circuit. When wired in this way, the LED is dark whenever the path between the IR LED and the phototransistor is clear. When something blocks the path and the phototransistor stops detecting IR light, the red LED comes on.

Project 27 shows how to build this circuit configured as an interrupter, and Figure 4-8 shows the finished project. When you connect this circuit to power, the red LED stays dark. But if you pass an object between the IR LED and the phototransistor, the red LED turns on.

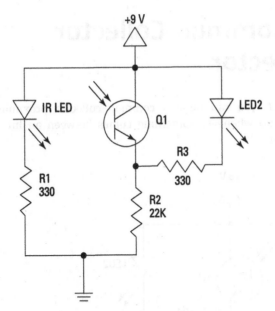

FIGURE 4-7:
A common-collector proximity detector circuit.

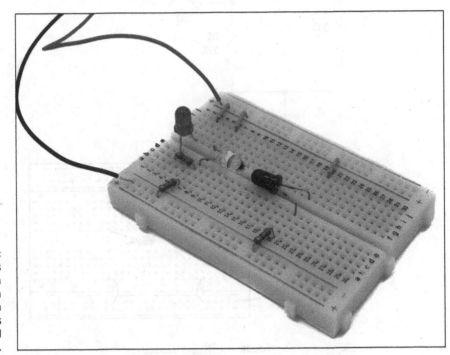

FIGURE 4-8:
This circuit lights a red LED when the path between an IR LED and a phototransistor is blocked (Project 27).

Project 27: A Common-Collector Proximity Detector

In this project, you can build a common-collector proximity detector that lights a red LED whenever something comes between an infrared LED and a phototransistor.

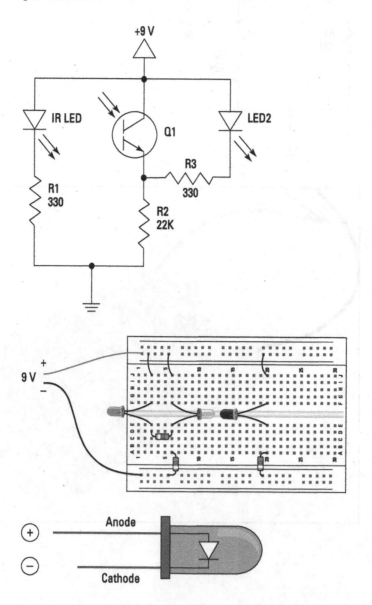

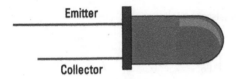

Phototransistor

Emitter

Collector

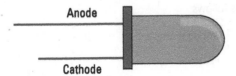

Infrared LED

Anode

Cathode

Parts

>> One 9 V battery snap holder

>> One 9 V battery

>> One IR phototransistor

>> One IR LED

>> One red LED

>> Two 330 Ω ¼ W resistors (orange-orange-brown)

>> One 22 kΩ ¼ W resistor (orange-orange-orange)

>> Three short lengths of jumper wire

Steps

Throughout these steps, use the negative (–) bus strip on the left side of the board for ground and the positive (+) bus strip on the right side of the board as the positive voltage.

1. **Insert the red LED.**

   Insert the LED as follows:

Cathode (Short)	Anode (Long)
E3	F3

2. Insert the phototransistor.

Insert the phototransistor as follows:

Collector (Long)	Emitter (Short)
F6	E6

3. Insert the IR LED.

Insert the IR LED as follows:

Cathode (Short)	Anode (Long)
E20	F20

4. Insert the resistors.

Resistor	From	To
R1 (330 Ω)	A20	Ground bus
R2 (22 kΩ)	A6	Ground bus
R3 (330 Ω)	D3	D6

5. Insert the jumper wires.

Insert the jumper wires as follows:

From	To
J3	Positive bus
J6	Positive bus
J20	Positive bus

6. Connect the battery.

Attach the red lead to the positive bus and the black lead to the negative bus.

7. Use a piece of paper or other flat object to block the light path between the IR LED and the phototransistor.

The red LED will light up when you interrupt the light path between the IR LED and the phototransistor. When you remove the obstacle, the red LED will go dark.

5
Doing Digital Electronics

Contents at a Glance

Chapter **1**

Understanding Digital Electronics

Welcome to the world of digital electronics. Beginning with this chapter, you start to learn about the fundamental building-block circuits that you'll ultimately use to create computers and other advanced electronic devices.

Digital electronics is a complex, almost-overwhelming topic, but the building blocks are actually quite simple. In this chapter, you learn some basic principles, such as how to tell the difference between analog and digital circuits, how the binary system works, and how basic logic operations work. In the remaining chapters in this part, you build on that knowledge as you explore the amazing world of digital electronics.

Hang on!

Distinguishing Analog and Digital Electronics

All electronics can be divided into two broad categories: analog and digital.

Analog refers to circuits in which quantities such as voltage or current vary at a continuous rate. When you turn the dial of a potentiometer, for example, you change the resistance by a continuously varying rate. The resistance of the potentiometer can be any value between the minimum and maximum allowed by the pot.

If you create a voltage divider by placing a fixed resistor in series with a potentiometer, the voltage at the point between the fixed resistor and the potentiometer increases or decreases smoothly as you turn the knob on the potentiometer.

In *digital* electronics, quantities are counted rather than measured. There's an important distinction between counting and measuring. When you *count* something, you get an exact result. When you *measure* something, you get an approximate result.

Consider a cake recipe that calls for 2 cups of flour, 1 cup of milk, and 2 eggs. To get 2 cups of flour, you scoop some flour into a 1-cup measuring cup, pour the flour into the bowl, and then do it again. To get a cup of milk, you pour milk into a liquid measuring cup until the top of the milk lines up with the 1-cup line printed on the measuring cup and then pour the milk into the mixing bowl. To get 2 eggs, you count out 2 eggs, crack them open, and add them to the mixing bowl.

The measurements for flour and milk in this recipe are approximate. A teaspoon too much or too little won't affect the outcome. But the eggs are precisely counted: exactly 2. Not 3, not 1, not 1½, but 2. You can't have a teaspoon too many or too few eggs. There will be exactly 2 eggs, because you count them.

One of the most common examples of the difference between analog and digital devices is a clock. Figure 1-1 shows two clocks: one analog and the other digital. On the analog clock, the time is represented by hands that spin around a dial and point to a location on the dial that represents the approximate time. On a digital clock, a numeric display indicates the exact time.

FIGURE 1-1:
Analog and digital clocks.

Another example is a thermometer. A traditional glass–mercury thermometer has a small amount of liquid mercury inside a glass column. Mercury expands when it gets warm, so the warmer the mercury is, the higher it climbs in the glass column. Little tick marks are printed on the column to help you read the temperature. On a digital thermometer, the exact temperature is indicated by a numeric display.

So which is more accurate — analog or digital? In one sense, digital circuits are more accurate because they count with complete precision. You can precisely count the number of jelly beans in a jar, for example. But if you weigh the jar by putting it on an analog scale, your reading may be a bit imprecise because you can't always judge the exact position of the needle. Say that the needle on the scale is about halfway between 4 pounds and 5 pounds. Does the jar weigh 4.5 pounds or 4.6 pounds? You can't tell for sure, so you settle for approximately 4.5 pounds.

On the other hand, digital circuits are inherently limited in their precision because they must count in fixed units. Most digital thermometers, for example, have only one digit to the right of the decimal point. Thus, they can indicate a temperature of 98.6 or 98.7 but can't indicate 98.65.

Here are a few other thoughts to ponder concerning the differences between digital and analog systems:

>> The term *digital* is actually a reference to your digits — that is, your fingers. The most natural way for us humans to count is with our fingers.

>> The term *analog* is a variation on the word *angler* — a reference to the ancient practice of estimating the size of a fish by spreading out your arms while saying, "He was at least this big!" That's the original analog measurement.

>> Saying that a system is digital isn't the same as saying that it's binary. *Binary* is a particular type of digital system in which the counting is all done with the binary number system. Nearly all digital systems are also binary systems, but the two words aren't interchangeable. (For more information about the binary number system, see the next section, "Understanding Binary.")

>> Many systems are a combination of binary and analog systems. In a system that combines binary and analog values, special circuitry is required to convert from analog to digital, or vice versa. An input voltage (analog) might be converted to a sequence of pulses, one for each volt; then the pulses can be counted to determine the voltage.

>> I made up that item about the angler, of course. Made you think about it for a second, though, didn't I?

Understanding Binary

Most digital electronic circuits work with the binary number system. Thus, before you can understand the details of how digital circuits work, you need to understand how the binary numbering system works.

Knowing your number systems

A *number system* is simply a way of representing numeric values. Number systems use symbols called *numerals* to represent numeric quantities. The numerals 1, 2, and 3 represent the numeric quantities commonly known as one, two, and three.

In most number systems, numerals can be strung together to create larger numeric values, and the position of each numeral in the string determines its relative value. In the number 12, the numeral 1 represents the quantity ten, and the numeral 2 represents the quantity two. In the number 238, the numeral 2 represents the quantity two hundred, the numeral 3 represents the quantity thirty, and the numeral 8 represents the quantity eight.

You learned all this stuff in elementary school, so it's completely intuitive. Unfortunately, it's so intuitive that it's easy to miss its brilliance — and to overlook the fact that it's completely arbitrary.

The fact that we have ten numerals in our everyday counting system — which is called the *decimal system*, or *base 10* — is a simple result of the fact that humans have ten fingers. If humans had evolved with 12 fingers, we would've learned how to count in base 12, and we would've invented two more numerals.

Different number bases may seem strange, but we actually encounter them every day without thinking about it. One common example is the system we use for keeping time. Our system for measuring time actually works in several number bases. An hour contains 60 minutes, for example. We recognize that 1:30 is halfway between 1:00 and 2:00 without even thinking about it. You may not realize it, but you're thinking in base 60 when you tell time.

Counting by ones

Binary is one of the simplest of all number systems because it has only two numerals: 0 and 1. In the decimal system (with which most people are accustomed), you use 10 numerals: 0 through 9. In an ordinary decimal number, such as 3,482, the rightmost digit represents ones; the next digit to the left, tens; the next, hundreds; the next, thousands; and so on. These digits represent powers of ten: first 10^0 (which is 1); next, 10^1 (10); then 10^2 (100); then 10^3 (1,000); and so on.

In binary, you have only two numerals rather than ten, which is why binary numbers look somewhat monotonous, as in 110011, 101111, and 100001.

The positions in a binary number (called *bits* rather than *digits*) represent powers of two rather than powers of ten: 1, 2, 4, 8, 16, 32, and so on. To figure the decimal value of a binary number, you multiply each bit by its corresponding power of two and then add the results. The decimal value of binary 10111, for example, is calculated as follows:

$$
\begin{aligned}
1 \times 2^0 &= 1 \times 1 = 1 \\
+1 \times 2^1 &= 1 \times 2 = 2 \\
+1 \times 2^2 &= 1 \times 4 = 4 \\
+0 \times 2^3 &= 0 \times 8 = 0 \\
+1 \times 2^4 &= 1 \times 16 = \underline{16} \\
& 23
\end{aligned}
$$

Fortunately, converting a number between binary and decimal is something that a computer is good at — so good, in fact, that you're unlikely ever to need to do any conversions yourself. The point of learning binary is not to be able to look at a number such as 1110110110110 and say instantly, "Ah! Decimal 7,606!" (If you could do that, Barbara Walters would probably interview you, and Hollywood would even make a movie about you. You'd probably be played by Dustin Hoffman.)

Instead, the point is to have a basic understanding of how computers store information and — most importantly — of how the binary counting system works, which I describe in the following section.

If you do find that you need to convert binary numbers to decimal, or vice versa, and you have access to a computer, you can use the Calculator program that comes free with Windows to do the conversion for you. For more information, see the nearby sidebar "Using Windows Calculator for binary conversions."

Here are some of the most interesting characteristics of binary, which explain how the system is similar to and different from the decimal system:

>> **In decimal, the number of decimal places allotted for a number determines how large the number can be.** If you allot six digits, for example, the largest number possible is 999,999. Because 0 is itself a number, however, a 6-digit number can have any of 1 million different values.

Similarly, the number of bits allotted for a binary number determines how large that number can be. If you allot 8 bits, the largest value that number can store is 11111111, which happens to be 255 in decimal. Thus, a binary number that is 8 bits long can have any of 256 different values (including 0).

TIP

» **To quickly figure how many different values you can store in a binary number of a given length, use the number of bits as an exponent of two.** An 8-bit binary number, for example, can hold 2^8 values. Because 2^8 is 256, an 8-bit number can have any of 256 different values. This is why a *byte* — 8 bits — can have 256 different values.

TECHNICAL STUFF

» **This "powers of two" thing is why digital systems don't use nice, even round numbers for measuring such values as memory capacity.** A value of 1k, for example, isn't an even 1,000 bytes: It's actually 1,024 bytes, because 1,024 is 2^{10}. Similarly, 1MB isn't an even 1,000,000 bytes, but 1,048,576 bytes, which happens to be 2^{20}.

REMEMBER

One basic test of digital geekdom is knowing your powers of two because they play such an important role in binary numbers. Just for the fun of it, but not because you really need to know, Table 1-1 lists the powers of 2 up to 32.

Table 1-1 also shows the common shorthand notation for various powers of 2. The abbreviation *k* represents 2^{10} (1,024). The *M* in *MB* stands for 2^{20}, or 1,024k, and the *G* in *GB* represents 2^{30}, which is 1,024MB.

TABLE 1-1   **Powers of 2**

Power	Bytes	Kilobytes	Power	Bytes	k, MB, or GB
2^1	2	2^{17}	131,072		128k
2^2	4	2^{18}	262,144		256k
2^3	8	2^{19}	524,288		512k
2^4	16	2^{20}	1,048,576	1MB	
2^5	32	2^{21}	2,097,152	2MB	
2^6	64	2^{22}	4,194,304	4MB	
2^7	128	2^{23}	8,388,608	8MB	
2^8	256	2^{24}	16,777,216	16MB	
2^9	512	2^{25}	33,554,432	32MB	
2^{10}	1,024	1k	2^{26}	67,108,864	64MB
2^{11}	2,048	2k	2^{27}	134,217,728	128MB
2^{12}	4,096	4k	2^{28}	268,435,456	256MB
2^{13}	8,192	8k	2^{29}	536,870,912	512MB
2^{14}	16,384	16k	2^{30}	1,073,741,824	1GB
2^{15}	32,768	32k	2^{31}	2,147,483,648	2GB
2^{16}	65,536	64k	2^{32}	4,294,967,296	4GB

Doing the logic thing

One of the great things about binary is that it's very efficient at handling special operations called logical operations. *Logical operations* compare two binary bits and render a third binary bit as a result. There are 16 possible logical operations. For now, I want to introduce you to three of them: AND, OR, and XOR.

The following list summarizes these three basic logical operations:

>> **AND:** An AND operation compares two binary values. If both values are 1, the result of the AND operation is 1. If one value is 0 or both of the values are 0, the result is 0.

>> **OR:** An OR operation compares two binary values. If at least one of the values is 1, the result of the OR operation is 1. If both values are 0, the result is 0.

>> **XOR:** An XOR operation compares two binary values. If exactly one of them is 1, the result is 1. If both values are 0 or if both values are 1, the result is 0.

Table 1-2 summarizes how AND, OR, and XOR work.

TABLE 1-2

Logical Operations for Binary Values

First Value	Second Value	AND	OR	XOR
0	0	0	0	0
0	1	0	1	1
1	0	0	1	1
1	1	1	1	0

You can apply logical operations to binary numbers that have more than one binary digit by applying the operation one bit at a time. The easiest way to do this manually is to line the two binary numbers on top of one another and then write the result of the operation below each binary digit. The following example shows how you calculate 10010100 AND 11011101:

```
        10010100
AND 11011101
        10010100
```

As you can see, the result is 10010100.

USING WINDOWS CALCULATOR FOR BINARY CONVERSIONS

If you have a computer, you can use the free Calculator program that comes with all versions of Windows to work with binary numbers. The Calculator program has a special Programmer mode that many users don't know about. When you flip the Calculator into this mode, you can do instant binary and decimal conversions, which occasionally come in handy when you're working with IP addresses.

To use the Windows Calculator in Programmer mode, launch the Calculator (click the Start button and type **Calculator** to find it). Then click the menu icon near the upper left of the Calculator and choose View ⇨ Programmer. The Calculator changes to a fancy programmer model — the kind that I paid $150 for when I was a rookie computer programmer 40 years ago. All kinds of buttons appear:

You can select the Hex, Dec, Oct, and Bin radio buttons to switch from decimal to the different numbering systems commonly used in digital electronics: hexadecimal, octal, and binary. For example, to find the binary equivalent of decimal 155, enter **155** and then select the Bin radio button. The value in the display changes to 10011011.

Here are a few other things to note about the Programmer mode of the Calculator:

- The Programmer Calculator has several features that are designed specifically for binary calculations, such as AND, XOR, NOT, and NOR.

- The Programmer Calculator can also handle hexadecimal conversions. Hexadecimal doesn't come into play when you're dealing with IP addresses, but it's used for other types of binary numbers, so this feature sometimes proves to be useful.

- The calculator in earlier versions of Windows (prior to Windows 7) doesn't have Programmer mode. However, the older versions do have a Scientific mode which included features for working with binary numbers.

Using Switches to Build Gates

To give you an idea of how basic gates work, Projects 28, 29, and 30 show you how to assemble an AND gate, an OR gate, and an XOR gate using simple DPDT knife switches. In actual practice, gate circuits are built with transistors or with integrated circuits. But these three projects will give you a good idea of how gates work.

Figure 1-2 shows these three projects fully assembled. The AND gate (Project 28) is shown in the top photo. As you can see, the AND gate is implemented by connecting two switches together in series. Both switches must be closed for the lamp to light.

The circuit for the OR gate (Project 29) is shown in the middle photo of Figure 1-2. To implement an OR gate with switches, you simply wire the switches in parallel. Then, the lamp lights if either of the switches is closed.

The XOR gate circuit (Project 30) is shown in the bottom photo of Figure 1-2. This wiring is a bit trickier. The DPDT switches are cross-connected in such a way that the circuit to the lamp is completed if one of the switches is in the A position and the other is in the B position, or vice versa. If both switches are in the same position, the circuit is not completed so the lamp does not light.

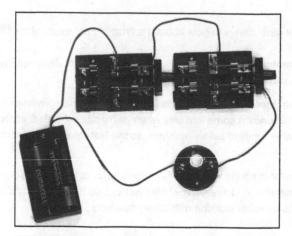

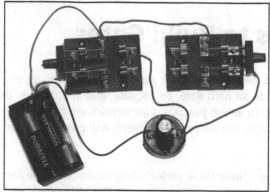

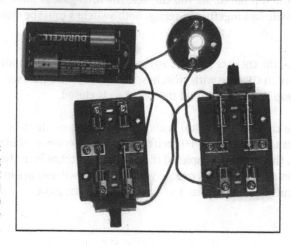

FIGURE 1-2:
Implementing
AND (top), OR
(middle), and
XOR (bottom)
gates with knife
switches.

Project 28: A Simple AND Circuit

In this project, you use two switches and a lamp to create a simple circuit that performs a logical AND operation.

The two switches represent the two binary values that are input to the AND operation. A closed switch represents the binary value 1, and an open switch represents the binary value 0.

The lamp represents the binary output of the AND operation. When lit, the lamp represents the binary value 1. When not lit, the lamp represents the binary value 0.

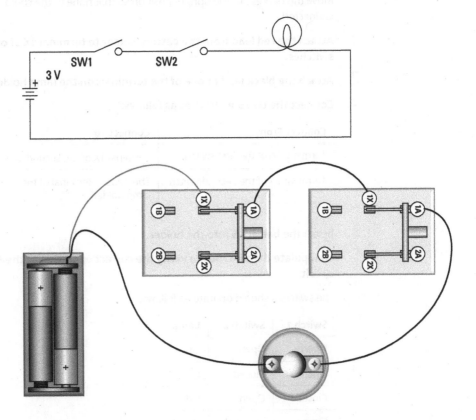

Parts

>> Two AA batteries

>> One battery holder

>> One lamp holder

>> One 2.33 V flashlight lamp

>> Two DPDT knife switches

>> Two 5-inch lengths of 22-gauge stranded wire, stripped ½ inch on each end

Steps

1. **Open both switches.**

   Move the handles to the upright position so that none of the contacts is connected.

2. **Attach the red lead from the battery holder to terminal 1X of one of the switches.**

3. **Attach the black lead to one of the terminals on the lamp holder.**

4. **Connect the two 5-inch wires as follows:**

Connect From	Connect To
Terminal 1A of the first switch	Terminal 1X of the second switch
Terminal 1A of the second switch	The unused terminal of the lamp holder

5. **Insert the batteries into the holder.**

6. **Manipulate the switches to verify the correct operation of the AND circuit.**

   The switches should operate as follows:

Switch 1	Switch 2	Lamp
Open	Open	Off
Open	Closed	Off
Closed	Open	Off
Closed	Closed	On

   Notice that the lamp comes on only when both switches are closed, which is exactly how an AND circuit should work.

   You're done! Give yourself a pat on the back, because you've just built your first logic circuit.

Project 29: A Simple OR Circuit

In this project, you build a simple OR circuit by wiring two switches in parallel to control a lamp. The lamp will be lit when either one of the two switches is closed. The switches represent the two binary inputs to the OR operation, and the lamp represents the binary output.

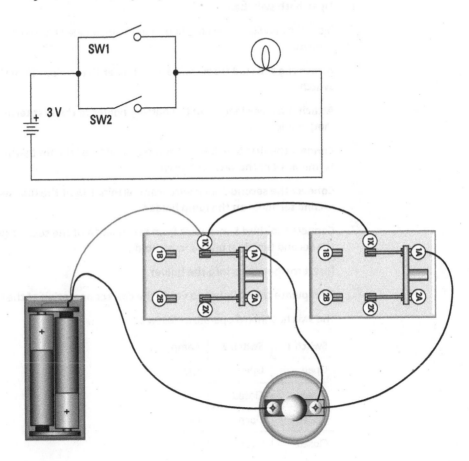

Parts

» Two AA batteries

» One battery holder

» One lamp holder

» One 2.33 V flashlight lamp

» Two DPDT knife switches

» Three 5-inch lengths of 22-gauge stranded wire

Steps

1. **Open both switches.**

   Move the handles to the upright position so that none of the contacts is connected.

2. **Attach the red lead from the battery holder to terminal 1X of the first switch.**

3. **Attach the black lead from the battery holder to the first terminal on the lamp holder.**

4. **Connect the first 5-inch wire from terminal 1X of the first switch to terminal 1X of the second switch.**

5. **Connect the second 5-inch wire from terminal 1A of the first switch to the second terminal on the lamp holder.**

6. **Connect the third 5-inch wire from terminal 1A of the second switch to the second terminal on the lamp holder.**

7. **Insert the batteries into the holder.**

8. **Manipulate the switches to verify the correct operation of the OR circuit.**

   The switches should operate as follows:

Switch 1	Switch 2	Lamp
Open	Open	Off
Closed	Closed	On
Closed	Open	On
Closed	Closed	On

   Notice that the lamp is on whenever at least one of the two switches is closed. The lamp is off only when both switches are open. That's the way an OR circuit should operate.

   You're done! Congratulate yourself on a job well done. You've completed your second logic circuit.

Project 30: A Simple XOR Circuit

In this project, you build a simple XOR circuit by wiring two switches to control a lamp. In this circuit, a lamp will light when one or the other switch is closed. If both switches are open or both are closed, the lamp won't light. The switches represent the two binary inputs to the XOR operation, and the lamp represents the binary output.

Unlike the other two projects in this chapter, the switches in this project use both the A and B positions. Thus, a switch represents a binary 1 when it's in the A position. In the B position, the switch represents a binary 0.

Notice how in the schematic diagram the A and B terminals are cross-connected between the two switches. That way, the circuit to the lamp will be closed only if the switches are in opposite positions. In other words, the lamp will light if SW1 is in the A position and SW2 is in the B position, or if SW1 is in the B position and SW2 is in the A position. If both switches are in the A position or if both switches are in the B position, the lamp won't light.

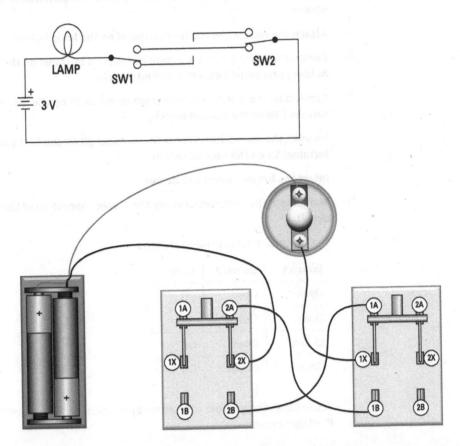

Understanding Digital Electronics

Parts

» Two AA batteries

» One battery holder

» One lamp holder

» One 2.33 V flashlight lamp

» Two DPDT knife switches

» Three 5-inch lengths of 22-gauge stranded wire

Steps

1. **Open both switches.**

   Move the handles to the upright position so that none of the contacts is connected.

2. **Attach the black lead from the battery holder to terminal 2X of the first switch.**

3. **Attach the red lead to the first terminal on the lamp holder.**

4. **Connect the first 5-inch wire from the second terminal on the lamp holder to terminal 1X on the second switch.**

5. **Connect the second 5-inch wire from terminal 2A on the first switch to terminal 1B on the second switch.**

6. **Connect the third 5-inch wire from terminal 2B on the first switch to terminal 1A on the second switch.**

7. **Insert the batteries into the holder.**

8. **Manipulate the switches to verify the correct operation of the XOR circuit.**

   The switches should operate as follows:

Switch 1	Switch 2	Lamp
Open	Open	Off
Open	Closed	On
Closed	Open	On
Closed	Closed	Off

   You're done! Congratulate yourself on a job well done. You've completed your third logic circuit!

Chapter **2**

Getting Logical

I grew up on *Star Trek*. Captain Kirk and his valiant crew sparked my imagination as they journeyed through the galaxy on their amazing adventures. So it's only logical that I'm inclined to begin a chapter on logic with a reference to Mr. Spock. In fact, it would be illogical to begin this chapter in any other way.

In this chapter, I look at the basic principles of logic, which are the underpinnings of digital electronics. In particular, I explore *logic gates*, which are handy devices that perform a logical operation on two binary inputs and produce a single binary output result.

It's incredible to realize that the overwhelming complexity of modern computer systems is built on the simple concept of logic gates. A modern computer processor consists of billions of individual logic gates connected in a way that enables the processor to perform complicated operations at amazing speed.

So back to Mr. Spock: My favorite Spock quote concerning logic is in an episode entitled "The Changeling." At the very end of the episode, Spock congratulates

Captain Kirk for getting the *Enterprise* crew out of a real bind by using sophisti-cated logic on a destruction-bent robot:

Spock: My congratulations, Captain. A dazzling bit of logic.

Kirk: You didn't think I had it in me, did you, Spock?

Spock: No.

Well, Kirk did have it in him, and I think you have it in you, too. So let's get started!

Introducing Boolean Logic and Logic Gates

In digital electronics, *Boolean logic* refers to the manipulation of binary values in which a 1 represents the concept of *true* and a 0 represents the concept of *false*. In electronic circuits that implement logic, binary values are represented by voltage levels. In the most common convention, a binary value of one is represented by +5 V (also called *HIGH*), and a binary zero is represented by 0 V (also called *LOW*).

This type of logic is called *Boolean* because it was invented in the 19th century by George Boole, an English mathematician and philosopher. In 1854, he published a book titled *An Investigation of the Laws of Thought*, which laid out the initial con-cepts that eventually came to be known as Boolean algebra, also called Boolean logic. Boolean logic is among the most important principles of modern computers. Thus, most people consider Boole to be the father of computer science.

As I mention at the start of this section, in Boolean logic, *true* is represented by the binary digit 1 and *false* by the binary digit 0. *Logical operations* (also called *logical functions)* are functions that can be applied to one or more logic inputs and produce a single logic output. One of the most common types of logic operations is NOT, which simply inverts the state of its input. In other words, with the NOT opera-tion, if the input is true, the output is false; if the input is false, the output is true.

A *gate* is a circuit or device that implements a logical function. Thus, a NOT gate is a circuit or device that implements the logical NOT operation. NOT gates are very common in digital circuits.

You can create gate switches in a variety of ways. The most common method uses transistors as switches, arranged in such a way that the correct output is generated based on the logical inputs and the type of gate being implemented. In Chapter 3 of this minibook, you see how gates are implemented. For the rest of this chapter, I discuss what the various types of gates do and how they can be combined in real-world circuits.

Regardless of the method used to create gate circuits, all logic circuits depend on different voltage ranges to represent 1 and 0. As I've already mentioned, the most common voltage convention is to represent 1 by approximately +5 V and 0 by approximately 0 V. The +5 V signal is usually referred to as *HIGH*, and the 0 V signal is usually called simply *LOW*.

In the rest of this chapter, I look at seven of the most common types of logic gates: NOT, AND, OR, NAND, NOR, XOR, and NXOR. All these gates except NOT use at least two inputs; the NOT gate has just one input. To help you get your bearings, Table 2-1 provides a brief overview of the distinctions among the gate types:

TABLE 2-1

The Most Common Types of Logic Gates

Gate	Description
NOT	Inverts the input (HIGH becomes LOW, LOW becomes HIGH)
AND	Outputs HIGH if all the inputs are HIGH; otherwise, outputs LOW
OR	Outputs HIGH if at least one of the inputs is HIGH; otherwise, outputs LOW
NAND	Outputs HIGH if all the inputs are LOW; otherwise, outputs LOW
NOR	Outputs HIGH if at least one of the inputs is LOW; otherwise, outputs LOW
XOR	Outputs HIGH if one, and only one, of the inputs is HIGH; otherwise, outputs LOW
NXOR	Outputs HIGH if one, and only one, of the inputs is LOW; otherwise, outputs LOW

Looking at NOT Gates

The simplest of all gates is the *NOT gate*, which is also called an *inverter*. A NOT gate has just one input, and its output is the opposite of the input. If the input is LOW, the output is HIGH. If the input is HIGH, the output is LOW.

Table 2-2 shows the truth table of an inverter. A *truth table* is simply a table that lists every possible combination of input values and shows the resulting output for each combination. For an inverter, the truth table is simple. Because only one input exists, only two possibilities exist: The input is LOW, or it is HIGH. As you can see in the table, the output is simply the opposite of the input.

TIP

Note that in truth tables, it's common to use 0s and 1s to represent logic values rather than the terms *HIGH* and *LOW*.

TABLE 2-2

The Truth Table for an Inverter

Input	Output
0	1
1	0

Figure 2-1 shows the standard logic symbol for a NOT gate. Symbols such as this are often used in schematic diagrams for circuits that use gates. The NOT symbol is simply a triangle with the input at one end and the output at the other. The small circle on the output is called a *negation bubble*, which indicates that the output is inverted.

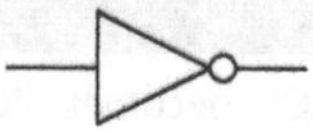

FIGURE 2-1:
The symbol for a NOT gate.

Looking at AND Gates

A *two-input AND gate* is a gate that has two inputs and one output. The output is HIGH only if both of the inputs are HIGH. Any other combination of inputs results in the output's being LOW. Table 2-3 shows the truth table for an AND gate with two inputs.

TABLE 2-3

The Truth Table for a Two-Input AND Gate

Input A	Input B	Output
0	0	0
0	1	0
1	0	0
1	1	1

The standard symbol for an AND gate is shown in Figure 2-2. The inputs are on the left, and the output is on the right.

Note that you can create AND gates with more than two inputs. For each additional input that you add to a gate, the number of possible input combinations doubles. A two-input gate has four possible input combinations; for a three-input gate, there are eight possible combinations; a four-input gate has 16 possible input combinations, and so on.

FIGURE 2-2:
The symbol for
a two-input AND
gate.

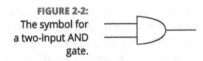

Table 2-4 shows the truth table for a three-input AND gate. As you can see, the output is HIGH only if all the inputs are HIGH. Any other combination of inputs produces LOW output.

TABLE 2-4

The Truth Table for a Three-Input AND Gate

Input A	Input B	Input C	Output
0	0	0	0
0	0	1	0
0	1	0	0
0	1	1	0
1	0	0	0
1	0	1	0
1	1	0	0
1	1	1	1

You can combine gates in a circuit to create logic networks that are more complicated than a single gate could produce. For example, you can create a three-input AND gate by using two two-input AND gates as shown in Figure 2-3. In that figure, the first AND gate produces a HIGH output only if the inputs A and B are true. Then the output of the first AND gate is used as one of the inputs to the second AND gate; the other input is the input C.

FIGURE 2-3:
A pair of two-
input AND gates
can be used to
create a logic
network that
operates like a
three-input AND
gate.

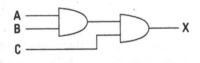

Because the output of the second gate will be HIGH only if both of its inputs are HIGH, and because the first input to the second gate is the output from the first gate, which is HIGH only if both of its inputs are HIGH, the output of the entire circuit (designated as X) is HIGH only if all three inputs (A, B, and C) are HIGH.

TECHNICAL STUFF

If you want to, you can think of an AND gate as being a form of multiplication. To understand why, first consider that when you multiply any quantity of single-bit binary numbers, only two possible results exist: zero or one. Then, consider that any value times zero is zero. So, no matter how many inputs there are, if any one of them is zero (LOW), the result of the multiplication will be zero.

Look back at the truth tables in Table 2-3 and Table 2-4, and try multiplying the binary inputs in each row. In each case, the answer will be zero for any combination that contains a zero in any of the inputs. The answer will be 1 only if all the inputs are 1.

The following paragraphs describe just two of the many situations in which you might use an AND gate in a real-life electronic circuit:

>> Figure 2-4 shows how AND gates might be used in a home alarm system. Here, the inputs for the various sensors placed on the home's doors and windows are processed by the sensor circuit, which sends a HIGH signal to one of the inputs of the AND gate if any of the sensors indicates an intrusion anywhere in the house. Then the arming circuit sends a 1 to the other input of the AND gate if the system is armed. Finally, the alarm circuit sounds an audible alarm if the AND gate's output is 1. As a result, the alarm will sound if an intrusion is detected and the system is armed.

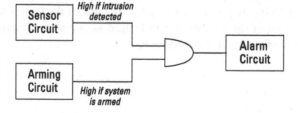

FIGURE 2-4:
An AND gate used in a home alarm system.

When used in this way, an AND gate is often called an *enable input*, as one of the inputs to the AND gate enables the other input to be processed. When the enable input is HIGH, the controlled input is allowed to pass through the AND gate. When the enable is LOW, the controlled input is inhibited.

Figure 2-5 shows a more developed version of the home alarm system that uses an enable input. In this version, the alarm doesn't sound immediately when an intrusion is detected. Instead, the alarm sounds 30 seconds after it is triggered. This gives you time to disable the alarm before waking up the neighbors. When the sensor circuit detects an intrusion, it sends a trigger pulse to a 555 timer circuit, which then generates a 30-second HIGH pulse. The HIGH signal from the 555 timer output is routed through a NOT gate, which inverts the signal to LOW. The LOW signal is sent to one of the inputs to the second AND gate. The other input to the second AND gate is the output of the first AND gate, which indicates that an intrusion has been detected, and the system is armed.

For the first 30 seconds after the intrusion is detected, the first input to the second AND gate is LOW, and the second input is HIGH, so the output from the second AND gate is LOW. Thus, the alarm doesn't sound. In effect, the timer output inhibits the alarm from sounding. But when the output from the 555 timer circuit goes LOW after the 30-second pulse ends, the NOT gate inverts the signal, sending a HIGH output to the second AND gate. This situation causes the second AND gate's output to go HIGH, so the alarm sounds. Thus, the inverted pulse from the timer circuit enables the alarm circuit.

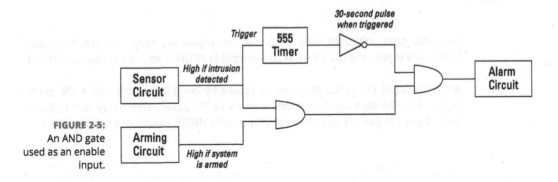

FIGURE 2-5:
An AND gate used as an enable input.

Looking at OR Gates

An *OR gate* produces a HIGH output if any of the inputs is HIGH. The output from an OR gate is LOW only if all the inputs are LOW. Table 2-5 shows the truth table for a two-input OR gate.

TABLE 2-5

The Truth Table for a Two-Input OR Gate

Input A	Input B	Output
0	0	0
0	1	1
1	0	1
1	1	1

It's important to note that in an OR gate, it doesn't matter how many of the inputs are HIGH. If at least one input is HIGH, the output will be HIGH. Thus, in a two-input OR gate, the output will be HIGH if either input is HIGH or both inputs are HIGH. In a three-input OR gate, the output will be HIGH if any one, any two, or all three of the inputs are HIGH.

The standard symbol for an OR gate is shown in Figure 2-6. The inputs are on the left, and the output is on the right.

FIGURE 2-6:
The symbol for a two-input OR gate.

Like AND gates, OR gates with more than two inputs are easy to create. No matter how many inputs the OR gate has, the output is HIGH if any of the inputs is HIGH.

Multiple-input OR gates are easy to create by combining two-input OR gates. Figure 2-7 shows a logic network with three OR gates, effectively acting like a four-input OR gate: If any of the four inputs is HIGH, the output will be HIGH.

FIGURE 2-7:
Three OR gates used to create a four-input OR gate.

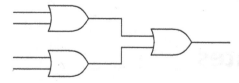

You can combine two-Input OR gates in a circuit to create OR networks that have more than two inputs. For example, Figure 2-8 shows how OR gates might be used in the sensor circuit of a home alarm system that has more than two inputs. In this circuit, eight separate alarm sensors are fed into a network of OR gates. If any one of the inputs is HIGH, the output from the sensor circuit is HIGH.

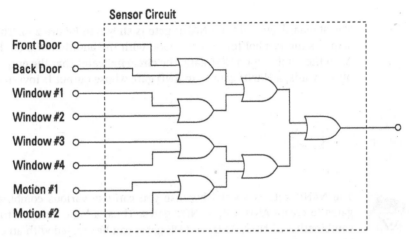

Sensor Circuit

Front Door
Back Door
Window #1
Window #2
Window #3
Window #4
Motion #1

FIGURE 2-8:
OR gates used in
a sensor circuit.   Motion #2

The sensor circuit uses seven OR gates to create an eight-input OR gate. Study the way that the OR gates are arranged in this network for a few moments to make sure you understand how it works. Each of the eight inputs is routed to one of the four OR gates in the first tier of gates. These four OR gates reduce the eight inputs to four outputs, which are then sent to the two OR gates in the second tier. These two OR gates reduce their four inputs to two outputs, which are sent to the final OR gate. Then the output of the last OR gate becomes the output of the entire sensor circuit.

Looking at NAND Gates

A *NAND* gate is a combination of an AND gate and a NOT gate. In fact, the name *NAND* is a contraction of *NOT* and *AND*. As you can see in Table 2-6, the output of a NAND gate is LOW when both of the inputs are HIGH. Otherwise, the output of the NAND gate is HIGH.

TABLE 2-6

The Truth Table for a Two-Input NAND Gate

Input A	Input B	Output
0	0	1
0	1	1
1	0	1
1	1	0

The standard symbol for a NAND gate is shown in Figure 2-9. This symbol is the same as the symbol for an AND gate, with the addition of a circle at the output. As in the symbol for a NOT gate, the circle indicates that the output is inverted. In other words, a NAND gate is an AND gate whose output is inverted.

FIGURE 2-9:
The symbol for a
two-input NAND
gate.

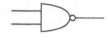

TECHNICAL STUFF

The NAND gate is special because you can use various combinations of NAND gates to create AND, OR, or NOT gates. Thus, a logic network that consists of a combination of NOT, AND, and OR gates can be created with an equivalent combination of just NAND gates. For this reason, the NAND gate is called a *universal gate*. You learn more about this characteristic of NAND gates in "Universal NAND Gates," later in this chapter.

Another interesting thing about the NAND gate is that it can be used as a kind of OR gate that tests for LOW inputs instead of HIGH inputs. In other words, the output of a NAND gate is HIGH whenever either of the inputs are LOW.

This characteristic of NAND gates is useful in many situations. For example, consider the alarm sensor circuit that was presented in the previous section and in Figure 2-8. Suppose all the alarm sensors produce a HIGH signal when there is no intrusion, then go LOW when there is an intrusion. In the real world, many alarm sensors work exactly in this way. For example, a sensor that detects whether a door is open is essentially a simple switch that is closed when the door is closed and open when the door is open. When the door is closed, current flows through the switch so the signal from the sensor is HIGH. When the door is opened, current stops flowing and the signal from the sensor goes LOW.

Figure 2-10 shows how NAND gates can be used to test for a LOW input on any of eight sensors. This circuit is identical to the circuit that was shown in Figure 2-8, except that all the OR gates have been replaced by NAND gates.

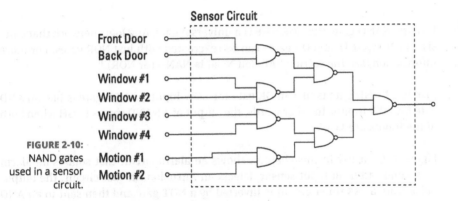

FIGURE 2-10:
NAND gates
used in a sensor
circuit.

Sensor Circuit

Front Door
Back Door
Window #1
Window #2
Window #3
Window #4
Motion #1
Motion #2

Looking at NOR Gates

A NOR gate is a combination of an OR gate and a NOT gate. As with NAND, the name NOR is a contraction of NOT and OR. The truth table for a NOR gate is shown in Table 2-7. As you can see, the output of a NOR gate is LOW when any of its input are HIGH. Otherwise, the output of the NAND gate is HIGH.

TABLE 2-7

The Truth Table for a Two-Input NOR Gate

Input A	Input B	Output
0	0	1
0	1	0
1	0	0
1	1	0

As you can see in Figure 2-11, the standard symbol for a NOR gate is the same as the symbol for an OR gate, with a negation circle added to the output. The circle simply indicates that the output is inverted. A NOR gate is essentially a combination of an OR gate and a NOT gate.

FIGURE 2-11:
The symbol for
a two-input NOR
gate.

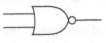

TECHNICAL STUFF

Like the NAND gate, the NOR gate is a universal gate. Any logic network that consists of NOT, AND, and OR gates can be re-created with just NOR gates. For more information, see the section "All You Need Is NAND (or NOR)."

Just as a NAND gate is like an OR gate for LOW inputs, a NOR gate is like an AND gate for LOW inputs. In other words, the output of a NAND gate is HIGH when both of the inputs are LOW.

Figure 2-5, earlier in this chapter, shows an alarm circuit that sounds an alarm 30 seconds after an input sensor detects an intrusion. In that circuit, the output pulse from a 555 timer circuit is inverted by a NOT gate and then sent to an AND gate, which sends a HIGH output to an audible alarm circuit when the output from a sensor circuit is HIGH and when the 30-second timer pulse ends.

Figure 2-12 shows a version of the same circuit that uses a NOR gate instead of an AND gate to feed the audible alarm circuit. In this circuit, the alarm sounds when the output from the sensor circuit is LOW and the output from the timer circuit is also LOW. A NOT gate isn't needed for the 555 timer output, but now a NOT gate is required on the output from the AND gate to invert its signal, so that it emits LOW when the system is armed and the alarm is tripped.

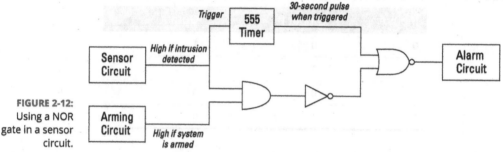

FIGURE 2-12: Using a NOR gate in a sensor circuit.

Looking at XOR and XNOR Gates

I have but two gate types remaining for your approval in this chapter: *XOR*, which stands for *Exclusive OR*, and *XNOR*, which stands for *Exclusive NOR*.

In an XOR gate, the output is HIGH if one, and only one, of the inputs is HIGH. If both inputs are LOW or both are LOW, the output is LOW. The truth table for an XOR gate is shown in Table 2-8.

TABLE 2-8

The Truth Table for a Two-Input XOR Gate

Input A	Input B	Output
0	0	0
0	1	1
1	0	1
1	1	0

Another way to explain an XOR gate is as follows: The output is HIGH if the inputs are different; if the inputs are the same, the output is LOW.

The XOR gate has a lesser-known cousin called the XNOR gate. An *XNOR gate* is an XOR gate whose output is inverted. Table 2-9 lists the truth table for an XNOR gate.

TABLE 2-9

The Truth Table for a Two-Input XNOR Gate

Input A	Input B	Output
0	0	1
0	1	0
1	0	0
1	1	1

Figure 2-13 shows the symbols used for both XOR and XNOR gates. As you can see, the only difference between these two symbols is that the XNOR has a circle on its output to indicate that the output is inverted.

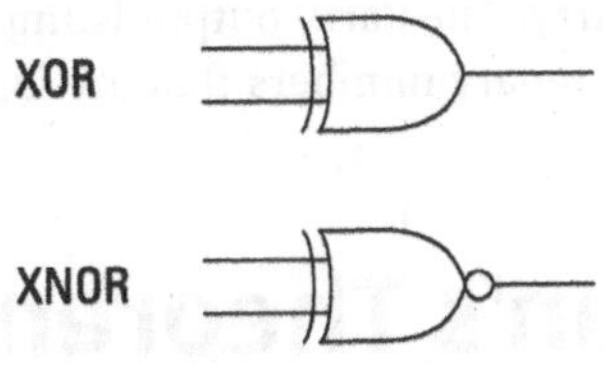

XOR

XNOR

FIGURE 2-13:
The symbols for
two-input XOR
and XNOR gates.

One of the most common uses for XOR gates is to add two binary numbers. For this operation to work, the XOR gate must be used in combination with an AND gate, as shown in Figure 2-14.

Getting Logical

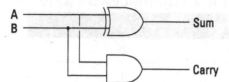

FIGURE 2-14:
An XOR gate and
an AND gate
can be used to
add two binary
numbers.

A

B

Sum

Carry

To understand how the circuit shown in Figure 2-14 works, review how binary addition works:

0 + 0 = 0

0 + 1 = 1

1 + 0 = 1

1 + 1 = 10

If you wanted, you could write the results of each of the preceding addition statements by using two binary digits, like this:

0 + 0 = 00

0 + 1 = 01

1 + 0 = 01

1 + 1 = 10

When results are written with two binary digits, as in this example, you can easily see how to use an XOR and an AND circuit in combination to perform binary addition. If you consider just the first binary digit of each result, you'll notice that it looks just like the truth table for an AND circuit and that the second digit of each result looks just like the truth table for an XOR gate.

The adder circuit, shown in Figure 2-14, has two outputs. The first is called the Sum, and the second is called the Carry. The Carry output is important when several adders are used together to add binary numbers that are longer than 1 bit.

De Marvelous De Morgan's Theorem

It's time for a theorem! Although I present some pretty sophisticated logic concepts in this chapter, I've avoided mentioning any that go by the name *theorem* — until now, anyway.

De Morgan's Theorem was created by Augustus De Morgan, a 19th-century British mathematician who developed many of the concepts that make Boolean logic work. Among De Morgan's most important work are two related theorems that have to do with how NOT gates are used in conjunction with AND and OR gates:

» An AND gate with inverted output behaves the same as an OR gate with inverted inputs.

» An OR gate with inverted output behaves the same as an AND gate with inverted inputs.

An AND gate with inverted output is also called a NAND gate, of course, and an OR gate with inverted output is also called a NOR gate. Thus, De Morgan's laws can also be stated like this:

» A NAND gate behaves the same as an OR gate with inverted inputs.

» A NOR gate behaves the same as an AND gate with inverted inputs.

TIP

An OR gate with inverted inputs is called a *negative OR gate*, and an AND gate with inverted inputs is called a *negative AND gate*.

In case you're not persuaded, review for a moment the truth table for a NAND gate:

A	B	X
0	0	1
0	1	1
1	0	1
1	1	0

Now look at the truth table for an OR gate, with an extra set of columns added to show the inverted inputs:

A	B	NOT A	NOT B	X
0	0	1	1	1
0	1	1	0	1
1	0	0	1	1
1	1	0	0	0

Here, the A and B columns represent the inputs. The NOT A and NOT B columns are the inputs after they've been inverted. Finally, the X column represents an OR operation applied to the NOT A and NOT B values.

As you can see, the final output column of these truth tables is the same. Thus, a NAND gate is equivalent to a negative OR gate. Any time you see a NAND gate in a circuit diagram, you can substitute a negative OR gate.

Now take a look at the other side of De Morgan's Theorem. Here's a truth table for a NOR gate:

A	B	X
0	0	1
0	1	0
1	0	0
1	1	0

And here's the output of a negative AND gate:

A	B	NOT A	NOT B	X
0	0	1	1	1
0	1	1	0	0
1	0	0	1	0
1	1	0	0	0

Again, you can see that these two truth tables give the same output.

Just as a circle is used on the output of a NAND or NOR gate to indicate that the output is inverted, you can use a circle on the inputs to an OR or AND gate to indicate that the inputs are inverted. Figure 2-15 shows these symbols. The figure also shows that the negative OR and AND gates are interchangeable with the NAND and NOR gates.

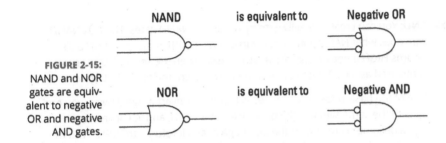

FIGURE 2-15:
NAND and NOR
gates are equiv-
alent to negative
OR and negative
AND gates.

All You Need Is NAND (Or NOR)

In "Looking at NAND Gates" and "Looking at NOR Gates," earlier in this chapter, I mention that the NAND gate is a universal gate because any other type of gate (such as AND, OR, XOR, and NXOR) can be constructed solely from combinations of NAND gates.

This fact is incredibly useful because it enables you to build any logic circuit, simple or complex, by using just NAND gates. When you begin to build your own digital circuits, you can stock up on integrated circuits that contain just NAND gates and be assured that you can build even the most complex circuits with your stock of NAND gates.

In "Looking at NOR Gates," I also mention that the NOR gate is a universal gate. Thus, you can also build any logic circuit by using nothing but NOR gates.

In the following sections, I first explain how you can use NAND gates to build other types of gates; then I show you how to do the same thing with NOR gates.

Universal NAND gates

Figure 2-16 shows how you can use NAND gates in various combinations to create NOT, AND, OR, and NOR gates. The following paragraphs describe how the circuits work:

>> **NOT:** You can create a NOT gate from a NAND gate simply by tying the two inputs of the NAND gate together. Because the two inputs of the NAND gate are tied together, only two input combinations are possible: both HIGH or both LOW. If both inputs are HIGH, the NAND gate will output a LOW. If both inputs are LOW, the NAND gate will output HIGH. Thus, the circuit behaves exactly as a NOT gate would.

>> **AND:** You can create an AND gate by using two NAND gates. The first NAND gate does what NAND gates do: returns LOW if both inputs are HIGH and returns HIGH if both inputs are anything else. Then the second NAND gate is configured as a NOT gate to invert the output from the first NAND gate.

One of the basic rules of NOT gates is that if you invert a signal twice, you end up with the same signal. If the original input is HIGH, and you invert it, the signal becomes LOW. Invert the input again, and it returns to HIGH. With this rule in mind, you should be able to see how the two NAND gates work together to create an AND gate.

>> **OR:** You need three NAND gates to create an OR gate. First, you use a pair of NAND gates configured as NOT gates to invert the two inputs. Then the third NAND gate produces a LOW output if both of the original inputs are LOW. If one of the original inputs is HIGH, or if both of the original inputs are HIGH, the output of the third gate is HIGH.

>> **NOR:** It takes four NAND gates to create a NOR gate. As you can see in Figure 2-16, this circuit is the same as the OR gate circuit, with the addition of another NOT gate to invert the output from the third NAND gate. Inverting the output changes the overall function of the circuit from OR to NOR.

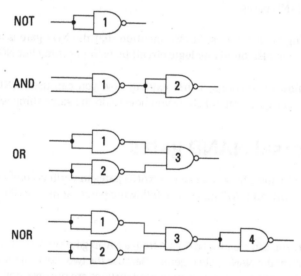

FIGURE 2-16:
Creating NOT, AND, OR, and NOR gates by using nothing but NAND gates.

Universal NOR gates

Like NAND, NOR is a universal gate. Figure 2-17 shows how NOR gates can be combined in various and sundry ways to create NOT, AND, OR, and NAND gates.

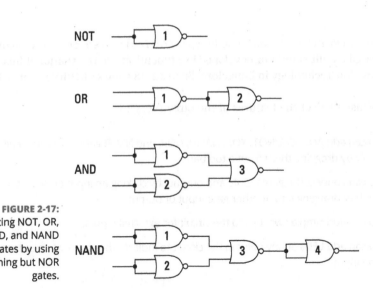

FIGURE 2-17:
Creating NOT, OR, AND, and NAND gates by using nothing but NOR gates.

The following paragraphs describe how the circuits work:

>> **NOT:** Creating a NOT gate from a NOR gate is the same as creating a NOT gate from a NAND gate: You simply tie the two inputs of the NOR gate together. If both inputs are LOW, the NOR gate will output a HIGH. Otherwise, the output is LOW.

>> **OR:** You need two NOR gates to create an OR gate. The first NOR gate returns LOW if either input is HIGH or both inputs are HIGH. Then the second NOR gate is configured as a NOT gate to invert the output of the first NOR gate.

>> **AND:** You need three NOR gates to create an AND gate. The first two are configured as NOT gates, so they invert the inputs. Then the third NOR gate produces a HIGH output if both of the original inputs are HIGH.

>> **NAND:** It takes four NOR gates to create a NAND gate. The first three NOR gates are configured just as they are for an AND gate. Then a fourth NOR gate configured as a NOT gate inverts output from the third NOR gate.

Using Software to Practice with Gates

TIP

If you really want to learn how logic gates work, one of the best ways is to use one of the many software logic gate simulators that are available free on the Internet. You can find these programs by firing up Google and searching for keywords such as *"logic gate simulator."*

One of my favorites is CircuitVerse (https://circuitverse.org/simulator), a cloud-based circuit simulator developed by students at the International Institute of Information Technology in Bangalore. Figure 2-18 shows CircuitVerse in action.

Here are just a few of the features of the CircuitVerse:

>> You can add AND, OR, NOT, XOR, NAND, NOR, and NXOR gates to your circuit simply by dragging them from a toolbar.

>> You can connect the gate inputs and outputs by clicking an input or output and then dragging it to another gate input or output.

>> You can add simple switches to the circuit for external inputs.

>> You can add LEDs at any point in the circuit to indicate the status of an output or an input.

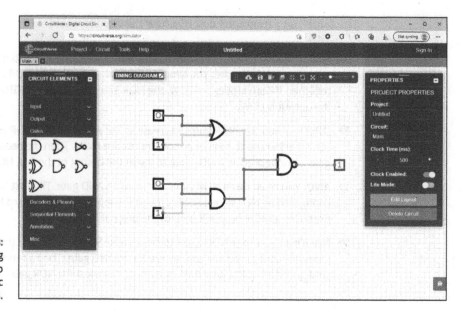

FIGURE 2-18: Using CircuitVerse to simulate logic circuits.

Chapter **3**

Working with Logic Circuits

n Book 5, Chapter 2, you learn all about logic gates, including the seven most popular kinds of gates: NOT, AND, OR, NAND, NOR, XOR, and XNOR.

In this chapter, you learn how to create actual circuits that use logic gates. I start you off by showing you the basics of how logic gates can be created from simple transistor circuits. Then I look at two popular integrated circuit families that provide prebuilt logic gates.

If you haven't already read Chapter 2 of this minibook, I suggest that you go back and do so now before you read this chapter. Little of this chapter will make sense if you aren't familiar with the different types of gates described in that chapter.

Creating Logic Gates with Transistors

In Book 2, Chapter 6, you see how transistors can be used as switches. In a nutshell, a voltage applied to the base of a transistor allows current to flow from the collector to the emitter. Thus, by applying an input signal to a transistor's base, you can control an output signal taken from the collector-emitter path.

You can build any logic gate you want by cobbling together a few transistors and resistors in just the right way. In this section, I look at simple transistor circuits for five gate types — NOT, AND, OR, NAND, and NOR — and I present projects to show you how to build three of them: NOT, NAND, and NOR.

If you need a refresher on how transistors work, see Book 2, Chapter 6.

TIP

All the circuits in this chapter assume that a HIGH signal (logical one) is represented by a DC voltage of +5 V or more. LOW (logical zero) is represented by near-zero voltage.

TECHNICAL
STUFF

Note that you won't often actually build your own logic gates by using transistors and resistors. Instead, you'll use integrated circuits that contain prefabricated logic gates. Before you use logic ICs in your circuits, however, you should have a basic understanding of how the gates that they contain work. After you take a look at simple transistor circuits for basic logic gates, you'll take a look at the logic ICs.

A transistor NOT gate circuit

A NOT gate simply inverts its input. If the input is HIGH, the output is LOW, and if the input is LOW, the output is HIGH. Such a circuit is easy to build, using a single transistor and a pair of resistors. Figure 3-1 shows the schematic.

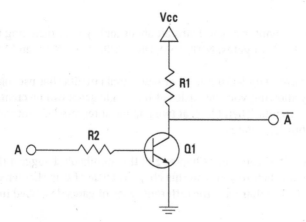

FIGURE 3-1:
A transistor
NOT gate.

The operation of this circuit is simple. The input is connected through resistor R2 to the transistor's base. When no voltage is present on the input, the transistor turns off. When the transistor is off, no current flows through the collector-emitter path. Thus, current from the supply voltage (V_{cc} in the schematic, typically between +5 V and +9 V) flows through resistor R1 to the output. In this way, the circuit's output is HIGH when its input is LOW.

When voltage is present at the input, the transistor turns on, allowing current to flow through the collector-emitter circuit directly to ground. This ground path creates a shortcut that bypasses the output, which causes the output to go LOW.

In this way, the output is HIGH when the input is LOW and LOW when the input is HIGH.

Project 31 shows how to assemble a simple transistor NOT gate on a solderless breadboard. For this project, a normally open push button is used as the input. When the button isn't pressed, the input is LOW and the output is HIGH, which causes the LED to light. When you press the button, the input goes HIGH, the output goes LOW, and the LED goes out. The assembled project is shown in Figure 3-2.

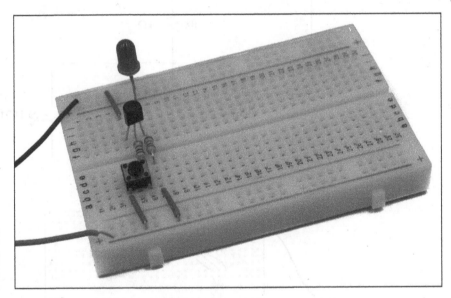

FIGURE 3-2:
A transistor NOT gate assembled on a breadboard (Project 31).

Project 31: A Transistor NOT Gate

In this project, you build a simple NOT gate by using a bipolar transistor. A NOT gate, shown in Table 3-1, is also known as an inverter. It simply reverses the logic level of its input. So, if the input is HIGH, the output of a NOT gate is LOW. If the input is LOW, the output is HIGH.

The input to this gate is controlled by a push-button switch (SW1). When the switch is open (not pressed), the input is LOW. When the switch is pressed, the input is HIGH.

TABLE 3-1

NOT Gate Truth Table

Input	Output
0	1
1	0

The output from this gate is sent through an LED, so the LED is on when the output is HIGH and off when the output is LOW.

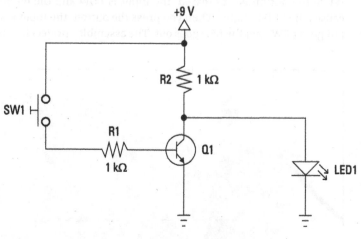

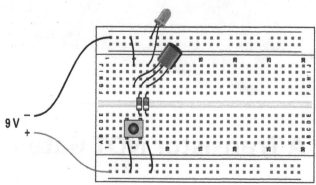

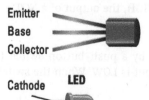

2N2222 Transistor

Emitter
Base
Collector

Cathode **LED**

Anode

Parts

» One 9-volt battery

» One 9-volt battery clip

» One small solderless breadboard

» One normally open DIP breadboard push button

» One NPN switching transistor, 2N2222A or similar

» One 5 mm red LED

» Two 1 kΩ resistors (brown-black-red)

» Three short jumper wires

Steps

For the following steps, orient the breadboard so that the hole at location A1 is near the bottom-left corner of the breadboard. The project will use the positive ground bus at the bottom of the breadboard and the negative ground bus at the top of the breadboard.

1. **Insert the transistor Q1.**

Emitter: G5

Base: G6

Collector: G7

2. **Insert resistors R1 and R2.**

R1 (1 kΩ: E7 to F7).

R2 (1 kΩ: E6 to F6).

3. **Insert LED1.**

Cathode (short lead): Ground bus at the top of breadboard

Anode (long lead): I7

4. **Insert SW1.**

The pins should be inserted in B4, D4, B4, and D6 such that the switch opens and closes across rows 4 and 6.

5. Insert the jumper wires.

1: Positive bus at the bottom of the breadboard to A4.

2: Positive bus at the bottom of the breadboard to A6.

3: Ground bus at the top of the breadboard to J5.

6. Connect the battery clip leads.

Red lead: Positive bus at the bottom of the breadboard.

Black lead: Ground bus at the top of the breadboard.

7. Test the circuit.

Connect the battery to the battery holder. The LED will come on, indicating a HIGH output because the input is LOW (that is, the switch is open so there is no voltage on the transistor's base).

When you press the button, the input is HIGH and there is voltage on the base. The LED turns off, indicating that the output is LOW.

Release the button to change the input back to LOW; the LED comes back on, indicating that the output is HIGH.

TIP

If the circuit doesn't behave as it should, double-check your wiring. Make sure the transistor is oriented correctly; if it's inserted into the breadboard backward, the circuit won't work. Also, confirm that the LED is oriented correctly; if the cathode and anode are reversed in the circuit, the LED won't come on. Finally, verify that the battery holder leads are connected to the correct negative and positive buses.

A transistor AND gate circuit

A two-input AND gate produces a HIGH output if both of its inputs are HIGH. You can create a two-input AND gate by using two transistors and three resistors, as shown in Figure 3-3. In this circuit, the output current must flow from the Vcc supply voltage through the collector-emitter circuits of both transistors to reach the output. Thus, current will flow to the output only if both of the transistors are on.

The bases of both transistors are fed through R2 and R3 from the two inputs. Thus, if both inputs are HIGH, current will flow through the base-emitter path of both transistors, turning both transistors on and allowing current to flow through to the output. If either input is LOW, the corresponding transistor turns off, and the output goes LOW.

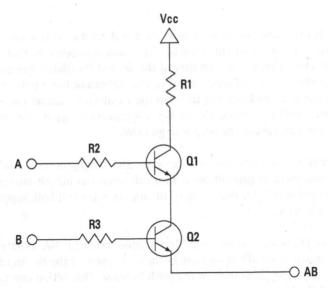

FIGURE 3-3:
A transistor
AND gate.

A transistor NAND gate circuit

A two-input NAND gate produces a LOW output if both of its inputs are HIGH. You could create a NAND gate by combining the circuits shown in Figure 3-1 and Figure 3-3 so that the output from the AND gate is used as input to the NOT gate, but that combination would require three transistors. It's easy enough to create a NAND gate by using just two transistors, as shown in Figure 3-4.

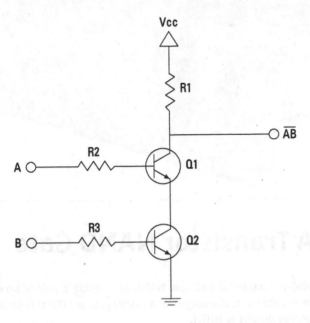

FIGURE 3-4:
A transistor
NAND gate.

This NAND gate circuit is almost identical to the AND gate circuit shown in Figure 3-3, earlier in this chapter. The only difference is that instead of connecting the output to the emitter of the second transistor, the output is obtained before the collector of the first transistor. If both of the inputs are HIGH, both of the transistors will conduct through their collector-emitter paths, which creates a short circuit to ground. This causes the current to bypass the output altogether, which in turn causes the output to go LOW.

If either transistor turns off, however, the supply current can't flow through the transistors to ground, so it flows through the output circuit instead. Thus, the output is HIGH if either one of the inputs is LOW. If both inputs are HIGH, the output is LOW.

Project 32 shows how to assemble a simple transistor NAND gate on a solderless breadboard. Normally open push buttons are used for the two inputs. The LED will be on until you press both of the push buttons. This action causes both inputs to go HIGH, which causes the output to go LOW and the LED to go dark. The completed project is shown in Figure 3-5.

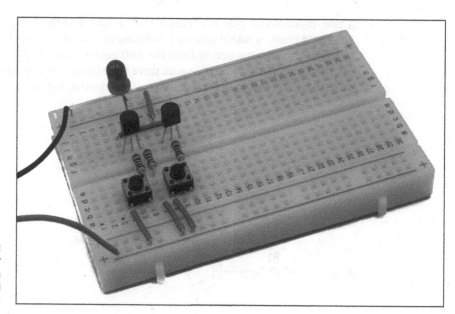

FIGURE 3-5:
A two-transistor NAND gate on a breadboard (Project 32).

Project 32: A Transistor NAND Gate

In this project, you build a simple NAND gate using a pair of bipolar transistors. As shown in Table 3-2, the output of a NAND gate is LOW if both inputs are HIGH; otherwise, the output is HIGH.

TABLE 3-2

NAND Gate Truth Table

Input A	Input B	Output
0	0	1
0	1	1
1	0	1
1	1	0

The inputs in this project are provided by two breadboard-mounted push buttons, and the output is indicated by an LED. The LED is lit unless both of the push buttons are pressed.

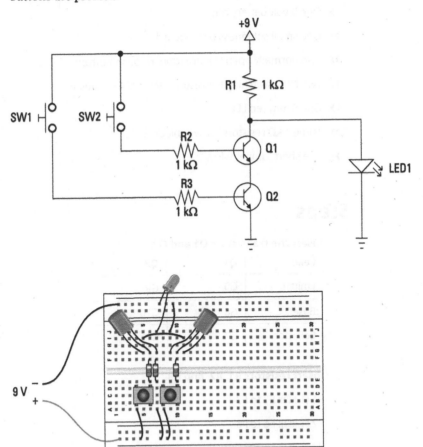

2N2222 Transistor

Emitter
Base
Collector

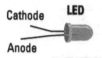

LED

Cathode

Anode

Parts

>> One 9-volt battery

>> One 9-volt battery clip

>> One small solderless breadboard

>> Two normally open DIP breadboard push buttons

>> Two NPN switching transistors, 2N2222A or similar

>> One 5 mm red LED

>> Three 1 kΩ resistors (brown-black-red)

>> Five short jumper wires

Steps

1. **Insert the transistors Q1 and Q2.**

Lead	Q1	Q2
Emitter	G5	G9
Base	G6	G10
Collector	G7	G11

2. **Insert resistors R1, R2, and R3.**

    R1 (1 kΩ): E7 to F7

    R2 (1 kΩ): E6 to F6

    R3 (1 kΩ): E10 to F10

3. **Insert LED1.**

Cathode (short lead): Ground bus

Anode (long lead): J7

4. **Insert SW1.**

The pins should be inserted in B4, D4, B6, and D6 such that the switch opens and closes across rows 4 and 6.

5. **Insert SW2.**

The pins should be inserted in B8, D8, B10, and D10 such that the switch opens and closes across rows 8 and 10.

6. **Insert the jumper wires.**

1: Positive bus to A4

2: Positive bus to A7

3: Positive bus to A8

4: Ground bus to J9

5: I5 to I11

7. **Connect the battery clip leads.**

Red lead: Positive bus

Black lead: Ground bus

8. **Test the circuit.**

Insert the battery. The LED will come on. This is because both of the switches are in the open position (not pushed). Because both inputs are LOW, the output is HIGH.

Press the first button; the LED will remain on. The two inputs are now HIGH and LOW, and the output is still HIGH.

Release the first button and press the second button; the LED again remains on. In this case, the inputs are LOW and HIGH, so the output is HIGH.

Finally, press both buttons at the same time. Now, both inputs are HIGH. The LED goes out, confirming that the output is LOW.

TIP

If the circuit doesn't behave as it should, double-check your wiring. Make sure the transistors are oriented correctly; if either one is inserted into the breadboard backward, the circuit won't work. Also confirm that the LED is oriented correctly; if the cathode and anode are reversed in the circuit, the LED won't come on. Finally, verify that the battery holder leads are connected to the correct negative and positive buses.

A transistor OR gate circuit

A two-input OR gate produces a HIGH output if either of its inputs is HIGH or both of its inputs are HIGH. Figure 3-6 shows a schematic for an OR gate created with two transistors and three resistors.

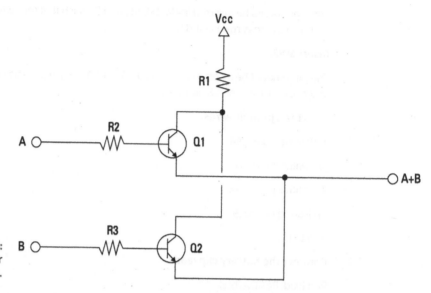

FIGURE 3-6: A transistor OR gate.

In the OR gate circuit, the supply voltage is connected to the collector of each transistor. Then the emitters of both transistors are connected to the output. That way, if voltage is applied to the base of either one of the transistors, that transistor will turn on and pass current through to the output.

Thus, the output is HIGH if one input is HIGH or both inputs are HIGH. The output is LOW only if both inputs are LOW.

A transistor NOR gate circuit

A NOR gate is an inverted OR gate. If at least one of the inputs is HIGH, the output is LOW. If both inputs are LOW, the output is HIGH.

Figure 3-7 shows a schematic for a NOR gate. This circuit is similar to the circuit shown in Figure 3-6, earlier in this chapter, except that the output is connected to the collector of both transistors, and the emitter of each transistor is connected to ground. If either one of the transistors is on, current from Vcc will be short-circuited to ground, bypassing the output. As a result, the output will be HIGH only when both inputs are LOW. If either input is HIGH or both inputs are HIGH, the output will be LOW.

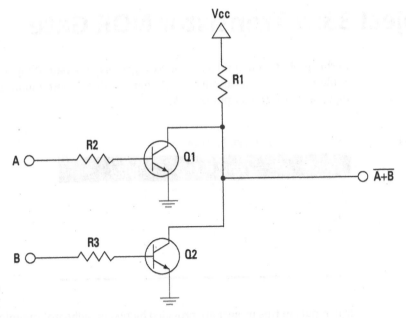

FIGURE 3-7:
A transistor
NOR gate.

You can build a two-transistor NOR gate by following the steps outlined in Project 33. As with the other projects in this chapter, Project 33 uses normally open push buttons to control the input circuits. When power is applied to this circuit, both inputs initially will be LOW, and the output will be HIGH. Pressing either one of the switches causes that switch's input to go HIGH, which in turn causes the output to go LOW. The assembled project is shown in Figure 3-8.

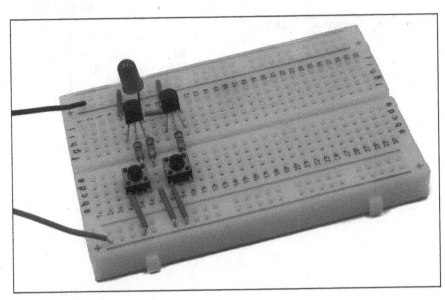

FIGURE 3-8:
A two-transistor
NOR gate on
a breadboard
(Project 33).

Project 33: A Transistor NOR Gate

In this project, you build a simple NOR gate using a pair of bipolar transistors. As Table 3-3 shows, the output of a NOR gate is HIGH if both inputs are LOW. If either input is HIGH, the output is LOW.

TABLE 3-3

NOR Gate Truth Table

Input A	Input B	Output
0	0	1
0	1	0
1	0	0
1	1	0

The inputs in this project are provided by two breadboard-mounted push buttons, and the output is indicated by an LED. The LED is lit unless at least one of the push buttons is pressed.

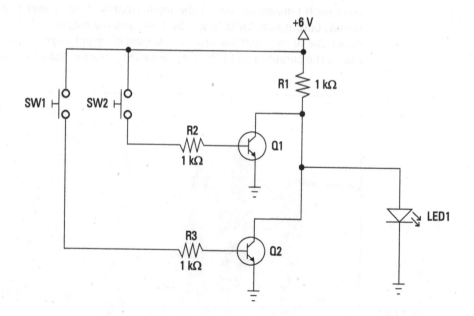

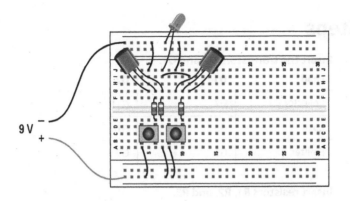

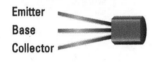

2N2222 Transistor

Emitter
Base
Collector

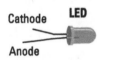

LED

Cathode

Anode

Parts

>> One 9-volt battery

>> One 9-volt battery clip

>> One small solderless breadboard

>> Two normally open DIP breadboard push buttons

>> Two NPN switching transistors, 2N2222A or similar

>> One 5 mm red LED

>> Three 1 kΩ resistors (brown-black-red)

>> Six short jumper wires

Steps

1. **Insert the transistors Q1 and Q2.**

Lead	Q1	Q2
Collector	G7	G11
Base	G6	G10
Emitter	G5	G9

2. **Insert resistors R1, R2, and R3.**

   R1 (1 kΩ): E7 to F7

   R2 (1 kΩ): E6 to F6

   R3 (1 kΩ): E10 to F10

3. **Insert LED1.**

   Cathode (short lead): Ground bus

   Anode (long lead): J7

4. **Insert SW1.**

   The pins should be inserted in B4, D4, B6, and D6 such that the switch opens and closes across rows 4 and 6.

5. **Insert SW2.**

   The pins should be inserted in B8, D8, B10, and D10 such that the switch opens and closes across rows 8 and 10.

6. **Insert the jumper wires.**

   1: Positive bus to A4

   2: Positive bus to A7

   3: Positive bus to A8

   4: Ground bus to J5

   5: Ground bus to J9

   6: I7 to I11

7. **Connect the battery clip leads.**

   Red lead: Positive bus

   Black lead: Ground bus

8. **Test the circuit.**

Insert the battery. The LED will come on. This is because both of the switches are in the open position (not pushed). Because both inputs are LOW, the output is HIGH.

Press the first button. The LED turns off because the first input has gone HIGH and the output has gone LOW.

Release the first button. The LED comes back on because both inputs are now LOW.

Now press the second button. The LED turns off again because the second input is HIGH.

Finally, press both buttons at the same time. The LED turns off again, properly reflecting the truth table of a NOR gate: When either or both inputs are HIGH, the output is LOW.

TIP

If the circuit doesn't behave as it should, double-check your wiring. Make sure the transistors are oriented correctly; if either one is inserted into the breadboard backward, the circuit won't work. Also confirm that the LED is oriented correctly; if the cathode and anode are reversed in the circuit, the LED won't come on. Finally, verify that the battery holder leads are connected to the correct negative and positive buses.

Introducing Integrated Circuit Logic Gates

Although you can build your own logic gates by using transistors and resistors as described so far in this chapter, it's far easier to buy prepackaged integrated circuits that implement logic gates. The main advantage of using integrated circuit logic gates is that you don't have to design the individual gates yourself or waste time assembling them.

The logic gate circuits you've learned about so far in this book are among the simplest circuits for creating logic gates, but they're by no means the only — and not necessarily the best — ways to create logic gates. Over the 50 years or so that circuit designers have been working on semiconductor-based logic circuits, many designs have been developed for creating logic gates.

Because each approach to designing logic circuits results in an entire family of logic circuits for the various types of gates (NOT, AND, OR, NAND, NOR, XOR, and XNOR), the different designs are often referred to as *design families*. Also, because these design families were developed by engineers far nerdier than you and me, each family has a nice three- or four-letter acronym. Here are the most popular:

>> **RTL:** *Resistor-Transistor Logic,* which uses resistors and bipolar transistors. The circuits presented so far in this chapter are examples of RTL circuits.

>> **DTL:** *Diode-Transistor Logic,* which is similar to RTL but adds a diode to each input circuit.

>> **TTL:** Not Tinkertoy Logic, but *Transistor-Transistor Logic.* It uses two transistors, one configured to work as a switch and the other configured to work as an amplifier. The switching transistor is used in the input circuits, and the amplifier transistor is used in the output circuits. The amplifier allows the gate's output to be connected to a larger number of inputs than RTL or DTL circuits.

In a TTL circuit, the switching transistors are actually special transistors that have two or more emitters. These transistors are called *multiple-emitter transistors.*

Each input is connected to one of the emitters so that the separate inputs all control the same collector-emitter circuit. The switching transistor's base is connected to the Vcc supply voltage, and the collector is connected to the base of the amplifying transistor. If voltage is present at any or all of the emitter inputs, the base-collector circuit is pulled LOW, effectively creating a NAND gate. Figure 3-9 shows a typical TTL circuit.

Although you can build TTL circuits by using individual transistors, ICs with TTL circuits are readily available. One of the most popular types of TTL ICs is designated by numbers in the form 74*nn.* In all, a few hundred types of 7400-series integrated circuits are available. Many of them provide advanced logic circuits that you aren't likely to use for home electronics projects. The ICs listed in Table 3-4 provide several basic logic gates in a single package.

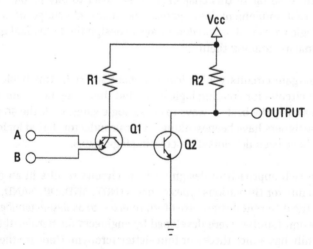

FIGURE 3-9:
A typical TTL gate.

TABLE 3-4

7400-Series TTL Logic Gates

Number	Description
7400	Quad two-input NAND gate (four NAND gates)
7402	Quad two-input NOR gate (four NOR gates)
7404	Hex inverter (six NOT gates)
7408	Quad two-input AND gate (four AND gates)
7432	Quad two-input OR gate (four OR gates)
7486	Quad two-input XOR gate (four XOR gates)

TIP

There actually is such a thing as Tinkertoy Logic. In fact, a team of MIT students actually built a complete computer that plays tic-tack-toe by using Tinkertoys. To learn more, search the Internet for *"Tinkertoy Computer."*

>> **CMOS:** *Complementary Metal-Oxide Semiconductor Logic,* which refers to logic circuits built with a special type of transistor called a MOSFET. *MOSFET* stands for *Metal Oxide Semiconductor Field Effect Transistor,* but that won't be on the test. The physics of how a MOSFET differs from a standard bipolar transistor aren't all that important unless you want to become an IC designer. What is important is that MOSFETs use much less power, can switch states much faster, and are significantly smaller than bipolar transistors. These differences make MOSFETs ideal for modern integrated circuits, which often require millions of transistors on a single chip.

Apart from drawing less power and operating faster than TTL circuits, CMOS circuits work much like TTL circuits. In fact, CMOS chips are designed to be interchangeable with comparable TTL chips.

CMOS logic chips have a four-digit part number that begins with the number 4 and are often called 4000-series chips. As with the 7400 series of TTL logic chips, several hundred types of 4000-series chips are available. Table 3-5 lists the 4000-series chips that provide basic logic gates.

WARNING

CMOS logic circuits are very sensitive to static electricity. As a result, you should take special precautions when you handle them. Make sure that you discharge yourself properly by touching a grounded metal surface before you touch a CMOS chip. For maximum protection, wear an antistatic wrist band. For more information about static precautions, see Book 1, Chapter 4.

The remaining sections of this chapter describe several popular 4000-series ICs and present a few projects that use them.

TABLE 3-5

4000-Series CMOS Logic Gates

Number	Description
4001	Quad two-input NOR gate (four NOR gates)
4009	Hex inverter (six NOT gates)
4011	Quad two-input NAND gate (four NAND gates)
4030	Quad two-input XOR gate (four XOR gates)
4071	Quad two-input OR gate (four OR gates)
4077	Quad two-input XNOR gate (four XNOR gates)
4081	Quad two-input AND gate (four AND gates)

Introducing the Versatile 4000-Series Logic Gates

The 4000-series CMOS logic circuits include several ICs that provide several logic gates in a single package. Figure 3-10 shows the pinout connections for six popular 4000-series chips. Each of these six chips contains 4 2-input logic gates in a 14-pin DIP package. Power, which can range from +3 V to +15 V, is connected to pin 14, and ground is connected to pin 7.

Note that the 4001 and 4011 chips contain four NOR and NAND gates. Because NOR and NAND are universal gates, you can use them in combination to create other types of gates. As a result, if you stock up on 4001 or 4011 chips, you'll be able to create any type of logic circuit you may need.

Before you start building circuits with CMOS logic chips, here are a few tips for working with 4000-series chips:

>> The outputs can source as much as 10 mA with a 9 V power supply. At 6 V, the maximum is about 5 mA, which is just enough to light up an LED. If your output circuit requires more current, you can always use a transistor. Just connect the transistor's collector to the positive voltage source and the base to the logic gate output. Then connect your output circuit to the transistor's emitter. The output circuit should eventually lead to ground, of course, to complete the circuit.

>> The input pins of CMOS logic chips are notorious for picking up stray signals in the form of electrical noise. Although it isn't necessary to do so in experimental breadboard circuits, in an actual circuit you should connect all unused

input pins to the positive supply voltage or to ground. (You don't have to connect the unused outputs to anything — just the inputs.)

» In addition to connecting unused inputs to positive voltage or ground, it's a good idea to place a small capacitor (47 μF is typical) across the power leads (pins 7 and 14). The capacitor helps ensure that the supply voltage stays constant.

» Don't forget that CMOS chips are very susceptible to damage from small amounts of static electricity. Be sure to ground yourself by touching a metal object before you touch CMOS chips.

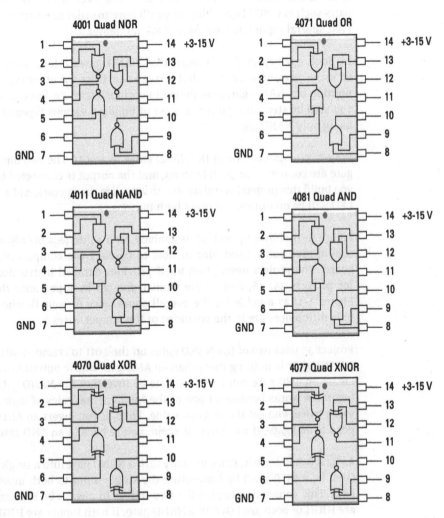

FIGURE 3-10:
Pinout chart for 4000-series Quad Two-Input Logic Gate chips.

Building Projects with the 4011 Quad Two-Input NAND Gate

The 4011 Quad Two-Input NAND Gate is a popular CMOS logic gate IC. As its name suggests, this IC contains four two-input NAND gates. The pinouts for this IC are shown in Figure 3-10, earlier in this chapter, along with the pinouts for several other quad two-input gate chips.

Unlike most of the components used in this book, a 4011 IC isn't available at your local RadioShack. Thus, you'll have to find a local electronics store that carries parts such as CMOS logic chips, or you'll have to order them from an online source. I list several reputable sources in Book 1, Chapter 3.

In Book 5, Chapter 2, you learn that NAND gates (along with NOR gates) are *universal gates,* which means that you can construct any other type of gate by using nothing but NAND gates combined in various ways. Projects 34 through 37 walk you step by step through the process of building various types of gate circuits by using only NAND gates.

Project 34 uses just one of the NAND gates in a 4011. The two inputs of the NAND gate are connected to push buttons, and the output is connected to an LED. When you build this project, you'll be able to visualize the operation of a NAND gate: The LED will be on unless you press both buttons.

Figure 3-11 shows Project 34 assembled on a solderless breadboard. This figure should give you a good idea of how to connect the components to the breadboard. The project description itself provides detailed instructions. The circuits for projects 35, 36, and 37 are similar enough in appearance that you can use Figure 3-11 as a guide for the overall appearance of your finished projects. (The only differences are in the positions of the jumper wires.)

Project 35 uses two of the NAND gates on the 4011 to create an AND gate. Because a NAND gate is nothing more than an AND gate whose output is inverted, you can create an AND gate from a NAND gate by inverting the NAND gate's output. This inversion works because of one of the fundamental rules of logic: If you invert a value twice, you get the original value. Thus, if you invert an AND gate once, you get a NAND gate; if you invert it again, you're back to an AND gate.

As luck would have it, you can easily turn a NAND gate into a single-input inverter (that is, a NOT gate) by connecting the single input to both inputs of the NAND gate. This connection causes the two inputs to always be the same: Either both are HIGH or both are LOW. In a NAND gate, if both inputs are HIGH, the output is

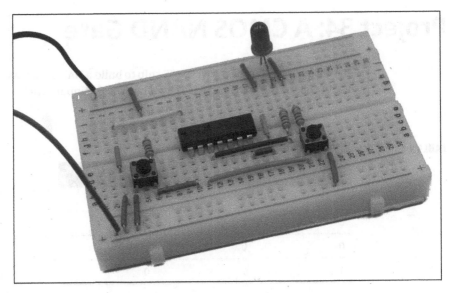

FIGURE 3-11:
A NAND gate
circuit that uses a
CMOS logic chip
(Project 34).

LOW, and if both inputs are LOW, the output is HIGH. Thus, wiring the inputs of a NAND gate together has the effect of inverting the input.

In Project 36, you see how to create an OR gate by using three NAND gates. Book 5, Chapter 2 shows that a NAND gate is the same as an OR gate whose inputs have been inverted. Thus, to create an OR gate by using NAND gates, you invert the two inputs with NAND gates configured as inverters (that is, with their inputs wired together). The output from these inverters is sent to the inputs of the third NAND gate.

The final project in this chapter, Project 37, uses all four of the NAND gates on the 4011 chip to create a NOR gate. A NOR gate is nothing more than an OR gate whose output has been inverted. Thus, you first use three NAND gates to create an OR gate by using the technique in Project 36; then you configure the fourth NAND gate on the 4011 chip as an inverter to invert the output of the OR gate.

Because these projects build on one another, I recommend that you don't tear down your breadboard after completing each project. Instead, you can use each assembled project as a starting point for the next one. If you choose to build the projects in this way, you can just scan the steps to see what resistors and jumper wires need to be moved for each project. (The 4011 IC itself, the LED, and the two switches are in the same locations for all four projects.)

Project 34: A CMOS NAND Gate

In this project, you use a 4011 CMOS chip to build a NAND gate. As Table 3-6 shows, the output of the NAND gate is LOW if both of the inputs are HIGH. Otherwise, the output is HIGH.

TABLE 3-6

NAND Gate Truth Table

Input A	Input B	Output
0	0	1
0	1	1
0	0	1
1	1	0

The inputs to this gate are fed through two breadboard-mounted push buttons, and the output is indicated by an LED.

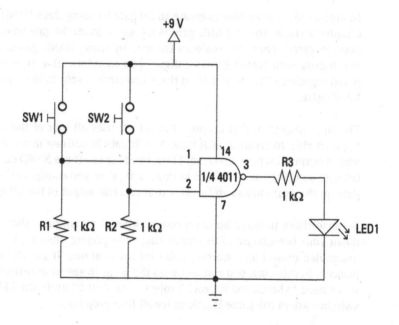

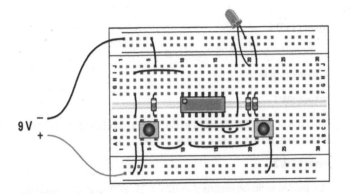

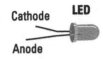

Cathode **LED**

Anode

Parts

>> One 9-volt battery

>> One 9-volt battery clip

>> One small solderless breadboard

>> Two normally open DIP breadboard push buttons

>> One 4011 CMOS Quad Two-Input NAND Gate

>> One 5 mm red LED

>> Three 1 kΩ resistors

>> 13 jumper wires

Steps

1. **Insert the 4011 IC.**

   Pin 1 should be in E10.

2. **Insert resistors R1, R2, and R3.**

   R1 (1 kΩ): E6 to F6

   R2 (1 kΩ): E21 to F21

   R3 (1 kΩ): E20 to F20

3. Insert LED1.

Cathode (short lead): Ground bus

Anode (long lead): J20

4. Insert SW1.

The pins should be inserted in B4, D4, B6, and D6 such that the switch opens and closes across rows 4 and 6.

5. Insert SW2.

The pins should be inserted in B21, D21, B23, and D23 such that the switch opens and closes across rows 21 and 23.

6. Insert the jumper wires.

1: Positive bus to A3

2: Positive bus to A4

3: Positive bus to A23

4: Ground bus to J6

5: Ground bus to J18

6: Ground bus to J21

7: E3 to F3

8: E18 to F18

9: A6 to A10

10: A11 to A21

11: C16 to C18

12: D12 to D20

13: I3 to I10

7. Connect the battery clip leads.

Red lead: Positive bus

Black lead: Ground bus

8. Test the circuit.

Insert the battery. The LED will come on. This is because both of the push buttons are open (not pushed), so both inputs are LOW. Therefore, the output is HIGH.

Press the first button. The LED remains on. With only one of the inputs HIGH, the output of a NAND gate remains HIGH.

Release the first button and press the second button. Again, the LED remains on because only one of the inputs is HIGH.

Now press both buttons at the same time. This makes both inputs HIGH, which causes the output to go LOW and turns off the LED.

Project 35: A CMOS AND Gate

In this project, you use two of the NAND gates in a 4011 CMOS chip to build an AND gate. As Table 3-7 shows, the AND gate's output is HIGH if both inputs are HIGH; otherwise, the output is LOW.

TABLE 3-7

AND Gate Truth Table

Input A	Input B	Output
0	0	0
0	1	0
0	0	0
1	1	1

To create the AND gate, you send the output from the first NAND gate to the second NAND gate, which you configure as an inverter by tying its inputs together. Inverting the output of a NAND gate creates an AND gate.

The inputs to this gate are fed through two breadboard-mounted push buttons, and the output is indicated by an LED.

Note that the only difference between this project and Project 34 is the placement of a few of the jumper wires.

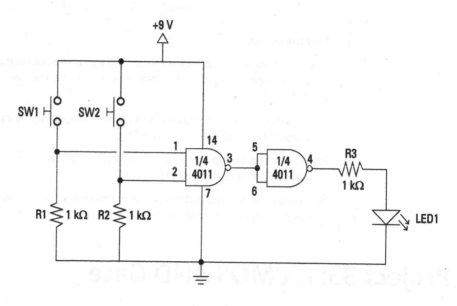

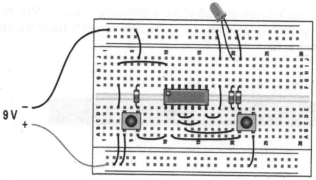

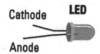

Cathode   **LED**

Anode

Parts

» One 9-volt battery

» One 9-volt battery clip

» One small solderless breadboard

» Two normally open DIP breadboard push buttons

» One 4011 CMOS Quad Two-Input NAND Gate

- » One 5 mm red LED
- » Three 1 kΩ resistors (brown-black-red)
- » 15 Jumper wires

Steps

1. **Insert the 4011 IC.**

   Pin 1 should be in E10.

2. **Insert resistors R1, R2, and R3.**

   R1 (1 kΩ): E6 to F6

   R2 (1 kΩ): E21 to F21

   R3 (1 kΩ): E20 to F20

3. **Insert LED1.**

   Cathode (short lead): Ground bus

   Anode (long lead): J20

4. **Insert SW1.**

   The pins should be inserted in B4, D4, B6, and D6 such that the switch opens and closes across rows 4 and 6.

5. **Insert SW2.**

   The pins should be inserted in B21, D21, B23, and D23 such that the switch opens and closes across rows 21 and 23.

6. **Insert the jumper wires.**

   1: Positive bus to A3

   2: Positive bus to A4

   3: Positive bus to A23

   4: Ground bus to J6

   5: Ground bus to J18

   6: Ground bus to J21

   7: E3 to F3

   8: E18 to F18

   9: A6 to A10

10: A11 to A21

11: B13 to B20

12: C12 to C15

13: D16 to D18

14: D12 to D14

15: I3 to I10

7. **Connect the battery clip leads.**

   Red lead: Positive bus

   Black lead: Ground bus

8. **Test the circuit.**

   Insert the battery. The LED will remain dark because neither input is HIGH.

   Press the first button. The LED remains off. A HIGH on one input is not enough to turn on the output of an AND gate.

   Release the first button and press the second button. Again, the LED remains dark because only one of the inputs is HIGH.

   Now press both buttons at the same time. This makes both inputs HIGH, which causes the output to go HIGH and turns on the LED.

Project 36: A CMOS OR Gate

In this project, you use two of the NAND gates in a 4011 CMOS chip to build an OR gate. As Table 3-8 shows, the OR gate's output is HIGH if either one (or both) of the inputs are HIGH. If both inputs are LOW, the output is also LOW.

TABLE 3-8

OR Gate Truth Table

Input A	Input B	Output
0	0	0
0	1	1
0	0	1
1	1	1

To create the OR gate, you simply invert the two inputs to a NAND gate. Thus, this circuit uses two of the NAND gates on the 4011 to invert the two inputs. It then sends these inverted inputs to the inputs of a third NAND gate to complete the OR gate.

The inputs to this gate are fed through two breadboard-mounted push buttons, and the output is indicated by an LED.

Note that the only difference between this project and Project 35 is the placement of one of the resistors and a few of the jumper wires.

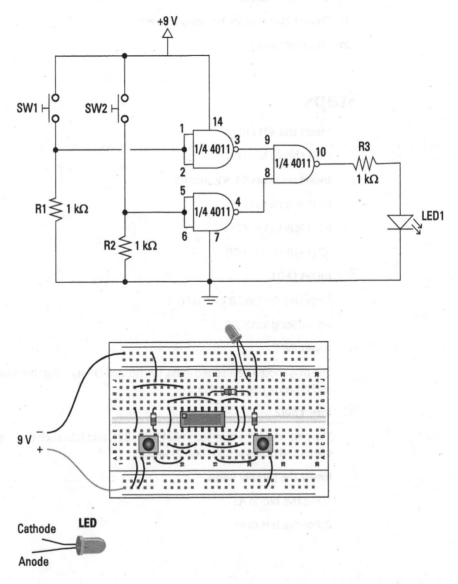

Parts

>> One 9-volt battery

>> One 9-volt battery clip

>> One small solderless breadboard

>> Two normally open DIP breadboard push buttons

>> One 4011 CMOS Quad Two-Input NAND Gate

>> One 5 mm red LED

>> Three 1 kΩ resistors (brown-black-red)

>> 20 jumper wires

Steps

1. **Insert the 4011 IC.**

   Pin 1 should be in E10.

2. **Insert resistors R1, R2, and R3.**

   R1 (1 kΩ): E6 to F6

   R2 (1 kΩ): E21 to F21

   R3 (1 kΩ): H14 to H20

3. **Insert LED1.**

   Cathode (short lead): Ground bus

   Anode (long lead): J20

4. **Insert SW1.**

   The pins should be inserted in B4, D4, B6, and D6 such that the switch opens and closes across rows 4 and 6.

5. **Insert SW2.**

   The pins should be inserted in B21, D21, B23, and D23 such that the switch opens and closes across rows 21 and 23.

6. **Insert the jumper wires.**

   1: Positive bus to A3

   2: Positive bus to A4

3: Positive bus to A23

4: Ground bus to J6

5: Ground bus to J18

6: Ground bus to J21

7: E3 to F3

8: E8 to F8

9: E18 to F18

10: E19 to F19

11: A6 to A10

12: A15 to A21

13: B10 to B11

14: B14 to B15

15: C8 to C12

16: C13 to C19

17: D16 to D18

18: G8 to G15

19: G16 to G19

20: I3 to I10

7. **Connect the battery clip leads.**

Red lead: Positive bus.

Black lead: Ground bus.

8. **Test the circuit.**

Insert the battery. The LED will remain dark because neither input is HIGH.

Press the first button. The LED comes on. A HIGH on one of the two inputs raises the output to HIGH, turning on the LED.

Release the first button and press the second button. Again, the LED comes on, this time because the other input is HIGH.

Now press both buttons at the same time. With both inputs HIGH, the LED comes on, reflecting the truth table of an OR gate.

Project 37: A CMOS NOR Gate

In this project, you use all four of the NAND gates in a 4011 CMOS chip to build a NOR gate. As Table 3-9 shows, the NOR gate's output is LOW if either one (or both) of the inputs are HIGH. If both inputs are LOW, the output is HIGH.

TABLE 3-9

NOR Gate Truth Table

Input A	Input B	Output
0	0	1
0	1	0
0	0	0
1	1	0

A NOR gate is simply an OR gate whose output is inverted. Thus, to create a NOR gate, you use three of the NAND gates on the 4011 to create an OR gate, as described in Project 36. Then you use the fourth NAND gate to invert the output of the OR gate.

The inputs to this gate are fed through two breadboard-mounted push buttons, and the output is indicated by an LED.

Note that the only difference between this project and Project 36 is the placement of the third resistor and the addition of two jumpers.

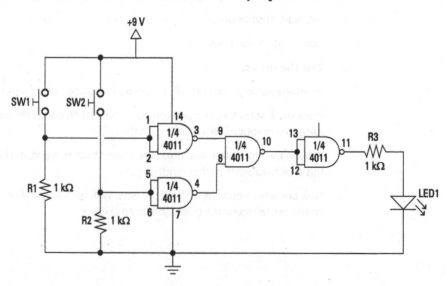

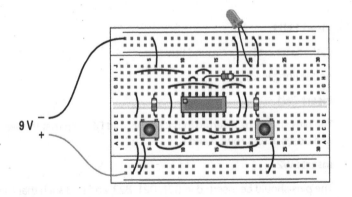

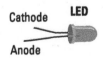

Cathode **LED**

Anode

Parts

>> One 9-volt battery

>> One 9-volt battery clip

>> One small solderless breadboard

>> Two normally open DIP breadboard push buttons

>> One 4011 CMOS Quad Two-Input NAND Gate

>> One 5 mm red LED

>> Three 1 kΩ resistors (brown-black-red)

>> 22 jumper wires

Steps

1. **Insert the 4011 IC.**

    Pin 1 should be in E10.

2. **Insert resistors R1, R2, and R3.**

    R1 (1 kΩ): E6 to F6

    R2 (1 kΩ): E21 to F21

    R3 (1 kΩ): H13 to H20

3. Insert LED1.

Cathode (short lead): Ground bus

Anode (long lead): J20

4. Insert SW1.

The pins should be inserted in B4, D4, B6, and D6 such that the switch opens and closes across rows 4 and 6.

5. Insert SW2.

The pins should be inserted in B21, D21, B23, and D23 such that the switch opens and closes across rows 21 and 23.

6. Insert the jumper wires.

1: Positive bus to A3

2: Positive bus to A4

3: Positive bus to A23

4: Ground bus to J6

5: Ground bus to J18

6: Ground bus to J21

7: E3 to F3

8: E8 to F8

9: E18 to F18

10: E19 to F19

11: A6 to A10

12: A15 to A21

13: B10 to B11

14: B14 to B15

15: C8 to C12

16: C13 to C19

17: D16 to D18

18: G8 to G15

19: G16 to G19

20: H11 to H12

21: I3 to I10

22: I12 to I14

7. Connect the battery clip leads.

Red lead: Positive bus

Black lead: Ground bus

8. Test the circuit.

Insert the battery. The LED immediately comes on. With both inputs LOW, the output of a NOR gate is HIGH.

All other combinations of inputs in a NOR gate cause the output to go LOW, turning off the LED.

Press the first button; the LED goes off.

Release the first button and press the second button. Again, the LED goes off.

Now press both buttons at the same time. With both inputs HIGH, the output goes LOW, so the LED once again goes out.

Chapter **4**

Working with Flip-Flops

This chapter is about flip-flops, but not the kind you wear on your feet. I wish it were about the kind you wear on your feet, especially if I happened to be wearing a pair while I wrote this because that might mean I'd be writing this at the beach, parked in a beach chair watching children play with the waves.

Don't you love watching children play with waves? They laugh uncontrollably as they chase after a retreating wave. But when the next wave comes in, the children turn and run, screaming with delight for fear that the wave might catch them, and they might — of all things — get wet.

Alas, this chapter isn't about the kind of flip-flops you wear to the beach. Instead, it's about the electronic kind of flip-flop. A *flip-flop* is a circuit that stores data. As such, flip-flops are the basis of modern computers.

Don't you think it's odd that one of the fundamental building blocks of modern thinking machines has a name that suggests it can't make up its mind?

In this chapter, you learn how to work with simple flip-flop circuits. Before I tell you about flip-flops, however, I first show you a simpler type of circuit called a *latch*.

Looking at Latches

A *latch* is a logic circuit that has two inputs and one output. One of the inputs is called the *SET input*; the other is called the *RESET input*.

Latch circuits can be either active-high or active-low. The difference is determined by whether the operation of the latch circuit is triggered by HIGH or LOW signals on the inputs. (See Chapter 2 of this minibook for an explanation of concepts such as HIGH and LOW.)

» **Active-high circuit:** Both inputs are normally tied to ground (LOW), and the latch is triggered by a momentary HIGH signal on either of the inputs.

» **Active-low circuit:** Both inputs are normally HIGH, and the latch is triggered by a momentary LOW signal on either input.

In an active-high latch, both the SET and RESET inputs are normally connected to ground. When the SET input goes HIGH, the output also goes HIGH. When the SET input returns to LOW, however, the output remains HIGH. The output of the active-high latch stays HIGH until the RESET input goes HIGH. Then, the output returns to LOW and will go HIGH again only when the SET input is triggered once more.

In other words, the latch remembers that the SET input has been activated. If the SET input goes HIGH for even a moment, the output goes HIGH and stays HIGH, even after the SET input returns to LOW. The output returns to LOW only when the RESET input goes HIGH.

On the other hand, in an active-low latch the inputs are normally held at HIGH. When the SET input momentarily goes LOW, the output goes HIGH. The output then stays HIGH until the RESET input momentarily goes LOW.

Note that most latch circuits actually have a second output that is simply the first output inverted. In other words, whenever the first output is HIGH, the second output is LOW, and vice versa. These outputs are usually referred to as Q and $\bar{Q}$.

TECHNICAL STUFF

The notation $\bar{Q}$ is usually pronounced either "bar Q" or "Q bar," though some people pronounce it "not Q." The horizontal bar symbol over a label is a common logical shorthand for inversion. That is, $\bar{Q}$ is the inverse of Q. If Q is HIGH, $\bar{Q}$ is LOW, and if Q is LOW, $\bar{Q}$ is HIGH.

You can easily create an active-high latch from a pair of NOR gates, as shown in Figure 4-1. (In Chapter 2 in this minibook you see that the output of a NOR gate is HIGH if both inputs are LOW; otherwise, the output is LOW.) In this circuit, the RESET input is connected to one of the inputs of the first NOR gate, and the SET input is connected to one of the inputs of the second NOR gate. The trick of the latch circuit is that the output of the NOR gates are cross-connected to the remaining NOR gate inputs. In other words, the output from the first NOR gate is connected to one of the inputs of the second NOR gate, and the output from the second NOR gate is connected to one of the inputs of the first NOR gate.

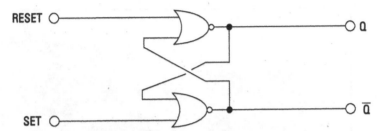

FIGURE 4-1:
Schematic diagram for an active-high latch.

It may seem a bit confusing, but in an active-high latch, the RESET input is at the top of the circuit and the SET input is connected to the gate at the bottom of the circuit. At first, this makes the circuit look like it's wired upside-down. But for an active-high latch, this is the standard way to depict the circuit, with RESET at the top and SET at the bottom.

The schematic for an active-low latch is shown in Figure 4-2. As you can see, there are several differences between this circuit schematic and the one shown in Figure 4-1:

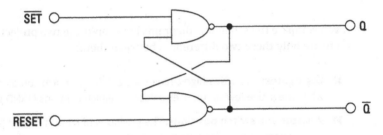

FIGURE 4-2:
Schematic diagram for an active-low latch.

>> The active-low latch uses NAND gates instead of NOR gates.

>> The inputs are active-low rather than active-high and are referred to as $\overline{SET}$ and $\overline{RESET}$ rather than SET and RESET

>> The position of the inputs within the circuit is reversed: $\overline{SET}$ is connected to the first NAND gate, at the top of the circuit, and $\overline{RESET}$ is connected to the second NAND gate, at the bottom of the circuit.

Projects 38 and 39 show you how to build simple active-high and active-low latch circuits using a 4001 quad 2-input NOR gate IC and a 4011 quad 2-input NAND gate IC. Both the Q and $\bar{Q}$ outputs are used to drive LEDs so you can see the state of the latch, and both inputs are controlled by normally open push buttons so that you can trigger the latch by momentarily pressing the buttons. Figure 4-3 shows the assembled active-high latch, and Figure 4-4 shows the assembled active-low latch.

FIGURE 4-3:
The assembled active-high latch (Project 38).

If you compare the schematic diagrams between these two projects, you'll see that there are only these two differences between them:

>> **Logic gates:** The active-high circuit uses a 4001 IC, which contains NOR gates, while the active-low circuit uses a 4011 IC, which contains NAND gates.

>> **Resistor and switch positions:** The positions of R1 and R2 and SW1 and SW2 are reversed. In the active-high circuit, the resistors connect the two gate inputs to ground and the switches short the gate inputs to +6 V. In the active-low circuit, the resistors connect the gate inputs to +6 V and the switches short the gate inputs to ground.

Both of these circuits use simple push-button switches to provide the trigger inputs. However, you can easily imagine other sources for the trigger pulse. For example in a home alarm system, the SET input in an active-low latch might come from a window switch that breaks contact when the window is open, and the RESET input may come from a key lock on the alarm system's control panel to reset the alarm.

FIGURE 4-4:
The assembled
active-low latch
(Project 39).

TIP

Here are a few other things you should know about latches before we move on:

>> A latch with a SET and RESET input is often called an *SR latch*. The term *RS latch* is also used.

>> In some cases, you may need a latch in which one of the inputs is active-high and the other is active-low. For example, your home alarm system might send HIGH to the SET input when a window is opened, but it may send LOW when the system needs to be reset.

You can easily accomplish that by adding an inverter to one of the inputs, as shown in Figure 4-5. Here, I've used NOR gates to create an active-high latch, but I've added a NOT gate to invert the RESET input. Strictly speaking, the RESET input of the gate is active-high, but because of the inverter, the circuit behaves as if the RESET input were active-low.

FIGURE 4-5:
A latch in which
$\overline{\text{SET}}$ is active-low
and RESET is
active-high.

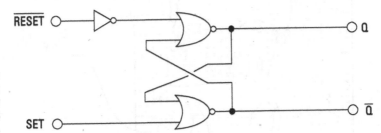

Project 38: An Active-High Latch

In this project, you use two of the NOR gates in a 4001 CMOS chip to build an active-high latch. The inputs to this latch are controlled by two breadboard-mounted push buttons; SW1 is the RESET input and SW2 is the SET input. The outputs are indicated by LEDs; LED1 is Q and LED2 is $\bar{Q}$.

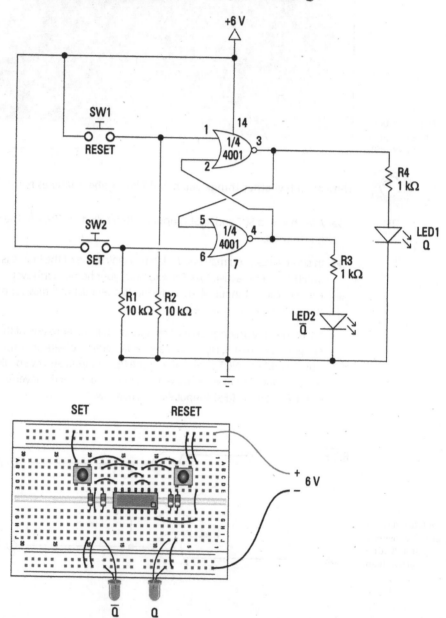

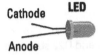
Cathode **LED**

Anode

Parts

>> Four AA batteries

>> One four AA battery holder

>> One small solderless breadboard

>> Two normally open DIP breadboard push buttons

>> One 4001 CMOS Quad Two-Input NOR Gate

>> Two 5 mm red LEDs

>> Two 10 kΩ resistors (brown-black-orange)

>> Two 1 kΩ resistors (brown-black-red)

>> 16 jumper wires

Steps

1. **Insert the 4001 IC.**

   Pin 1 should be in E10.

2. **Insert resistors R1, R2, R3, and R4.**

   R1 (10 kΩ): E20 to F20

   R2 (10 kΩ): E7 to F7

   R3 (1 kΩ): E18 to F18

   R4 (1 kΩ): E8 to F8

3. **Insert LED1 and LED2.**

   LED1: J18 to ground bus

   LED2: J8 to ground bus

   Connect the cathode (short lead) to the ground bus.

4. **Insert SW1.**

   The pins should be inserted in B5, D5, B7, and D7 such that the switch opens and closes across rows 5 and 7.

5. Insert SW2.

The pins should be inserted in B20, D20, B22, and D22 such that the switch opens and closes across rows 20 and 22.

6. Insert the jumper wires.

1: Positive bus to A4

2: Positive bus to A5

3: Positive bus to A22

4: Ground bus to J7

5: Ground bus to J19

6: Ground bus to J20

7: E4 to F4

8: E19 to F19

9: A7 to A10

10: A15 to A20

11: B8 to B12

12: B13 to B18

13: C12 to C14

14: D11 to D13

15: D16 to D19

16: G4 to G10

7. Connect the battery holder jumper leads.

Red lead: Positive bus

Black lead: Ground bus

8. Test the circuit.

Insert the batteries. In its initial state, Q will most likely be Low and $\bar{Q}$ will be High, so LED1 will be dark and LED2 will be lit. (In some cases, the reverse may be true. The initial state of an SR latch is not predictable.)

Whether LED1 or LED2 is on when you power up the circuit, press and release the Reset button (SW1) to reset the latch. Q goes Low and $\bar{Q}$ goes High, so LED1 turns off and LED2 turns on. At this point, you can press the Reset button (SW1) repeatedly without changing the state of the output.

Now press and release the Set button (SW2). *Q* now goes HIGH and $\bar{Q}$ goes LOW, so LED1 lights up and LED2 goes dark. The output status remains the same when you release the Set button (SW2). And you can press and release the Set button repeatedly without affecting the output.

Finally, press and release the Reset button (SW1) again to reset the latch. Q goes LOW and $\bar{Q}$ goes HIGH so LED1 turns off and LED2 turns on, indicating that the latch has been reset. You can repeat the process to verify that pressing Set (SW2) sets the Q output to HIGH and pressing Reset (SW1) resets the latch so that Q is LOW.

Project 39: An Active-Low Latch

In this project, you use two of the NAND gates in a 4011 CMOS chip to build an active-low latch. The inputs to this latch are controlled by two breadboard-mounted push buttons (SW1 is $\overline{\text{SET}}$ and SW2 is $\overline{\text{RESET}}$), and the outputs are indicated by LEDs (LED1 is Q and LED2 is $\bar{Q}$).

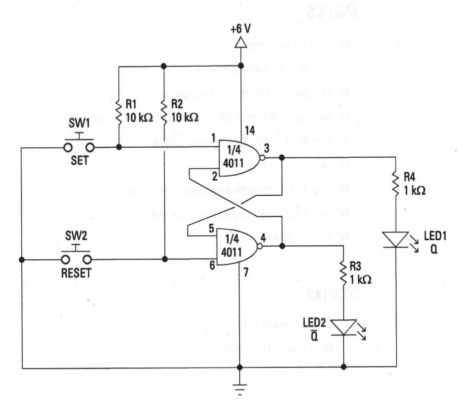

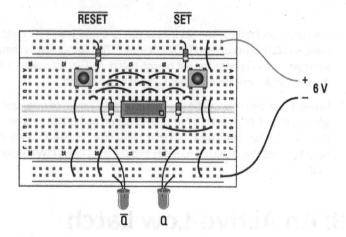

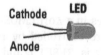

Parts

>> Four AA batteries

>> One four AA battery holder

>> One small solderless breadboard

>> Two normally open DIP breadboard push buttons

>> One 4011 CMOS Quad Two-Input NAND Gate

>> Two 5 mm red LEDs

>> Two 10 kΩ resistors (brown-black-orange)

>> Two 1 kΩ resistors (brown-black-red)

>> 18 jumper wires

Steps

1. **Insert the 4011 IC.**

   Pin 1 should be in E10.

2. Insert resistors R1, R2, R3, and R4.

R1 (10 kΩ): A7 to positive bus

R2 (10 kΩ): A20 to positive bus

R3 (1 kΩ): E18 to F18

R4 (1 kΩ): E8 to F8

3. Insert LED1 and LED2.

LED1: J8 to ground bus

LED2: J18 to ground bus

Connect the cathode (short lead) to the ground bus.

4. Insert SW1.

The pins should be inserted in A4, C4, A6, and C6 such that the switch opens and closes across rows 4 and 6.

5. Insert SW2.

The pins should be inserted in A21, C21, A23, and C23 such that the switch opens and closes across rows 21 and 23.

6. Insert the jumper wires.

1: Positive bus to A3

2: Ground bus to J4

3: Ground bus to J19

4: Ground bus to J23

5: E3 to F3

6: E4 to F4

7: E19 to F19

8: E23 to F23

9: B8 to B12

10: B13 to B18

11: C12 to C14

12: C16 to C19

13: D7 to D10

14: D11 to D13

15: D15 to D20

16: E6 to E7

17: E20 to E21

18: G3 to G10

7. **Connect the battery holder leads.**

   Red lead: Positive bus

   Black lead: Ground bus

8. **Test the circuit.**

   Insert the batteries. In its initial state, Q is Low and $\bar{Q}$ is High, so LED2 is lit and LED1 is dark.

   Press and release the Set button (SW1) to set the latch. Q goes High and $\bar{Q}$ goes Low, so LED1 turns on and LED2 turns off. You can press and release SW1 repeatedly without changing the state of either output.

   Now press and release the Reset button (SW2). Q now returns to Low and $\bar{Q}$ goes High. You can now press and release SW2 repeatedly, again without affecting the output. But if you press SW1, the latch sets, so LED1 turns on and LED2 turns off.

Looking at Gated Latches

A *gated latch* is a latch that has a third input that must be active in order for the SET and RESET inputs to take effect. This third input is sometimes called ENABLE because it enables the operation of the SET and RESET inputs.

The ENABLE input can be connected to a simple switch. Then, when the switch is closed, the SET and RESET inputs are enabled; when the switch is open, any changes in the SET and RESET inputs are ignored.

Alternatively, the ENABLE input can be connected to a clock pulse. For example, you could connect the output of a 555 timer circuit to the ENABLE input. Then, the latch inputs will be operational only when the 555 timer's output is HIGH. Note that the ENABLE input is often called the *CLOCK input*. (For more information about 555 timer circuits, see Book 3, Chapter 2.)

You can easily add an ENABLE input to a latch by adding a pair of NAND gates as shown in Figure 4-6. Here, the SET and RESET inputs (SR latch) are connected to one input of each of the two NAND gates. The ENABLE input is connected to the other input of each NAND gate. Then, the outputs from these gates are used as the inputs to the basic latch circuit.

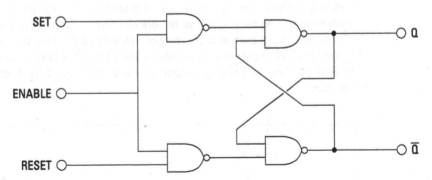

FIGURE 4-6:
A gated SR latch.

Another common type of gated latch is called a *gated D latch*, which has just two inputs: DATA and ENABLE. When a HIGH is received at the ENABLE input, the DATA input is copied to the output. Even if the ENABLE input then goes low, the output remains unchanged. The output cannot be changed until the ENABLE input goes high.

To create a gated D latch from a gated SR latch, you simply connect the SET and RESET inputs together through an inverter, as shown in Figure 4-7. Thus, the SET and RESET inputs will always be opposite of one another. When the DATA input is HIGH, the SET input is HIGH and the RESET input is LOW. When the DATA input is LOW, the SET input is LOW and the RESET input is HIGH.

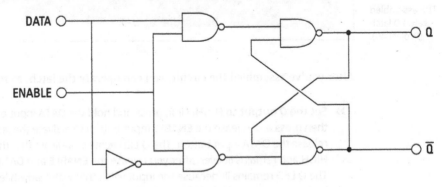

FIGURE 4-7:
A gated D latch.

Project 40 shows how to build a gated D latch using two 4011 Quad 2-Input NAND gates. Two 4011 chips are required because the latch requires a total of five gates (four NAND gates and one NOT gate), and each 4011 provides just four gates. In Chapter 2 of this minibook, you learn that you can create a NOT gate from a NAND gate simply by tying the two inputs of the NAND gate together. In this project, you use that technique to convert one of the 4011's NAND gates to a NOT gate.

Figure 4-8 shows the assembled gated D latch. In this circuit, the ENABLE input push button is on the top side of the breadboard, in columns B through D, and the DATA input push button is on the bottom side of the breadboard, in columns F through G. The Q output is indicated by the LED at the bottom of the breadboard, in H30 and I29, and the $\bar{Q}$ output LED is at the top of the breadboard, in A28 and B29.

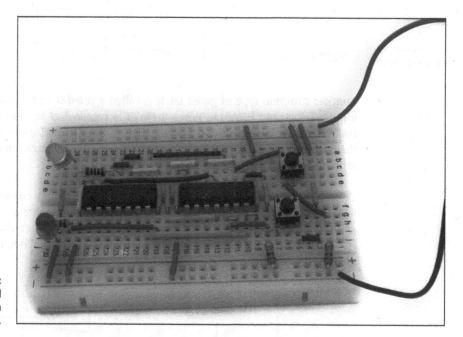

FIGURE 4-8:
The assembled
gated D latch
(Project 40).

After you've assembled the circuit, you can operate the latch, as follows:

>> **Set the Q output to HIGH.** First, press and hold the DATA input button, and then press and release the ENABLE input button to activate the latch. Then release the DATA input button. The Q LED lights to indicate that the output is HIGH and remains lit even after you release the ENABLE and DATA buttons. The Q LED remains lit because the input value that was read while the ENABLE button was pressed is held by the latch.

>> **Set the Q output to LOW (which sets the $\bar{Q}$ output to HIGH).** Just press and release the ENABLE button without pressing the DATA button. The Q LED goes out and the $\bar{Q}$ LED comes on when you pressed the ENABLE button, the

input on the DATA switch was LOW. Therefore, the Q output becomes LOW and is held at LOW when the ENABLE button is released.

The operation of the gated D latch can seem confusing, because pressing the ENABLE button by itself seems to reset the latch. That's not really what's happening. Instead, each time you press the ENABLE button, the Q output is set to match the input.

To get an idea of how a gated D latch would work when the ENABLE input is connected to a clock, press and quickly release the ENABLE button repeatedly about once per second, as if it were connected to a metronome. Each press of the ENABLE button represents a clock tick. While you keep clicking ENABLE, press and hold the DATA button for a few ENABLE clicks and note that the Q LED turns on, indicating that the output is HIGH and remains high. Now release the DATA button and continue clicking the ENABLE button. The Q LED will turn off, indicating that the LOW value on the DATA input has been read

Project 40: A Gated D Latch

In this project, you use five NAND gates provided by two 4011 CMOS chips to build a gated D latch. The inputs are controlled by two breadboard-mounted push buttons, and the outputs are indicated by LEDs.

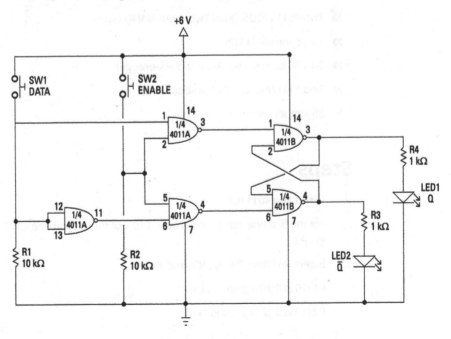

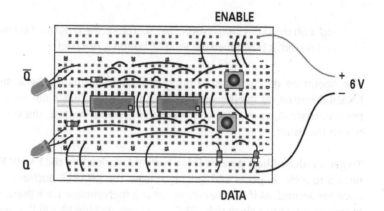

ENABLE

$\overline{Q}$

6 V

Q

DATA

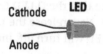

Cathode **LED**

Anode

Parts

>> Four AA batteries

>> One four AA battery holder

>> One small solderless breadboard

>> Two normally open DIP breadboard push buttons

>> Two 4011 CMOS Quad Two-Input NAND Gates

>> Two 5 mm red LEDs

>> Two 10 kΩ resistors (brown-black-orange)

>> Two 1 kΩ resistors (brown-black-red)

>> 35 jumper wires

Steps

1. **Insert the 4011 ICs.**

   For the first one, pin 1 should be in E10. For the second one, pin 1 should be in E20.

2. **Insert resistors R1, R2, R3, and R4.**

   R1 (10 kΩ) J8 to ground bus

   R2 (10 kΩ): J2 to ground bus

R3 (1 kΩ): C23 to C28

R4 (1 kΩ): G28 to G30

3. Insert LED1 and LED2.

LED1: Cathode in I29, anode in H30

LED2: Cathode in B29, anode in A28

4. Insert SW1.

The pins should be inserted in B4, D4, B6, and D6 such that the switch opens and closes across rows 4 and 6.

5. Insert SW2.

The pins should be inserted in F5, H5, F7, and H7 such that the switch opens and closes across rows 5 and 7.

6. Insert the jumper wires.

1: Positive bus to A3

2: Positive bus to A4

3: Positive bus to A9

4: Ground bus to J17

5: Ground bus to J27

6: Ground bus to J29

7: E3 to F3

8: E8 to F8

9: E9 to F9

10: E17 to F17

11: E18 to F18

12: E19 to F19

13: E27 to F27

14: E29 to F29

15: A6 to C11

16: A12 to A20

17: A21 to A23

18: B11 to B14

19: B15 to B18

20: B22 to B24

21: C16 to C17

22: C19 to C22

23: D8 to D10

24: D13 to D25

25: D26 to D27

26: G2 to E6

27: G9 to G10

28: G11 to G12

29: G13 to G18

30: H8 to H12

31: H19 to H28

32: I7 to I8

33: H19 to H28

34: J3 to J5

35: I10 to I20

7. **Connect the battery leads.**

   Red lead: Positive bus.

   Black lead: Ground bus.

8. **Test the circuit.**

   Connect the batteries. In its initial state, no input has been read. The Q output is Low, so the Q LED is dark and the $\bar{Q}$ LED is on.

   Press and hold the INPUT button and tap the ENABLE button. When you tap the ENABLE button, the input will be read and the Q LED will come on. (The $\bar{Q}$ LED will turn off to show the inverse of the input.)

   Tap the ENABLE button again without holding the INPUT button. The input will be read again, and the Q LED will go off to indicate the output is zero. (The $\bar{Q}$ LED will turn on to show the inverse of the output.)

Introducing Flip-Flops

A *flip-flop* is a special type of gated latch. The difference between a flip-flop and a gated latch is that in a flip-flop, the inputs aren't enabled merely by the presence of a HIGH signal on the CLOCK input. Instead, the inputs are enabled by the *transition* of the CLOCK input. Thus, at the moment that the clock input transitions from low to high, the inputs are briefly enabled. Once the clock stabilizes at the HIGH setting, the output state of the flip-flop is latched until the next clock pulse.

Flip-flops are often said to be *edge-triggered* because it's the edge of the clock signal that triggers the flip-flop. When used in clock-driven computer circuits, edge-triggering is an important characteristic because it helps circuit designers maintain better control over the timing in circuits that contain hundreds or perhaps thousands of flip-flops.

The circuitry that enables a flip-flop to respond to just the leading edge can be pretty complicated. One of the simplest methods is to feed the clock input into a NAND gate, passing one of the legs through an inverter, as shown in Figure 4-9. This works because in all logic gates, there is a very small delay between the time a signal arrives at the input and the correct signal arrives at the output.

FIGURE 4-9:
A circuit that
detects a clock
transition.

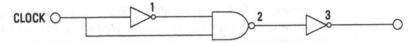

Here I step you through what happens when the clock transitions from LOW to HIGH in Figure 4-9:

1. Initially, the clock input is LOW. The inverter causes the first input to the NAND gate (marked *1* in the figure) to be HIGH, while the second input is LOW. Because the inputs aren't both HIGH, the output from the NAND gate at point 2 in the figure is HIGH. The second inverter inverts the NAND gate output so the final output from the circuit at point 3 is LOW, just like the clock input.

2. When the clock input goes high, the second input to the NAND gate goes high immediately. However, it takes a few milliseconds for the inverter to respond, so for those few milliseconds, the output from the inverter is still HIGH. Thus, both inputs to the NAND gate are HIGH for a few milliseconds, which causes the output from the NAND gate at point 2 in the figure to go LOW. Then, the second NOT gate inverts the NAND gate output, causing the output at point 3 in the signal to go HIGH for a brief moment.

3. Once the first NOT gate catches up and its output goes LOW (at point 1 in the figure), the NAND gate responds to the LOW and HIGH input by setting its output to HIGH at point 2 in the figure. The second NOT gate then inverts that output at point 3 in the figure.

The net result of the circuit in Figure 4-9 is that long clock pulses are turned into short clock pulses. The duration between the pulses remains the same, but the HIGH part of the pulse becomes much shorter.

Flip-flops are designed for use in circuits that use steady clock pulses. An easy way to provide clock pulses for a flip-flop circuit is to use a 555 timer IC, as described in Book 3, Chapter 2. However, the input source for the CLOCK input of a flip-flop doesn't have to be an actual clock; it can also be a one-shot input triggered by a push button.

As with latches, there are several different types of flip-flops. The most common are

» **SR flip-flop:** Is similar to an SR latch. Besides the CLOCK input, an SR flip-flop has two inputs, labeled SET and RESET. If the SET input is HIGH when the clock is triggered, the Q output goes HIGH. If the RESET input is HIGH when the clock is triggered, the Q output goes LOW.

Note that in an SR flip-flop, the SET and RESET inputs shouldn't both be HIGH when the clock is triggered. This is considered an invalid input condition, and the resulting output isn't predictable if this condition occurs.

» **D flip-flop:** Has just one input in addition to the CLOCK input. This input is called the DATA input. When the clock is triggered, the Q output is matched to the DATA input. Thus, if the DATA input is HIGH, the Q output goes HIGH, and if the DATA input is LOW, the Q output goes LOW.

Most D-type flip-flops also include S and R inputs that let you set or reset the flip-flop. Note that the S and R inputs in a D flip-flop ignore the CLOCK input. Thus, if you apply a HIGH to either S or R, the flip-flop will be set or reset immediately, without waiting for a clock pulse.

» **JK flip-flop:** A common variation of the SR flip-flop. A JK flip-flop has two inputs, labeled J and K. The J input corresponds to the SET input in an SR flip-flop, and the K input corresponds to the RESET input.

The difference between a JK flip-flop and an SR flip-flop is that in a JK flip-flop, both inputs can be HIGH. When both the J and K inputs are HIGH, the Q output is *toggled,* which means that the output alternates between HIGH and LOW. For example, if the Q output is HIGH when the clock is triggered and J and K are both HIGH, the Q output is set to LOW. If the clock is triggered again while J and K both remain HIGH, the Q output is set to HIGH again, and so forth, with the Q output alternating from HIGH to LOW at every clock tick.

» **T flip-flop:** This is simply a JK flip-flop in which the output alternates between HIGH and LOW with each clock pulse. Toggles are widely used in logic circuits because they can be combined to form counting circuits that count the number of clock pulses received.

You can create a T flip-flop from a D flip-flop by connecting the $\bar{Q}$ output directly to the D input. Thus, whenever a clock pulse is received, the current state of the Q output is inverted (that's what the $\bar{Q}$ output is) and fed back into the D input. This causes the output to alternate between HIGH and LOW.

You can also create a T flip-flop from a JK flip-flop simply by hard-wiring both the J and K inputs to HIGH. When both J and K are HIGH, the JK flip-flop acts as a toggle.

Although you can construct your own flip-flop circuits using NAND gates, it's much easier to use ICs that contain flip-flops. One common example is the 4013 Dual D Flip-Flop. This chip contains two D-type flip-flops in a 14-pin DIP package. The pinouts are listed in Table 4-1.

TABLE 4-1 **Pinouts for the 4013 Dual D Flip-Flop IC xxx**

Pin	Name	Explanation	Pin	Name	Explanation
1	Q1	Flip-flop 1 Q output	8	SET2	Flip-flop 2 SET input
2	$\overline{Q1}$	Flip-flop 1 $\bar{Q}$ output	9	DATA2	Flip-flop 2 DATA input
3	CLOCK1	Flip-flop 1 CLOCK input	10	RESET2	Flip-flop 2 RESET input
4	RESET1	Flip-flop 1 RESET input	11	CLOCK2	Flip-flop 2 CLOCK input
5	DATA1	Flip-flop 1 DATA input	12	$\overline{Q2}$	Flip-flop 2 $\bar{Q}$ output
6	SET	Flip-flop 1 SET input	13	Q2	Flip-flop 2 Q output
7	GND	Ground	14	V_{DD}	+3 to 15 V

TIP

When you use one of the flip-flops in a 4013 IC, be sure to connect any unused inputs to ground. All unused inputs in CMOS logic chips should be connected to ground, but for simple breadboard circuits, the ground connections aren't usually required. However, the DATA and CLOCK inputs of a 4013 flip-flop won't work properly if you don't ground the SET and RESET inputs.

Project 41 shows how to use a 4013 IC to create a basic D flip-flop. This circuit works much the same as the D-type latch you create in Project 40. However, it requires only one IC rather than two, and the wiring is much simpler. That's because the engineers who designed the 4013 IC crammed all the wiring between the individual NAND gates in the IC, so you don't have to wire the gates together on the breadboard. Instead, all you have to do is hook up the inputs and the outputs and watch the circuit work. Figure 4-10 shows the assembled circuit.

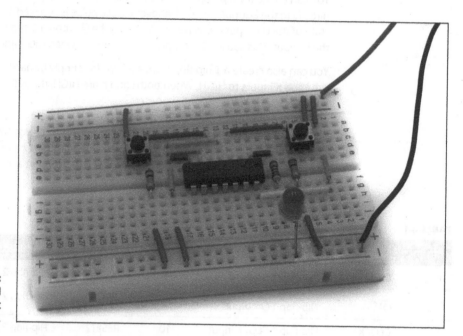

FIGURE 4-10:
The assembled
D flip-flop circuit
(Project 41).

Project 41: A D Flip-Flop

In this project, you build a circuit that demonstrates the operation of a D flip-flop. This circuit uses one of the two flip-flops on a 4013 Dual D Flip-Flop IC. The Data and Clock inputs are controlled via breadboard-mounted push buttons, and the Q output is connected to an LED. The $\bar{Q}$ output isn't connected.

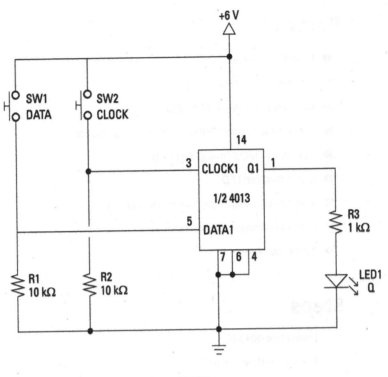

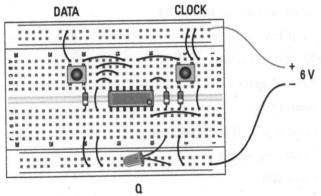

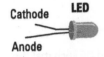

Parts

>> Four AA batteries

>> One four AA battery holder

>> One small solderless breadboard

>> Two normally open DIP breadboard push buttons

>> One 4013 CMOS Dual D Flip-Flop

>> One 5 mm red LED

>> Two 10 kΩ resistors (brown-black-orange)

>> One 1 kΩ resistor (brown-black-red)

>> 15 jumper wires

Steps

1. **Insert the 4013 IC.**

   Pin 1 should be in E10.

2. **Insert resistors R1, R2, and R3.**

   R1 (10 kΩ): E20 to F20

   R2 (10 kΩ): E6 to F6

   R3 (1 kΩ): E8 to F8

3. **Insert LED1.**

   Cathode (short lead): Ground bus

   Anode (long lead): J8

4. **Insert SW1.**

   The pins should be inserted in B20, D20, B22, and D22 such that the switch opens and closes across rows 20 and 22.

5. **Insert SW2.**

   The pins should be inserted in B4, D4, B6, and D6 such that the switch opens and closes across rows 4 and 6.

6. Insert the jumper wires.

1: Positive bus to A3

2: Positive bus to A4

3: Positive bus to A22

4: Ground bus to J6

5: Ground bus to J18

6: Ground bus to J20

7: E3 to F3

8: E18 to F18

9: A6 to A12

10: A14 to A20

11: B13 to B18

12: C15 to C18

13: D8 to D10

14: D16 to D18

15: G3 to G10

7. Connect the battery leads.

Red lead: Positive bus

Black lead: Ground bus

8. Test the circuit.

Insert the batteries. Press and hold the Data button to indicate a HIGH input; then press and release the Clock button to read the input. The LED will turn on, indicating that the output has been set to HIGH.

Now press and release the Clock button without pressing the Data button. This time, the input is LOW, so the LED is turned off to indicate the new output state.

Project 42 shows how to build a T flip-flop in which each press of a button causes an output LED to alternate between on and off. For this project, the $\bar{Q}$ output from the flip-flop is connected to the DATA input. Then, each time the Clock input goes HIGH, the inverted output from the $\bar{Q}$ output is fed into the DATA input. This causes the Q output to invert. Figure 4-11 shows the assembled circuit.

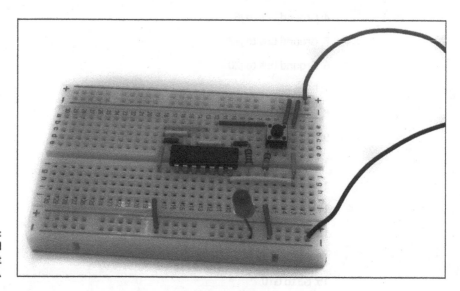

FIGURE 4-11:
The assembled
T flip-flop circuit
(Project 42).

Project 42: A Toggle Flip-Flop

In this project, you build a toggle flip-flop using a 4013 Dual D Flip-Flop IC. The clock input is connected to a breadboard-mounted push button, and the Q output is connected to an LED. The $\bar{Q}$ output is connected to the DATA input, which causes the Q output to invert at each clock pulse.

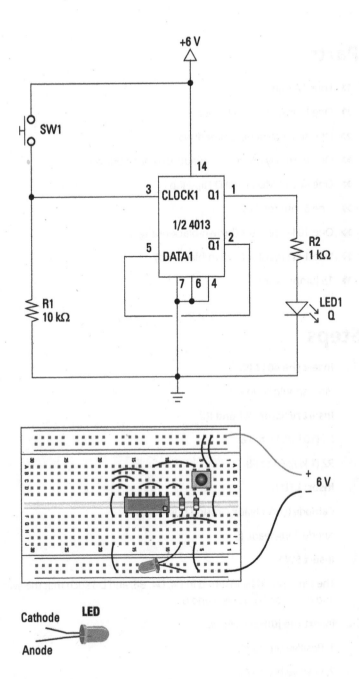

Parts

>> Four AA batteries

>> One four AA battery holder

>> One small solderless breadboard

>> One normally open DIP breadboard push button

>> One 4013 CMOS Dual D Flip-Flop

>> One 5 mm red LED

>> One 10 kΩ resistor (brown-black-orange)

>> One 1 kΩ resistor (brown-black-red)

>> 15 jumper wires

Steps

1. **Insert the 4013 IC.**

   Pin 1 should be in E10.

2. **Insert resistors R1 and R2.**

   R1 (10 kΩ): E6 to F6

   R2 (1 kΩ): E8 to F8

3. **Insert LED1.**

   Cathode (short lead): Ground bus

   Anode (long lead): J8

4. **Insert SW1.**

   The pins should be inserted in B4, D4, B6, and D6 such that the switch opens and closes across rows 4 and 6.

5. **Insert the jumper wires.**

   1: Positive bus to A3

   2: Positive bus to A4

   3: Ground bus to J6

   4: Ground bus to J18

   5: E3 to F3

   6: E18 to F18

7: A6 to A12

8: B13 to B18

9: C15 to C18

10: D8 to D10

11: D11 to D14

12: D16 to D18

13: G3 to G10

6. **Connect the battery leads.**

   Red lead: Positive bus

   Black lead: Ground bus

7. **Test the circuit.**

   Insert the batteries. Press and release the button. The LED will come on to indicate that the output is HIGH. Press and release the button again to turn the LED off. Each time you press the button, the output turns from LOW to HIGH or from HIGH to LOW.

Debouncing a Clock Input

When you use a mechanical switch to trigger the clock input of a flip-flop, the switch will very likely have some mechanical bounce. This bounce happens when the switch contacts don't close completely cleanly; instead, the contacts bounce a little bit when they first touch each other. Even though these bounces are usually just a few milliseconds apart, they can end up confusing the flip-flop, as it thinks that each bounce of the switch contacts is actually a separate press of the button. So instead of just turning the LED attached to the Q output from off to on, a single press of the button might turn it from off to on, and then back off, then on, then off again, and so on until the switch settles down into its fully closed position.

There are several different ways you can *debounce* a mechanical switch — that is, eliminate the bounce effect. The easiest is to connect the mechanical switch to a one-shot timer circuit that uses an RC network to create a very short time interval such as 10 or 20 ms. Though short, this interval is enough to eliminate the negative bouncing effect.

For more information about how to build a one-shot circuit using a 555 timer IC, see Book 3, Chapter 2.

Chapter **5**

Introducing Microcontrollers

I first learned about microcontrollers years ago, when I wanted to create an automated prop for a Halloween display. The prop consisted of a Frankenstein creature that was attached to a pneumatic (air-powered) pop-up mechanism. Unsuspecting trick-or-treaters would walk into Frankenstein's lab, where the creature was sitting on a big work table. After a few moments, a voice would cry out, "It's alive! It's *alive! It's alive!*" and lights would flash on and off. Then the creature would start to twitch, and then pop up screaming "puttin' on the Ritz!" As the creature popped up, a bright floodlight would illuminate him. After a few seconds, the light would go out and the creature would lie back down.

I pondered how I could create the electronic circuit to control my prop. It could be done with an old computer and relay interface that would connect to the computer's parallel port. But I then realized it would be more fun and cost about the same to do it with a microcontroller instead of an old laptop computer and a parallel-port relay board. So I decided to build the prop using a microcontroller instead. The microcontroller would handle all the sequencing of lights and motion for the prop, and it could even play the sounds when the creature begins to awake and then finally pops up.

In a nutshell, a *microcontroller* is a small computer on a single board, which you can purchase typically for less than $50. This chapter introduces you to some of the basic concepts of microcontrollers. Then the next two minibooks show you how to work with two popular types of microcontrollers: Arduino and Raspberry Pi.

In this chapter, you learn several basic concepts that apply to microcontrollers of all types: what they are, how they're programmed, and how they can be connected to the outside world via digital I/O pins. You learn the specifics of working with Arduino and Raspberry Pi microcontrollers in Books 6 and 7.

Introducing Microcontrollers

A *microcontroller* is a complete computer on a single chip. Like all computer systems, microcomputers consist of several basic subsystems:

>> **Central processing unit (CPU):** The brains of the microprocessor. A CPU carries out the instructions provided to it by a program. The CPU can do basic arithmetic as well as other operations necessary to the proper functioning of the computer, such as moving data from one location of memory to another or receiving data as input from the outside world.

The CPU of a microcontroller is usually much simpler than the CPU found in a desktop computer. However, it is conceptually very similar. In fact, the CPUs found in many modern microcontrollers are as advanced as CPUs used in desktop computers just a few years ago.

>> **Clock:** The CPU and other components of the microcontroller are driven by a clock that provides timing pulses that control the pacing of program instructions as they are executed one at a time by the CPU. For most microcontrollers, the clock ticks along at a pace of a few million ticks per second. In contrast, the clock that drives a typical desktop computer ticks along at a few *billion* ticks per second.

>> **Random access memory (RAM):** Provides a scratchpad area where the computer can store the data it's working on. For example, if you want the computer to determine the result of a calculation (such as 2 + 2), you need to provide a location in RAM where the computer can store the result.

In a desktop computer, the amount of available RAM is measured in billions of bytes (GB for gigabytes). Microcontrollers typically have much less RAM available to them. For example, a typical Arduino processor has just 2KB of RAM. And some versions of the BASIC Stamp have just 32 bytes of RAM. One of the challenges of programming a microcontroller is often figuring out how to get your program to run with the small amount of memory available to it.

» **EEPROM:** A special type of memory that holds the program that runs on a microcontroller. *EEPROM* stands for *Electrically Erasable Programmable Read-Only Memory*, but that won't be on the test.

EEPROM is *read-only*, which means that once data has been stored in EEPROM, the data can't be changed by a program running on the microcontroller's CPU. However, it's possible to write data to EEPROM memory by connecting the EEPROM to a computer via a USB port. Then, the computer can send data to the EEPROM.

This is how microcontrollers are programmed. You use special software on a PC to create the program that you want to run on the microcontroller. Then, you connect the microcontroller to the PC and transfer the program from the PC to the microcontroller. The microcontroller then executes the instructions set forth in the program.

Most microcontrollers have a few thousand bytes of EEPROM memory, which is enough to store relatively complicated programs downloaded from a PC.

One of the most important features of EEPROM memory is that it doesn't lose its data when you turn off the power. Thus, once you transfer a program from a PC to a microcontroller's EEPROM, the program remains in the microcontroller until you replace it with some other program. You can turn the microcontroller off and put it on a closet shelf for years, and when you turn the microcontroller back on, the program that was recorded years ago will run again.

» **I/O pins:** One of the most important features of a microcontroller is its *I/O pins*, which enable the microcontroller to communicate with the outside world. Although some microcontrollers have separate input pins and output pins, most have shared I/O pins that can be used for both input and output.

I/O pins usually use the basic TTL logic interface that I describe throughout this minibook: HIGH (logic 1) is represented by +5 V, and LOW (logic 0) is represented by 0 V.

Most microcontrollers can handle only a small amount of current directly through the I/O pins. 20 mA to 25 mA is typical. That's enough to light up an LED, but circuits that require more current should isolate the higher current load from the microcontroller I/O pins. This is usually done by using a transistor driver.

Programming a Microcontroller

If you've never done any form of computer programming before, you're in for a fun and fascinating adventure, during which you'll learn a lot about how computers really work. In a nutshell, a computer program is a set of written instructions that a computer knows how to read, interpret, and carry out. The instructions are written in a language that both humans and computers can read. The instructions aren't quite English, but they resemble English enough that we English-speaking people can understand what they mean. (There are programming languages based on spoken languages other than English, but English-based programming languages are the most popular throughout the world.)

Computer programs are stored in text files that consist of one or more lines of written instructions. In most cases, each line of the computer program contains one instruction. Each instruction tells the computer to do something specific, such as add two numbers together or read the status of a particular input pin.

The trick of computer programming is to put the right instructions together in the right sequence to get the program to do exactly what you want it to do. Of course, to do that, you need to have a solid understanding of what you want the program to do, and you need a solid understanding of the variety of instructions that are available to you. You also must take into consideration the resources of the particular microcontroller you're working with and what circuits or devices are connected to the microcontroller.

The most common way to program a microcontroller is to connect the microcontroller to a PC via a USB cable, write the program using a special program called an *integrated development environment* (IDE), and then using the IDE to download the program to the microcontroller.

When you download the program to the microcontroller, the program is transferred into the microcontroller's EEPROM and then executed. Once it's in the EEPROM, you can disconnect the microcontroller from the computer. If you turn the microcontroller off or cut power to it, the program will be retained in EEPROM until you turn the microcontroller on again. In other words, the microcontroller will retain the program in its EEPROM so it can be run over and over again, even when the microcontroller is not connected to the computer. I explain how to do this for Arduino and Raspberry Pi in Books 6 and 7, respectively.

Working with I/O Pins

Most microcontrollers have several I/O pins, which you can connect external circuits to. The exact number of I/O pins varies depending on the specific microcontroller that you are using.

I/O pins use the standard conventions of digital logic, which I cover in the previous chapters of this minibook. Thus, a positive voltage level of 5 V is considered HIGH, and a level of 0 V is considered LOW.

The I/O pins for most microcontrollers can handle about 20 mA of current, so you need to be sure you use current-limiting resistors in your circuits to avoid overrunning the circuit and possibly damaging the microcontroller.

Figure 5-1 shows a simple schematic diagram for a circuit that drives an LED from an I/O pin of a microcontroller. Notice in this schematic that the pin output is represented by a simple five-sided shape. It's common in schematics *not* to draw the microcontroller as a single rectangle as you would other integrated circuits. Instead, each I/O connection in the circuit is shown using the five-sided connector shape. This convention gives you plenty of flexibility in how you lay out the schematic for your circuit.

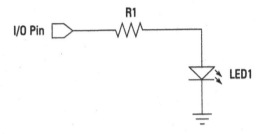

FIGURE 5-1:
Schematic for an LED circuit connected to a microcontroller I/O pin.

6

Working with Arduino Microprocessors

Contents at a Glance

Chapter **1**

Introducing Arduino

A *rduino* (pronounced ar-*dwee*-no) is one of the most popular microprocessor systems in use today. Arduino originally started in Ivrea, Italy, in 2005 and has become a global phenomenon. Arduino boards can be purchased online for less than $10.

Arduino is formally known as the Arduino Project, reflecting the fact that Arduino is an open-source project developed by an online community. As a result, several different companies manufacture Arduino-compatible microprocessor boards. You can purchase Arduino boards directly from the Arduino website (www.arduino.cc) or from other online retailers such as Amazon or Newegg.

Introducing the Arduino UNO

As of this writing, there are about three dozen different variations of Arduino boards, each providing different capabilities. For example, several Arduino boards come with built-in wired or wireless network abilities. Others are designed to be used in robots or game controllers.

The most popular Arduino board is called the *UNO* (see Figure 1-1). This mini-book focuses on the UNO, but just about everything you learn here about the UNO will apply to other Arduino models as well. In addition, Chapter 6 of this minibook delves into another Arduino model called the *Leonardo.*

FIGURE 1-1:
An Arduino UNO.

TIP

Several manufacturers make compatible versions of the UNO board. You can construct the projects in this book using any of these compatible UNO boards.

Figure 1-2 shows the layout of components on an UNO board. Here are its major features:

>> An Atmel ATmega328P single-chip microprocessor with 2KB of RAM for data storage, 1KB of EEPROM memory for program storage, 32KB of flash memory, and a variety of useful input and output capabilities that are exploited by the UNO board.

>> A 6 V to 20 V power connector that can be used to power the device from a battery or wall adapter. Built-in 5 V and 3.3 V regulators drop the voltage to the levels required by the Arduino. (Note that the device can also be powered from the USB Type B connector.)

>> A power LED that indicates the UNO's power status.

>> A USB Type B connector that is used to connect the UNO to a computer for programming. (The Type B connector is the squarish type of USB connector that is often used to connect a printer.) When connected, the UNO draws power directly from the USB connection, so a separate power connection is not required.

» Fourteen digital I/O pins, identified on the board as D0 through D13. (On all Arduino boards, the digital pins are numbered beginning with zero.)

» An onboard LED connected to pin 13. This LED can be controlled from programs that you upload from your computer to the UNO.

» Six analog input pins, identified as A0 through A5. (Like the digital pins, the analog pins are numbered starting with zero.) Note that if you want, the analog pins can be used as additional digital pins. In that case, the analog pins are identified as digital pins 14 through 19. (A0 becomes D14, A1 becomes D15.)

» Power pins that provide access to 5 V, 3.3 V, and ground for external circuits.

» Onboard TX and RX LEDs that indicate when the board is communicating with the computer via the USB port.

» A Reset button that interrupts the program currently running and restarts the UNO.

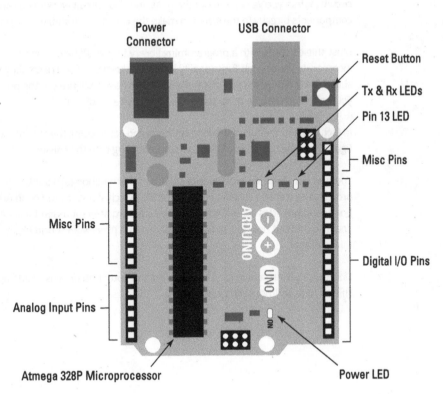

FIGURE 1-2: Components of an Arduino UNO.

RAISE THE SHIELDS!

One of the most interesting and useful features of Arduino microcontrollers are add-on boards known as *shields*. These add-on boards can piggyback on top of an Arduino board to provide additional capabilities. For example, you can get shields that enable your Arduino project to connect to the Internet, to control motors, or to provide an audible voice for your Arduino.

The best thing about Arduino shields is that they're stackable. In other words, you can combine two or more shields in a stack on top of an Arduino board to provide multiple functions. For example, you could use a wireless networking shield along with a motor driver shield and a GPS shield to create an autonomous robot. Or you could combine a voice-generator shield with a voice-recognition shield to create a project that can both listen and talk.

You can also get prototyping shield boards that allow you to incorporate your own circuits as shields. Some of these boards come with breadboards for prototyping your circuits. When you've worked out the bugs and your circuit works, you can solder the components directly to the shield to make the circuit more stable.

Most shields come with a programming library that you'll have to learn in order to incorporate the shield's functionality into your sketches. If you're creating your own shield using one of the prototype shields, you'll have to figure out the programming yourself, which can be quite challenging but also a lot of fun!

In fact, most of the fun of working with Arduino lies in using the shields. So, after you've figured out the basics, you'll want to start dabbling with the shields.

You find out about several types of shields in this minibook. To start, most of the projects use a shield called a *prototype shield*, which stacks on top of an Arduino UNO board and provides a small solderless breadboard area for assembling basic circuits. You also find out how to use an audio shield that plays MP3 sound files in Chapter 5 of this minibook.

Do a quick online search for *Arduino shield* and you'll find dozens of additional types of shields to experiment with. Have fun!

Buying an UNO Starter Kit

TIP

Although you can purchase an Arduino UNO board by itself, you may want to purchase your first UNO board as part of a starter kit that includes a variety of other components that will be helpful as you build your projects. These starter kits typically cost anywhere from $30 to $100 or more, depending on the additional components they include.

I suggest you start with an inexpensive kit such as the Elegoo UNO Project Super Starter Kit, which comes with a compatible UNO board, a small tutorial book, and a variety of goodies such as a breadboard, an LCD display, servo and stepper motors, some jumper wires, LEDs, resistors, and push buttons, and so on. In all, it contains enough parts to build a variety of UNO projects.

If you already have breadboards, resistors, LEDs, jumper cables, and the like, you can get a genuine Arduino UNO board for about $25, or you can get a compatible UNO board for less than $10.

I also suggest you pick up one or more Prototype Shields to build the circuits for the projects in the chapters that follow. The most popular is made by Adafruit and is called a *Proto Shield*. I explain more about this useful shield in Chapter 2 of this minibook. You can buy the Proto Shield directly from Adafruit (www.adafruit.com) or you can purchase one from online retailers such as Amazon or Walmart.

Don't forget that you'll also need a short mini-B USB cable to connect the UNO board to your computer to program the board.

Installing the Arduino IDE

The Arduino IDE is the software that you use on your computer to create programs that can be uploaded from your computer to an Arduino board. This software is available free from the Arduino website. Just point your browser to www.arduino.cc and click the Software menu. Then download the IDE package for the operating system you're using. (The IDE comes in separate versions for Windows, Mac, and Linux.)

If you're using Windows, you should choose the Windows Installer rather than the zip version. The Windows Installer version will install itself directly onto your computer so that you can then run it from the Start menu.

If you're a Mac user, download the Arduino IDE program. Then copy the downloaded program to your Applications folder. Finally, to make it easier to run the program, drag it from the Applications folder onto the Dock at the bottom of the screen.

When you get the software installed, run it by choosing Arduino IDE from the Start menu (Windows) or from the Dock (Mac). Figure 1-3 shows how the IDE appears when you first run it on a Windows PC.

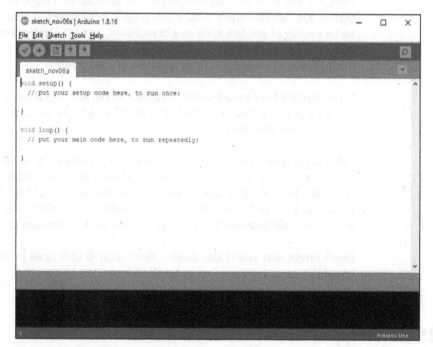

```
sketch_nov06a | Arduino 1.8.16                                    —    □    ×
File Edit Sketch Tools Help

sketch_nov06a

void setup() {
    // put your setup code here, to run once:

}

void loop() {
    // put your main code here, to run repeatedly:

}

                                                               1                  Arduino Uno
```

FIGURE 1-3:
The Arduino IDE.

Connecting to an UNO

Before you can use the IDE to program an Arduino, you must first connect the Arduino to your computer. With an UNO, this is an easy task: Simply plug the squarish end of the USB cable into the USB receptacle on the UNO board, and then plug the normal USB connector on the other end of the cable into any available USB port on your computer. The power LED lights up on the UNO, and the Arduino IDE automatically recognizes the UNO on the USB port.

To verify that the Arduino is connected properly, open the IDE software and perform the following steps:

1. **In the IDE, choose Tools ⇨ Board and verify that Arduino UNO is selected.**

   If not, choose Tools ⇨ Board ⇨ Arduino UNO.

2. **Choose Tools ⇨ Port and verify that the USB port is selected.**

   If it isn't, manually select the USB port.

   After you've verified that the UNO is connected, you can proceed with uploading your first program to the UNO.

Looking at a Simple Arduino Sketch

In the Arduino world, a program is called a *sketch*. Contrary to what this gentle-sounding term implies, however, there is nothing artistic or visual about an Arduino sketch: It's simply a program written in an advanced programming language called C++. Calling the program a *sketch* instead of a program is simply a way of helping lower your defenses against the daunting task of learning how to program.

In just about every book on programming languages, the first program presented is called Hello World. This simple program displays the string "Hello, World!" as a way of demonstrating what the simplest possible program looks like.

Such a program is possible in Arduino, but it isn't really the best sample program to start with. That's because unlike most computers, microprocessors such as the Arduino do not have a built-in console that can show the "Hello, World!" text.

So instead, we'll start with a simple program that flashes the built-in LED that is established on pin 13. This program simply flashes the LED on and off repeatedly, as long as the program is allowed to run — in other words, until you turn the UNO off. The program turns the LED on for one second, then off for one second, then on for one second, and so on, indefinitely.

Before I show you the code for this program, let's walk through the steps that the program must take:

1. First, the program must designate I/O pin 13 as an output pin. It will do this using a command called pinMode.

2. Next, it must turn the LED on. It does this by using a command called digitalWrite, which is used to write the value HIGH to pin 13.

3. Then the program waits for one second to elapse. It does this by using a command called delay. During this delay, the LED remains lit because the output status of pin 13 is HIGH.

4. After the one-second delay, digitalWrite is used again, this time to turn the LED off by writing the value LOW to pin 13.

5. Then the program uses delay again to wait for one second. During this delay, the LED remains off because the output status of pin 13 is LOW.

6. Now the program repeats Steps 2 through 5, in sequence over and over again until the UNO is turned off.

The actual Arduino program to implement these steps is shown in Listing 1-1.

LISTING 1-1: **The Blink Program**

```
void setup() {
pinMode(13, OUTPUT);
}


void loop() {
digitalWrite(13, HIGH);
delay(1000);
digitalWrite(13, LOW);
delay(1000);
}
```

If you're new to programming, there are a lot of details within this listing that I don't expect you to understand yet. Don't worry — I explain them all in the next chapter. For now, I just want you to peruse the code casually and note that the program uses the pinMode, digitalWrite, and delay commands as described in the steps listed earlier.

The commands in the Blink program are grouped together into two *functions*, which are simply sections of code that are collected together in a unit. These two functions are

>> setup: The commands in the setup function are executed once at the very beginning of the program. In other words, the setup function is run once whenever the UNO is turned on or when the UNO's Reset button is pressed. In the Blink program, the setup function has just one command — pinMode — which sets pin 13 to OUTPUT mode.

>> loop: After the setup function completes, the commands in the loop function are executed repeatedly until the UNO is turned off or reset. These commands turn the LED on, wait for one second, turn the LED off, then wait for one more second. As a result, each time the loop function is executed, one blinking cycle occurs — the LED goes on for one second, and then goes off for one second.

REMEMBER

All Arduino programs, from the simplest to the most complex, are organized into these two basic functions: setup, which is executed once when the program starts, and loop, which is executed repeatedly until the program stops.

Running the Blink Program

To actually run the Blink program described in the previous section on your Arduino UNO, follow these steps:

1. Connect the Arduino UNO to your computer.

Use the USB cable to connect the USB port on the UNO to a USB port on your computer.

2. Open the Arduino IDE program on your computer.

On a Windows computer, choose Arduino IDE from the Start menu. On a Mac, double-click the Arduino IDE icon from the Dock or open it from the Applications folder.

3. Choose the File ⇨ New Command.

A window for the new sketch appears. Notice that this sketch already contains the outline for the setup and loop functions.

4. Edit the program so that it looks exactly like Listing 1-1.

The IDE editor works like any other text editor you've worked with. If you want, you can start by selecting the entire contents of the editor window and pressing the Delete key. Then simply type the program exactly as shown in Listing 1-1 into the editing window.

5. Save the sketch.

Choose File ⇨ Save, enter the name "Blink-1," and click the Save button.

When you've created and saved the program, the editor window should look like Figure 1-4.

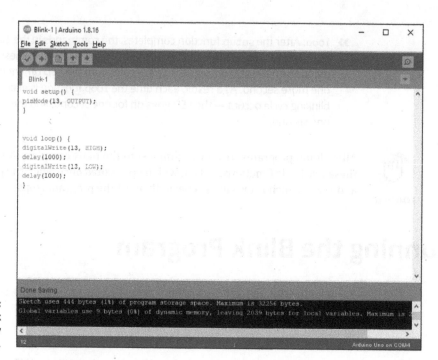

FIGURE 1-4:
The Blink
program ready
to run.

6. **Choose Sketch ⇨ Upload.**

The program is uploaded to the UNO. The TX and RX LEDs on the UNO board will flash for a few seconds while the program is being uploaded to the UNO. After the program has finished uploading, the onboard LED will begin to flash on and off at one-second intervals.

TIP

The onboard LED is very small — you may need to squint your eyes to see it. But the steady on/off of its blinking is a sure sign that your program is running.

Congratulations! You've written your first Arduino program, and you've now stepped into the world of microcontroller programming!

Using a Digital I/O Pin to Control an LED

Now that you've written an Arduino sketch that can flash the onboard LED that's connected to pin 13, the next step is to tap into pin 13 on the UNO board to light an external LED.

The UNO board has a total of 14 digital I/O pins, which are labeled on the board for your convenience. The pins are accessible via two sets of *pin headers* (usually just called *headers*). Headers are a common way of connecting to digital circuit boards.

The headers on an UNO board are female, with a pin spacing of 0.1 inch. If you hold the board with the power and USB connectors at the top, you'll see two rows of female headers — one on the left edge of the board, the other on the right edge. Figure 1-5 identifies each of the pins on these headers.

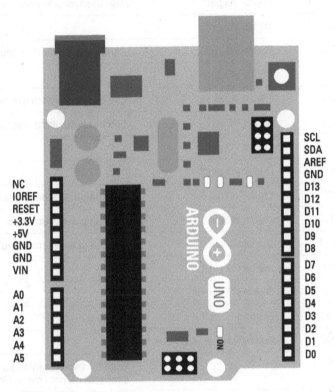

FIGURE 1-5:
The headers on an UNO board.

The digital pins 0 through 13 are on the header on the right-hand side of the UNO. From the bottom, they're labeled D0 through D13. Immediately above the D13 header is a ground connection, labeled GND. This header pin is used to complete the circuit for each digital I/O pin. There are a few additional header pins above the GND pin, but I won't be using them in this minibook.

For your reference, Table 1-1 lists all the pin connections available via the UNO boards headers.

As with most microcontrollers, the digital I/O pins on the UNO operate at standard logic-level voltage, so +5 V is present at the pin when it is HIGH, and 0 V is present when the pin is LOW. On the UNO, the I/O pins are capable of sourcing 20 mA, so you'll need to provide an appropriate current-limiting resistor to avoid burning out the LED. For this example, we use a 470 Ω resistor.

TABLE 1-1 **UNO Board Header Pins**

Label	Purpose	Additional Information
D0→RX	Digital I/O pin 0	Also used for serial communication (RX)
D1←TX	Digital I/O pin 1	Also used for serial communication (TX)
D2	Digital I/O pin 2	
~D3	Digital I/O pin 3	Can use pulse width modulation (PWM)
D4	Digital I/O pin 4	
~D5	Digital I/O pin 5	Can use pulse width modulation (PWM)
~D6	Digital I/O pin 6	Can use pulse width modulation (PWM)
D7	Digital I/O pin 7	
D8	Digital I/O pin 8	
~D9	Digital I/O pin 9	Can use pulse width modulation (PWM)
~D10	Digital I/O pin 10	Can use pulse width modulation (PWM)
~D11	Digital I/O pin 11	Can use pulse width modulation (PWM)
D12	Digital I/O pin 12	
D13	Digital I/O pin 13	
GND	Ground	
AREF	Analog voltage reference	
SDA		Used for two-wire interface (TWI) connections, not covered in this book
SCL		Used for two-wire interface (TWI) connections, not covered in this book
A0	Analog input pin 0	Can also be used as digital I/O pin 14
A1	Analog input pin 1	Can also be used as digital I/O pin 15
A2	Analog input pin 2	Can also be used as digital I/O pin 16
A3	Analog input pin 3	Can also be used as digital I/O pin 17
A4	Analog input pin 4	Can also be used as digital I/O pin 18
A5	Analog input pin 5	Can also be used as digital I/O pin 19
VIN	Input power (6–20 V)	Can be used to power the Arduino

Label	Purpose	Additional Information
GND	Ground	
GND	Ground	
+5 V	+5 V	Provides regulated +5 V for external circuits (maximum 400 mA)
+3.3 V	+3.3 V	Provides regulated +3.3 V for external circuits (maximum 50 mA)
RESET	Reset	Not covered in this book
IOREF	Voltage reference	Not covered in this book

Figure 1-6 shows the schematic for this circuit. As you can see, the circuit is very simple: Pin 13 connects to the 470 Ω resistor, which in turn connects to the LED's anode. The cathode then connects to ground. Although the schematic doesn't indicate it, we use one of the two ground pins on the UNO card for this ground connection.

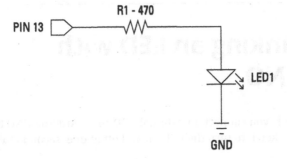

FIGURE 1-6:
Adding an external LED to an UNO board.

Project 43 shows how to assemble this project by using a breadboard to hold the LED and resistor, with jumpers to connect the breadboard to the UNO board. Figure 1-7 shows how the project appears when completed.

Note that the project uses the same Arduino sketch used in Project 42. This works because the internal LED that was used in Project 42 is connected to digital I/O pin 13, which is used in Project 43. So no reprogramming of the Arduino is required for Project 43.

FIGURE 1-7:
The assembled
Arduino
LED circuit
(Project 43).

Project 43: Blinking an LED with an Arduino UNO

In this project, you connect an external LED to an Arduino UNO board. Then you use a simple sketch to turn the LED on and off at one-second intervals.

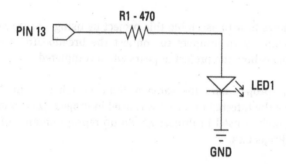

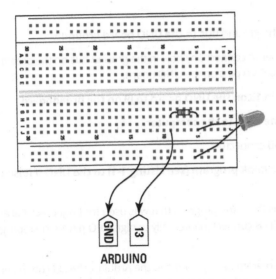

ARDUINO

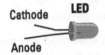

Cathode **LED**

Anode

Parts

- » One computer with Arduino IDE installed
- » One Arduino UNO board
- » One USB cable
- » One small solderless breadboard
- » One red LED
- » One 470 Ω resistor
- » Two jumper wires (male to male)

Steps

1. **Insert resistor R1.**

   R1 (470 Ω): H5 to H9

2. **Insert LED1.**

   Cathode (short lead): Ground bus

   Anode (long lead): J5

3. **Connect the ground bus to the UNO board ground.**

   Use a jumper to connect any hole in the ground bus on the breadboard to any of the three GND pins on the UNO board.

4. **Connect pin 13 on the UNO board to J9 on the breadboard.**

5. **Connect the UNO to the computer.**

   Use the USB connector.

6. **Upload the Blink program (see Listing 1-1) to the UNO if it isn't already uploaded.**

   Again, this is the same program that was used for Project 42. Because the onboard LED is connected internally to digital I/O pin 13, no reprogramming is needed.

   When the program has uploaded to the Arduino, the LED on the breadboard will flash on and off at one-second intervals. Note that the LED on the breadboard will flash in sync with the built-in LED on the UNO board.

TIP

If the LED on the breadboard doesn't flash, double-check your connections. In all likelihood, you've inserted the LED backward. Remember that the short lead (the cathode) must be connected to the ground rail on the breadboard.

IN THIS CHAPTER

» **Learning the essentials of the Arduino language**

» **Working with variables**

» **Adding `if` statements to your programs**

» **Using `while` and `for` loops**

» **Creating your own functions**

» **Using a prototyping board**

Chapter **2**

Creating Arduino Sketches

Welcome to programming! This chapter is a very brief introduction to the programming language you use to develop Arduino sketches. If you've never done any kind of computer programming before, you're in for an interesting and fun journey. The nuances of computer programming can be difficult to grasp at first, but once you get your mind around some of the basic concepts, you'll discover all sorts of things you can coerce your Arduino to do.

TECHNICAL STUFF

The Arduino language is actually a standard programming language known as C++. C++ happens to be an advanced version of a somewhat simpler language called C. Because most beginning Arduino programs don't actually use any of the advanced features, you can think of the Arduino language as simply C for now. In fact, I refer to the Arduino language simply as C rather than as C++ throughout this chapter. At some point, you may want to venture into the world of C++ programming for your Arduino, but programming in C++ is beyond the scope of this humble book.

Introducing C

As you learn in Chapter 1 of this minibook, you use the Arduino IDE to create sketches and upload them to your Arduino. The term *sketch* is simply Arduino's word for a C program.

WARNING

If you're unclear on how to create and upload a sketch to your Arduino, please read Chapter 1 of this minibook. You can't progress very far in Arduino programming without writing some actual programs, downloading them to an Arduino, and observing how they work.

Before we get going, I recap the program that was presented in Chapter 1 of this minibook. For your convenience, the program is repeated in Listing 2-1.

LISTING 2-1:	The Blink Program

```
void setup() {                                                        →1
  pinMode(13, OUTPUT);                                                →2
}

                                                                      →3

void loop() {                                                         →4
  digitalWrite(13, HIGH);                                            →5
  delay(1000);                                                        →6
  digitalWrite(13, LOW);                                             →7
  delay(1000);                                                        →8
}                                                                     →9
```

Here are some important details about this program — going through it line by line — that you need to understand in order to create other, more complicated sketches:

>> →1 The first line of the program declares a function named setup. As you find in Chapter 1 of this minibook, a *function* is a set of programming instructions that are collected together and given a name. In this case, there is just a single instruction (the pinMode instruction in line 2). The setup function is called by the Arduino once, as soon as the Arduino starts up or is reset. This function is usually used to execute instructions that need to be executed only once, such as the pinMode instruction that establishes whether a particular pin will operate as an input or output pin.

There are a few other details about this line to note:

- The line begins with the word void. In C, all functions can return a value. That is, when a function finishes executing, it can return a value that can be used elsewhere in the program. Not all functions actually do return values, however. When a function does not return a value, the declaration must begin with the word void. This simply indicates that the setup function does not return a value.

- The function name setup is followed by a pair of parentheses with nothing in between the parentheses. In C, data can be passed to a function using what are called *arguments*. These arguments must be listed in parentheses following the function name. In the case of the setup function, no arguments are required. However, the parentheses are always required; hence, a set of empty parentheses must always follow setup.

- The left curly brace character follows the parentheses. This marks the beginning of the programming instructions that are contained within the setup function. Everything between this left curly brace and a matching right curly brace (which is found in line 3) belongs to the function. In this case, that is just one instruction, the one found in line 2.

» →2 This line sets the input/output mode of pin 13. It does so by calling a built-in function named pinMode. The pinMode function is a function just like the setup function, except that it is predefined by Arduino as part of the standard *function library* — a collection of many standard functions that you can use in your Arduino programs. The pinMode function simply designates whether a particular I/O pin will be used as an input pin or an output pin.

Like most standard library functions, and unlike the setup function, the pinMode function *does* require arguments. In fact, it requires two arguments. These arguments must be enclosed in parentheses:

- The first argument specifies the pin for which you want to set the mode. In this case, we are setting the mode of pin 13.

- The second argument specifies what the mode should be. In this case, we're setting the pin to operate as an output pin.

You may be wondering why the word OUTPUT is written in all capital letters. That's because the word OUTPUT is what is known in C as a *constant*. A constant is a word that is defined within the C program as a symbol that stands for some predefined value. In this case, the constant OUTPUT is defined by the same standard library that defines the pinMode function. Another constant defined by this library is INPUT, which you could use instead of OUTPUT to designate the pin as an input pin.

One final and very important point to note about this line: It ends with a semicolon. In C, all *statements* must end with a semicolon. In general, a statement is an instruction that directs the program to do something. In this case, the statement tells the program to execute the pinMode function to define the mode of pin 13. The first line of the program doesn't require a semicolon because it doesn't tell the program to do anything; it simply marks the start of the setup function.

» →3 This closing (right) curly brace is matched with the opening (left) curly brace at the end of line 1 to mark the end of the setup function. Thus, this brace and the one in line 1 enclose the instructions of the setup function.

» →4 This line marks the beginning of the loop function. As you can see, it closely resembles line 1, which marks the beginning of the setup function. Like the setup function, the loop function doesn't return a value, so it begins with the word void. And like the setup function, the loop function doesn't use any arguments, so it includes an empty set of parentheses. And this line ends with an opening curly brace to mark the beginning of the statements that will be executed when the Arduino calls the loop function.

» →5 This line calls the digitalWrite function, another standard Arduino library function. This function sends output data to a pin. The two arguments specify the pin to which the data should be sent and the data value to send to the pin. In this case, the value HIGH is sent to pin 13, which causes the voltage level at pin 13 to go to +5 V, thus lighting the LED. As you might guess, HIGH is another constant defined by the standard library. Notice also that this line ends with a semicolon to mark the end of the statement.

» →6 This line calls yet another built-in function called delay. The delay function takes a single argument, which represents the number of milliseconds that the program should pause. In this case, 1,000 milliseconds is specified as the argument; this pauses the program for one full second. Notice that the comma is not used in the argument value 1000. Notice also that this statement ends with a semicolon.

» →7 After the LED on pin 13 has been allowed to stay lit for one second, this line calls digitalWrite again to turn off the LED. To accomplish that, it uses the constant LOW as the data to be sent to the pin. This sets the output at pin 13 to low, which turns off the LED.

» →8 Now the program calls delay again to pause the program for another 1,000 milliseconds, allowing the LED to stay dark for one second.

» →9 The final closing (right) brace marks the end of the loop function. Thus, the loop function includes four statements: two that write data to pin 13 and two that pause the program. Because the loop function is called repeatedly by the Arduino, the loop function causes the LED to flash on and off repeatedly as long as the program continues to run.

Working with a Prototyping Shield

In the next section, I show you how to build a test circuit that you can then use to experiment with the basics of Arduino programming. The test circuit and all the remaining UNO projects in this minibook use an Arduino prototyping shield that snaps into the top of an Arduino UNO and provides a small breadboard on which you can assemble simple circuits.

Many manufacturers make prototyping shields. The one I use is made by HiLetgo (www.hiletgo.com) and can be purchased directly from Amazon. If you prefer, you can use prototyping boards made by other companies, but they may have a different layout than the HiLetgo shields used in this book.

Here are the key features of a prototyping shield:

>> **Stackable headers that snap into the headers on the UNO board to pass along the functions of each of the UNO's header pins.**

>> **A prototyping area with pre-drilled holes and solder pads to add components.**

>> **A small breadboard (called a *tiny breadboard*) on which you can assemble simple circuits.** The tiny breadboard has an adhesive backing so it can be permanently placed over the prototyping area. For all the examples in this minibook, I use the breadboard rather than the prototyping area to assemble the circuits.

Figure 2-1 shows the prototyping shield in action, attached to the top of an UNO card. In this photo, you can see how the prototyping shield's header pins provide access to the UNO's header pins. And you can see the tiny breadboard, which can be used to assemble circuits.

Figure 2-2 is a diagram that provides a better look at the elements of the prototype shield. As you can see, headers at the left and right of the prototype shield provide access to the I/O pins on the UNO card. In addition, headers at the bottom of the card provide a row of five GND pins and five +5 V pins.

The breadboard itself has 17 rows of ten holes divided into two groups of five. In each row, the five holes on the left are internally connected and the five holes on the right are internally connected. As with other breadboards, the holes are not internally connected across the center gap.

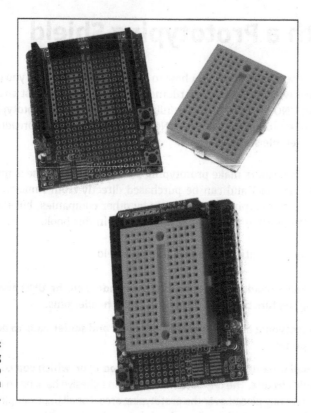

FIGURE 2-1:
The prototyping
shield snaps into
the top of an
Arduino UNO.

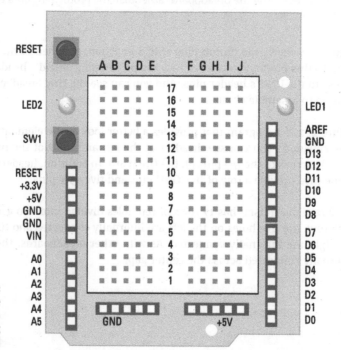

FIGURE 2-2:
Identifying the
connections on
a prototyping
shield

TIP

The rows and columns on most tiny breadboards are not identified in any way. For convenience, I refer to the rows using the numbers 1 through 17 and the columns using the letters A through J, as indicated in Figure 2-2.

Building a Test Circuit

Project 44 provides a simple testing environment that will be useful to you as you learn the basics of Arduino programming. This project simply places eight LEDs and eight current-limiting resistors on a breadboard and connects them to digital pins 0 through 7 on the Arduino. After you've built this circuit, you'll create several Arduino sketches that use various programming techniques to flash these LEDs in interesting patterns.

Figure 2-3 shows the finished circuit, ready to be used for testing the programs you'll write as you work your way through this chapter.

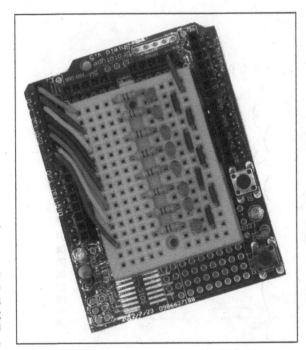

FIGURE 2-3:
An Arduino UNO with eight LEDs on a prototyping shield (Project 44).

Project 44: An Arduino LED Test Circuit

In this project, you assemble a circuit with eight LEDs on a prototyping shield, connecting the LEDs to digital ports 0 through 7. You can use this test circuit to test the programs shown in this chapter.

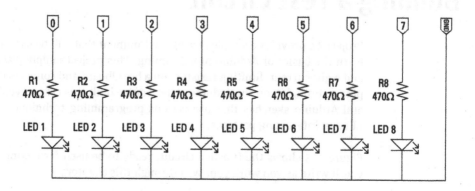

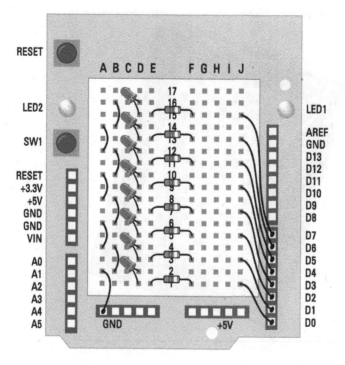

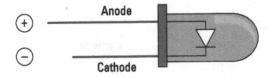

Parts

>> One computer with Arduino IDE installed

>> One A-B USB cable (the kind used for printers)

>> One Arduino UNO board

>> One prototyping shield

>> Eight red LEDs

>> Eight 470 Ω resistors

>> Sixteen jumper wires

Steps

1. **Insert the resistors.**

Resistor	From	To
R1 (470 Ω)	E1	F1
R2 (470 Ω)	E3	F3
R3 (470 Ω)	E5	F5
R4 (470 Ω)	E7	F7
R5 (470 Ω)	E9	F9
R6 (470 Ω)	E11	F11
R7 (470 Ω)	E13	F13
R8 (470 Ω)	E15	F15

2. Insert the LEDs.

LED	Anode	Cathode
LED1	D1	D2
LED2	D3	D4
LED3	D5	D6
LED4	D7	D8
LED5	D9	D10
LED6	D11	D12
LED7	D13	D14
LED8	D15	D16

3. Use jumpers to connect digital pins to the breadboard.

Digital Pin	Breadboard Pin
D0	J1
D1	J3
D2	J5
D3	J7
D4	J9
D5	J11
D6	J13
D7	J15

4. Use jumpers to complete the ground circuit.

From	To
GND 1	A2
B2	B4
A4	A6
B6	B8
A8	A10
B10	B12

From	To
A12	A14
B14	B16

5. **Connect the prototyping shield to the Arduino UNO.**

6. **Connect the UNO to the computer.**

    Use the type-B USB connector.

7. **Upload the LED Flasher program (see Listing 2-2, in the next section) to the UNO.**

    The LEDs on the breadboard will flash on and off at one-second intervals.

Flashing the LEDs

Earlier in this chapter, you see a program that flashes a single LED on pin 13 on and off. In the rest of this chapter, I show several variants of that program, which flash the eight LEDs in the test project (Project 44) in various sequences. Along the way, you can add more C statements to your repertoire to provide more and more complex ways to control the flashing.

REMEMBER

Keep in mind throughout this chapter that if you can turn an LED on or off with an Arduino sketch, you can control *anything* that can be connected to an Arduino's digital I/O port. The Arduino itself doesn't know or care what kind of circuit you connect to a digital I/O pin. All it knows is that when you tell it to, the Arduino makes the I/O port HIGH or LOW. It's the external circuitry connected to the pin that determines what happens when the pin goes HIGH.

The only limitation is that the Arduino itself can swing only about 20 mA through its I/O pins. If the circuit connected to the pin requires more current than 20 mA, you must isolate the higher-current portion of the circuit from the Arduino. The easiest way to do that is to use a transistor driver or a relay.

Listing 2-2 shows a simple program that flashes all eight of the LEDs on and off at half-second intervals. This program uses nothing more than the `pinMode`, `digitalWrite`, and `delay` functions that you already know about. The program turns all eight LEDs on, pauses 500 ms (half a second), turns the LEDs off, waits another half-second, and then jumps back to the `Main` label to start the whole process over.

LISTING 2-2: **The LED Flasher Program**

```
void setup() {                                              →1

  pinMode(0, OUTPUT);                                       →2
  pinMode(1, OUTPUT);
  pinMode(2, OUTPUT);
  pinMode(3, OUTPUT);
  pinMode(4, OUTPUT);
  pinMode(5, OUTPUT);
  pinMode(6, OUTPUT);
  pinMode(7, OUTPUT);

}

void loop() {                                              →3

  digitalWrite(0, HIGH);                                   →4
  digitalWrite(1, HIGH);
  digitalWrite(2, HIGH);
  digitalWrite(3, HIGH);
  digitalWrite(4, HIGH);
  digitalWrite(5, HIGH);
  digitalWrite(6, HIGH);
  digitalWrite(7, HIGH);

  delay(500);                                              →5

  digitalWrite(0, LOW);                                    →6
  digitalWrite(1, LOW);
  digitalWrite(2, LOW);
  digitalWrite(3, LOW);
  digitalWrite(4, LOW);
  digitalWrite(5, LOW);
  digitalWrite(6, LOW);
  digitalWrite(7, LOW);

  delay(500);                                              →7

}
```

The following paragraphs summarize the operation of this program:

» →1 The start of the setup function, run once when the UNO starts up or is reset.

» →2 The next eight lines call the setMode function eight times to set the I/O mode for pins 0 through 7 to OUTPUT.

» →3 The start of the loop function, called repeatedly while the UNO runs.

» →4 The next eight lines call the digitalWrite function eight times to set the status of pins 0 through 7 to HIGH, thus turning on all eight LEDs.

» →5 This line calls the delay function to pause the program for 500 milliseconds.

» →6 The next eight lines call the digitalWrite function to set the status of pins 0 through 7 to LOW, thus turning off all eight LEDs.

» →7 This line calls the delay function to pause the program for 500 milliseconds.

Using Comments

A *comment* is a bit of text that provides an explanation of your code. The Arduino completely ignores comments, so you can put any text you want in a comment. Using plenty of comments in your programs to explain what your program does and how it works is a good idea.

REMEMBER

To create a comment, place two forward slashes at the beginning of the line. When C sees the two slashes, it ignores the rest of the line. Thus, if you place the slashes at the beginning of a line, the entire line is considered to be a comment. If you place a comment in the middle of a line (for example, after a statement), everything after the slashes is ignored.

It's a common programming practice to begin a program with a group of comments that indicates what the program does, who wrote it, and when. This block of comments may also indicate what I/O devices are expected to be connected to the Arduino. Listing 2-3 shows a version of the LED Flasher program that includes both types of comments.

LISTING 2-3: **LED Flasher with Comments**

```
// The LED Flasher Program
// Doug Lowe
// December 2, 2021
//
// This program flashes LEDs connected to digital pins 0
// through 7 at half-second intervals.

void setup() {

  // Set pins 0 through 7 to OUTPUT
  pinMode(0, OUTPUT);
  pinMode(1, OUTPUT);
  pinMode(2, OUTPUT);
  pinMode(3, OUTPUT);
  pinMode(4, OUTPUT);
  pinMode(5, OUTPUT);
  pinMode(6, OUTPUT);
  pinMode(7, OUTPUT);

}

void loop() {

  // Turn all LEDs on
  digitalWrite(0, HIGH);
  digitalWrite(1, HIGH);
  digitalWrite(2, HIGH);
  digitalWrite(3, HIGH);
  digitalWrite(4, HIGH);
  digitalWrite(5, HIGH);
  digitalWrite(6, HIGH);
  digitalWrite(7, HIGH);

  // Wait a bit
  delay(500);

  // Turn all LEDs off
  digitalWrite(0, LOW);
  digitalWrite(1, LOW);
  digitalWrite(2, LOW);
  digitalWrite(3, LOW);
  digitalWrite(4, LOW);
  digitalWrite(5, LOW);
```

```
digitalWrite(6, LOW);
digitalWrite(7, LOW);

// Wait a bit
delay(500);

}
```

Creating Identifiers

An *identifier* is a word that you make up to refer to a programming element by name. Although you can assign identifiers to many types of elements, they're most commonly used for the following:

>> Function names, such as setup and pinMode

>> Variables and fields, which hold data used by your program

>> Arguments, which pass data values to functions

REMEMBER

You must follow a few simple rules when you create identifiers:

>> **Identifiers are case-sensitive.** As a result, SalesTax and salesTax are distinct identifiers.

>> **Identifiers can be made up of upper- or lowercase letters, numerals, underscore characters (_), and dollar signs ($).** Thus, identifiers have names such as Port1, SalesTax$, and Total_Sales.

>> **All identifiers must begin with a letter.** Thus, a15 is a valid identifier, but 13Unlucky isn't (because it begins with a numeral).

Using Variables

A *variable* is a name that is assigned to a particular bit of data in your program. Variables are essential to almost all serious Arduino programs, because they provide a way to keep track of data, perform calculations on data, and make decisions based on data.

In C, a variable can store four distinct types of data:

>> `int`: The `int` type stores integer values — that is, whole numbers.

>> `float`: The `float` type stores floating point numbers — that is, numbers with decimal points.

>> `double`: The `double` type stores what are called *double precision* floating-point numbers — that is, numbers with decimal points that store twice the number of digits as a regular `float` variable.

>> `char`: The `char` type stores a single byte of data.

You must define a variable before you can use it in an Arduino program. The most basic form for declaring a variable is this:

```
type name;
```

Here are some examples:

```
int time;
float reading;
char status;
```

Notice that a variable declaration is considered to be a statement in C, so it must be followed by a semicolon.

Once you've created a variable, you can use it in an *assignment statement* to assign it a value. An assignment statement consists of a variable name followed by an equals sign, followed by the value to be assigned. For example, this assignment statement assigns the value 500 to a variable named `time`:

```
time = 500
```

You can combine the declaration of a variable and the assignment of a value into a single statement, like this:

```
int time = 500;
```

Here, the variable `time` is both declared and assigned the value 500 in a single statement.

The value on the right side of the equals sign can be an arithmetic calculation. For example:

```
time = 500 + 10
```

In this example, the value 510 is assigned to the variable named time.

There's not much point in doing arithmetic using only numerals. After all, you could just do the calculation yourself. Thus, the previous example could be written like this:

```
time = 510
```

The real power of variable assignments happens when you use variables on the right side of the equals sign. For example, the following statement increases the value of the time variable by 10:

```
time = time + 10
```

In this example, the previous value of time is increased by 10. For example, if the time variable's value was 150 before this statement executed, it will be 160 after.

An important thing to know about variables is that you can declare them either inside of a function or outside of a function. If you declare a variable inside of a function, the variable can be used only within that function. However, if you declare a variable outside of a function, the variable can be used in any of the functions in the program. Such a variable is sometimes called a *global variable* because it can be used globally throughout the program.

Although not required, it is customary to declare all variables within a function at the very beginning of the function, before any other statements in the function. For global variables, it is customary to define them near the beginning of the program, before any functions.

Doing Math

As you've already seen, C lets you perform addition, subtraction, multiplication, and division using the symbols (called *operators*) +, −, *, and /. For example:

```
int x = 10;
int y;
y = x * 3;
```

In this example, the value 30 will be assigned to the variable y.

Here are a few things you need to know about how C does math:

» **C uses the normal order of operations found in math.** So, multiplication and division are done before addition and subtraction. For example:

```
int x = 10;
int y;
y = x + 5 * 3
```

This statement assigns the value 25 to y, because 5 is first multiplied by 3 and then the result is added to 10.

» **You can use parentheses to force calculation of a certain part of the formula first.** For example:

```
int x = 10;
int y;
y = (x + 5) + 3
```

Here, C first does the calculation inside the parentheses, giving a result of 15. It then adds the 3 to the 15 to give the final result, 18.

» **When you do division with integers, the fractional part of the answer is discarded and the result is an integer.** For example:

```
int x = 8 / 3
```

This statement assigns the value 2 to x. That's because 8 divided by 3 is 2 with a remainder of 2. The remainder is discarded.

A Program That Uses Variables and Math

Listing 2-4 shows a program that uses a global variable to change the speed at which the LEDs flash each time the loop function is called. As you can see, a variable named Time is used to provide the number of milliseconds that the delay function is paused each time it is called. Each time through the loop, the value of the Delay variable is increased by 10. Thus, the LEDs flash very fast when the program first starts, but the flashing gets progressively slower as the program loops.

LISTING 2-4: The LED Flasher Program with a Variable

```
// The LED Flasher Program
// Doug Lowe
// December 2, 2021
```

```
//
// This program flashes LEDs connected to digital pins 0
// through 7 at half-second intervals.

// The Time variable
int Time;

void setup() {

  // Set pins 0 through 7 to OUTPUT
  pinMode(0, OUTPUT);
  pinMode(1, OUTPUT);
  pinMode(2, OUTPUT);
  pinMode(3, OUTPUT);
  pinMode(4, OUTPUT);
  pinMode(5, OUTPUT);
  pinMode(6, OUTPUT);
  pinMode(7, OUTPUT);

  // Initialize the Time variable
  Time = 10;

}

void loop() {

  // Turn all LEDs on
  digitalWrite(0, HIGH);
  digitalWrite(1, HIGH);
  digitalWrite(2, HIGH);
  digitalWrite(3, HIGH);
  digitalWrite(4, HIGH);
  digitalWrite(5, HIGH);
  digitalWrite(6, HIGH);
  digitalWrite(7, HIGH);

  // Wait a bit
  delay(Time);

  // Turn all LEDs off
  digitalWrite(0, LOW);
  digitalWrite(1, LOW);
  digitalWrite(2, LOW);
  digitalWrite(3, LOW);
```

(continued)

LISTING 2-4:   *(continued)*

```
digitalWrite(4, LOW);
digitalWrite(5, LOW);
digitalWrite(6, LOW);
digitalWrite(7, LOW);

// Wait a bit
delay(Time);

// Increase the wait time
Time = Time + 10;

}
```

Using If Statements

An if statement lets you add conditional testing to your programs. In other words, it lets you execute certain statements only if a particular condition is met. This type of conditional processing is an important part of any but the most trivial of programs.

Every if statement must include a *conditional expression* that lays out a logical test to determine whether the condition is true or false. For example:

```
x > 5
```

This condition is true if the value of the variable x is greater than 5. If x has a value that is equal to or less than 5, the condition is false.

One of the oddities about the C programming language is that it requires you to use two equal sign to test for equality. For example:

```
x == 5
```

This condition is true if the value of the variable x is exactly 5. If the value is anything other than 5, the condition is false.

The basic form of an if statement looks like this:

```
if (condition)
  statement
```

Note that the condition must be enclosed in parentheses.

Here's a simple example:

```
if (Time == 500)
  Time = 1000;
```

This if statement looks at the value of the Time variable and changes the value to 1000 if the current value is 500.

The *statement* part of the if statement can be either a single C statement or a group of statements. If you use a group of statements, you must enclose them in curly braces. For example:

```
if (Time == 500)
{
  Time = 1000;
  Counter = 0;
}
```

In this example, two statements are executed if the value of the Time variable is 500.

The if statement can also include an else component that provides one or more statements that are executed if the condition is *not* true. For example:

```
if (Time == 500)
  Time = 1000;
else
  Time = 500;
```

In this example, one of two things happens: If the Time variable has a value of 500, the Time variable's value is set to 1000; but if the Time variable has a value other than 500, the Time variable's value is set to 500.

As in the if component, the else component can have more than one statement. In that case, the statements must be bounded by braces, like this:

```
if (Timer == 500)
{
  Timer = 1000;
  Counter = 0;
}
Else
```

```
{
  Timer = 500;
  Counter = Counter + 1;
}
```

One final thing to know about if statements is that one if statement can be contained within another. This arrangement is called *nesting*, and it's very useful for implementing complex decisions.

Listing 2-5 shows a program that cleverly uses a nested if statement to flash the LEDs in an alternating pattern of three quick flashes followed by three longer flashes. The program uses a variable named Time to control the delay interval (250 for the quick flashes, 500 for the longer flashes) and a second variable named Counter to count how many times the LEDs have been flashed. The nested if statement looks like this:

```
if (Counter == 3)
{
  if (Time = 250)
  {
    Time = 500;
  }
  else
  {
    Time = 250;
  }
  Counter = 0;
}
```

Here, the first if statement is used to change the Time variable and reset the Counter variable to zero whenever the Counter variable reaches 3. Within this if statement, a second if statement checks the value of the Time variable and sets it to 500 if the variable's current value is 250; otherwise, the Time variable is set to 500. In essence, this inner if statement toggles the Time variable between 250 and 500 each time the Counter variable reaches 3.

LISTING 2-5: The LED Flasher Program with an IF statement

```
// The LED Flasher Program
// Doug Lowe
// December 2, 2021
//
// This program flashes LEDs connected to digital pins 0
// through 7. The LEDs are alternately flashed quickly 3
// times, then more slowly 3 times.
```

```
// The Time and Counter variables
int Time;
int Counter;

void setup() {

  // Set pins 0 through 7 to OUTPUT
  pinMode(0, OUTPUT);
  pinMode(1, OUTPUT);
  pinMode(2, OUTPUT);
  pinMode(3, OUTPUT);
  pinMode(4, OUTPUT);
  pinMode(5, OUTPUT);
  pinMode(6, OUTPUT);
  pinMode(7, OUTPUT);

  // Initialize the Time and Counter variables
  Time = 500;
  Counter = 0;

}

void loop() {

  // Turn all LEDs on
  digitalWrite(0, HIGH);
  digitalWrite(1, HIGH);
  digitalWrite(2, HIGH);
  digitalWrite(3, HIGH);
  digitalWrite(4, HIGH);
  digitalWrite(5, HIGH);
  digitalWrite(6, HIGH);
  digitalWrite(7, HIGH);

  // Wait a bit
  delay(Time);

  // Turn all LEDs off
  digitalWrite(0, LOW);
  digitalWrite(1, LOW);
  digitalWrite(2, LOW);
  digitalWrite(3, LOW);
  digitalWrite(4, LOW);
```

(continued)

LISTING 2-5: *(continued)*

```
digitalWrite(5, LOW);
digitalWrite(6, LOW);
digitalWrite(7, LOW);

// Wait a bit
delay(Time);

// Determine the correct wait time
if (Counter == 3)
{
  if (Time == 250)
  {
    Time = 500;
  }
  else
  {
    Time = 250;
  }
  Counter = 0;
}

// Increment the counter
Counter = Counter + 1;

}
```

Using While Loops

As you already know, the main part of any Arduino program is a function called loop, which the Arduino calls repeatedly as long as the program runs. So, you're already familiar with the basic concept of *looping*, which is an essential element of nearly all computer programs, regardless of the programming language used or the type of device on which the program is running.

In addition to the basic looping mechanism provided by the loop method, you can provide additional forms of looping within the setup or loop methods. The C language provides several ways to create program loops. The most basic is called a *while* loop, which is simply a loop that repeats continuously as long as some condition is met.

The basic form of a while loop is this:

```
while (condition)
  statement
```

Note that the *statement* part of the while loop can be a single statement or a series of statements enclosed in braces.

Here's an example of a while loop that initializes pins 0 through 7 as OUTPUT pins:

```
int pin = 0;
while (pin < 8)
{
  pinMode(pin, OUTPUT);
  pin = pin + 1;
}
```

On each execution of this loop, the setMode method is called to set the mode of the pin indicated by the pin variable to OUTPUT. Then the value of the pin variable is increased by 1. So, the first time through the loop, the mode of pin 0 is set. The second time, pin 1 is set. And so on, until pin 7 is set. On that execution of the loop, the pin variable is increased to 8. Then, on the next execution of the loop, the condition is false and the loop ends.

TECHNICAL STUFF

This business of adding one to a variable in a loop is so common that the C programming language has a helpful shortcut called the *increment operator*. The increment operator is simply two plus signs in a row that you write *after* a variable name instead of before it. For example:

```
pin++;
```

Here's what the complete loop looks like using the increment operator:

```
int pin = 0;
while (pin < 8)
{
  setMode(pin, OUTPUT);
  pin++;
}
```

Listing 2-6 shows a version of the LED flasher that flashes the LEDs one at a time for half a second each, starting with the LED on pin 0 and proceeding through the LED on pin 7. Notice that because of the while statement inside of the loop function, the program actually flashes all eight of the LEDs on each execution of the loop function.

LISTING 2-6: **LED Flasher with a while Loop**

```
// The LED Flasher Program
// Doug Lowe
// December 2, 2021
//
// This program flashes LEDs connected to digital pins 0
// through 7. The LEDs on each pin are flashed one at a
// time in sequence from pin 0 to pin 7 for half a
// second each.

// The Time and Pin variables
int Time;
int Pin;

void setup() {

  // Set pins 0 through 7 to OUTPUT
  Pin = 0;
  while (Pin < 8)
  {
    pinMode(Pin, OUTPUT);
    Pin++;
  }

  // Initialize the Time variable
  Time = 500;

}

void loop() {

  Pin = 0;
  while (Pin < 8)
  {
    digitalWrite(Pin, HIGH);
    delay(Time);
    digitalWrite(Pin, LOW);
    Pin++;
  }
}
```

Using For Loops

A for loop is a special type of looping statement that automatically keeps a counter variable. For loops are ideal when you want to execute a loop a certain number of times or when you want to perform an action on multiple I/O pins.

The basic structure of a for loop looks like this:

```
for (initialize; test; increment)
  statement
```

The following paragraphs describe what the three parts in the parentheses do:

» The *initialize* part initializes the loop and establishes the counter variable that will be used. The variable can already exist or it can be declared in the initialize part. Typically, the variable is given a starting value here.

» The *test* part is a condition test that is used to determine when the loop ends. Typically, the condition will test for a certain value of the counter variable. The loop keeps executing as long as the condition is true. When the condition tests false, the loop ends.

» The *increment* part is used to increment the counter variable.

Here's a simple example that sets the mode of pins 0 through 7 to OUTPUT:

```
for (int Pin = 0; Pin < 8; Pin++)
{
  pinMode(Pin, OUTPUT);
}
```

In this example, Pin is declared within the for loop as the counter variable and is given an initial value of zero. The loop continues to execute as long as the Pin variable is less than 8, and after each execution of the loop, the Pin variable is incremented by 1. In this way, the loop sets the mode for pins 0 through 7.

One interesting feature of a for loop is that you can count backward. The easiest way to do this is to use the *decrement operator*, which is similar to the increment operator you learn about earlier in this chapter but which uses two minus signs instead of two plus signs. As you might guess, the decrement operator subtracts one rather than adds one.

To count backward, you should set the initial value of the counter variable to the value you want to start with and test for the lower limit value in the condition. For example:

```
for (int Pin = 7; Pin >= 0; Pin--)
{
  digitalWrite(Pin, HIGH);
  delay(500);
  digitalWrite(Pin, LOW);
}
```

This example flashes the LEDs in reverse order, from pin 7 to pin 0. (Notice the >= condition test, which tests True if the value of the Pin variable is either greater than or equal to zero.)

You can also skip-count with a for loop. For example, the following loop flashes every other LED (that is, the LEDs on pins 0, 2, 4, and 6):

```
for (int Pin = 0; Pin < 8; Pin = Pin + 2)
{
  digitalWrite(Pin, HIGH);
  delay(500);
  digitalWrite(Pin, LOW);
}
```

Listing 2-7 shows a version of the LED Flasher program that uses a pair of for loops to flash the LEDs first in one direction and then in the opposite direction. This creates an effect similar to the spooky electronic eyes on the evil Cylons in the old TV series *Battlestar Galactica*. (To achieve the menacing Cylon effect, I sped up the loop by decreasing the Time variable from 500 to 50 milliseconds.)

The setup function uses a for loop to initialize the output pins. Then, in the loop function, two for loops are used. The first for loop flashes the LEDs in sequence from pins 0 to 7. The second for loop flashes the LEDs in reverse order, from pins 6 through 1. Notice that the second for loop doesn't flash all eight pins. If it did, pins 0 and 7 would actually flash twice with each call of the loop function, which would mar the timing of the effect.

LISTING 2-7: **The LED Flasher Program with for Loops**

```
// The LED Flasher Program
// Doug Lowe
// December 2, 2021
//
```

```
// This program flashes LEDs connected to digital pins 0
// through 7. The LEDs on each pin are flashed one at a
// time in forward sequence from pin 0 to pin 7 and then
// backward, each for half a second.

// The Time variable
int Time;

void setup() {

  // Set pins 0 through 7 to OUTPUT
  for (int Pin = 0; Pin < 8; Pin++)
  {
    pinMode(Pin, OUTPUT);
  }

  // Initialize the Time variable
  Time = 50;

}

void loop() {

  // Flash pins 0 through 7 forward
  for (int Pin = 0; Pin < 8; Pin++)
  {
    digitalWrite(Pin, HIGH);
    delay(Time);
    digitalWrite(Pin, LOW);
  }

  // Flash pins 6 through 1 backward
  for (int Pin = 6; Pin > 0; Pin--)
  {
    digitalWrite(Pin, HIGH);
    delay(Time);
    digitalWrite(Pin, LOW);
  }
}
```

Crafting Your Own Functions

In the previous sections, I explain how to create the setup and loop functions that are required by every Arduino program, and how to use standard library functions such as pinMode, digitalWrite, and delay. In the last section of this chapter, I tell you how to create your own functions that you can call from the setup or loop functions to simplify your code.

Why would you want to do this? Here's a simple example: In Listing 2-7, you may have noticed that the same three lines of code were repeated in the two for loops used in the loop function. Rather than repeat those three lines of code, you could create your own function that contains those lines. You might call this function flashLED. Then all you'd have to do in the for loop is call the flashLED function.

To create a function, you must declare the function much the same as you declare the setup and loop functions. The main difference is that when you create your own functions, you'll likely need to pass data to the functions via arguments. For example, you'll need to pass the pin number to the flashLED function. You may also want to pass the delay time as an argument.

Here's an example of a function that turns an LED on, waits for a certain period of time, and then turns the LED off:

```
void flashLED(int Pin, int Time)
{
  digitalWrite(Pin, HIGH);
  delay(Time);
  digitalWrite(Pin, LOW);
}
```

As you can see, the first line of the function declares the arguments in a manner that's similar to how variables are declared. The first argument is of type int and is named Pin, while the second argument is also of type int but is named Time.

Listing 2-8 shows a version of the Cylon-eyes program that uses a function to flash each LED. Notice that this version of the program eliminates the global Time variable, because the duration of the flash is passed to the flashLED function as a parameter.

LISTING 2-8: **The LED Flasher Program with a Function**

```
// The LED Flasher Program
// Doug Lowe
// November 7, 2021
//
// This program flashes LEDs connected to digital pins 0
// through 7. The LEDs on each pin are flashed one at a
// time in forward sequence from pin 0 to pin 7 and then
// backward, each for half a second.

void setup() {

  // Set pins 0 through 7 to OUTPUT
  for (int Pin = 0; Pin < 8; Pin++)
  {
    pinMode(Pin, OUTPUT);
  }

}

void loop() {

  // Flash pins 0 through 7 forward
  for (int Pin = 0; Pin < 8; Pin++)
  {
    flashLED(Pin, 50);
  }

  // Flash pins 6 through 1 backward
  for (int Pin = 6; Pin > 0; Pin--)
  {
    flashLED(Pin, 50);
  }
}

void flashLED(int Pin, int Time)
{
  digitalWrite(Pin, HIGH);
  delay(Time);
  digitalWrite(Pin, LOW);
}
```

Besides accepting arguments, a function can also return a value. There's not much use for functions that return values in programs that simply flash LEDs, so let's consider a different example. Suppose your Arduino project has a need to calculate the area of a rectangle, where the height and the width of the rectangle are represented by int values. You could calculate the area as follows:

```
int width = 5;
int height = 10;
int a;
a = width * height;
```

Another way to do this would be to create a function that accepts the width and the height as arguments and returns the area of the rectangle. Here's how you can write the function:

```
int area(int width, int height)
{
  int area;
  area = width * height;
  return area;
}
```

In this example, the function area is given a return type of int, meaning that the function will return an integer value when called. Then, in the body of the function, a return statement is used to end the function and provide the return value.

Here's how you could use this function to calculate the area:

```
int x = 5;
int y = 10;
int a;
a = area(x, y);
```

It's important to note here that the names of the variables you specify as arguments when you call the function do not have to match the names of the arguments specified within the function. Instead, the arguments are positional: The first argument you pass to the function will be used as the width, and the second will be used as the height.

Chapter **3**

More Arduino Programming Tricks

I n this chapter, you learn some additional Arduino programming techniques that will become invaluable in your Arduino projects. Specifically, you learn how to handle input data in the form of push buttons, how to generate random numbers that will make your programs more interesting by adding a degree of randomness, and how to read the value of a potentiometer.

Using a Push Button with an Arduino

In Chapter 2 of this minibook, you learn how to connect an LED to an Arduino I/O pin and turn the LED on or off by using `digitalWrite` function in an Arduino program. The `digitalWrite` function uses the I/O pins as output pins by setting the status of an I/O pin to HIGH or LOW so that external circuitry (such as an LED) can react to the pin's status.

But what if you want to use an I/O pin as an input instead of an output? In other words, what if you want the Arduino to react to the status of an external circuit instead of the other way around? The easiest way to do that is to connect a push button to an I/O pin. Then, you can add commands to your Arduino program to detect whether the push button is pressed.

There are two ways to connect a push button to a Arduino I/O pin:

» **Active-high:** This connection places +5 V on the I/O pin when the push button is pressed. When the button is released, the I/O pin sees 0 V.

» **Active-low:** This connection sees +5 V when the push button is not pressed. When you press the push button, the +5 V is removed, and the I/O pin sees no voltage.

Figure 3-1 shows examples of both active-high and active-low push buttons. In the Active-high circuit, the I/O pin is connected to ground through resistor R1 when the push button is not pressed. Thus, the voltage at the I/O pin is 0. When the push button is pressed, the I/O pin is connected to +5 V through R1, causing the I/O pin to see +5 V. As a result, the I/O pin is LOW when the button is not pressed and HIGH when the button is pressed.

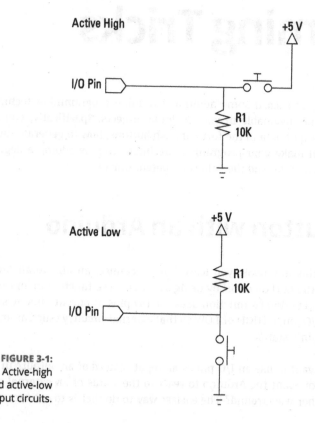

FIGURE 3-1:
Active-high and active-low input circuits.

In the active-low circuit, the I/O pin is connected to +5 V through R1, causing the I/O pin to go HIGH. But when the button is pressed, the current is shorted to ground through R1, causing the voltage at the I/O pin to drop to zero. Thus, the I/O pin is HIGH when the button isn't pressed and LOW when the button is pressed.

WARNING

Note that in both circuits, R1 is connected between the +5 V and ground to prevent a short circuit. Without this resistor, the Arduino would likely be damaged when the button is pressed.

TECHNICAL STUFF

In an active-high circuit, R1 is called a *pull-down* resistor because it pulls the current from the I/O pin down to zero when the push button isn't depressed. In an active-low circuit, R1 is called a *pull-up resistor* because it pulls the voltage at the I/O pin up to Vdd (+5 V) when the push button isn't depressed.

Checking the Status of a Switch in Arduino

Once you've connected a switch to an Arduino I/O pin, you need to know how an Arduino program can determine whether a switch is open or closed. The easiest way to do that is to first designate the pin as an input pin by calling the pinMode function, and then checking the status of the input pin by calling the digitalRead function. The digitalRead function returns 0 or 1 depending on the current status of the input pin.

For example, you can designate pin 10 as an input pin like this:

```
pinMode(10, INPUT);
```

Then, to get the status of the pin, you could call digitalRead like this:

```
int Pin10Status;
Pin10Status = digitalRead(10);
```

The digitalRead function returns the current status of pin 10 and stores it in an int variable named Pin10Status. You can then test the status of the pin in an if statement:

```
if (Pin10Status == 1)
{
  // The switch status is high
}
```

In many cases, you can shorten your code by skipping the need for a variable to store the result of the pinRead function, like this:

```
if (digitalRead(10))
{
  // The switch status is high
}
```

Notice in the preceding if statement that two consecutive right parentheses are required. The first one closes the list of arguments passed to the digitalRead function; the second one closes the expression in the if statement.

TIP

If you prefer, you can use bool variables to determine the pin status. A bool variable can have a value of true or false. (The word bool is short for Boolean.) Using a bool variable, you can test the status of a switch like this:

```
bool Pin10Status;
Pin10Status = digitalRead(10);
if (Pin10Status == true)
{
  // The switch status is true (high)
}
```

Here's an example that lights an LED on pin 0 if the user presses the button on pin 10 and turns the LED off when the button is released:

```
if (digitalRead(10))
{
  digitalWrite(0, HIGH);
}
Else
{
  digitalWrite(0, LOW);
}
```

Here, the LED on pin 0 is made HIGH if the button is pressed and LOW if the button isn't pressed.

REMEMBER

Remember that whenever you test the status of an input pin that's connected to a switch, determining whether the user has pressed the switch depends on whether the switch is configured in an Active High or Active Low circuit. In an Active High circuit, the digitalRead function will return HIGH if the user presses the switch. In an Active Low circuit, LOW will be returned when the switch is pressed.

Listing 3-1 shows an interesting program that works with an Arduino that has a push button switch connected to pin 10 and LEDs connected to pins 0 and 1. The program flashes the LED connected to pin 0 on and off at quarter-second intervals until the push button switch is depressed. Then, it flashes the LED on pin 1. The program assumes the button is connected in an Active High circuit.

LISTING 3-1:	**The Push Button Program**

```
// The LED Flasher Program with a Push Button
// Doug Lowe
// December 3, 2021
//
// This program flashes one of two LEDs connected to
// pins 0 and 1, depending on whether the push button on
// pin 10 is pressed.

void setup() {
  pinMode(0, OUTPUT);
  pinMode(1, OUTPUT);
  pinMode(10, INPUT);
}

void loop() {
  if (digitalRead(10) == 1)  // Button is pressed
  {
    digitalWrite(1, HIGH);
    delay(250);
    digitalWrite(1, LOW);
    delay(250);
  }
  else                       // Button is not pressed
  {
    digitalWrite(0, HIGH);
    delay(250);
    digitalWrite(0, LOW);
    delay(250);
  }
}
```

Project 45 shows how to build a simple circuit you can use to test the program in Listing 3-1, and Figure 3-2 shows the completed circuit.

FIGURE 3-2:
A circuit for
testing an
active-high push
button switch
(Project 45).

Project 45: A Push-Button Controlled Arduino LED Flasher

In this project, you connect a breadboard with two LEDs and a push button to an Arduino UNO board.

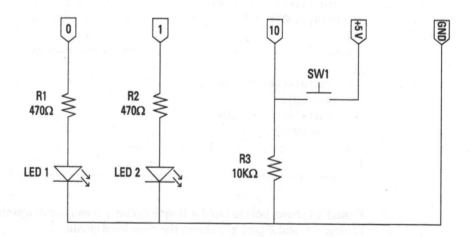

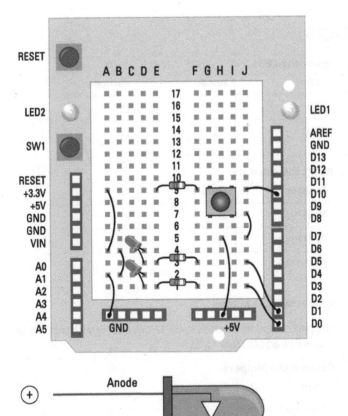

RESET
LED2
SW1
RESET
+3.3V
+5V
GND
GND
VIN
A0
A1
A2
A3
A4
A5

A B C D E   F G H I J

17
16
15
14
13
12
11
10
9
8
7
6
5
4
3
2

GND   +5V

LED1
AREF
GND
D13
D12
D11
D10
D9
D8
D7
D6
D5
D4
D3
D2
D1
D0

Anode
(+)
(−)
Cathode

Parts

» One computer with Arduino IDE installed

» One Arduino UNO board

» One A-B USB cable

» One UNO prototyping shield

» Two red LEDs

» One normally open DIP breadboard push button

» Two 470 Ω resistors (yellow-violet-brown)

» One 10 kΩ resistor

» Eight jumper wires

Steps

1. **Insert the LEDs.**

LED	Anode	Cathode
LED1	D1	D2
LED2	D3	D4

2. **Insert the resistors.**

Resistor	From	To
R1 (470 Ω)	E1	F1
R2 (470 Ω)	E3	F3
R3 (10 kΩ)	E9	F9

3. **Insert the push button.**

   The pins should be inserted in G7, I7, G9, and F9 such that the switch opens and closes across rows 7 and 9.

4. **Connect the jumpers.**

From	To
GND header	A2
B2	B4
A4	A9
+5 V header	H5
J5	J7
Digital pin D0	J1
Digital pin D1	J3
Digital pin D10	J9

5. **Connect the assembled prototyping shield to the Arduino UNO.**

6. **Connect the UNO to the computer.**

   Use the type-B USB connector.

7. **Upload the Flashing LED program (see Listing 3-1) to the UNO.**

   LED1 will flash on and off a quarter-second intervals. If you press the push button, LED2 will flash instead.

Randomizing Your Programs

Many computer-controlled applications require a degree of randomness to their operation. A classic example is the *Indiana Jones* ride at Disneyland and Disney World. Every time you go on this ride, the adventure is slightly different. At the start, there are three doors that your vehicle can drive through; the exact door chosen for your ride is determined randomly. Many other details of the ride are randomly varied in an effort to make the adventure slightly different every time you ride it.

You can add a bit of randomness to your own Arduino programs by using the random function. This function returns a random number whose value lies within a range that you specify via a pair of arguments.

The first argument indicates the minimum value you want returned. Somewhat confusingly, the second argument is one greater than the maximum value you want. For example, if you want a value within the range of 1 to 10, you would call the random function like this:

```
random(1,11);
```

Listing 3-2 shows a sample program that lights LED1 (pin 0) until the push button on pin 14 is pressed. Then it turns off LED 1 and flashes LED2 (pin 1) a random number of times (between 1 and 10). It then switches back to LED1. This program will work with the circuit you built for Project 45.

LISTING 3-2: **The Random Program**

```
// Random Program
// Doug Lowe
// December 3, 2021
//
// This program turns on the LED at pin 0. When the user
// presses the button at pin 10, the LED at pin 0 is
// turned off and the LED at pin 1 is flashed a
// random number of times (between 1 and 10).
// Then the LED at pin 0 is turned on again.

void setup() {
  pinMode(0, OUTPUT);
  pinMode(1, OUTPUT);
  digitalWrite(0, HIGH);
  digitalWrite(1, LOW);
}
```

```
void loop() {
  if (digitalRead(10) == 1)
  {
    digitalWrite(0, LOW);
    int rnd = random(1, 11);
    for (int i = 1; i < rnd; i++)
    {
      digitalWrite(1, HIGH);
      delay(250);
      digitalWrite(1, LOW);
      delay(250);
    }
    digitalWrite(0, HIGH);
  }
}
```

Each time you press the push button in this program, the LED on pin 1 lights up for a different length of time between 1 and 5 seconds.

TECHNICAL STUFF It turns out that the random function isn't really all that random. The random function applies a complex mathematical calculation to create a sequence of numbers that appear to be random. However, the numbers aren't actually random. That's because when you start the program, the random number generator is seeded with a default starting value, and the random function will generate the same sequence of numbers.

Thus, the program in Listing 3-2 always generates the same sequence of random delays. In particular, the sequence for the first ten button presses will always be as follows:

>> **First press:** 7 flashes

>> **Second press:** 9 flashes

>> **Third press:** 3 flashes

>> **Fourth press:** 8 flashes

>> **Fifth press:** 2 second

>> **Sixth press:** 4 flashes

>> **Seventh press:** 8 flashes

>> **Eighth press:** 3 flashes

>> **Ninth press:** 9 flashes

>> **Tenth press:** 5 flashes

This sequence appears fairly random, but every time you reset the program and start over, the sequence will be identical.

You can get around this lack of true randomness by calling the randomSeed function to provide a *seed value* for the random function. The seed value provides a starting point within the sequence of random numbers generated by the random function. If you use a different seed value each time the program runs, you'll get a different random sequence each time.

So, how do you use a different seed value? You can do that easily enough by using the millis function, which returns the number of milliseconds that the program has been running. If you call the millis function when the user presses a button, and use its return value to seed the random number calculation, you'll get a truly random result.

Listing 3-3 shows an improved version of the random program that uses this technique to create a truly random delay. It turns out that only one additional line of code is needed to properly randomize the delay:

```
randomSeed(millis());
```

This line reseeds the random number generator each time the button is pressed. Thus, a true random value is generated.

LISTING 3-3: **An Improved Version of the Random Program**

```
// Improved Random Program
// Doug Lowe
// December 3, 2021
//
// This program turns on the LED at pin 0. When the user
// presses the button at pin 10, the LED at pin 0 is
// turned off and the LED at pin 1 is flashed a
// random number of times (between 1 and 10).
// Then the LED at pin 0 is turned on again.
// The random seed is reset each time the
// button is pressed.
```

```
void setup() {
  pinMode(0, OUTPUT);
  pinMode(1, OUTPUT);
  digitalWrite(0, HIGH);
  digitalWrite(1, LOW);
}

void loop() {
  if (digitalRead(10) == 1)
  {
    randomSeed(millis());
    digitalWrite(0, LOW);
    int rnd = random(1, 11);
    for (int i = 1; i < rnd; i++)
    {
      digitalWrite(1, HIGH);
      delay(250);
      digitalWrite(1, LOW);
      delay(250);
    }
    digitalWrite(0, HIGH);
  }
}
```

Reading a Value from a Potentiometer

As you know, a *potentiometer* (often called a *pot*) is simply a variable resistor with a knob you can turn to vary the resistance. Pots of various types are often used as input devices for Arduino projects. For example, you might use a simple pot to control the speed of a pair of flashing LEDs: As you turn the pot's knob, the rate at which the LEDs flash changes.

The Arduino can read the value of a pot directly using any of its six *analog input ports*. On the Uno board, these ports are located on a six-pin header that's on the opposite side of the digital ports. Each of these ports converts a voltage level between 0 V and a reference voltage to a integer between 0 and 1023. Although you can change the reference voltage for an analog port, we'll use the standard reference voltage of +5 V.

Figure 3-3 shows how simple it is to connect a potentiometer to an Arduino. Here, the two outside terminals of the potentiometer are connected to ground and +5 V, respectively, and the middle terminal is connected to an analog input port.

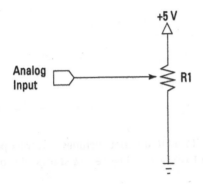

+5 V

Analog
Input
→ R1

FIGURE 3-3:
Connecting a pot
to an Arduino
I/O pin.

To read the value of an analog input in an Arduino program, you use the analogRead function. This function accepts the number of the port to be read as an argument and returns a number between 0 and 1023 to indicate the voltage at the input relative to the reference voltage of +5 V. Thus, if the voltage is 2.5 V, the analogRead function will return 511.

Listing 3-4 shows a simple program that alternately flashes LEDs connected to pins 0 and 1. The rate at which the LEDs flash is set by a potentiometer on analog pin 0. As you can see, the program simply uses the value returned by the analogRead function as the delay between flashes. Thus, as the user turns the potentiometer's knob, the flash rate varies between 0 and a tad over one second.

| LISTING 3-4: | An LED Flashing Program That Uses a Pot |

```
// Variable Rate Flasher
// Doug Lowe
// December 3, 2011
//
// This program flashes a pair of LEDs on pins 0 and 1
// at a rate set by a potentiometer on analog input pin
// 0.

int Rate;

void setup() {
  pinMode(4, OUTPUT);
  pinMode(6, OUTPUT);
}

void loop() {
  Rate = analogRead(0);
  digitalWrite(4, LOW);
  digitalWrite(6, HIGH);
```

```
  delay(Rate);
  digitalWrite(6, LOW);
  digitalWrite(4, HIGH);
  delay(Rate);
}
```

Project 46 shows how to build a circuit that includes a 10 MΩ potentiometer so that you can test the code in Listing 3-4. Figure 3-4 shows the completed circuit.

FIGURE 3-4:
A circuit that uses
a potentiometer
to control
flashing LEDs
(Project 46).

Project 46: A Variable-Rate LED Flasher

In this project, you connect a breadboard with two flashing LEDs and a potentiometer that controls the rate at which the LEDs flash.

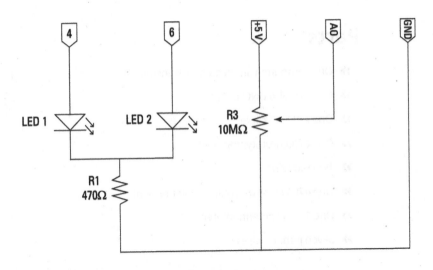

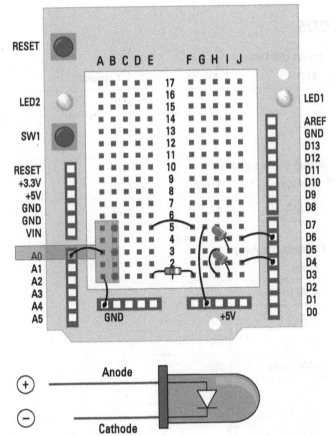

Parts

>> One computer with Arduino IDE installed

>> One Arduino UNO board

>> One A-B USB cable

>> One UNO prototyping shield

>> Two red LEDs

>> One 470 Ω resistor (yellow-violet-brown)

>> One 10 MΩ potentiometer

>> Seven jumper wires

Steps

1. **Insert the LEDs.**

LED	Anode	Cathode
LED1	I2	I1
LED2	I4	I3

2. **Insert the resistor.**

Resistor	From	To
R1 (470 Ω)	E1	F1

3. **Connect the jumpers.**

From	To
GND header	A1
Analog pin A0	A3
E5	F5
+5 V header	G5
H1	H3
Digital pin D6	J4
Digital pin D4	J2

4. **Insert the potentiometer.**

The pins should be inserted in B1, B3, and B5.

5. **Connect the assembled prototyping shield to the Arduino UNO.**

6. **Connect the UNO to the computer.**

Use the type-B USB connector.

7. **Upload the Variable-Rate Flasher (see Listing 3-4) to the UNO.**

The LEDs will alternately flash at a rate determined by the potentiometer. When you turn the knob on the potentiometer, the rate of flashing will increase or decrease.

Chapter **4**

An Arduino Proximity Sensor

I n this chapter, you learn how to work with two new devices that can be easily connected to an Arduino. The first is an ultrasonic range finder, which uses a short ultrasonic pulse to determine the distance to a nearby object. The second is a liquid crystal display (LCD) device that lets you display readable output from your Arduino.

Then you build a project that uses these two devices together to display the distance between the Arduino and a nearby object. Have fun!

Using an Ultrasonic Range Finder

An *ultrasonic range finder* is a device that determines the distance to a nearby object by bouncing ultrasonic sound waves off of the object. In short, the range finder emits a short burst of ultrasonic sound (that is, sound in a frequency range that cannot be heard by humans) and then listens for the sound reflected off the object. The amount of time that elapses between the initial burst and the detected reflection can be used to determine how far away the object is from the detector.

Looking at the HC-SR04 Range Finder

Figure 4-1 shows a commonly used ultrasonic range finder called an HC-SR04. You can purchase this range finder online for just a few dollars from a variety of online retailers, including Amazon and Newegg (www.newegg.com).

FIGURE 4-1:
An HC-SR04
Range Finder.

The HC-SR04 is built on a small circuit board (approximately 2 inches long and ¾ inch tall) with a four-pin connector that can easily attach to a breadboard. The HC-SR04 includes a transmitter that can generate ultrasonic sounds, a receiver that can listen for ultrasonic sounds, and circuitry that measures how much time elapses between transmission of a sound pulse and the reception of the sound's echo.

Table 4-1 defines the four pins on the HC-SR04.

TABLE 4-1   ## Pinouts for the HC-SR04 Ultrasonic Range Finder

Pin	Name	Function
1	VCC	+5 V power
2	TRIG	Triggers the generation of an ultrasonic pulse to measure the distance to a nearby object
3	ECHO	Following a trigger pulse, generates an echo pulse whose duration represents the distance to the object
4	GND	Ground

To connect the HC-SR04 to an Arduino, first connect pins 1 and 4 to power and ground, respectively. Then, connect pins 2 and 3 to any of the Arduino's digital I/O pins.

After the HC-SR04 is connected to the Arduino, you can initiate a range check by sending a short HIGH pulse to the TRIG pin. This causes the HC-SR04 to emit a short ultrasonic pulse over its transmitter and then listen for the echo on its receiver. When the echo has been received, the HR-SR04 sends a pulse over its ECHO pin. The duration of this pulse is the same as the amount of time it took for the ultrasonic echo to be received.

You can then use the duration of the ECHO pulse to determine the distance to the object that reflected the ultrasonic sound pulse. You'll have to do a little math, but nothing too complicated.

In the next few sections, I show you how to generate a trigger pulse for an HC-SR04, how to read the length of the echo pulse, and how to calculate the distance to the object based on the echo pulse length.

Generating a trigger pulse

The HC-SR04 requires a short pulse on its TRIG pin to initiate the range-finding process. This pulse must be at least 10 microseconds, but it's typically 100 microseconds.

To generate a pulse on an Arduino digital I/O pin, you simply set the output pin to HIGH, delay for the desired length of the pulse, and then set the pin to LOW. Unfortunately, the delay function (see Chapter 1 of this minibook) works in milliseconds (thousands of a second) rather than in microseconds (millionths of a second). You need to use the delayMicroseconds function to create a delay shorter than 1 millisecond.

Here's a complete bit of code that generates a 100-microsecond pulse on pin 9:

```
digitalWrite(9, LOW);
delayMicroseconds(20);
digitalWrite(9, HIGH);
delayMicroseconds(100);
digitalWrite(9, LOW);
```

This code begins by setting the output pin to LOW and waiting for 20 microseconds. This action gives the pulse a clean start. Then pin 9 is set to HIGH for 100 microseconds before it's returned to LOW.

Assuming that pin 9 is connected to the TRIG pin of the HC-SR04, this code will initiate the range check.

Reading pulse input

The Arduino programming library includes a function called pulseIn, which reads an input pulse on a digital input pin and returns the duration of the pulse as an integer. The pulseIn function accepts two arguments. The first argument indicates the pin to read (which must be previously configured as an input pin). The second argument indicates whether you want to listen for a HIGH or LOW pulse.

An optional third argument specifies a time-out value; if a pulse hasn't been received before the time-out value elapses, the pulseIn function returns 0.

The pulseIn function waits for the pulse to begin, and begins counting microseconds until the pulse ends. It then returns the counted length of the pulse.

Here's an example of how you can use it, assuming that pin 9 is connected to the TRIG pin and pin 8 is connected to the ECHO pin:

```
int duration;
pinMode(8, INPUT);     // ECHO pin
pinMode(9, OUTPUT);    // TRIG pin

// Trigger the HC-SR04
digitalWrite(9, LOW);
delayMicroseconds(20);
digitalWrite(9, HIGH);
delayMicroseconds(100);
digitalWrite(9, LOW);

// Read the ECHO pulse
duration = pulseIn(8, HIGH)
```

The preceding code starts by setting the I/O mode for the TRIG and ECHO pins. It then sends a 100-microsecond trigger pulse to the TRIG pin, and then reads the echo pulse on the ECHO pin. When this code completes, the duration variable will contain the length of the echo pulse in microseconds.

Doing the math

When you've obtained the length of the echo pulse from the HC-SR04, you can use the result to calculate the distance to the object that reflected the range finder's ultrasonic sound.

You first need to decide what unit of measure you want to use. For this chapter, measure the distance in centimeters. You could work in inches and feet, but the math will be much simpler if you work in centimeters.

To convert the duration of the echo pulse to a distance in centimeters, you need to know a key fact: Sound travels at a speed of approximately 340 meters per second. This is assuming the sound is traveling through ordinary air at an ambient temperature of 68°F, but the variance for higher or lower temperatures isn't significant for our purposes here.

There are 100 centimeters in a meter, so sound travels about 34,000 centimeters per second. The pulse length is measured in microseconds (millionths of a second), so divide 34,000 by 1 million to get 0.034. So, each microsecond of the echo pulse indicates that the sound has traveled 0.034 centimeter. If you multiply the pulse length by 0.034, the result indicates how far the sound has traveled.

Keep in mind, though, that the sound has traveled from the pulse transmitter to the nearby object, and then the echo has traveled back to the pulse detector. So, the actual distance between the range detector and the object whose echo is received is half the distance indicated by the pulse length. Rather than multiply the pulse length by 0.034, you multiply the pulse length by half that — 0.017 — to determine how far away the closest object is from the range finder.

So, here's what the math looks like, after you've read the pulse duration into the pulseLength **variable**:

```
distance = pulseLength * 0.017;
```

The preceding statement will save the distance in centimeters to the object in distance.

Before you put all of this together into a single program, I want to direct your attention to the second major component of this chapter's range finder project, the LCD that you'll use to display the distance detected by the range finder.

Using an LCD

Many different types of display devices are available for Arduino projects. Figure 4-2 shows one of the most common: a small liquid crystal display (LCD) that can show two lines of 16 characters, controlled by a popular LCD controller called the Hitachi HD44780. You can purchase one of these displays for just a few dollars from online retailers such as Amazon or Newegg (www.newegg.com).

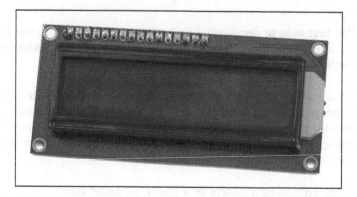

FIGURE 4-2:
A typical
2 x 16 LCD.

In this section, I show you how to connect this type of LCD to an Arduino board and how use the LCD in your Arduino sketches.

Connecting an LCD to Arduino

An HD44780-type LCD includes a 16-pin connection that you can plug directly into a breadboard. Table 4-2 lists the name and function of each of the 16 pins in the LCD's interface.

The LCD module can operate in two basic modes:

>> **8-bit:** In 8-bit mode, all eight of the data bit pins are used to send a single byte of data to the LCD. This mode is efficient and simple to use, but it requires that you connect all eight of these data pins to digital I/O pins on the Arduino, leaving only a few pins available for other purposes.

>> **4-bit:** In 4-bit mode, only four of the data bit pins are used (pins 11 through 14) to send each byte. Because each byte of data requires eight bits, it takes two reads of the four data bits to send a complete byte. This complicates the programming but reduces the number of Arduino pins that must be used. (Note that in 4-bit mode, the last four data bits are used, not the first four data bits. Thus, in 4-pin mode, data bits 4 through 7 are used.)

Fortunately, the complications of using the LCD in 4-bit mode are not your problem. The Arduino comes with a special library of functions that handle all those complexities. As a result, you'll almost always use the LCD in 4-bit mode.

In addition to the four data bits, two additional pins must be connected to Arduino digital I/O pins: the RS pin (pin 4) and the CLOCK pin (pin 6). Once again, the LCD programming library takes care of all the details of using these pins properly to display data on the LCD, so you don't need to worry about the details of how these pins work.

TABLE 4-2

Pinouts for an HD44780-Type LCD

Pin	Name	Function
1	GND	Ground.
2	VSS	+5 V.
3	VO	Controls the contrast on the display. Typically connected to the center pole of a potentiometer.
4	RS	Used to indicate whether data or a command is being sent to the device.
5	RW	Read/write, almost always connect to ground.
6	CLOCK	Also called Enable or just E. Used to trigger a read of the data pins.
7	D0	Data bit 0. (Not used in four-bit mode.)
8	D1	Data bit 1. (Not used in four-bit mode.)
9	D2	Data bit 2. (Not used in four-bit mode.)
10	D3	Data bit 3. (Not used in four-bit mode.)
11	D4	Data bit 4.
12	D5	Data bit 5.
13	D6	Data bit 6.
14	D7	Data bit 7.
15	LED+	Backlight +5 V.
16	LED-	Backlight ground.

Pins 1 and 2 must be connected to ground and +5 V, respectively, to provide basic power for the LCD, and pins 15 and 16 are also usually connected to power and ground, respectively, to energize the backlight, which makes the LCD easier to read.

Pin 3 is also commonly used to adjust the contrast of the display. You'll generally connect this pin to the center pole of a potentiometer (1 MΩ or so is usually appropriate), with the outside poles of the pot connected to +5 V and ground.

Programming the LCD

Arduino provides an entire library of functions designed to work directly with an HD44780 LCD. Table 4–3 lists a few of the functions that are available in this library. (For a complete list, go to www.arduino.cc/en/Reference/LiquidCrystal).

TABLE 4-3

Pinouts for an HD44780-Type LCD

Function	Function
LiquidCrystal()	Creates a variable of type LiquidCrystal, which you can then use to call other functions. The arguments passed to this function specify the Arduino pins that are used for the various LCD features.
begin()	Must be called to initialize the LCD before any information can be written to it. The arguments indicate how many columns and rows of characters can be displayed on the LCD.
clear()	Erases the LCD.
home()	Moves the cursor to the top left of the display.
setCursor()	Moves the cursor to a particular row and column position.
print()	Displays text on the LCD.

In order to use any of the functions listed in Table 4-3, your program must include the following statement:

```
#include LiquidCrystal.h
```

This statement is required to make the functions of the LiquidCrystal library available to your program.

Next, your program should create a variable of type LiquidCrystal. Creating a variable of a library type is similar to creating a variable of a built-in type such as int or float, except that you can provide arguments in the variable declaration. In the case of the LiquidCrystal data type, you provide a total of six arguments, representing the digital I/O pins that will be used for the RS, ENABLE, D4, D5, D6, and D7 pins. For example:

```
LiquidCrystal lcd(6, 7, 9, 10, 11, 12);
```

In this example, the variable will be named lcd; Arduino pins 6 and 7 will be used for RS and ENABLE; and pins 9, 10, 11, and 12 will be used for the data input.

After you've assigned the digital I/O pins to the LCD, you don't need to directly access those pins within your program. Instead, the other LiquidCrystal library functions will use those pins to control the LCD.

Having created an lcd variable, the next step is to initialize it using the begin function, like this:

```
lcd.begin(16, 2);
```

Here, the LCD is initialized for a display of two rows of 16 characters each.

You can then print data to the LCD, like this:

```
lcd.print("Hello, World!");
```

Here, the text "Hello, World!" is displayed on the LCD.

In some cases, you'll want to control exactly where text should be displayed. To do that, you can use the setCursor function, followed by a print function. For example, if you want to display the text "Hello, World!" beginning in column 5 of the second row, use this code:

```
lcd.setCursor(4, 1);
lcd.print("Hello, World!");
```

Note that both rows and columns are numbered starting with 0, not 1. So the fifth column is column 4, and the second row is row 1.

Finally, you may need to erase the entire contents of the LCD. To do that, use the clear function:

```
lcd.clear();
```

Here, the entire LCD screen is erased.

Now that I've explained the details of connecting an LCD to an Arduino board and controlling it from a sketch, let's look at a project that combines a range finder with an LCD board to display the distance from a nearby object and light an LED if the object gets too close.

Building a Proximity Sensor

In this section, you build a simple proximity sensor that uses an HC-SR04 ultrasonic range finder and an HD44780-compatible LCD. The ultrasonic range finder is used to track the distance to nearby objects; that distance is displayed on the LCD. If an object gets closer than 10 centimeters, an LED lights up to indicate that an object is close by.

Figure 4-3 shows the assembled project. The details for building the project, including the schematic diagram, are shown in Project 47, and the complete program to run the project is shown in Listing 4-1. Have fun!

FIGURE 4-3:
The assembled
Arduino
proximity sensor
(Project 47).

Project 47: An Arduino Proximity Sensor

In this project, you build a proximity sensor that displays the distance to a nearby object on an LCD panel and flashes an LED when the object is closer than 10 cm.

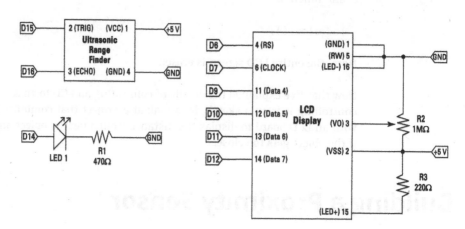

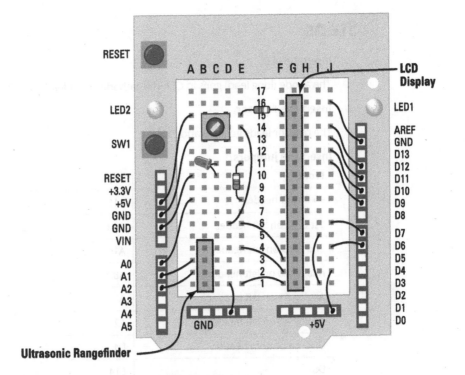

RESET

A B C D E   F G H I J   LCD Display

LED2   17 16 15 14 13 12 11 10 9 8 7 6 5 4 3 2 1   LED1

SW1   AREF
GND
D13
RESET   D12
+3.3V   D11
+5V   D10
GND   D9
GND   D8
VIN   D7
A0   D6
A1   D5
A2   D4
A3   D3
A4   D2
A5   GND   +5V   D1
D0

Ultrasonic Rangefinder

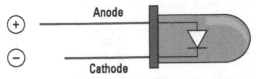

(+)   Anode

(−)   Cathode

Parts

>> One computer with Arduino IDE installed

>> One Arduino UNO board

>> One A-B USB cable

>> One UNO prototyping shield

>> One red LED

>> One HC-SR04-type Ultrasonic Range Finder

>> One HD44780-compatible 16 x 2 LCD

>> One 470 Ω resistor (yellow-violet-brown)

>> One 1 MΩ trim pot

>> 20 jumper wires

Steps

1. **Insert the LED.**

   The anode (long lead) goes in C11, the cathode (short lead) in C10.

2. **Insert the resistor.**

   The resistor leads go in E8 and E11.

3. **Insert the jumpers.**

From	To
+5 on left header	A15
1st GND on left header	A13
2nd GND on left header	A10
Analog A0 on left header	A8
Analog A1 on left header	A3
Analog A2 on left header	A2
GND on bottom header	D1
D6	E14
E6	F3
E4	F2
E1	F1
I1	I5
+5 V on bottom header	J2
Digital pin 6 on right header	J4
Digital pin 7 on right header	J6
Digital pin 9 on right header	J11
Digital pin 10 on right header	J12
Digital pin 11 on right header	J13
Digital pin 12 on right header	J14
GND on right header	J16

4. **Insert the trim pot.**

   The trim pot has three pins: two on the edges of one side and one in the middle of the opposite side. The two pins that are on the same side should go in B13 and B15. The single pin on the opposite side will go in D14.

5. **Attach the prototype shield to the UNO.**

6. **Insert the LCD.**

   Orient the LCD so that pin 1 is inserted into G1 and pin 16 is inserted into G16. The LCD will extend over the top of columns H, I, and J on the tiny breadboard and past the right side of the prototype board.

7. **Insert the Ultrasonic Range Finder.**

   Orient the range finder so that pin 1 (Vcc) is inserted into B4 and pin 4 (Gnd) is inserted into B1. The transmitter and receiver of the range finder should face to the left, away from the breadboard.

8. **Connect the UNO to the computer.**

   Use the type- B USB connector.

9. **Upload the Proximity Sensor program (see Listing 4-1) to the UNO.**

   The LCD will display the distance of the nearest object. If necessary, adjust the trim pot so you can read the display. When the distance is less than 10 cm, the LED will come on.

LISTING 4-1:	**The Proximity Sensor (Project 47)**

```
// Proximity Sensor
// Doug Lowe
// December 3, 2021


// This program uses an HC-SR04 Ultrasonic Range
// Detector and an HD44780-compatible LCD to detect and
// display proximity of nearby objects. The distance to
// the object is displayed on the LCD, and the cycle repeats
// every 50 milliseconds. If the object is closer than 10 cm, the
// LED comes on.

#include <LiquidCrystal.h>


// Arduino Digital Pin Constants

int US_Trig = 15;       // Ultrasonic Trigger
int US_Echo = 16;       // Ultrsonic Echo
```

```
int LED = 14;              // LED
int LCD_Reset = 6;         // LCD Reset
int LCD_Enable = 7;        // LCD Enable/Clock
int LCD_D4 = 9;            // LCD Data Pin 4
int LCD_D5 = 10;           // LCD Data Pin 5
int LCD_D6 = 11;           // LCD Data Pin 6
int LCD_D7 = 12;           // LCD Data Pin 7

// Create the LCD and assign the interface pins
LiquidCrystal lcd(LCD_Reset, LCD_Enable, LCD_D4, LCD_D5, LCD_D6,
   LCD_D7);

int close = 10;                    // The distance that triggers the LED

void setup() {

   // Initialize the LCD and display the headings
   lcd.begin(16, 2);
   lcd.print("Proximity Sensor");
   lcd.setCursor(0, 1);
   lcd.print("Range: ");


   // Set the pin modes
   pinMode(US_Trig,OUTPUT);   // Trigger pin
   pinMode(US_Echo, INPUT);   // Echo pin
   pinMode(LED, OUTPUT); // LED pin

}

void loop() {

   float distance;
   int pulseLength;

   // Send the trigger pulse to start the echo ranging
   digitalWrite(US_Trig, LOW);
   delayMicroseconds(20);
   digitalWrite(US_Trig, HIGH);
   delayMicroseconds(100);
   digitalWrite(US_Trig, LOW);
```

```
// Get the echo pulse and calculate the distance
pulseLength = pulseIn(US_Echo, HIGH);
distance=pulseLength * 0.017;

// Show the distance and erase the rest of the line
lcd.setCursor(7,1);
lcd.print(distance);
lcd.print("       ");

// Turn on the LED if the object is close
if (distance < close)
{
  digitalWrite(LED, HIGH);
}
else
{
  digitalWrite(LED, LOW);
}

delay(50);

}
```

Chapter 5

Adding Sound and Motion to Your Arduino Projects

In this chapter, you learn how to work with devices that add sound and motion to your Arduino projects. To create sound, you can add a speaker to create audible output tones. This is useful in situations where your Arduino program might need to get someone's attention or when you want to create a sound effect. For more realistic sound effects and music, you can add a shield to your Arduino to play MP3 files that are stored on a micro-SD card. To create motion, you can add a very useful device called a *servo*, which lets you control mechanical motion with an Arduino program.

Using a Speaker with an Arduino

An inexpensive way to add sound to an Arduino project is to simply add a small 8 Ω speaker to a digital output pin. You can then rapidly alternate the pin between HIGH and LOW to create a square-wave signal on the pin, which the speaker will render as sound. The frequency of this square wave will determine the pitch of the sound.

As an alternative to a speaker, you can use a *piezoelectric buzzer*, sometimes referred to as a *piezo buzzer*. A piezo buzzer is similar to a speaker but isn't capable of playing complex waveforms such as voice or music. Piezo buzzers and small speakers are essentially interchangeable when connected directly to a digital output pin. Figure 5-1 shows a small speaker (on the right) and a piezo buzzer (on the left). Either one of these devices will work for the project that's coming up (Project 48).

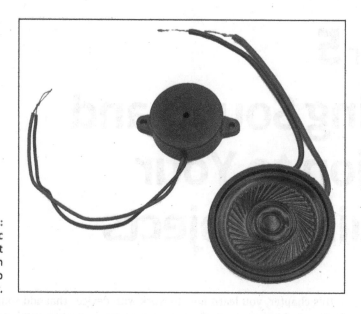

FIGURE 5-1:
The piezoelectric speaker that comes with the Arduino Activity Kit.

Note that the speaker is polarized, so when you connect it to an I/O pin, be sure to connect the + terminal to the I/O pin and the other terminal to Vss (ground), as shown in the schematic in Figure 5-2.

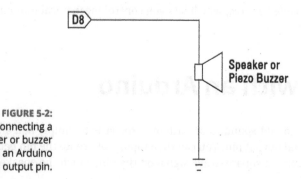

FIGURE 5-2:
Connecting a speaker or buzzer to an Arduino digital output pin.

Using the tone function

Programming a speaker is remarkably simple. Arduino C includes a command called tone that sends a frequency of your choice to an output pin. Thus, you can create an audible tone on a speaker by using the tone command, using the following syntax:

```
tone(pin, frequency, duration)
```

Here's how that syntax works:

>> *pin* is simply the pin number that you want to send the frequency to.

>> *frequency* is the frequency in hertz (Hz) that you want to generate.

>> *duration* is simply the length of time in milliseconds you want the frequency to play. Note that if you omit *duration,* the tone plays indefinitely until you tell it to stop by calling the noTone function (more on that later).

For example, the following command generates a 2,000 Hz frequency for five seconds on pin 8:

```
tone(8, 2000, 5000);
```

Note that the tone function returns immediately; it does not wait for the tone to be completed. Thus, your program can do other work while the tone is playing.

You can easily create a beeping sound by alternately sending short bursts of a frequency to the speaker followed by a brief pause. For example, here's a function you can call to beep a speaker on a specified digital output pin a certain number of times:

```
void beep(int pin, int count)
{
  for(int i=0; i < count; i++)
  {
    tone(8, 1000, 250);
    delay(500);
  }
}
```

To send three beeps to digital pin 8, call the function like this:

```
beep(8, 3);
```

In the beep function, a for loop handles the counting of the beeps. For each pass through the for loop, the tone sounds for 250 milliseconds. The delay of 500 milliseconds accounts for the 250 milliseconds during which the tone plays plus another 250 milliseconds of silence. The result is a nice *beep beep beep* sound.

Earlier, I mention that you can omit the *duration* argument when you call the tone function. If you do, the tone continues indefinitely until you call the noTone function, which silences the tone. Here's a version of the beep function that omits the *duration* argument:

```
void beep(int pin, int count)
{
  for(int i=0; i < count; i++)
  {
    tone(8, 1000);
    delay(250);
    noTone(8);
    delay(250);
  }
}
```

In the for loop, a 1,000 Hz tone is started on pin 8. Then, the program is delayed for 250 milliseconds and the tone is stopped. Then the program is delayed again for 250 milliseconds. The result is a beep tone that's on for a quarter of a second and then off for a quarter of a second.

Connecting a speaker to an Arduino

As I've already mentioned and as the schematic diagram that was presented in Figure 5-2 shows, a speaker or piezo buzzer connects to an Arduino via a GND pin and a digital output pin. Conceptually, this is easy. But in practice, it's a little more complicated because speakers and piezo buzzers usually have leads constructed of stranded wire rather than solid wire. That makes the leads more difficult to insert into header pins on an Arduino or into the holes in a solderless breadboard.

Here are a few ways you can work around this challenge:

>> Strip off ⅜ inch or so of insulation and cleanly twist the individual wires. Then carefully insert the twisted ends into the appropriate Arduino header or a breadboard hole. This method will work only if the strands are twisted cleanly and tightly together.

» Before twisting the wires, separate out about one-third of the strands and cut them off. Then carefully and cleanly twist the remaining wires. When you have a clean twist, apply a small amount of solder to hold the wires together. This takes a bit of soldering skill, as the resulting soldered end of the wire must be smooth to fit nicely into a header or breadboard hole. (This is the method I use for the projects in this chapter.)

» Attach a header pin to the ends of the speaker or buzzer leads. You can purchase kits with header pin parts and a crimping tool to create header pins.

» Use a two-terminal screw connector block. You can snap the connector block into a breadboard and then use the screw connector to attach the stranded speaker wires.

» Use an Arduino *screw-terminal breakout block* such as the ones pictured in Figure 5-3. These shields provide a screw terminal for each of the header pins on the UNO. Just attach the shield to your Arduino UNO, and then connect your stranded speaker leads to the appropriate screw terminal.

FIGURE 5-3:
Screw-terminal breakout shields that make it easy to connect stranded wires to Arduino headers.

In the next section, you connect a speaker directly to an Arduino UNO so that you can test various ways to generate tones. Figure 5-4 shows the assembled project.

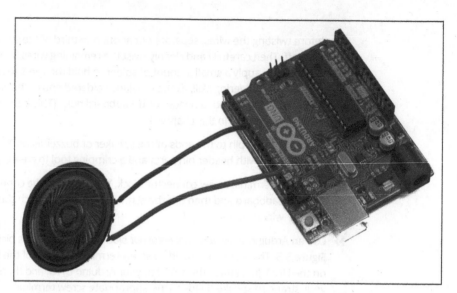

FIGURE 5-4:
A speaker
connected to
an Arduino
(Project 48).

Project 48: Creating Sound with a Speaker

In this project, you connect a speaker to an Arduino UNO to run programs that generate sound. The project is very simple, because the only component other than the UNO is a speaker.

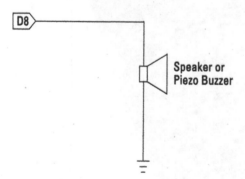

UNO

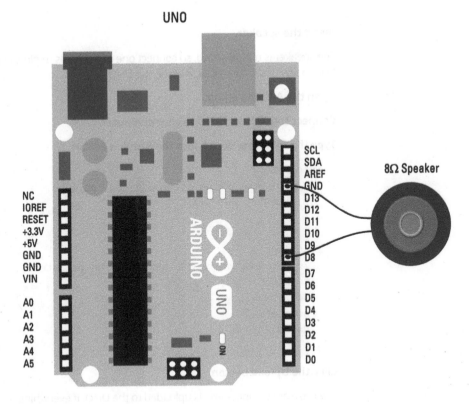

Parts

» One computer with Arduino IDE software installed

» One Arduino UNO

» One USB type-B cable to connect the UNO to your computer.

» One 8 Ω speaker (alternatively, you can use a piezo buzzer)

Steps

1. **Prepare the speaker leads.**

   Strip off about ⅜ inch of insulation. Separate out about one-third of the individual wires in each lead and snip them off. Then carefully twist the remaining wire strands together. Apply a very small amount of solder to hold the individual strands together.

2. **Insert the speaker.**

The speaker is not polarized, so connect one of the leads to pin 8 and the other lead to any of the three GND pins on the UNO.

3. **Open the Arduino editor.**

4. **Connect the Arduino UNO to your computer.**

5. **Type the following sketch into the Arduino editor:**

```
void setup() {
    // put your setup code here, to run once:
    pinMode(8, OUTPUT);
}

void loop() {
    // put your main code here, to run repeatedly:
    tone(8, 1000);
    delay(250);
    noTone(8);
    delay(250);
}
```

6. **Click the Upload button.**

The program compiles and is uploaded to the UNO. If everything works, you'll hear a steady *beep-beep-beep* from the speaker.

7. **Now try the programs in Listings 5-1, 5-2, and 5-3.**

These programs vary the tone in different ways to demonstrate the versatility of the tone function.

Sounding the alarm with Morse code

Listing 5-1 is an Arduino program that sounds the familiar SOS emergency call in Morse code. The Morse code for SOS is three dots, three dashes, and three dots. This program simply sounds a 1,000 Hz tone in the familiar pattern: dot-dot-dot-dash-dash-dash-dot-dot-dot.

TECHNICAL
STUFF

The proper way to pronounce Morse code is *dit* for dots and *dah* for dashes, so the correct way to say "SOS" in Morse code *dit dit dit dah dah dah dit dit dit.*

According to the rules of Morse code, a dash is three times longer than a dot, the gap between letters is three times the length of a dot, and the gap between words is seven times as long as a dot. However, for the special case of the code SOS, the gaps between letters are skipped.

The program uses a variable named pace to establish the pace of the Morse code sounded. This variable is set to 60, which is a pretty good pace for reading Morse code. (A pace of 60 milliseconds per dot translates to 20 words per minute. The world record is 75.2 words per minute, which is a pace of 16 milliseconds per dot. Just for fun, try the program with the pace variable set to 16. The result is gibberish to me. If you can make out individual dots and dashes, your hearing is much better than mine!)

The program uses four functions to sound each of the four elements of Morse code:

» soundDot: Sounds a dot using a single pace unit and then pauses for a single pace unit.

» soundDash: Sounds a dot by using three pace units and then pauses for a single pace unit.

» soundLetterGap: Delays for three pace units. (Note that this function isn't called in the program, because SOS doesn't require inter-letter gaps.)

» soundWordGap: Delays for seven pace units.

| LISTING 5-1: | SOS! |

```
// SOS Program
// Doug Lowe
// December 3, 2021

// This program sounds the alert by playing SOS in Morse Code
// on a speaker or piezo buzzer connected to pin 8.

int pin = 8;            // The speaker pin.

int pace = 60;          // Sets the speed at which SOS is sounded.
                        // Morse code standards spell out the duration
                        // of each element of the coded signal:
                        // A dot is one unit.
                        // A dash is three units.
                        // The gap between dots and dashes is one unit.
                        // The gap between letters is three units.
                        // However the gap is skipped for the
                        // specific sequence "SOS."
                        // The gap between words is seven units.
void setup() {
  pinMode(8, OUTPUT);
}
```

(continued)

Adding Sound and Motion to Your Arduino Projects

LISTING 5-1: *(continued)*

```
void loop() {
  // Sound SOS
  soundDots(3);
  soundDashes(3);
  soundDots(3);
  soundWordGap();
}

void soundDots(int count) {
  for(int i=0; i < count; i++)
  {
    tone(pin, 1000);
    delay(pace);
    noTone(pin);
    delay(pace);
  }
}

void soundDashes(int count) {
  for(int i=0; i < count; i++)
  {
    tone(pin, 1000);
    delay(pace * 3);
    noTone(pin);
    delay(pace);
  }
}

void soundLetterGap() {
  delay(pace * 3);
}

void soundWordGap() {
  delay(pace * 7);
}
```

Playing a siren

Listing 5-2 shows how you can use tone within a pair of for loops to create a continuously rising and falling tone, much like a police siren. The first for loop varies the pitch from a low value to a high value, and the second for loop varies the pitch from the high value back down to the low value.

The sound characteristics of the siren are set by four variables:

>> low: Sets the low end of the frequency range. This variable is set to 300, but you can change it to start from a lower or higher frequency.

>> high: Sets the high end of the frequency range. This variable is set to 1,000, but you can change it if you want.

>> increment: Sets the amount that is added or subtracted to or from the frequency at each pass through the for loops. Increasing this value speeds up the siren.

>> pace: Sets the length at which each pitch is played. Increase this value to make the siren slower. If you increase the value enough (say, to 100 or 150), you'll hear the individual pitches separately.

LISTING 5-2:   **Generating a Siren Effect**

```
// Siren Program
// Doug Lowe
// December 3, 2021

// This program generates a siren sound on a speaker or piezo buzzer
// connected to Pin 8.

int pin = 8;              // The speaker pin.
int low = 300;            // Frequency of the lowest pitch of the siren.
int high = 1000;          // Frequency of the highest pitch of the siren.
int increment = 5;        // The amount added or subtracted from the pitch
                          // each time through the for loop.
int pace = 10;            // The speed at which the pitch changes.
void setup() {
  pinMode(8, OUTPUT);
}

void loop() {
  for (int f = low; f < high; f = f+increment)
  {
    tone(pin, f);
    delay(pace);
  }
  for (int f = high; f > low; f = f-increment)
  {
    tone(pin, f);
    delay(pace);
  }
}
```

Playing a song

Listing 5-3 shows a program that plays "Mary Had a Little Lamb."

To simplify the code that generates the musical notes, the program defines several constants that represent the frequency for each of the notes required by the songs. For example, the constant NoteC4 is 262, the frequency in hertz of middle C on a piano keyboard. The constants span one full octave, which is plenty of range for "Mary Had a Little Lamb."

The program also sets up constants for the duration of a quarter note, half note, and whole note. The constants make it easy to specify a particular pitch for a particular duration in a tone command.

Two functions are used to actually play the music:

» **playNote:** Accepts the note's pitch and a duration as arguments. A little trick is used to ensure that the notes are distinguishable from one another to ensure that 10 percent of each note's duration is used as a gap between notes.

» **playRest:** Accepts a duration value and simply delays for the indicated duration.

Thus, playing a melody is simply a matter of writing a sequence of calls to these two functions to play the correct notes for the correct durations in the correct order. As you can see, I've added a comment to each of the calls to playNote so you can see what part of the song is being played.

LISTING 5-3: **Making Music with an Arduino**

```
// Mary Had a Little Lamb
// Doug Lowe
// December 3, 2021


// This program plays Mary Had a Little Lamb on a speaker or piezo buzzer
// connected to Pin 8.

int pin = 8;            // The speaker pin.
int tempo = 100;        // The speed at which the pitch changes.

// Note pitch constants for the Key of C (Starting at Middle C)
const int Note_C4 = 262;
const int Note_D4 = 294;
const int Note_E4 = 330;
const int Note_F4 = 349;
const int Note_G4 = 392;
```

```
const int Note_A4 = 440;
const int Note_B4 = 494;
const int Note_C5 = 523;

// Note duration constants
const int quarter = 300;
const int half = quarter * 2;
const int whole = quarter * 4;

void setup() {
  pinMode(8, OUTPUT);
}

void loop() {
  playNote(Note_E4, quarter);     // Mar-
  playNote(Note_D4, quarter);     // y-
  playNote(Note_C4, quarter);     // Had
  playNote(Note_D4, quarter);     // a
  playNote(Note_E4, quarter);     // Lit-
  playNote(Note_E4, quarter);     // le
  playNote(Note_E4, quarter);     // Lamb
  playRest(quarter);
  playNote(Note_D4, quarter);     // Lit-
  playNote(Note_D4, quarter);     // le
  playNote(Note_D4, quarter);     // Lamb
  playRest(quarter);
  playNote(Note_E4, quarter);     // Lit-
  playNote(Note_G4, quarter);     // le
  playNote(Note_G4, quarter);     // Lamb
  playRest(quarter);
  playNote(Note_E4, quarter);     // Mar-
  playNote(Note_D4, quarter);     // y-
  playNote(Note_C4, quarter);     // Had
  playNote(Note_D4, quarter);     // a
  playNote(Note_E4, quarter);     // Lit-
  playNote(Note_E4, quarter);     // le
  playNote(Note_E4, quarter);     // Lamb
  playNote(Note_E4, quarter);     // Its-
  playNote(Note_D4, quarter);     // Fleece
  playNote(Note_D4, quarter);     // Was
  playNote(Note_E4, quarter);     // White
  playNote(Note_D4, quarter);     // As
  playNote(Note_C4, half);        // Snot
  playRest(half);
  playRest(whole);
}
```

(continued)

Adding Sound and Motion to Your Arduino Projects

CHAPTER 5 Adding Sound and Motion to Your Arduino Projects   693

LISTING 5-3:   *(continued)*

```
void playNote(int pitch, int duration){
  int phrasing = quarter * 0.1;
  tone(pin, pitch);
  delay(duration - phrasing);
  noTone(pin);
  delay(phrasing);
}

void playRest(int duration) {
  delay(duration);
}
```

Using an MP3 Shield

The tone function used in combination with a speaker or piezo buzzer is adequate for simple sounds, but what if your project needs more sophisticated sound effects? The Arduino doesn't have the ability to play audio files in the common MP3 audio format, but you can easily add that capability to an Arduino by adding an MP3 shield to your project.

There are many MP3 shields available — just search the web for "Arduino MP3 shield," and you'll find plenty of options. The one I use in this chapter is a popular MP3 shield from Adafruit (www.adafruit.com) called the Music Maker shield (see Figure 5-5).

The Music Maker shield comes in two versions — one that has just a line-out signal level, which you can use with headphones or as input to an external amplifier, and another that has a built-in 3-watt stereo amplifier that lets you connect small speakers. I'm using the one that has the built-in amplifier. You can purchase it directly from Adafruit or from other sources for about $35.

The Music Maker shield includes the following features:

» A micro-SD card reader into which you can insert a card loaded with audio files.

» An on-board audio decoder chip that can play MP3 files as well as several other file formats, including WAV and MIDI.

» A standard ⅛-inch stereo audio jack to connect headphones or an external amplifier.

FIGURE 5-5:
The Adafruit
Music Maker
shield.

>> Screw terminals to connect external speakers. The shield will work with either 4 Ω or 8 Ω speakers. Don't expect the output to be very loud — 3 W is not a powerful amplifier, but it is suitable when a lot of volume isn't required.

>> A programming library that makes it easy to play audio files stored on the micro-SD card.

>> Seven additional digital I/O pins that you can access via the Music Maker library. (To access these pins, you'll need to solder additional header pins to the Music Maker shield.)

Assembling the Music Maker shield

When you purchase the Music Maker shield, you'll find that a bit of assembly is required: The shield's main board is fully assembled, but you'll have to solder the header pins (included) onto the board. The soldering is pretty simple and should only take about 15 minutes.

Preparing the micro-SD card

The Music Maker shield reads audio files from its on-board micro-SD card reader. Before you can use the Music Make shield, you'll need to copy at least one audio file to a micro-SD card.

If you've never worked with micro-SD cards, you'll be surprised at how small they are. Most computers that have multi-format card readers don't have a slot that's small enough for micro-SD cards. However, micro-SD cards usually come with an adapter that lets you use a micro-SD card in a standard SD card reader. Simply insert the micro-SD card into the adapter, and then slide the adapter into your computer's SD card reader. In a few moments, Windows will recognize the micro SD card so that you can copy files to it.

There's one important caveat to know about saving your audio files to the micro-SD card: The Music Maker shield does not recognize file names longer than eight characters, and file extensions can be only three characters long. So, a filename like Puff the Magic Dragon.mp3 won't work. You'll need to shorten the name to something like Puff.mp3.

You can put as many audio files on the micro-SD card as will fit. And you can, if you want, create folders to organize them — provided that the folder names are no longer than eight characters in length.

For the purposes of this chapter, copy just one MP3 file to the card and name it TRACK01.mp3. This can be any audio file you want — your favorite song, or perhaps a sound effect file with an owl screeching or a werewolf howling. As long as the file is named TRACK01.mp3, the program in the project that's coming up will be able to play the file.

Programming the Music Maker shield

Adafruit supplies a programming library with the Music Maker shield, making it simple to play individual audio files from your Arduino sketches. Before you can use this library, you have to download and install it into the Arduino IDE. Here are the steps:

1. **Open the Arduino IDE.**

2. **Choose Tools⇨Manage Libraries.**

   The dialog box, shown in Figure 5-6, appears.

3. **Enter** Adafruit_VS1053 **in the search box at the top of the Library Manager.**

   The Library Manager locates this library in the cloud.

4. **Select the most recent version in the Select Version drop-down list.**

5. **Click Install.**

   The library is installed.

   You're done!

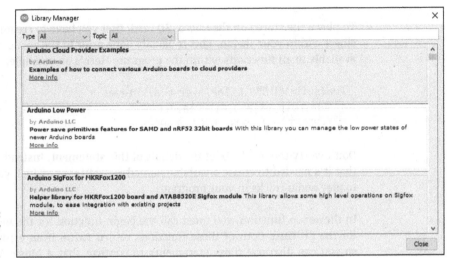

FIGURE 5-6:
The Library
Manager
dialog box.

After you've installed the library, you can use it to play sound files on the micro-SD card that's loaded into the shield.

The Music Maker shield uses four pins to play audio files:

Name	What it Does	Default Digital Pin
MCS	Chip select pin	D7
DCS	Data select pin	D6
CCS	Card select pin	D4
DREQ	Request interrupt pin	D3

There's usually no reason to change these pins from their defaults, and you don't need to worry about the details of what the pins actually do. But you do need to remember that these pins are used by the Music Maker shield, so you can't use them for other purposes in your program.

It's common to define these pin settings as constants at the beginning of an Arduino sketch that uses the Music Maker shield. For example:

```
#define CS      7       // Select pin (output)
#define DCS     6       // Data/command select pin (output)
#define CARDCS  4       // Card chip select pin
#define DREQ    3       // Data request
```

To play a file stored on the micro-SD card, first you have to create a media player object within your sketch. This is usually done prior to the setup function, so it's available to all functions within the program. Here's an example:

```
Adafruit_VS1053_FilePlayer mediaPlayer =
  Adafruit_VS1053_FilePlayer(
    RESET, CS, DCS, DREQ, CARDCS);
```

Don't worry too much about the details of this statement. Instead, just remember that it's needed to create a variable named mediaPlayer, which you can then use to play audio tracks in your program.

In the setup function, you must call the begin function for the media player and for the SD card. Both of these functions return false if an error has occurred, so you can also use these statements to confirm that a Music Maker shield is installed onto the UNO and that a card is inserted into the micro-SD card reader:

```
if (mediaPlayer.begin() == false)
{
  while (true);     // Don't do anything -- Shield is missing
}
if (SD.begin(CARDCS) == false)
{
  while (true);     // Don't do anything -- SD card is missing
}
```

Notice that both if statements execute the command while (true) if an error is detected. The while (true) statement simply loops forever, preventing the program from actually doing anything if the shield or the card is missing.

After you've established that the shield and micro-SD card are present, you can play an audio file on the card like this:

```
mediaPlayer.playFullFile("TRACK01.mp3");
```

This statement supplies the name of the audio file. As long as there's a file on the card named TRACK01.mp3, the Music Player shield will play the file.

One other statement you may want to use is setVolume, which adjusts the left and right channel volume of the playback. Use zeros to set the maximum volume:

```
mediaPlayer.setVolume(0, 0);
```

Building an Arduino Music Player

In this section, you build a simple Arduino project that uses a Music Maker shield to play MP3 files. Figure 5-7 shows the completed project.

FIGURE 5-7:
An Arduino
Music Player
(Project 49).

Project 49: Using a Music Maker Shield to Play Sounds

This project uses an Adafruit Music Maker shield so you can add sound effects or music to your Arduino projects.

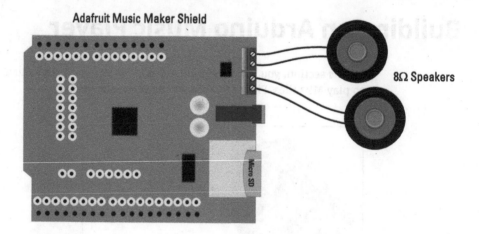

Adafruit Music Maker Shield

8Ω Speakers

Micro SD

Parts

>> One computer with Arduino IDE software installed

>> One Arduino UNO

>> One USB type-B cable to connect the UNO to your computer

>> One Adafruit Music Maker shield

>> One micro-SD card

>> Two 8 Ω speakers (alternatively, you can use a headset with a ⅛-inch audio connector)

Steps

1. **Assemble the Adafruit Music Maker shield.**

   Carefully follow the instructions that come with the shield. (You'll need a soldering iron and some solder to perform this step. A lighted magnifying viewer or goggles will help as well.)

2. **Prepare the micro-SD card.**

   Copy any MP3 file to the root folder of the SD card. Change the name of the MP3 file to TRACK01 .mp3.

3. **Open the Arduino editor.**

4. **Attach the Music Maker shield to the Arduino UNO.**

5. **Connect the leads from the two speakers to the speaker terminal blocks on the Music Maker shield.**

6. **Connect the Arduino UNO to your computer.**

7. **Enter the sketch shown in Listing 5-4 into the Arduino editor.**

8. **Click the Upload button.**

   The program compiles and is uploaded to the UNO. You'll soon hear your MP3 file play on the speakers. When the MP3 file finishes, it will play again after a pause of 5 seconds.

LISTING 5-4: | **Playing an MP3 Track**

```
// MP3 Program
// Doug Lowe
// December 3, 2021

// This program uses an Adafruit Music Player shield
// to play an MP3 file named TRACK01.mp3.

#include <SPI.h>
#include <Adafruit_VS1053.h>
#include <SD.h>

#define CS      7        // Select pin (output)
#define DCS     6        // Data/command select pin (output)
#define CARDCS  4        // Card chip select pin
#define DREQ    3        // Data request

Adafruit_VS1053_FilePlayer mediaPlayer =
    Adafruit_VS1053_FilePlayer(RESET, CS, DCS, DREQ, CARDCS);

void setup() {
  if (mediaPlayer.begin() == false)
  {
    while (true);     // Don't do anything -- Music Maker shield is missing
  }

  if (SD.begin(CARDCS) == false)
  {
    while (true);     // Don't do anything -- SD card is missing
  }

  mediaPlayer.setVolume(0,0);
}

void loop() {
  mediaPlayer.playFullFile("TRACK01.mp3");
  delay(5000);
}
```

Adding Sound and Motion to Your Arduino Projects

Using a Servo with an Arduino

The previous sections in this chapter look at two ways to create Arduino projects that make noise. In this section, you add an element of movement by utilizing a device called a servo.

A *servo* is a special type of motor that is designed to rotate to a particular position and hold that position until it's told to rotate to a different position. Hobby servos are frequently used in radio-controlled vehicles such as airplanes, boats, and cars, but there are many other uses for servos. For example, I often use them in Halloween props to add movement such as eyes or a mouth.

Servos are readily available from many sources. Electronics suppliers, such as Adafruit (www.adafruit.com) and Jameco (www.jameco.com), sell these, as does the giant retailer Amazon. You may also be able to purchase servos from a local hobby store. Figure 5-8 shows a typical hobby servo.

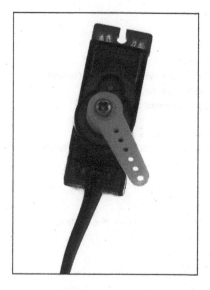

FIGURE 5-8:
A typical
hobby servo.

Connecting a servo to an Arduino

Servos use a special three-conductor cable to provide power and a *control signal* that tells the servo what position it should move to and hold. The cable's three wires are colored red, black, and white and have the following functions:

>> **Red:** Supplies the voltage required to operate the servo. For most servos, this voltage can be anywhere between +4 V and +9 V. On an Arduino, you should connect this to one of the +5 V pins.

>> **Black:** The ground connection. On the Arduino, you should connect it to one of the GND pins.

>> **White:** The control wire; it connects to one of the Arduino's digital I/O pins.

Figure 5-9 shows how these wires should be connected in an Arduino circuit.

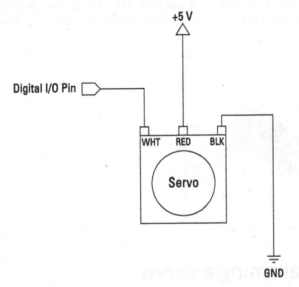

FIGURE 5-9:
Connecting
a servo to an
Arduino.

The control wire controls the position of the servo by sending the servo a series of pulses approximately 20 ms apart. The length of each one of these pulses determines the position that the servo rotates to and holds.

Most hobby servos have a range of motion of 180 degrees — that is, one-half of a complete revolution. The complete range of pulse durations is 0.5 ms to 2.5 ms, where 0.5 ms pulses move the servo to its minimum position (0 degrees), and 2.5 ms pulses move the servo to its maximum position (180 degrees). To hold the servo at the center point of this range (90 degrees), the pulses should be 1.5 ms in duration.

Fortunately, Arduino provides a library of functions for programming servos that removes all the details of pulse duration. With the servo library, you just tell the servo what position you want to move it to, and the library moves the servo to that position and holds it there until you tell it otherwise. Because of this library, programming an Arduino to control a servo is easy.

To connect a servo to an Arduino, you must use a three-pin header adapter such as the one shown in Figure 5-10. A header like this is required because the leads from the servo have female pin connections, just like the female connections in the Arduino headers or on a breadboard. The header simply adapts the female header on the servo leads so it can properly connect. You simply plug the three-pin adapter into the breadboard and then plug the servo leads into the adapter.

FIGURE 5-10:
A three-pin
header adapter
for connecting
the servo to an
Arduino.

Programming a servo

Arduino provides a servo library that makes it easy to work with servos. Because servos are so commonly used, the servo library is automatically installed in the Arduino integrated development environment (IDE), so you don't have to do anything special to use the library.

For most programs, there are four steps you must take to program a servo:

1. **Include the servo library so your program can use its functions.**

To do this, add the following line near the top of your program:

```
#include <Servo.h>
```

REMEMBER

Case is important in Arduino C, so you must capitalize the word *Servo* here.

2. **Create a servo variable.**

Servo is a special variable type defined by the servo library. You'll need a variable of this type to refer to your servo. The easiest way to do this is to create the variable outside of the setup( ) or loop( ) functions, like this:

```
Servo Servo1;
```

Here, I've created a servo variable named Servo1.

3. **Attach a pin to the servo variable.**

You'll usually do that in the setup( ) function with a line like this:

```
Servo1.attach(2);
```

If you've create a constant for the servo pin number, you'll use something like this instead:

```
Servo1.attach(ServoPin);
```

where ServoPin is the constant you've created.

4. **Move the servo.**

To move the servo to a specific position, use a line such as this:

```
Servo1.write(90);
```

This example moves the servo to the 90-degree position. After it's there, the servo will remain there until you use the write function to move the servo to a new position.

Listing 5-5 shows a complete program that moves the servo based on the status of two push-button switches that are available to the program. The switches are identified as SW1, on pin 10, and SW2 on pin 11. The servo itself is on pin 2.

When you press SW1, the servo moves to 0 degrees. When you press SW2, the servo moves to 180 degrees. If you press both buttons simultaneously, the program moves the servo to the middle position (90 degrees).

LISTING 5-5: | **A Servo Control Program**

```
// Servo Program
// Doug Lowe
// December 5, 2021
```

(continued)

LISTING 5-5: *(continued)*

```
#include <Servo.h>

#define ServoPin   2
#define SW1        10
#define SW2        11
#define LowPos      0
#define MidPos     90
#define HighPos   180

Servo Servo1;

int ServoPos = MidPos;

void setup() {
  pinMode(SW1, INPUT);
  pinMode(SW2, INPUT);

  Servo1.attach(ServoPin);
  Servo1.write(ServoPos);

}

void loop() {

    bool sw1HIGH = digitalRead(SW1);
    bool sw2HIGH = digitalRead(SW2);

    if (sw1HIGH & sw2HIGH)
      ServoPos = MidPos;
    else if (sw1HIGH)
      ServoPos = LowPos;
    else if (sw2HIGH)
      ServoPos = HighPos;

    Servo1.write(ServoPos);
    delay(100);
}
```

Building a servo project

Project 50 shows you how to build a complete circuit that uses a servo as well as two push buttons. This circuit is capable of running the program that was shown in Listing 5-4. Figure 5-11 shows the assembled project.

FIGURE 5-11:
An Arduino
project that
controls a servo
(Project 50).

Project 50: Using a Servo with an Arduino

In this project, you connect a servo to an Arduino. The circuit also includes two push buttons (on pins 10 and 11) that you can use to control the action of the servo.

Adding Sound and Motion
to Your Arduino Projects

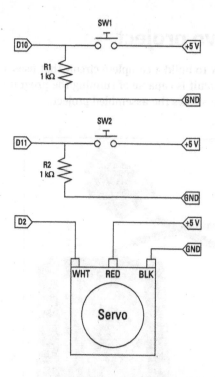

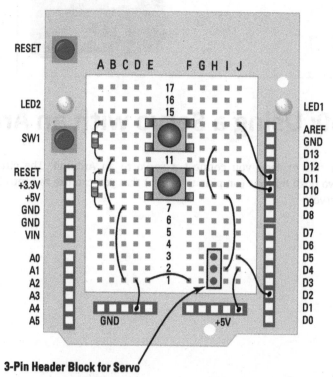

3-Pin Header Block for Servo

Parts

» One computer with Arduino Editor software installed

» One Arduino UNO

» One prototyping shield

» One USB cable

» One hobby servo

» One three-pin male-to-male header

» Two normally open push buttons

» Two 1 k4 Ω resistors

» Ten jumper wires

Steps

1. Insert the jumper wires.

From	To
B11	B7
C7	C1
GND on bottom left header	D1
E1	F1
H12	H8
I8	I2
Digital pin 11 on right header	J14
Digital pin 10 on right header	J10
Digital pin 2 on right header	J3
+5 V on bottom right header	J2

2. Insert the resistors.

Resistor	From	To
R1 (470 Ω)	A14	A11
R2 (470 Ω)	A10	A7

3. Insert the switches

Resistor	From	To
SW1	E8 and F8	E10 and F10
SW2	E12 and F12	E14 and F14

Be sure to orient the switches correctly, as shown in the breadboard diagram.

4. Insert the three-pin male-to-male header.

The three pins should go in H1, H2, and H3.

5. Connect the servo.

Plug the three-pin servo connector into the three-pin male-to-male header. Ensure that the control wire (usually white) connects to H3.

6. Open the Arduino Editor.

7. Connect your Arduino to the computer and identify it in Arduino Editor.

For more information about how to do this, see Chapter 1 of this minibook.

8. Enter and run the following program to test that the servo is working:

```
#include <Servo.h>

  Servo Servo1;

void setup() {
  Servo1.attach(2);
}

void loop() {
  Servo1.write(45);
  delay(500);
  Servo1.write(135);
  delay(500);
}
```

9. Now try the program in Listing 5-4.

This program uses the switches to control the servo's action.

IN THIS CHAPTER

» **Considering keypads**

» **Discovering how keypads work**

» **Using the Keypad library for easier keypad programming**

» **Teaching an Arduino how to pretend to be a computer keyboard**

Chapter **6**

Keypads and Keyboards

This chapter shows you how to use a common input device called a *keypad*. It also shows you how to use a unique feature of some Arduino boards, including the popular Leonardo, to send information to a computer as if the Arduino were a keyboard.

Using a Keypad

A keypad is a common form of input for devices built with microcontrollers such as the Arduino. You're probably most familiar with keypads from devices such as calculators and phones — they consist of an array of push buttons arranged in rows and columns. Figure 6-1 shows a typical keypad, with four rows of four push buttons each.

You can purchase keypads such as this one from many different suppliers, including Amazon, Jameco Electronics (www.jameco.com), and others. Just search the web for "Arduino keypad," and you'll find a wide variety to choose from. Simple keypads such as the one shown in Figure 6-1 cost just a few dollars apiece. You can get nicer ones for a bit more.

FIGURE 6-1:
A typical keypad.

How a keypad works

If each push button in the keypad used two leads, a 4 x 4 keypad would have 32 leads. Even if the keypad used a common ground lead for all 16 buttons, a total of 17 leads would be required, taking up 16 digital I/O ports on the Arduino. Unfortunately, a 4 x 4 keypad would use nearly all of an UNO's available I/O ports if it is wired in this way.

Fortunately, some clever folks have figured out a more efficient way to wire a keypad. The key (pun intended) is to realize that the buttons in a keypad are arranged in an array, with rows and columns. They can be wired with a single lead for each row and a single lead for each column. So, a 4 x 4 keypad can be wired with just eight leads, and a 3 x 3 keypad requires just six leads.

Figure 6-2 shows how the wiring works for a 4 x 4 keypad. As you can see, one lead connects all the buttons in each separate row, and one lead connects all the buttons in each separate column. When any one of the push buttons is pressed, a circuit is completed between the lead that's connected to the button's row and the lead that's connected to the button's column.

For example, if the button labeled 6 is pressed, a circuit is completed between the lead on row 2 (R2) and the lead on column 2 (C3).

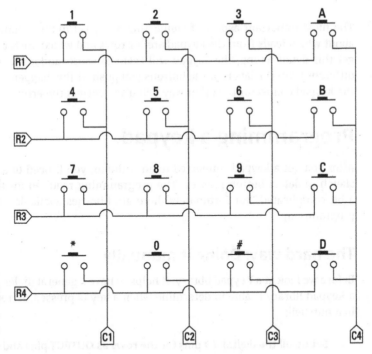

FIGURE 6-2:
How the push
buttons are
connected in a
keypad.

Connecting a keypad to an Arduino

Connecting a keypad to an Arduino is a simple task, but it can require a lot of digital I/O pins depending on the number of rows and columns in the keypad. A 4 x 4 keypad requires a total of eight digital I/O pins: four for the row pins and four for the column pins.

Although it's not required, consecutive pins are commonly used for the keypad, with the lower-numbered pins used for the columns and the higher-numbered pins used for the rows. Avoid pins 0 and 1, because those pins are often used for serial communications.

TIP

Unless you have other devices, the easiest way to connect a 4 x 4 keypad is to use pins 2, 3, 4, and 5 for the rows and pins 6, 7, 8, and 9 for the columns. For a 3 x 4 keypad, use pins 2, 3, and 4 for the rows and pins 5, 6, 7, and 8 for the columns.

The leads on a keypad typically have female connectors, which means you need male-to-male adapters to connect them to the female terminal strips on the Arduino or to a solderless breadboard. For a 4 x 4 keypad, you need an eight-pin adapter. For a 3 x 4 keypad, a six-pin adapter will do the trick.

REMEMBER

The most important aspect of connecting a keypad to an Arduino is being clear about which leads from the keypad are for rows and which are for columns. If you get this detail wrong, the keypad will seem to work randomly, returning values different from the labels of the buttons you push. If this happens, simply flipping the keypad connector the other way will often correct the error.

Programming a keypad

After you get a keypad connected to an Arduino, you'll need to add code to your sketch to detect button presses. The programming required for that task can be a bit complicated, but fortunately there are libraries available that simplify the programming.

The hard way: Doing it manually

Before we look at a Keypad library, it helps to have a general understanding of how a Keypad library is able to determine when a key is pressed. Here's how it works in a nutshell:

1. **Set up all the digital I/O pins for the rows as OUTPUT pins and all the I/O pins for the columns as INPUT_PULLUP pins.**

   Active-Low switch logic will be used, and the Arduino's internal pullup resistors eliminate the need for external resistors.

2. **Make sure all the output pins are set to HIGH.**

3. **Set the output value of the I/O pin for row 1 to LOW; then check each of the input pins for a LOW value.**

   If you find a LOW value, you've found the row and column for the button that's pressed.

4. **Repeat step 3 for each remaining row, setting the I/O pin for the row to LOW and looking for a LOW value on all the input pins.**

5. **If you never find a LOW value on any of the input pins, no button is pushed; start over again with row 1.**

Here's how this may work in actual Arduino code. Steps 1 and 2 are done in the setup function:

```
void setup() {
  for (int row = 2; row<6; row++)          // Set the row pins to OUTPUT
  {
    pinMode(row, OUTPUT);
    digitalWrite(row, LOW);
  }
```

```
    for (int col = 6; col<10; col++)        // Set the column pins to INPUT
    {
      pinMode(col, INPUT_PULLUP);           // Use INPUT_PULLUP so external
    }                                       // resistors are not required
  }
```

Steps 3 through 5 are done in the loop function:

```
void loop() {
  // Look for a completed circuit
  String ButtonPushed = "NONE";
  for (int row = 2; row<6; row++)
  {
    digitalWrite(row, LOW);                           // Set the row to LOW
    for (int col = 6, col<10)                         // Test each column
    {
      If (digitalRead(col) == LOW)                    // Button is pressed!
      {
        ButtonPushed = 'R' + row + 'C' + col;  // Save the row and column
        delay(50);                                    // Debounce the button
      }
    }
    digitalWrite(row, HIGH);                           // Set the row to HIGH
  }
```

After one pass through this loop function, the ButtonPushed String variable indicates the status of the keypad. If no button has been pressed, ButtonPushed will be NONE. But if a button has been pressed, ButtonPushed will indicate the button's row and column. For example, a value of R2C3 indicates that the button at row 2 and column 3 has been pressed.

The easy way: Using the Keypad library

Fortunately, you don't have to write all that code every time you want to use a keypad. Instead, you can use one of several libraries that handle the details for you. The one I present in this chapter is known simply as Keypad. You can install it by following these steps:

1. **Open the Arduino integrated development environment (IDE).**

2. **Choose Tools⇨Manage Libraries.**

   The Library Manager dialog box, shown in Figure 6-3, appears.

3. **Type** Keypad **in the search box at the top of the Library Manager and press Enter.**

   The Library Manager searches its vast store of libraries and displays every library that includes the word *keypad*.

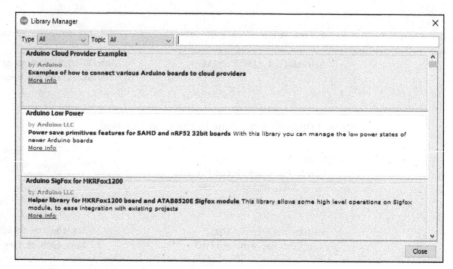

FIGURE 6-3:
The Library
Manager
dialog box.

4. **Select the Keypad library.**

Scroll through the list of libraries until you find the one identified simply as Keypad.

5. **Select the current version.**

Use the Select Version drop-down list to choose the version you want to install. At the time of this writing, the current version was 3.1.0.

6. **Click Install.**

The library is installed.

You're done!

After you've installed the Keypad library, you can use it to access matrix keypads. To do so, you'll need to include the following line near the top of your sketch:

```
#include <keypad.h>
```

Setting Up a Keypad

After you've installed and included the Keypad library, you can write the code necessary to define the characteristics of the keypad. You'll need to define five things:

>> The number of rows in the keypad

>> The number of columns in the keypad

>> The digital I/O pins that are connected to the keypad's rows

>> The digital I/O pins that are connected to the keypad's columns

>> A *keymap*, which provides a value for each button on the keypad

After you've set up these five things, you can create a variable of type Keypad, which you can then use to read key presses from the keypad.

TIP

All these items, including the Keypad variable, are usually best outside the setup or loop functions. That way, they're accessible throughout the sketch.

Defining the number of rows and columns

To define the number or rows and columns in the keypad, just create a pair of int variables that represent the number of rows and the number of columns in the keypad, like this:

```
int rows = 4;
int cols = 4;
```

Defining the row and column pins

To identify the pins used to connect the keypad's rows and columns, you must use an Arduino programming feature called an *array*. An array is a variable that can hold more than one value. The array of row pins will hold one value for each row on your keypad, and the array of column pins will hold one value for each column. The data type of these arrays should be byte.

The individual values in an array are identified using an integer value known as an *index*. The index value is enclosed in square brackets after the variable name, like this: rowPins[0]. Index values always begin with zero, so rowPins[0] refers to the first value in the array named rowPins.

To create an array, you must declare it using a data type, a variable name, and the number of values in the array. For example:

```
byte rowPins[4];
byte colPins[4];
```

In the preceding example, two arrays are defined — one to hold the row pins, and the other to hold the column pins. Both arrays will hold four byte values.

You can assign all the values as part of an array's declaration like this:

```
byte colPins[4] = {6, 7, 8, 9};
byte rowPins[4] = {2, 3, 4, 5};
```

In the preceding example, the colPins values are 6, 7, 8, and 9, and the rowPins values are 2, 3, 4, and 5. These values correspond to the Arduino digital I/O pins that the keypad's row and column leads will be connected to.

Defining the keymap

The keymap uses a more complex type of array, called a *two-dimensional array*, also known as a *matrix*. This type of array requires two indexes to access a value: The first index is a row index, and the second index is a column index.

TIP

The keymap uses char values rather than int or byte values. A char value is a single character enclosed in single quotes. It's important to remember that the char value '4' is not the same as the int value 4. The int value 4 represents the quantity four. A char value '4' represents the numeral 4, which is merely a symbol and does not represent a quantity.

You use a variation of the special syntax used to initialize the values of a simple array with a two-dimensional array as well, like this:

```
char keymap[rows][cols] = {
  {'1','2','3','A'},
  {'4','5','6','B'},
  {'7','8','9','C'},
  {'*','0','#','D'}
};
```

As you can see, the matrix of key values closely resembles the physical arrangement of keys on the 4 x 4 keypad.

Using a Keypad variable

The Keypad variable is what allows your program to detect key presses on a keypad. You'll need to define and initialize this variable before you can read the status of the keypad. Here's an example that defines and initializes a Keypad variable named kp:

```
Keypad kp = Keypad(makeKeymap(keymap), rowPins, colPins, rows, cols);
```

The Keypad variable is initialized using the keymap, rowPins, colPins, rows, and cols variables you've created, as described in the previous sections.

Detecting when a key has been pressed

After you've created a Keypad variable (see the preceding section), you can use that variable to read the status of the keypad by calling the Keypad variable's waitForKey function. This function waits until one of the keys on the keypad is pressed. It then returns the char value that corresponds to the key that was pressed, according to the Keymap array.

You'll want to call the waitForKey function somewhere in the loop function, like this:

```
void loop() {
  char key = kp.waitForKey();
  // Do something based on which key was pressed.
}
```

Here's a simple example that just echoes the key value on the serial port, which you can view by opening the Serial Monitor in the Arduino integrated development environment (IDE):

```
void loop() {
  char key = kp.waitForKey();
  Serial.println(key);
}
```

Note that for this to work, you'll need to initialize the serial port and provide a baud rate. This is usually done in the setup function with a line like this:

```
Serial.begin(9600);
```

If you open the serial monitor in the IDE (choose Tools⇨Serial Monitor), you'll see the character that corresponds to each button pressed on the keypad.

TECHNICAL STUFF

A major drawback of the waitForKey method is that your sketch can't do any other work while it's waiting for a key press. The Keypad library provides several alternative methods of detecting key presses that don't hold up the main program waiting for a key to be pressed. I encourage you to research the Keypad library documentation if you want to learn more (see www.arduino.cc/reference/en/libraries/keypad).

Building a Keypad Circuit

In this section, you build a simple project that connects a 4 x 4 keypad to an Arduino UNO. Key presses are echoed on the serial port, which you can monitor from the Arduino IDE via the Serial Port Monitor.

The details for building the project are presented in Project 51, and the complete program to run the project is shown in Listing 6-1.

Project 51: Using a Keypad

In this project, you connect a 4 x 4 keypad to an Arduino UNO and learn how to develop sketches that use it.

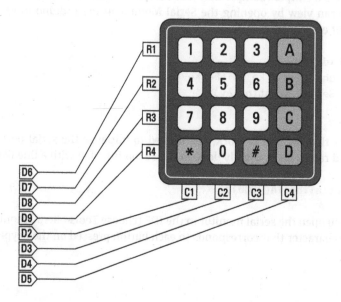

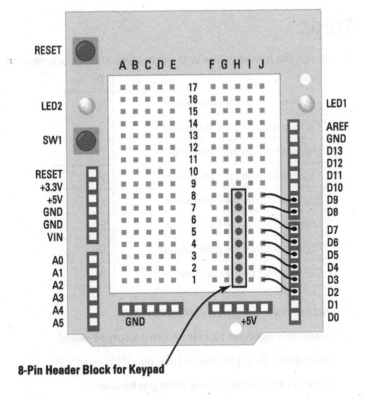

8-Pin Header Block for Keypad

Parts

» One computer with Arduino IDE installed

» One Arduino UNO board

» One USB-B cable

» One UNO prototyping shield

» One 4 x 4 matrix-style keypad

» One eight-pin male-to-male header adapter

» Eight short (⅜-inch) jumper wires.

Steps

1. **Insert the jumper wires in the prototype shield's breadboard.**

From	To
Digital pin D2 on the right header	J1
Digital pin D3 on the right header	J2
Digital pin D4 on the right header	J3
Digital pin D5 on the right header	J4
Digital pin D6 on the right header	J5
Digital pin D7 on the right header	J6
Digital pin D8 on the right header	J7
Digital pin D9 on the right header	J8

2. **Attach the prototype shield to the UNO.**

3. **Insert the eight-pin header on the prototype shield.**

   The header should be inserted into pins H1 through H8.

4. **Connect the keypad to the eight-pin header.**

   The pin for column 1 should be on the H1 side of the header.

5. **Connect the UNO to the computer.**

   Use the mini-B USB connector.

6. **Upload the keypad test program (see Listing 6-1) to the UNO.**

7. **Open the serial monitor by choosing Tools⇨Serial Monitor in the Arduino IDE.**

8. **Press each button on the keypad.**

   As you press each button, the serial monitor will show the button that was pressed (refer to Figure 6-4).

FIGURE 6-4:
Monitoring the
serial output
from the keypad
program.

LISTING 6-1: **Testing a Keypad**

```
// Keypad Tester
// Doug Lowe
// December 5, 2021

#include <Keypad.h>

const byte rows = 4;
const byte cols = 4;

char keymap[rows][cols] = {
  {'1','2','3','A'},
  {'4','5','6','B'},
  {'7','8','9','C'},
  {'*','0','#','D'}
};

byte colPins[rows] = {6, 7, 8, 9};
byte rowPins[cols] = {2, 3, 4, 5};

Keypad keypad = Keypad(makeKeymap(keymap), rowPins, colPins, rows, cols);

void setup() {
  Serial.begin(9600);
}

void loop() {
  char x = keypad.waitForKey();
  Serial.println(x);
}
```

Using an Arduino for Computer Keyboard Input

All Arduino boards have a USB port that you use to upload sketches from a Windows computer to the Arduino. Wouldn't it be great if you could write sketches that would reverse the flow of this USB port, sending data back to the Windows computer?

You can't with an UNO, but some Arduino boards *do* provide this capability. The most popular is the Arduino Leonardo, shown in Figure 6-5.

FIGURE 6-5:
An Arduino
Leonardo board.

In many respects, the Leonardo is very similar to the UNO. In particular, the I/O pin headers are the same, which means that most programs and circuits that will work on an UNO will also work on a Leonardo. All the projects presented so far in this minibook are interchangeable between UNO and Leonardo. In other words, any of the prototype boards built for the projects can be attached to a Leonardo instead of an UNO, and any of the sketches can be run on a Leonardo instead of an UNO.

One major difference, however, is that a Leonardo board contains the same USB circuitry that is found in most computer keyboards. As a result, the Leonardo can emulate a standard computer keyboard. That means you can write sketches that send keystrokes to a Windows computer. The program that happens to be running on the computer will receive those keystrokes and act as if you had typed them on a normal keyboard.

This type of Arduino project can be incredibly useful. For example, you could adapt Project 47 (see Book 6, Chapter 4) to send range information to a computer. If Excel is open on the computer, the range data will fill up the rows of the spreadsheet.

In this section, I show you how to create a Leonardo project that uses a 4 x 4 keypad to send Windows shortcut keys to the computer so you can use the keypad to lock or unlock the computer, minimize windows to reveal the desktop, or start programs from the taskbar.

One other difference between a Leonardo and an UNO is the type of USB port that's installed on the board. The UNO uses an old-fashioned USB-B connector, the same type that's commonly used on printers. The Leonardo, in contrast, uses a more modern and smaller USB-C connector. You'll need to pick up a USB-C cable to connect a Leonardo to a computer, both to upload sketches to the Leonardo and to enable the Leonardo to send keyboard input to the computer.

Using the Keyboard library

The secret to emulating a computer keyboard on a Leonardo lies in the Keyboard library. This library is a built-in part of the Arduino IDE, so you don't have to use the Library Manager to install it. However, you do have to tell the Arduino IDE that you're working with a Leonardo board rather than an UNO board. If you forget to do this, the compiler won't recognize any of the Keyboard calls.

The Keyboard library includes the following functions to help your sketches send keyboard input to a computer attached via the USB port:

>> Keyboard.begin: Starts a keyboard emulation session. You'll typically call this function from the sketch's setup function.

>> Keyboard.end: Ends a keyboard session. You may not need this in your sketch, but if you want the Leonardo to act like a keyboard at some times but not others, you can use Keyboard.end to turn off the keyboard.

>> Keyboard.print: Sends a complete string, as if you had typed the string on the computer.

>> Keyboard.println: Works just like Keyboard.print, but also sends the Enter key at the end of the string.

>> Keyboard.write: Sends a single keystroke.

>> Keyboard.press: Presses and holds a key. The key is not released until you call the Keyboard.release or Keyboard.releaseAll function.

>> Keyboard.release: Releases a specific key.

>> Keyboard.releaseAll: Releases all keys that are currently pressed.

WARNING

You must consider the ramifications of sending keyboard input to a computer from an Arduino sketch without ensuring that those keystrokes are triggered by some other form of input to your sketch. In particular, you should *never* use any of the keyboard functions that send keystrokes (such as Keyboard.println) unconditionally in the loop function of your sketch. If you do, the Leonardo will start sending keystrokes to your computer the moment you plug it in to a USB port on the computer. The only way to stop it will be to unplug the Leonardo. What's worse, you'll lose the ability to upload other sketches to the Leonardo. That's because the Leonardo will be too busy sending keystrokes to the computer to accept a new sketch. The Arduino IDE won't be able to communicate properly with the Leonardo. You'll effectively have ruined the Leonardo. (You can discover ways of fixing this by scouring the Internet, but the process is convoluted and you may be better just buying another Leonardo.)

With that warning duly noted, let's look at some code. To use the Keyboard library, you must include it. Use this command near the start of your sketch:

```
#include <Keyboard.h>
```

Then use a command such as this in the setup function to start a Keyboard session:

```
void setup() {
  Keyboard.begin();
}
```

Sending a line via the keyboard

Here's a snippet of code from a loop function that waits for a key to be pressed on a keypad. Then the program types one of four Shakespearean insults based on which key was pressed on the keypad:

```
char x = keypad.waitForKey();
if (x=='A')
{
  Keyboard.println("Thine face is not worth sunburning.");
}
if (x=='B')
```

```
{
  Keyboard.println("Would thou were clean enough to spit upon.");
}
if (x=='C')
{
  Keyboard.println("I'd beat thee, but I would infect my hands.");
}
if (x=='D')
{
  Keyboard.println("The tartness of his face sours ripe grapes.");
}
```

Using the switch statement to simplify choices

Input via a keypad cries out for a simpler way to determine which key was pressed than a bunch of if statements, one after the other. Fortunately, the Arduino language provides such a simplification via a statement called switch. The switch statement tests the value of a single variable and provides several different outcomes based on that value.

The snippet of code in the preceding section, which generated Shakespearean insults based on a key press, can be written with a switch statement like this:

```
char x = keypad.waitForKey();
switch (x)
{
  case 'A':
    Keyboard.println("Thine face is not worth sunburning.");
    break;

  case 'B':
    Keyboard.println("Would thou were clean enough to spit upon.");
    break;

  case 'C':
    Keyboard.println("I'd beat thee, but I would infect my hands.");
    break;
  case 'D':
    Keyboard.println("The tartness of his face sours ripe grapes.");
    break;
}
```

Here are the important rules to follow when composing switch statements:

>> The keyword switch names a variable in parentheses and is followed by one or more case statements, all of which are enclosed within a set of curly brackets.

>> Each case statement provides a value that is matched with the variable. The value is *not* enclosed in parentheses. The value is followed by a colon.

>> The statements after the colon are executed if the variable matches the value. Like other statements, these statements must end with a semicolon.

>> You can write as many statements as you want after a case statement. You can also use other structural statements, such as if statements or for loops.

>> Use the break statement to mark the end of a group of statements that should be executed for each case value. When the break statement is executed, the entire switch statement finished and no other case statements are processed.

>> If the variable doesn't match the value of the first case statement, the next case statement is evaluated. If that doesn't match, the case statement after that is evaluated, until all case statements have been checked.

TIP

Shakespeare truly was one of the greatest composers of insults ever to pick up a pen. Search the Internet for "Shakespearean insults" to see what I mean. You can even find Shakespearean insult generators that will create a customized Shakespearean insult just for you!

Pressing and holding keys

One of the important features of a standard keyboard is to use keys in combination. For example, pressing Ctrl+Alt+Del brings up a menu that lets you select actions such as locking the computer, signing out, switching users, or calling up the Windows Task Manager.

The Keyboard.press function lets you use keys in combination, by pressing and holding one or more keys and then using Keyboard.print or Keyboard.write to send keystrokes along with the modifier keys.

To facilitate this, the Arduino language defines a set of constants that represent the modifier keys. These constants are listed here. Note that some keys have two versions — one on the left side of the keyboard and the other on the right.

KEY_LEFT_CTRL	KEY_PAGE_UP
KEY_LEFT_SHIFT	KEY_PAGE_DOWN
KEY_LEFT_ALT	KEY_HOME
KEY_LEFT_GUI	KEY_END
KEY_RIGHT_CTRL	KEY_CAPS_LOCK
KEY_RIGHT_SHIFT	KEY_F1
KEY_RIGHT_ALT	KEY_F2
KEY_RIGHT_GUI	KEY_F3
KEY_UP_ARROW	KEY_F4
KEY_DOWN_ARROW	KEY_F5
KEY_LEFT_ARROW	KEY_F6
KEY_RIGHT_ARROW	KEY_F7
KEY_BACKSPACE	KEY_F8
KEY_TAB	KEY_F9
KEY_RETURN	KEY_F10
KEY_ESC	KEY_F11
KEY_INSERT	KEY_F12
KEY_DELETE	

Here's an example of a code snippet that types the Windows shortcut sequence Windows+L, which locks the computer:

```
Keyboard.press(KEY_LEFT_GUI);
Keyboard.print("l");
Keyboard.releaseAll();
```

In this example, the Keyboard.press function is called to press and hold the left graphical user interface (GUI) key, which is the Windows key. Then the letter *l* is sent. Finally, all keys are released. The result is that the computer is immediately locked.

Building a Windows Keypad Gadget

I want to wrap up this chapter with a project that uses a 4 x 4 keypad to provide handy shortcuts to several commonly used Windows shortcut key combinations. This project actually uses the same prototype breadboard built for Project 51. The only difference is that the breadboard is placed on a Leonardo rather than an UNO, and a different program is used.

The following functions are provided by this gadget:

Keypad Press	Windows Function
A	Locks the computer.
B	Unlocks the computer by entering a PIN or password.
C	Minimizes all windows.
D	Restores all windows.
1–9	Opens one of the programs pinned to the taskbar. (The programs are opened in the order in which they appear on the taskbar.)

Project 52: A Windows Keypad Gadget

In this project, you connect a 4 x 4 keypad to an Arduino Leonardo and learn how to use it to send keystrokes to a Windows computer.

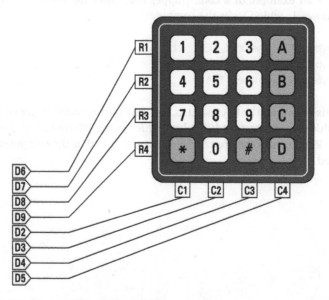

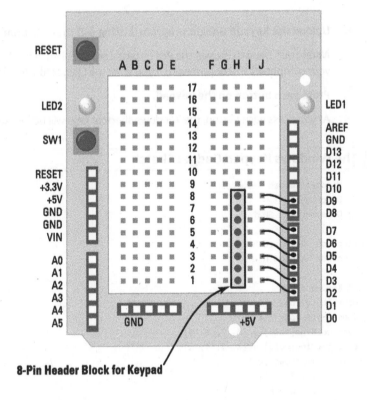

8-Pin Header Block for Keypad

Parts

>> One computer with Arduino IDE installed

>> One Arduino Leonardo

>> One USB-C cable

>> The assembled prototyping shield from Project 51, including the 4 x 4 keypad

Steps

1. If you haven't already done so, follow the steps in Project 51 to assemble the prototyping shield and the keypad.

2. Attach the assembled prototyping shield to the Leonardo.

3. Connect the Leonardo to the computer.

   Use the USB-C USB connector.

4. **Upload the Keypad Gadget program (Listing 6-2) to the Leonardo.**

   Note: You'll need to change the PIN or password supplied in the program to your computer's PIN or password if you want the Unlock function to work.

5. **Press each button on the keypad.**

   As you press each button, the Windows shortcut keys will be invoked.

LISTING 6-2: **The Windows Keypad Gadget Sketch**

```
// Windows Keypad Gadget
// Doug Lowe
// December 10, 2021

#include <Keyboard.h>
#include <Keypad.h>

// Keypad setup
const byte rows = 4;
const byte cols = 4;
byte colPins[rows] = {6, 7, 8, 9};
byte rowPins[cols] = {2, 3, 4, 5};
char keymap[rows][cols] = {
  {'1','2','3','A'},
  {'4','5','6','B'},
  {'7','8','9','C'},
  {'*','0','#','D'}
};

Keypad kp = Keypad( makeKeymap(keymap), rowPins, colPins, rows, cols);

void setup() {
  Keyboard.begin();

}

void loop() {
  char x = kp.waitForKey();
  int number = 0;
  switch(x) {
    case 'A':                           // Lock computer
      Keyboard.press(KEY_LEFT_GUI);
      Keyboard.print("l");
      Keyboard.releaseAll();
      break;
```

```
case 'B':                                    // Unlock computer
  Keyboard.press(KEY_LEFT_CTRL);
  Keyboard.press(KEY_LEFT_ALT);
  Keyboard.press(KEY_DELETE);
  Keyboard.releaseAll();
  delay(100);
  Keyboard.println("******");                // Use your PIN here!
  break;

case 'C':                                    // Minimize all windows
  Keyboard.press(KEY_LEFT_GUI);
  Keyboard.print("m");
  Keyboard.releaseAll();
  break;

case 'D':                                    // Restore all windows
  Keyboard.press(KEY_LEFT_GUI);
  Keyboard.press(KEY_LEFT_SHIFT);
  Keyboard.print("m");
  Keyboard.releaseAll();
  break;

case '1':
  OpenProgram(1);
  break;

case '2':
  OpenProgram(2);
  break;

case '3':
  OpenProgram(3);
  break;

case '4':
  OpenProgram(4);
  break;

case '5':
  OpenProgram(5);
  break;

case '6':
  OpenProgram(6);
  break;
```

(continued)

LISTING 6-2: *(continued)*

```
      case '7':
        OpenProgram(7);
        break;

      case '8':
        OpenProgram(8);
        break;

      case '9':
        OpenProgram(9);
        break;
      }

  delay(200);
}

void OpenProgram(int num) {
      Keyboard.press(KEY_LEFT_GUI);
      Keyboard.press(KEY_LEFT_SHIFT);
      Keyboard.print(num);
      Keyboard.releaseAll();
}
```

7

Working with Raspberry Pi

Contents at a Glance

IN THIS CHAPTER

» Finding out what a Raspberry Pi is

» Looking at various Raspberry Pi versions

» Connecting all the parts to make your Raspberry Pi work

» Installing the Raspberry Pi OS operating system

» Getting used to Raspberry Pi OS

» Writing your first Python program

Chapter **1**

Introducing Raspberry Pi

I n Book 6, you learn how to work with Arduino microprocessors, one of the most popular microprocessors available. In this minibook, you learn how to work with Raspberry Pi, a popular alternative to Arduino.

Physically, a Raspberry Pi resembles an Arduino. However, the Raspberry Pi is much more than a microcontroller; it is a full-blown computer system, implemented on a single small card. In fact, a Raspberry Pi has most of the features commonly found on a desktop or laptop computer. Yet besides its small size, a Raspberry Pi has other features not commonly found on a desktop, such as the ability to directly control digital I/O pins. Thus, you can use a Raspberry Pi with external devices such as LEDs, push buttons, potentiometers, various types of sensors, and servo or stepper motors.

The biggest difference between a Raspberry Pi and an Arduino is that the Raspberry Pi has an operating system. Arduino is really a single-program computer that doesn't have an operating system. You simply upload a program to an Arduino and the program runs. In contrast, a Raspberry Pi runs a variation of the Linux operating system known as the Raspberry Pi OS.

Raspberry Pi OS is similar in many ways to Microsoft Windows or the Apple macOS, so a Raspberry Pi can do many of the things that a desktop or laptop computer can do. Support for things like monitors, keyboards, mice, network, and sound are built-in features of a Raspberry Pi.

In this chapter, you learn exactly what a Raspberry Pi is, how to get one, how to hook it up, and how to get it running. Then, in the next few chapters, you learn how to write programs for the Raspberry Pi and how to add external components to it.

Introducing the Raspberry Pi

A Raspberry Pi (sometimes just called a Pi for short) is a very small computer. It contains most of the components found in a traditional desktop computer, but all squeezed onto a small board about the size of a deck of playing cards. The newest version of the Raspberry Pi, called the Raspberry Pi 4, is shown in Figure 1-1. This version of the Raspberry Pi includes all of the following packed onto the board:

» **CPU:** A quad-core 64-bit ARM Cortex-A72 microprocessor running at 1.5GHz.

» **RAM:** 2GB, 4GB, or 8GB.

» **USB ports:** Two USB 2.0 ports, two USB 3.0 ports. These standard-size USB-A ports can be used to connect any USB device, including a keyboard, a mouse, or a flash drive.

» **Video:** A built-in graphics processor that can support two monitors at 4K resolution (3,840 x 2,160).

» **HDMI:** Two micro HDMI connectors mounted on the board to connect video monitors.

» **Display serial interface (DSI):** A display interface designed to connect to small LCDs via a 15-pin ribbon cable.

» **MicroSDHC card:** The MicroSDHC card acts as the computer's disk drive. The operating system (Linux) is installed on the MicroSD card, along with any other software you want to use.

» **Ethernet networking:** A built-in RJ-45 connector for 1 Gbps networking.

» **802.11n wireless network:** A built-in wireless network connection. The antenna is actually built into the board itself, so no external antenna is needed.

» **Bluetooth:** Built-in Bluetooth networking for wireless devices such as a keyboard, a mouse, and headphones.

FIGURE 1-1:
A Raspberry Pi 4.

>> **Camera serial interface (CSI):** A special interface designed to connect to a camera device via a 15-pin ribbon cable.

>> **Audio:** A 3.5 mm audio jack for sound applications.

>> **Power:** The Raspberry Pi is powered by a 5 V supply connected to the board via a USB-C connection.

>> **GPIO header:** The most interesting thing about the Raspberry Pi from an electronics enthusiast's perspective is the 40-pin GPIO header, which provides access to a variety of features, including 26 general-purpose input-output (GPIO) pins. These pins work the same as the digital I/O pins found on Arduino microprocessors, and can be accessed via programs that you write for the Raspberry Pi. You can use these GPIO pins as output pins to connect to devices such as LEDs, servo or stepper motors, and so on. Or, you can use them as input pins to read input from external switches, potentiometers, or other types of sensors.

WARNING

Unlike the digital I/O pins found on Arduino microprocessors, the Raspberry Pi GPIO pins work at a voltage level of 3.3 V rather than 5 V to indicate HIGH signals. You'll need to adjust your circuits accordingly to deal with the smaller input and output voltage levels. In particular, if you apply a 5 V input to GPIO input pin, you run the risk of damaging your Raspberry Pi.

Considering Raspberry Pi Versions

Since the introduction of the original Raspberry Pi back in 2012, an almost dizzying array of versions have been released. The Raspberry Pi has been through four generations, known as 1, 2, 3, and 4. Each new generation has incorporated new features.

The Raspberry Pi 2 was introduced in 2015. It improved upon the Raspberry Pi 1 by using a more powerful processor and additional RAM, while holding the price the same. Importantly, the Raspberry Pi 2 also increased the number of GPIO ports from 17 to 26. Unfortunately, in doing so the GPIO pin configuration was changed, which means that projects created for the Raspberry Pi 1 are not compatible with projects created for the Raspberry Pi 2.

In 2016, the Raspberry Pi 3 was introduced. It improved upon the Raspberry Pi 2 by adding a more powerful processor, faster RAM, and built-in wireless networking. (Wireless networking with a wireless Raspberry Pi 2 was usually done using a wireless adapter inserted into one of the USB ports.)

The fourth generation Raspberry Pi 4 was introduced in 2019. It has a faster processor and more RAM — you can choose 2GB, 4GB, or 8GB. The Raspberry Pi 4 also supports dual monitors and 4K resolution.

The price of all four versions of the Raspberry Pi has been $35, though recent chip shortages have driven the prices up — especially for the Pi 4 versions. All the projects in this book use the Raspberry Pi 4, but they'll also work on a Raspberry Pi 3 or even a Pi 2.

The first-generation Raspberry Pi 1 was available in two models: Model A and Model B. The Model A version was a slimmed-down version of the Model B, with less memory and fewer I/O options but consuming less power. In 2014, these two models were replaced with the Model A+ and Model B+, which added additional capabilities.

TECHNICAL STUFF

The original Raspberry Pi 1 models had fewer GPIO pins than the newer versions. The original versions provided a 26-pin header block that provided access to 17 GPIO ports. On the newer versions, a 40-pin header provides access to 26 GPIO ports.

All the projects shown in this book will work with a Raspberry Pi 2, 3, or 4. If you don't yet own a Raspberry Pi, I suggest you purchase a Raspberry Pi 4 with 4GB of RAM as part of a kit that includes the materials you'll need to get started. These simple kits include a power supply to power your Raspberry Pi, a microSD card already loaded with the operating system software, and a small book to get you started. Some kits include additional goodies, such as an HDMI cable to connect your Raspberry Pi to a monitor and perhaps additional components such as LEDs, resistors, jumper cables, and a breadboard.

Setting Up a Raspberry Pi

Before you can fire up your Raspberry Pi and start building projects, you need to do some basic setup work. Start by setting up the hardware. You'll need the following to set up your Pi so that you can program it for the projects presented in this book:

>> **A Raspberry Pi 2, 3 or 4.**

>> **A suitable power supply:** The Raspberry Pi requires a 5 V power supply connected via the USB-C connection on the card. For a Raspberry Pi 4, you should use at least a 3 A power supply. Older versions, such as Pi 3 or Pi 2, can use smaller power supplies (2.5 A for Pi 3 or 2.0 A for Pi 2.)

>> **A monitor:** You don't need a huge monitor, but I suggest at least 17 inches. You also don't need 4K resolution, but I suggest at least 1,920 x 1,080.

>> **An HDMI cable:** If your monitor has an HDMI connection, you'll need a cable with a full-size HDMI connector on one end and a micro-HDMI connector on the other end. If your monitor has some other type of connection, such as DVI or VGA, you'll need an adapter to connect your monitor to the Pi's HDMI connector.

>> **A USB keyboard:** Any keyboard with a USB connector will do.

>> **A USB mouse:** Any mouse with a USB connector will do.

>> **A microSD card:** The microSD card acts as the disk drive for the Raspberry Pi. The minimum size is 8GB, but 16GB or even 32GB is better.

>> **A network connection:** A network connection is essential to download several of the support packages you'll need for the projects in this book. You can connect your Pi to a network in one of two ways:

- If you have a Raspberry Pi 3 or 4, you can use the built-in Wi-Fi to connect to a wireless network.

- You can use a standard Ethernet cable to plug a Raspberry Pi 2, 3, or 4 into a wired network, provided you have a nearby router or switch with an available network port.

Installing the Raspberry Pi Operating System

Although you can run any version of Linux on a Raspberry Pi, the official version that's designed specifically for Raspberry Pi is called *Raspberry Pi OS*. (It used to be called Raspian, but the name was changed a few years back.) Here are the steps for installing the Raspberry Pi OS:

1. **Insert a blank micro-SD card into your Windows computer.**

    You'll need a micro-SD card adapter to do this; most micro-SD cards come with the necessary adapter.

2. **Open a web browser (such as Google or Microsoft Edge) and go to** www.raspberrypi.com/software.

3. **Download the Raspberry Pi Imager program.**

    For Windows, click Download for Windows. (If you're a Mac user, you can choose Download for macOS instead. The remaining steps in this procedure are for Windows, but the steps for Mac users are similar.)

4. **When the download has completed, run the downloaded file.**

    Note that for all browsers, the file is downloaded to your Downloads folder. If you're not sure how to run the downloaded file on your browser, you can always open a File Explorer window and open your Downloads folder. Then, double-click the imager_1.6.2.exe file to run it. (The 1.6.2 is the version number; if your version is more recent, you'll see a different number.)

5. **Click Install to install the imager.**

    The installation should take less than a minute.

6. **When the installation is complete, click Finish.**

    The Raspberry Pi Imager program appears, as shown in Figure 1-2. As you can see, two buttons are visible: Choose OS and Choose Storage.

7. **Click Choose OS.**

    A list of operating system options appears, as shown in Figure 1-3. Feel free to explore the wide variety of operating systems you can install on a Raspberry Pi. After you've gained some experience with Raspberry Pi, you may want to try a different operating system.

8. **Click Raspberry Pi OS (32-bit).**

    This action selects the standard Raspberry Pi OS.

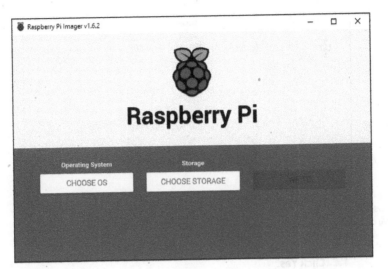

FIGURE 1-2:
The Raspberry Pi
Imager program.

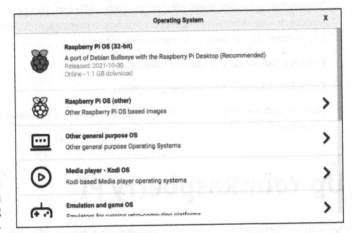

FIGURE 1-3:
Choose an
operating
system.

9. **Click Choose Storage.**

   A list of available storage devices appears, as shown in Figure 1-4. (In most cases, you'll see just one option.)

10. **Click the storage device you want to install the Raspberry Pi OS on.**

   You're returned to the Imager program. A third button labeled Write is now visible.

11. **Click Write.**

   You're asked whether you want to continue.

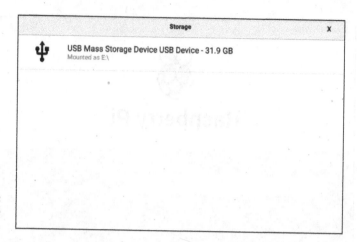

FIGURE 1-4:
Choose
your SD card.

12. **Click Yes.**

The operating system is installed on the micro-SD card. This process takes a minute or two. When the installation finishes, a message indicating that you can remove the SD card from the reader appears.

13. **Remove the SD card and click Continue.**

14. **Close the Raspberry Pi Imager program.**

You can find the Imager program on your Start menu if you need to use it again.

Starting Up Your Raspberry Pi

After you've installed the operating system on a micro-SD card, you can fire up your Raspberry Pi. Grab your Raspberry Pi and follow these steps:

1. **Insert the micro-SD card in the card slot.**

The card slot is located on the bottom of the Raspberry Pi, on the side opposite from the USB ports.

2. **Plug a mouse and keyboard into the USB ports.**

If you're using a wireless mouse/keyboard combo, just plug the USB wireless dongle into one of the Pi's USB ports.

3. Connect a monitor.

If you're using a Raspberry Pi 4, you'll need a micro-HDMI cable. For earlier Pi versions, you'll need a regular HDMI cable.

4. Connect the Raspberry Pi to a USB-C power source.

When you do, the Raspberry Pi will boot directly into the Raspberry Pi OS. After a few moments, you'll be greeted by the screen shown in Figure 1-5.

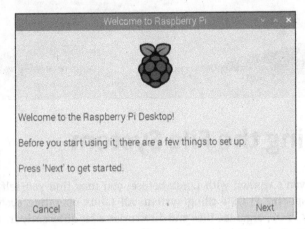

FIGURE 1-5:
The initial
configuration
for the
Raspberry
Pi OS.

5. Follow the steps for the initial configuration of the Raspberry Pi OS.

The configuration wizard walks you through the following configuration options:

- Country, language, and time zone
- Password for the default user account, which is named pi
- Screen setup, which confirms that the Linux task bar fits on the screen
- Wi-Fi network password so that the Pi can connect to the Internet
- Software updates

When the configuration completes, the Raspberry Pi OS prompts you to restart your Pi so that the changes can take effect.

6. Click Restart to reboot the Pi.

When the Pi has rebooted, you're shown the Raspberry Pi Desktop, as shown in Figure 1-6.

FIGURE 1-6:
The Raspberry Pi
Desktop.

Understanding the File System

If you haven't worked with Linux before, you may find yourself a bit confused about Raspberry Pi OS's filing system. All Linux operating systems, including Raspberry Pi OS, organize files and directories a bit differently from the Windows file system you're probably used to. Three of the most obvious differences are actually superficial, but they can trip you up if you're not mindful of them:

>> **Linux uses forward slashes rather than backward slashes to separate directories.** Thus, /home/pi is a valid path in Linux. (If you're not an experienced Linux user, you may have trouble finding the forward slash key. It's located on the question mark key next to the Shift key on the right side of the keyboard.)

>> **Linux filenames don't use extensions.** You can use periods within a filename, but unlike Windows, the final period doesn't identify a file extension.

>> **Linux is case-sensitive.** In Windows, upper- and lowercase letters don't matter in file or folder names. Thus, workfiles and WorkFiles are interchangeable. Not so in Linux: In Linux, workfiles and WorkFiles would refer to two different folders.

Apart from these superficial differences, the big difference between the Linux and Windows file system is that Linux treats everything in the entire system as a file, and it organizes everything into one gigantic tree of directories that begins at a single root directory.

When I say, "Everything is treated as a file," I mean that hardware devices such as floppy drives, serial ports, Ethernet adapters, even the GPIO pins on a Raspberry Pi are treated as files.

The root of the entire Linux file system is the top-level directory of the micro-SD card from which the Raspberry Pi boots. If you insert a USB flash drive or a USB disk drive into one of the Raspberry Pi's USB ports, that drive will be grafted into the tree as a directory called *mount points*. Thus, a directory in the Linux file system may actually be a device.

The root-level directories in a Linux filing system are listed in Table 1-1. If you open the File Manager by clicking the file cabinet icon at the top of the screen, these are the directories you'll see.

TABLE 1-1

Root-Level Directories in a Linux File System

Directory	Description
/bin	Essential command binaries
/boot	Static files of the boot loader
/dev	Devices
/etc	Configuration files for the local computer
/home	Home directories for users
/lib	Shared libraries and kernel modules
/media	Mount point for removable media such as USB flash drives
/mnt	Mount point for nonremovable file systems
/proc	Important kernel and process files
/opt	Add-on applications and packages
/root	Home directory for the root user
/sbin	Essential system binaries
/srv	Server data such as web pages, FTP files, and so on
/tmp	Temporary files
/usr	Read-only, shared files such as binaries for user commands and libraries
/var	Variable data files

Writing Your First Raspberry Pi Program

Raspberry Pi OS comes with Python built in, along with a fancy integrated development environment (IDE) called *Thonny*. To run this program, click the Raspberry Pi icon (shown in the margin), and then choose Programming ⇨ Thonny Python IDE. Thonny comes to life, as shown in Figure 1-7.

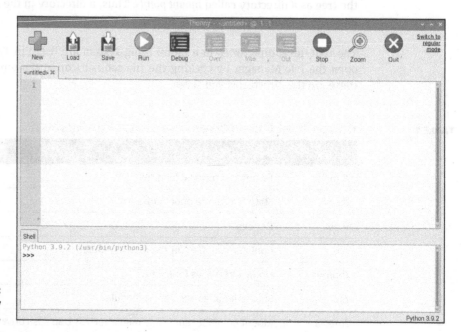

FIGURE 1-7:
The Thonny
Python IDE.

TIP

By default, Thonny opens in a simplified view called *simple mode*. Simplified mode includes a simple toolbar that provides single-click access to most of the functions you'll need to develop basic Python programs. If you prefer, you can switch to *regular mode*, which adds a menu with a full complement of Python programming options, as shown in Figure 1-8.

Near the bottom of the Thonny screen (in both Simple and Regular modes), you'll find an area labeled Shell. You can run Python statements directly within the shell without creating a program; when you type a Python statement in the shell and press the Enter key, the statement is executed immediately, and any results it generates are shown in the shell.

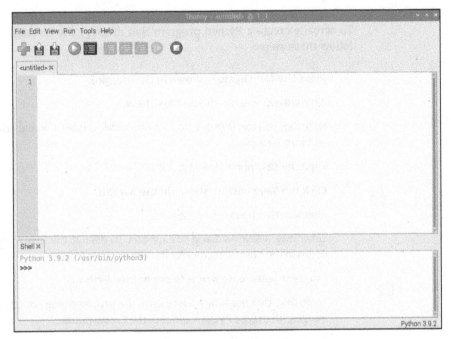

FIGURE 1-8:
Thonny in regular
mode, with
menus!

For example, type the following in the shell:

```
print("Hello, World!")
```

The shell responds by typing Hello, World!

If you make a mistake, the shell will let you know right away. For example, try typing this:

```
print("Hello, World!)
```

Notice that the closing quotation mark is missing after the text Hello, World!. The shell responds as follows:

```
    File "<phshell>", Line 1
        print("Hello, World!)
                            ^
    SyntaxError: EOL while scanning string literal
```

This somewhat cryptic response indicates that you made a Python mistake — specifically, Python reaches the end of the line (EOL) before it finds a closing quotation mark.

To actually create a Python program that displays the message `Hello, World!`, follow these steps:

New

1. **Press the New button (shown in the margin).**

   Alternatively, you can choose File ➪ New.

   Note that you can skip this step if a new window labeled <untitled> is already visible in Thonny.

2. **Type the text** print("Hello, World!").

Save

3. **Click the Save button (shown in the margin)**

   Alternatively, choose File ➪ Save As.

   Either way, a Save As dialog box appears. By default, the Save As dialog box opens in the home directory for the current user (/home/pi).

4. **I suggest you create a new folder named Python.**

   To do that, click the New Folder icon in the Save As dialog box (shown in the margin). Then type **Python** and click the Create button.

5. **Type** hello **as the filename, and then click OK.**

   The file is saved using the name hello. Figure 1-9 shows how the hello.py program appears after it has been saved.

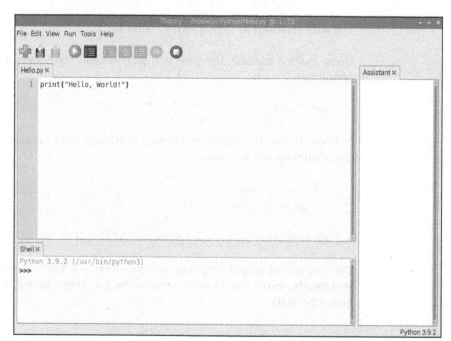

FIGURE 1-9:
The hello.py
program.

6. **Choose Run ⇨ Run Module (or press F5).**

   This action runs the hello.py program. The output from the program (that is, the message `Hello, World!`) appears in the Python Shell window, as shown in Figure 1-10.

7. **You're done!**

   Congratulate yourself! You have successfully written your first Python program.

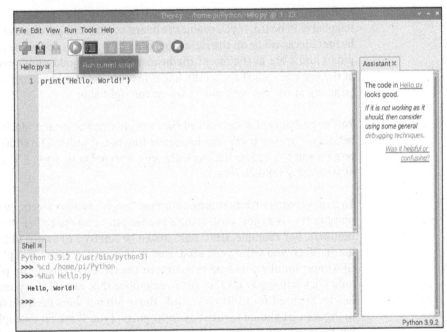

FIGURE 1-10:
The Hello, World!
Program in
action.

Examining GPIO Ports

General-purpose input/output (GPIO) ports make Raspberry Pi much more than just a very small computer. GPIO ports allow you to control your own electronic circuits directly from the Pi. The 40-pin GPIO header on the Raspberry Pi 2 and 3 provides access to a total of 17 GPIO ports that can be controlled from programs that you write in Python or other languages.

To use a GPIO port, you must first configure the port for input or output. When configured for output, a GPIO port can be set to a HIGH (+3.3 V)) or LOW (0 V) condition by your programs. When configured for input, a program can easily determine the current state (HIGH or LOW).

Figure 1-11 lists the function of each pin on the Raspberry Pi 40-pin header block. (Note that this diagram applies to Raspberry Pi versions 2, 3, and 4. The obsolete Raspberry Pi 1 had a different pin layout.)

Unfortunately, the pins on the header block are not labeled in any way on the Raspberry Pi board. If you orient the board so that USB ports are at the bottom, the header block will be on the right edge of the board, just above the USB ports. Then, pins 1 and 2 are at the top of the header block. The odd numbered pins (starting at the top with pin 1) are on the left side of the header block; even numbered pins (starting at the top with pin 2) are on the right side.

You've probably already noticed that the pins on the header block don't appear to be laid out in any particular logical or functional order. The GPIO ports are scattered about the header haphazardly, interspersed with 3.3 V, 5 V, and Ground pins at seemingly random locations.

To avoid confusion when programming Raspberry Pi projects, people commonly number the GPIO ports using their header pin numbers rather than the GPIO port numbers. For example, GPIO port 19 can be referred to as pin 10 in a Python program. When you write your programs in this way, you can completely ignore the GPIO port number, instead referring to the GPIO port using its pin number. The only trick when you do this is to remember that not all the pins on the 40-pin header are used for GPIO ports; only those pin numbers designated as GPIO ports in Figure 1-11 should be used as GPIO ports. (You'll see an example of this numbering in an actual program later in this chapter.)

WARNING

The GPIO ports on a Raspberry Pi are rated for about 16 mA of current flow each, with a total of 40 mA across all GPIO ports combined. When designing circuits that interface with the GPIO ports, keep this limit in mind. Make sure that you provide adequate current-limiting resistors to avoid damaging your Pi.

Pin	Function			Function	Pin
1	3.3V	◯	◯	5V	2
3	GPIO 2	◯	◯	5V	4
5	GPIO 3	◯	◯	Ground	6
7	GPIO 4	◯	◯	GPIO 14	8
9	Ground	◯	◯	GPIO 15	10
11	GPIO 17	◯	◯	GPIO 18	12
13	GPIO 27	◯	◯	Ground	14
15	GPIO 22	◯	◯	GPIO 23	16
17	3.3V	◯	◯	GPIO 24	18
19	GPIO 10	◯	◯	Ground	20
21	GPIO 9	◯	◯	GPIO 25	22
23	GPIO 11	◯	◯	GPIO 8	24
25	Ground	◯	◯	GPIO 7	26
27	ID_SD	◯	◯	ID_SC	28
29	GPIO 5	◯	◯	Ground	30
31	GPIO 6	◯	◯	GPIO 12	32
33	GPIO 13	◯	◯	Ground	34
35	GPIO 19	◯	◯	GPIO 16	36
37	GPIO 26	◯	◯	GPIO 20	38
39	Ground	◯	◯	GPIO 21	40

FIGURE 1-11: Raspberry Pi header pins.

Connecting an LED to a GPIO Port

If you configure a GPIO port as an output port, you can use the port to drive an LED. You'll need an LED and an appropriate current-limiting resistor. Figure 1-12 shows a typical schematic. In this case, the LED is connected to a GPIO pin on the Pi. A 330 Ω resistor is used to limit the current flowing through the LED, and the LED's cathode is connected to ground.

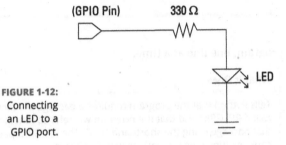

FIGURE 1-12: Connecting an LED to a GPIO port.

One detail that you should be aware of is that the header block on the Raspberry Pi uses male pins rather than female sockets as are found on Arduino and microcontrollers. As a result, to connect a breadboard circuit to pins on the Raspberry Pi, you'll need jumper wires that have male pins on one end (to plug into a breadboard hole) and female sockets on the other end (to plug into the male pins on the Raspberry Pi header). You can purchase these jumper wires from many Internet sources; just search the web for *Raspberry Pi jumper wires* and you'll find what you need. The wires you want will typically be called something like "female/male extension jumper wires" or "jumper wires (M/F)."

Flashing an LED in Python

Now that you know how to connect an LED to a Raspberry Pi GPIO port, let's examine how you can use Python to turn the LED on and off. In Chapter 2 of this minibook, I give you many more details about the Python programming language. For now, let's just take a quick look at a Python program that flashes an LED connected to GPIO port 2, which is physical pin 3 on the Pi. The LED will turn on and off alternately at half-second intervals. The complete program is shown in Listing 1-1.

LISTING 1-1: **An LED Flasher Program in Python**

```python
import RPi.GPIO as GPIO
import time

GPIO.setmode(GPIO.BOARD)
GPIO.setup(3, GPIO.OUT)

while True:
    GPIO.output(3, GPIO.HIGH)
    time.sleep(0.5)
    GPIO.output(3, GPIO.LOW)
    time.sleep(0.5)
```

Let's take a look at this program, one line at a time:

Program Line	What It Does
import RPi.GPIO as GPIO	Tells Python that the program requires a package of functions called RPi.GPIO and that the program will refer to functions in that package using the shorthand GPIO. The RPi.GPIO package contains functions for working with GPIO ports.

Program Line	What It Does
import time	Tells Python that the program requires a package of functions called time. This package is required so that the program can control the timing of the flashing LED.
GPIO. setmode(GPIO.BOARD)	Indicates that the program will refer to GPIO ports using the pin numbers on the Raspberry Pi board rather than using the internal GPIO port numbers.
GPIO. setup(3, GPIO.OUT)	Indicates that the program will use the GPIO port on pin 3 as an output port.
while True:	Sets up a loop that never ends. All the indented statements after this line are executed in sequence over and over again, until the program is stopped by the user.
GPIO.output(3, GPIO.HIGH)	Sets the status of the port on pin 3 to HIGH. This statement turns on the LED.
time.sleep(0.5)	Pauses the execution of the program for half a second. This allows the LED to stay on for half a second.
GPIO. output(3, GPIO.LOW)	Sets the status of the port on pin 3 to LOW. This turns the LED off.
time.sleep(0.5)	Pauses the execution of the program for half a second. This allows the LED to stay off for half a second.

For as long as the program is allowed to run, the four statements indented beneath the while statement will be executed over and over again, in the sequence list. First, the LED turns on for half a second; then the LED turns off for half a second. The flashing continues until the user stops the program.

Building a Raspberry Pi LED Flasher

Project 51 shows how to build a simple breadboard circuit that connects an LED to a Raspberry Pi via the GPIO port on pin 3 of the header. Figure 1-13 shows the completed project.

FIGURE 1-13:
The Raspberry
Pi LED Flasher
(Project 51).

Project 51: Blinking an LED with a Raspberry Pi

In this project, you connect an external LED to a Raspberry Pi, and then use a simple sketch to turn the LED on and off at 0.5-second intervals.

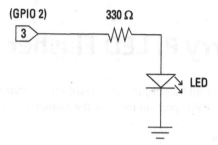

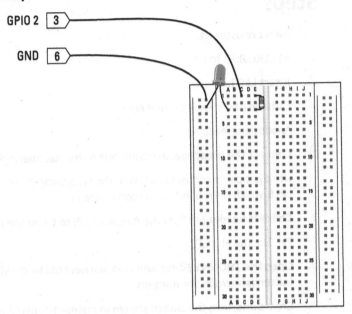

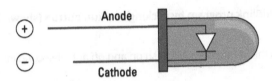

Anode (+)

Cathode (−)

Parts

>> One Raspberry Pi (4 is preferable, but a Pi 2 or 3 will work) with Raspberry Pi OS installed, connected to a monitor, keyboard, mouse, and power

>> One small solderless breadboard

>> One LED

>> One 330 Ω resistor (orange-orange-brown)

>> Two jumper wires (M/F)

Steps

1. **Insert resistor R1.**

   R1 (330 Ω): E1 to E3

2. **Insert LED1.**

   Cathode (short lead): Ground bus

   Anode (long lead): A3

3. **Connect the breadboard ground bus to the Raspberry Pi header pin 6.**

   Use a jumper to connect any hole in the breadboard's ground bus to header pin 6, which is a ground connection on the Pi.

4. **Connect header pin 3 on the Raspberry Pi to C1 on the breadboard.**

   Pin 3 on the Pi is GPIO 2.

5. **Open the THONNY Editor and create a new file by clicking the New button (shown in the margin).**

6. **Create and save the sketch shown in Listing 1-1, using the filename LedBlink.**

7. **Run the LedBlink program by clicking the Run button (shown in the margin).**

   The LED on the breadboard will flash on and off at half-second intervals.

IN THIS CHAPTER

» **Learning the essentials of the Python language**

» **Working with variables and constants**

» **Creating your own function definitions**

» **Working with if logic**

» **Using lists**

» **Looping with for loops**

Chapter **2**

Programming in Python

I n the preceding chapter, you learn how to create a basic Python program that lights up an LED connected to one of the Raspberry Pi's digital output ports. Getting the Python programming environment set up correctly is no small accomplishment, so by all means you should congratulate yourself. Well done!

However, you need to know a little more than how to turn on an LED to call yourself a Raspberry Pi programmer. In this chapter, you learn enough Python programming features to perform truly useful things with the circuits you connect to your Raspberry Pi. You learn how to work with basic Python programming features such as variables, function definitions, if statements, loops, and lists.

This chapter is by no means a comprehensive guide to Python programming. If you want a more complete tutorial on programming with Python, I suggest *Beginning Programming with Python For Dummies*, by John Paul Mueller (Wiley). You can also find plenty of good information about Python programming on the Internet. A great place to start is the official website of the Python Software Foundation (www.python.org).

Looking Closer at Python

As you learn in Chapter 1 of this minibook, Python is one of several programming languages you can use to create programs to run on a Raspberry Pi. You've already seen a short Python program that flashes an LED on and off every 0.5 second. This program is simple, but it introduces many of the basic concepts you need to know to create Python programs. In particular:

» **Python is what is known in the world of computer programming as an** *interpreted language.* That means that when you run a Python program, the Python shell reads the text of your Python program file, and then interprets and runs each statement one at a time. In fact, you can run Python statements without even creating a saved Python program by simply typing the statement into the Python shell. For example, when you type **print("Hello, World!")** in the shell and press Enter, the Python shell interprets the statement you entered and executes it, displaying the text Hello, World! on the console.

This is in contrast to most other programming languages, which are processed by a special program called a *compiler* before they're executed: The compiler converts the text of the program into machine instructions that the computer can execute. In this type of program, called a *compiled program,* the entire program is compiled all at once; then the compiled program is executed. In contrast, in an interpreted language, the text of the program is interpreted and executed one line at a time.

» **Unlike most other programming languages, Python uses indentation to indicate how programming statements are combined together to create blocks.** In a language such as C++ or Java, blocks of related statements are marked by opening and closing curly braces. Not so with Python. In Python, lines that are indented at the same level are considered part of the same block.

You've already seen this in the sample program shown in the previous chapter:

```
while True:
    GPIO.output(3, GPIO.HIGH)
    time.sleep(0.5)
    GPIO.output(3, GPIO.LOW)
    time.sleep(0.5)
```

Here, the four statements that are indented beneath the while statement are considered to be part of the while statement.

The preceding code would be interpreted quite differently if it had been written like this instead:

```
while True:
    GPIO.output(3, GPIO.HIGH)
    time.sleep(0.5)
    GPIO.output(3, GPIO.LOW)
  time.sleep(0.5)
```

In this case, the last line (`time.sleep(0.5)`) would *not* be considered a part of the same block as the three statements before it. As a result, it would not be executed in the loop indicated by the `while` statement.

>> **Python is very loose about how you use variables.** Unlike C++ or Java, you don't have to declare variables in your Python programs or assign data types to variables. Python will figure out based on the context what kind of data is stored in each variable you use.

In case you're wondering, Python is not named after the snake. Python's creators were very fond of the British TV series *Monty Python's Flying Circus* and named the language in honor of the comedy troupe. (IDLE, the standard development environment for Python, technically stands for *Integrated DeveLopment Environment*, but it's also a nod to one of Monty Python's founding members, Eric Idle.)

Building a Test Circuit

All the programs in this chapter are designed to work with the circuit shown in Project 56, which connects eight LEDs to the Raspberry Pi via eight GPIO ports.

All the programs in this chapter assume that an LED is connected to the following output pins:

LED	Pin	LED	Pin
LED1	29	LED5	37
LED2	31	LED6	36
LED3	33	LED7	38
LED4	35	LED8	40

Within the programs, these pins will be referred to using the variables LED1 through LED8.

Figure 2-1 shows the finished circuit, ready to be used for testing the programs you'll write as you work your way through this chapter.

FIGURE 2-1:
A Raspberry Pi
test circuit with
eight LEDs
(Project 52).

Project 52: A Raspberry Pi LED Test Board

In this project, you connect a breadboard with eight LEDs to GPIO ports on a Raspberry Pi. You can then use this board to test the programs shown in this chapter.

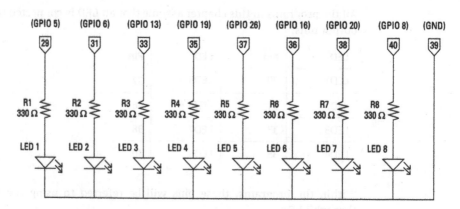

Raspberry Pi

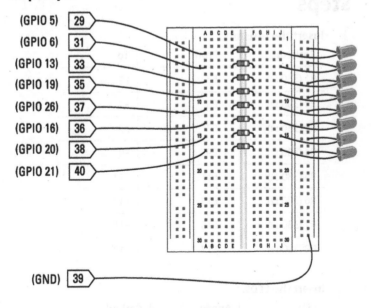

(GPIO 5) 29
(GPIO 6) 31
(GPIO 13) 33
(GPIO 19) 35
(GPIO 26) 37
(GPIO 16) 36
(GPIO 20) 38
(GPIO 21) 40

(GND) 39

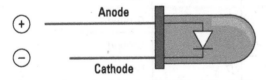

Anode

+

−

Cathode

Parts

>> One Raspberry Pi 4 with the Raspberry Pi OS installed and connected to a monitor, keyboard, and power (a Raspberry Pi 2 or 3 will also work)

>> One small solderless breadboard

>> Eight red LEDs

>> Eight 330 Ω resistors (orange-orange-brown)

>> Nine jumper wires (M/F)

Steps

1. **Insert the resistors.**

Resistor	From	To
R1 (330 Ω)	E3	F3
R2 (330 Ω)	E5	F5
R3 (330 Ω)	E7	F7
R4 (330 Ω)	E9	F9
R5 (330 Ω)	E11	F11
R6 (330 Ω)	E13	F13
R7 (330 Ω)	E15	F15
R8 (330 Ω)	E17	F17

2. **Insert the LEDs.**

LED	Anode	Cathode:
LED1	J3	Ground
LED2	J5	Ground
LED3	J7	Ground
LED4	J9	Ground
LED5	J11	Ground
LED6	J13	Ground
LED7	J15	Ground
LED8	J17	Ground

3. **Connect jumpers from the breadboard to the GPIO pins on the Raspberry Pi.**

Breadboard	GPIO Pin
A3	29
A5	31
A7	33
A9	35
A11	37
A13	36

Breadboard	GPIO Pin
A15	38
A17	40

4. **Connect the ground bus to pin 39 on the Raspberry Pi.**

   You can now use the programs in this chapter.

Flashing the LEDs

In the preceding chapter, I show you a program that flashes a single LED on and off. In this chapter, I show you several variants of that program, which flash the eight LEDs in the test project in various sequences. Along the way, you can add more Python statements to your repertoire to provide more and more complex ways to control the flashing.

REMEMBER

Keep in mind throughout this chapter that if you can turn an LED on or off with a Python program, you can control *anything* that can be connected to a GPIO port. The Raspberry Pi itself doesn't know or care what kind of circuit you connect to a GPIO port. All it knows is that when you tell it to, the Raspberry Pi makes the port HIGH or LOW. It's the external circuitry connected to the port that determines what happens when the port goes HIGH.

The only limitation is that the Raspberry Pi itself can swing only about 16 mA through its GPIO ports. If the circuit connected to the port requires more current than 16 mA, you must isolate the higher-current portion of the circuit from the Raspberry Pi. The easiest way to do that is to use a transistor driver or a relay.

Listing 2-1 shows a simple program that flashes all eight of the LEDs on and off at half-second intervals. This program is similar to the program that is presented in the preceding chapter; the only difference is that it flashes eight LEDs instead of just one.

LISTING 2-1: **Flashing LEDs**

```
import RPi.GPIO as GPIO
import time

GPIO.setmode(GPIO.BOARD)
GPIO.setup(29, GPIO.OUT)
GPIO.setup(31, GPIO.OUT)
```

(continued)

LISTING 2-1:   *(continued)*

```
GPIO.setup(33, GPIO.OUT)
GPIO.setup(35, GPIO.OUT)
GPIO.setup(37, GPIO.OUT)
GPIO.setup(36, GPIO.OUT)
GPIO.setup(38, GPIO.OUT)
GPIO.setup(40, GPIO.OUT)

while True:
    GPIO.output(29, GPIO.HIGH)
    GPIO.output(31, GPIO.HIGH)
    GPIO.output(33, GPIO.HIGH)
    GPIO.output(35, GPIO.HIGH)
    GPIO.output(37, GPIO.HIGH)
    GPIO.output(36, GPIO.HIGH)
    GPIO.output(38, GPIO.HIGH)
    GPIO.output(40, GPIO.HIGH)

    time.sleep(0.5)

    GPIO.output(29, GPIO.LOW)
    GPIO.output(31, GPIO.LOW)
    GPIO.output(33, GPIO.LOW)
    GPIO.output(35, GPIO.LOW)
    GPIO.output(37, GPIO.LOW)
    GPIO.output(36, GPIO.LOW)
    GPIO.output(38, GPIO.LOW)
    GPIO.output(40, GPIO.LOW)

    time.sleep(0.5)
```

ELIMINATING UNNECESSARY GPIO WARNINGS

You may notice when you run the program in Listing 2-1 that you see a bunch of warning messages in the Shell pane in Thonny. These messages look something like this:

```
/home/pi/Python/Listing_2_1.pi:5 RuntimeWarning: This channel is
    already in use, continuing anyway. Use GPIO.
    setwarnings(False) to disable warnings.
    GPIO.setup(29, GPIO.OUT)
```

```
/home/pi/Python/Listing_2_1.pi:6 RuntimeWarning: This channel is
    already in use, continuing anyway. Use GPIO.
    setwarnings(False) to disable warnings.
  GPIO.setup(31, GPIO.OUT)
/home/pi/Python/Listing_2_1.pi:7 RuntimeWarning: This channel is
    already in use, continuing anyway. Use GPIO.
    setwarnings(False) to disable warnings.
  GPIO.setup(33, GPIO.OUT)
/home/pi/Python/Listing_2_1.pi:8 RuntimeWarning: This channel is
    already in use, continuing anyway. Use GPIO.
    setwarnings(False) to disable warnings.
  GPIO.setup(35, GPIO.OUT)
/home/pi/Python/Listing_2_1.pi:9 RuntimeWarning: This channel is
    already in use, continuing anyway. Use GPIO.
    setwarnings(False) to disable warnings.
  GPIO.setup(37, GPIO.OUT)
/home/pi/Python/Listing_2_1.pi:10 RuntimeWarning: This channel
    is already in use, continuing anyway. Use GPIO.
    setwarnings(False) to disable warnings.
  GPIO.setup(36, GPIO.OUT)
/home/pi/Python/Listing_2_1.pi:11 RuntimeWarning: This channel
    is already in use, continuing anyway. Use GPIO.
    setwarnings(False) to disable warnings.
  GPIO.setup(38, GPIO.OUT)
/home/pi/Python/Listing_2_1.pi:12 RuntimeWarning: This channel
    is already in use, continuing anyway. Use GPIO.
    setwarnings(False) to disable warnings.
  GPIO.setup(40, GPIO.OUT)
```

If you read these warning messages carefully, you'll see that they're telling you that each of the eight digital output ports (called *channels* in the messages) is in use but that the program will continue anyway. Then each message indicates that you can disable warning messages by using the Python command GPIO.setwarnings(False).

These messages don't prevent your program from working, but they're annoying. So, I recommend you always include the following line in your startup code before you use GPIO.setup to initialize your GPIO ports:

```
GPIO.setwarnings(False)
```

That way, these warning messages won't be displayed. All the remaining programs in this chapter and the next use this command to suppress the warning messages.

Using Comments

A *comment* is text that provides an explanation of your code. Python completely ignores comments, so you can put any text you want in a comment. Using plenty of comments in your programs to explain what your program does and how it works is a good programming practice.

Python lets you use two different types of comments:

» **Single-line comments:** A single-line comment begins with a hash character (#) and extends to the end of the current line. Python ignores anything you type on a line following the hash character. Here's an example:

```
#Set the GPIO pin mode
```

You can also place a single-line comment at the end of a Python statement, as in this example:

```
GPIO.output(29, GPIO.HIGH) #LED1
```

Here, I used a comment to indicate that pin 29 is LED1.

» **Multi-line comments:** A *multi-line comment* is a comment that spans more than one line. It starts with a group of three adjacent quotation marks and continues until another group of three quotation marks is found. Here's an example:

```
"""Listing 2-1: Flash all eight LEDs.
This version of the program demonstrates
the use of comments.
"""
```

Although it isn't a requirement, it is customary for all Python programs to begin with a multi-line comment that provides basic information about what the program does.

Listing 2-2 shows a version of the LED Flasher program that includes both types of comments.

LISTING 2-2:   **LED Flasher with Comments**

```
"""Listing 2-1: Flash all eight LEDs.

This version of the program demonstrates the use of
comments.
"""
```

```
import RPi.GPIO as GPIO
import time

#Set up the GPIO mode and output pins.
GPIO.setmode(GPIO.BOARD)
GPIO.setwarnings(False)
GPIO.setup(29, GPIO.OUT) #LED1
GPIO.setup(31, GPIO.OUT) #LED2
GPIO.setup(33, GPIO.OUT) #LED3
GPIO.setup(35, GPIO.OUT) #LED4
GPIO.setup(37, GPIO.OUT) #LED5
GPIO.setup(36, GPIO.OUT) #LED6
GPIO.setup(38, GPIO.OUT) #LED7
GPIO.setup(40, GPIO.OUT) #LED8

while True:
    #Turn on the LEDs.
    GPIO.output(29, GPIO.HIGH) #LED1
    GPIO.output(31, GPIO.HIGH) #LED2
    GPIO.output(33, GPIO.HIGH) #LED3
    GPIO.output(35, GPIO.HIGH) #LED4
    GPIO.output(37, GPIO.HIGH) #LED5
    GPIO.output(36, GPIO.HIGH) #LED6
    GPIO.output(38, GPIO.HIGH) #LED7
    GPIO.output(40, GPIO.HIGH) #LED8

    #Wait half a second.
    time.sleep(0.5)

    #Turn off the LEDs.
    GPIO.output(29, GPIO.LOW) #LED1
    GPIO.output(31, GPIO.LOW) #LED2
    GPIO.output(33, GPIO.LOW) #LED3
    GPIO.output(35, GPIO.LOW) #LED4
    GPIO.output(37, GPIO.LOW) #LED5
    GPIO.output(36, GPIO.LOW) #LED6
    GPIO.output(38, GPIO.LOW) #LED7
    GPIO.output(40, GPIO.LOW) #LED8

    #Wait another half a second.
    time.sleep(0.5)
```

Creating Identifiers

In Python, an *identifier* is a name that is used to refer to items such as variables or function names. You'll often need to create your own identifiers when you write Python programs. When you create your own identifier, you must follow these simple rules:

>> **Identifier names can contain uppercase or lowercase letters, the digits 0 through 9, or underscore (_) characters.** Note that the underscore is the only special character allowed; you can't use other punctuation in an identifier.

>> **An identifier must begin with a letter or an underscore.** You cannot begin an identifier with a digit.

>> **Python is case-sensitive, which means that** ElapsedTime **and** elapsed-time **are two different identifiers.** This rule can be an annoying source of bugs, so if your program isn't working the way you expect it to, check the capitalization of any identifiers the program uses.

Using Constants

A *constant* is an identifier that has been assigned a fixed value that doesn't change throughout the program's execution. This allows you to use the constant name in your program rather than the value itself. Later, if you decide to change the value, you don't have to hunt through the program to find every occurrence of the constant. Instead, you simply change the line that defines the constant. In Python, constant names are usually written in all capital letters.

To create a constant in Python, you use the equals sign to assign a value to the constant, like this:

```
LED1 = 29
```

Here I've created a constant named LED1 that is given a value of 29. Then I can use the constant later in the program, like this:

```
GPIO.output(LED1, GPIO.HIGH)
```

Using a constant in this way makes it easier to keep track of the pin numbers used for each of the GPIO ports used by the program.

Listing 2-3 shows a complete program that uses constants to keep track of the pin numbers.

LISTING 2-3: **The LED Flasher Program with Constants**

```
"""Listing 2-3: Blink all eight LEDs.

This version of the program uses constants to identify
the LEDs.
"""

import RPi.GPIO as GPIO
import time

LED1 = 29
LED2 = 31
LED3 = 33
LED4 = 35
LED5 = 37
LED6 = 36
LED7 = 38
LED8 = 40

GPIO.setmode(GPIO.BOARD)
GPIO.setwarnings(False)
GPIO.setup(LED1, GPIO.OUT)
GPIO.setup(LED2, GPIO.OUT)
GPIO.setup(LED3, GPIO.OUT)
GPIO.setup(LED4, GPIO.OUT)
GPIO.setup(LED5, GPIO.OUT)
GPIO.setup(LED6, GPIO.OUT)
GPIO.setup(LED7, GPIO.OUT)
GPIO.setup(LED8, GPIO.OUT)

while True:
    GPIO.output(LED1, GPIO.HIGH)
    GPIO.output(LED2, GPIO.HIGH)
    GPIO.output(LED3, GPIO.HIGH)
    GPIO.output(LED4, GPIO.HIGH)
    GPIO.output(LED5, GPIO.HIGH)
    GPIO.output(LED6, GPIO.HIGH)
    GPIO.output(LED7, GPIO.HIGH)
    GPIO.output(LED8, GPIO.HIGH)

    time.sleep(0.5)
```

(continued)

LISTING 2-3: **(continued)**

```
GPIO.output(LED1, GPIO.LOW)
GPIO.output(LED2, GPIO.LOW)
GPIO.output(LED3, GPIO.LOW)
GPIO.output(LED4, GPIO.LOW)
GPIO.output(LED5, GPIO.LOW)
GPIO.output(LED6, GPIO.LOW)
GPIO.output(LED7, GPIO.LOW)
GPIO.output(LED8, GPIO.LOW)

time.sleep(0.5)
```

Using Variables

A *variable* is an identifier that refers to a memory location whose value can change throughout the program. For example, suppose I wanted to change the rate at which the LED Flasher program flashes, starting with a rapid flash but gradually flashing at an ever-slower rate. I could easily accomplish this by using a variable to represent the amount of time to wait between each flash and increasing the value of the variable with each repetition of the program's while loop.

The first step in doing this is to create the variable. You do that the same way you create a constant — in fact, a constant is just a variable whose value does not change during the program's execution. For example:

```
sleepTime = 0.05
```

Here, I create a variable named sleepTime and assign it an initial value of 0.05.

You could then use this variable when you call the sleep function, like this:

```
time.sleep(sleepTime)
```

This causes the program to delay the number of seconds indicated by the sleep-Time variable — in this case, five one-hundredths of a second.

Later in the program, you can change the value of a variable. For example, you can add 0.05 to the variable's value like this:

```
sleepTime = sleepTime + 0.05
```

The first time this statement is executed, the sleepTime variable is set to 0.1, which is 0.05 + 0.05. The second time the statement is executed, sleepTime is set to 0.15; each time, the value of sleepTime is increased by 0.05.

Note that you can do other types of math if you want. For example:

```
x = 10
y = x * 3
```

In this example, the value of the variable x is multiplied by 3 and the result is assigned to the variable y. Thus, 30 will be assigned to the variable y.

Here are a few things to know about math in Python:

>> **Python uses the normal order of operations found in math.** Thus, multiplication and division are done before addition and subtraction. For example:

```
x = 10
y = x + 5 * 3
```

This statement assigns the value 25 to y, because 5 is first multiplied by 3 and then the result is added to 10.

>> **You can use parentheses to change the order of calculation.** For example:

```
x = 10
y = (x + 5) * 3
```

Here, Python first does the calculation inside the parentheses, giving a result of 15. It then multiples the 3 by the 15 to give the final result, 45.

Listing 2-4 shows a program that uses a variable to change the speed at which the LEDs flash each time the while statement causes the program to loop. As you can see, the sleepTime variable is used to provide the delay time. Each time through the loop, the value of the sleepTime variable is increased by 0.05. Thus, the LEDs flash very fast when the program first starts, but the flashing gets progressively slower as the program continues.

LISTING 2-4: **The LED Flasher Program with a Variable**

```
"""Listing 2-4: Blink all eight LEDs.

This version of the program uses a variable to gradually
slow the flash rate.
"""

import RPi.GPIO as GPIO
import time

LED1 = 29
LED2 = 31
LED3 = 33
LED4 = 35
LED5 = 37
LED6 = 36
LED7 = 38
LED8 = 40

sleepTime = 0.05

GPIO.setmode(GPIO.BOARD)
GPIO.setwarnings(False)
GPIO.setup(LED1, GPIO.OUT)
GPIO.setup(LED2, GPIO.OUT)
GPIO.setup(LED3, GPIO.OUT)
GPIO.setup(LED4, GPIO.OUT)
GPIO.setup(LED5, GPIO.OUT)
GPIO.setup(LED6, GPIO.OUT)
GPIO.setup(LED7, GPIO.OUT)
GPIO.setup(LED8, GPIO.OUT)

while True:
    GPIO.output(LED1, GPIO.HIGH)
    GPIO.output(LED2, GPIO.HIGH)
    GPIO.output(LED3, GPIO.HIGH)
    GPIO.output(LED4, GPIO.HIGH)
    GPIO.output(LED5, GPIO.HIGH)
    GPIO.output(LED6, GPIO.HIGH)
    GPIO.output(LED7, GPIO.HIGH)
    GPIO.output(LED8, GPIO.HIGH)

    time.sleep(sleepTime)

    GPIO.output(LED1, GPIO.LOW)
    GPIO.output(LED2, GPIO.LOW)
```

```
GPIO.output(LED3, GPIO.LOW)
GPIO.output(LED4, GPIO.LOW)
GPIO.output(LED5, GPIO.LOW)
GPIO.output(LED6, GPIO.LOW)
GPIO.output(LED7, GPIO.LOW)
GPIO.output(LED8, GPIO.LOW)

time.sleep(sleepTime)

sleepTime = sleepTime + 0.05
```

Creating Your Own Functions

You have already seen how to use various functions that are part of function libraries that you can import into a Python program. For example, the statement time.sleep(1) calls a function named sleep to delay the execution of the program for one second.

TECHNICAL STUFF

The equivalent statements in any other programming languages, such as Arduino's sketch language and BASIC, measure time delays in milliseconds rather than microseconds.

You can also create your own functions in Python. This can be especially useful when you have a long sequence of statements that needs to be repeated in your program. Rather than just list the statements multiple times, you can place them in a function and call the function whenever you need the statements.

As an example, you could create a function that would turn on all eight LEDs. To do that, you write a def statement that specifies the name of the function. Then, you list all the statements that are part of the function with indentation after the def statement, like this:

```
def setLEDsOn():
    GPIO.output(LED1, GPIO.HIGH)
    GPIO.output(LED2, GPIO.HIGH)
    GPIO.output(LED3, GPIO.HIGH)
    GPIO.output(LED4, GPIO.HIGH)
    GPIO.output(LED5, GPIO.HIGH)
    GPIO.output(LED6, GPIO.HIGH)
    GPIO.output(LED7, GPIO.HIGH)
    GPIO.output(LED8, GPIO.HIGH)
```

Note that the `def` statement lists the name of the function (`setLEDsOn`) followed by an empty set of parentheses and a colon. I explain the empty parentheses in a just a moment. The colon indicates that indented lines that follow are to be executed whenever the function is called.

To call the function, you just type the name of the function as if it were any other Python statement. For example:

```
setLEDsOn()
```

When this statement is encountered, the statements that are defined for the `setLEDsOn` function are executed, which causes all eight LEDs to turn on.

There are two important details to note about defining and calling functions:

>> The statements that make up the function are not actually executed by the `def` statement. The `def` statement *defines* but does not *execute* the function.

>> The `def` statement for a function must appear in your program *before* any statement that calls the function.

Now to the empty parentheses that follow the function name in the `def` statement. A function can have one or more *arguments* that can be passed to it when the function is called. You've already seen several examples of this. For example, the `time.sleep` function accepts an argument that specifies the number of seconds of sleep. And the `GPIO.output` function accepts two arguments; the first indicates the output pin and the second indicates the output value.

The empty parentheses means that the function you're defining doesn't use any arguments. However, you can easily add one or more arguments to a function definition by listing argument names within the parentheses.

For example, suppose you want to create a function named `setLEDs` that can set the LED values to either HIGH or LOW. In that case, you could define the function like this:

```
def setLEDs(value):
    GPIO.output(LED1, value)
    GPIO.output(LED2, value)
    GPIO.output(LED3, value)
    GPIO.output(LED4, value)
    GPIO.output(LED5, value)
    GPIO.output(LED6, value)
    GPIO.output(LED7, value)
    GPIO.output(LED8, value)
```

Then, to turn the LEDs on, you would call the setLEDs function like this:

```
setLEDs(GPIO.HIGH)
```

To turn them off, call setLEDs like this:

```
setLEDs(GPIO.LOW)
```

Listing 2-5 shows a version of the LED flashing program that uses a function to flash the LEDs. As you can see, using the setLEDs function simplifies the program's while loop.

LISTING 2-5: **The LED Flasher Program with a Function**

```
"""Listing 2-5: Blink all eight LEDs.

This version of the program defines a function to turn
the LEDs on or off.
"""

import RPi.GPIO as GPIO
import time

LED1 = 29
LED2 = 31
LED3 = 33
LED4 = 35
LED5 = 37
LED6 = 36
LED7 = 38
LED8 = 40

GPIO.setmode(GPIO.BOARD)
GPIO.setwarnings(False)
GPIO.setup(LED1, GPIO.OUT)
GPIO.setup(LED2, GPIO.OUT)
GPIO.setup(LED3, GPIO.OUT)
GPIO.setup(LED4, GPIO.OUT)
GPIO.setup(LED5, GPIO.OUT)
GPIO.setup(LED6, GPIO.OUT)
GPIO.setup(LED7, GPIO.OUT)
GPIO.setup(LED8, GPIO.OUT)

def setLEDs(setValue):
    GPIO.output(LED1, setValue)
    GPIO.output(LED2, setValue)
    GPIO.output(LED3, setValue)
```

(continued)

LISTING 2-5: *(continued)*

```
        GPIO.output(LED4, setValue)
        GPIO.output(LED5, setValue)
        GPIO.output(LED6, setValue)
        GPIO.output(LED7, setValue)
        GPIO.output(LED8, setValue)

while True:
    setLEDs(GPIO.HIGH)
    time.sleep(0.5)
    setLEDs(GPIO.LOW)
     time.sleep(0.5)
```

Using If Statements

An if statement lets you add conditional testing to your programs. In other words, it lets you execute certain statements only if a particular condition is met. This type of conditional processing is an important part of any but the most trivial of programs.

Every if statement must include a *conditional expression* that lays out a logical test to determine whether the condition is true or false. For example:

```
x &gt; 5
```

This condition is true if the value of the variable x is greater than 5. If x has a value that is equal to or less than 5, the condition is false.

One thing that Python has in common with other languages such as C or Java is that it requires you to use two equal signs to test for equality. For example:

```
x == 5
```

This condition is true if the value of the variable x is exactly 5. If the value is anything other than 5, the condition is false.

The basic form of an if statement looks like this:

```
if condition:
    statements
```

Note that the condition is followed by a colon, and any statements that are executed if the condition is true must be indented (by the same amount) beneath the if statement.

Here's a simple example:

```
if sleepTime == 0.5:
    sleepTime = 1.0
```

This if statement looks at the value of the sleepTime variable and changes the value to 1.0 if the current value is 0.5.

The if statement can also include an else clause that provides one or more statements that are executed if the condition is *not* true. For example:

```
if sleepTime == 0.5:
    sleepTime = 1.0
else:
    sleepTime = 0.5
```

In this example, one of two things happens: If the sleepTime variable has a value of 0.5, the sleepTime variable's value is set to 1.0. But if the sleepTime variable has a value other than 0.5, the sleepTime variable's value is set to 0.5.

Keep in mind that both the if and else parts can have more than one statement. In that case, the indentation must be maintained, as in this example:

```
if Timer == 0.5:
    sleepTime = 1.0
    Counter = 0
else:
    sleepTime = 0.5
    Counter = Counter + 1
```

One final thing to know about if statements is that one if statement can be contained within another. This arrangement is called *nesting* and is very useful for implementing complex decisions. Indentation becomes especially important when if statements are nested.

Listing 2-6 shows a program that cleverly uses a nested if statement to flash the LEDs in an alternating pattern of three quick flashes followed by three longer flashes. The program uses a variable named sleepTime to control the delay interval (0.1 for the quick flashes, 0.5 for the longer flashes) and a second variable named `flashCount` to count how many times the LEDs have been flashed. The nested `if` statement looks like this:

```
if flashCount == 3:
    flashCount = 0
    if sleepTime == 0.1:
        sleepTime = 0.5
    else:
        sleepTime = 0.1
```

Here, the first `if` statement is used to check the `flashCount` variable's value. If it is 3, the `flashCount` variable is reset to zero. Then a second `if` statement is used to set the `sleepTime` variable to either 0.1 or 0.5. In essence, this inner `if` statement toggles the `sleepTime` variable between 0.1 and 0.5 each time the `flashCount` variable reaches 3.

LISTING 2-6:
The LED Flasher Program with an If Statement

```
"""Listing 2-6: Blink all eight LEDs.

This version of the program uses an If statement to
alternate between long and short flashes.
"""

import RPi.GPIO as GPIO
import time

LED1 = 29
LED2 = 31
LED3 = 33
LED4 = 35
LED5 = 37
LED6 = 36
LED7 = 38
LED8 = 40

sleepTime = 0.1
flashCount = 0

GPIO.setmode(GPIO.BOARD)
GPIO.setwarnings(False)
GPIO.setup(LED1, GPIO.OUT)
GPIO.setup(LED2, GPIO.OUT)
GPIO.setup(LED3, GPIO.OUT)
GPIO.setup(LED4, GPIO.OUT)
GPIO.setup(LED5, GPIO.OUT)
```

```
GPIO.setup(LED6, GPIO.OUT)
GPIO.setup(LED7, GPIO.OUT)
GPIO.setup(LED8, GPIO.OUT)

def setLEDs(setValue):
    GPIO.output(LED1, setValue)
    GPIO.output(LED2, setValue)
    GPIO.output(LED3, setValue)
    GPIO.output(LED4, setValue)
    GPIO.output(LED5, setValue)
    GPIO.output(LED6, setValue)
    GPIO.output(LED7, setValue)
    GPIO.output(LED8, setValue)

while True:
    setLEDs(GPIO.HIGH)
    time.sleep(sleepTime)
    setLEDs(GPIO.LOW)
    time.sleep(sleepTime)

    flashCount = flashCount + 1
    if flashCount == 3:
        flashCount = 0
        if sleepTime == 0.1:
            sleepTime = .5
        else:
            sleepTime = 0.1
```

Using While Loops

So far, all the Python programs you've seen have used the following while statement to cause the main portion of the program to loop repeatedly as long as the program is allowed to run:

```
while True:
```

It may not surprise you to learn that the full syntax for the while statement is this:

```
while condition:
    statements
```

Here, *condition* can be the same type of conditional test that is used in an if statement. Specifying just the word True as the conditional test is a special form of while statement that repeats forever, because the conditional test True, as its name suggests, *always* evaluates to True.

Note that the *statement* part of the while loop can be a single statement or a series of statements enclosed in curly brackets.

Here's an example of a while loop that flashes the LEDs three times for 0.1 second:

```
flashCount = 0
while flashCount < 3:
    setLEDs(GPIO.HIGH)
    time.sleep(.1)
    setLEDs(GPIO.LOW)
    time.sleep(.1)
    flashCount = flashCount + 1
```

On each execution of this loop, the setLEDs function is called twice to turn the LEDs on and off. Then, the flashCount variable is increased by 1. The flashCount variable starts with a value of zero, and the condition on the while statement says to repeat the loop as long as the flashCount variable is less than 3. Thus, the loop is run three times: The first time through, flashCount is zero; the second time, flashCount is 1; the third time, flashCount is 2.

Like if statements, while loops can be nested. Listing 2-7 shows a version of the LED flasher that uses two while loops to flash the LEDs in a pattern of three times for one-tenth of a second followed by three times for one-half of a second. These two while loops are nested inside a while True loop, so the pattern of three short, three long repeats indefinitely until the program is stopped.

LISTING 2-7: **LED Flasher with Nested While Loops**

```
"""Listing 2-7: Blink all eight LEDs.

This version of the program uses nested while loops to
alternate between long and short flashes.
"""

import RPi.GPIO as GPIO
import time

LED1 = 29
LED2 = 31
LED3 = 33
```

```
LED4 = 35
LED5 = 37
LED6 = 36
LED7 = 38
LED8 = 40

GPIO.setmode(GPIO.BOARD)
GPIO.setwarnings(False)
GPIO.setup(LED1, GPIO.OUT)
GPIO.setup(LED2, GPIO.OUT)
GPIO.setup(LED3, GPIO.OUT)
GPIO.setup(LED4, GPIO.OUT)
GPIO.setup(LED5, GPIO.OUT)
GPIO.setup(LED6, GPIO.OUT)
GPIO.setup(LED7, GPIO.OUT)
GPIO.setup(LED8, GPIO.OUT)

def setLEDs(setValue):
    GPIO.output(LED1, setValue)
    GPIO.output(LED2, setValue)
    GPIO.output(LED3, setValue)
    GPIO.output(LED4, setValue)
    GPIO.output(LED5, setValue)
    GPIO.output(LED6, setValue)
    GPIO.output(LED7, setValue)
    GPIO.output(LED8, setValue)

while True:
    flashCount = 0
    while flashCount < 3:
        setLEDs(GPIO.HIGH)
        time.sleep(.1)
        setLEDs(GPIO.LOW)
        time.sleep(.1)
        flashCount = flashCount + 1
    flashCount = 0
    while flashCount < 3:
        setLEDs(GPIO.HIGH)
        time.sleep(0.5)
        setLEDs(GPIO.LOW)
        time.sleep(0.5)
        flashCount = flashCount + 1
```

Using For Loops

Most programming languages have a for loop that is designed to loop through a series of statements while incrementing a counter variable through a given range, such as from 1 to 3.

In Python, this type of for loop looks like this:

```
for variable in range(count):
    statements
```

The statements indented beneath the for loops will be executed the number of times indicated by count. For each execution of the loop, variable will be increased by one, starting with zero for the first time through the loop.

For example, consider this for statement:

```
for x in range(3):
    print(x)
```

This will result in the following output in the Python Shell window:

```
0
1
2
```

The range function is actually more complex than this might lead you to believe. Range can actually take up to three arguments: start, stop, and stride.

If you use only one argument, it is taken to be the stop argument, and in practice the stop argument is a count of the number of values returned, starting with zero for the first time through the loop.

If you use two arguments, the first is the start argument and the second is the stop argument. In this case, the start argument specifies the value of the variable on the first iteration of the loop, and the stop argument is one greater than the value of the variable on the last trip through the loop.

For example, consider this for statement:

```
for x in range(5, 9):
    print(x)
```

When you run this statement, the following output will be displayed:

```
5
6
7
8
```

Here, the loop runs a total of four times. The starting value is 5 and the ending value is one less than 9.

The third argument lets you count by a value other than 1. For example:

```
for x in range(5, 25, 5):
    print(x)
```

The output created by this for loop is:

```
5
10
15
20
```

You can coax a for statement into counting backward by using a negative *stride* value. For example:

```
for x in range(5, 0, -1):
    print(x)
```

This creates the following output:

```
5
4
3
2
1
```

TIP

Every other programming language I've worked in (and I've worked in a lot!) uses the term *step* instead of *stride*. Perhaps this is simply a tribute to Monty Python's Ministry of Silly Walks sketch. The default value of 1 is a normal stride. Any other value could be a silly stride.

Listing 2-8 shows a version of the LED Flasher program that uses a pair of for statements to flash the LEDs three times with a short delay and then three times with a longer delay.

LISTING 2-8: **The LED Flasher Program with FOR Loops**

```
"""Listing 2-8: Blink all eight LEDs.

This version of the program uses a for statement to
alternate between long and short flashes.
"""

import RPi.GPIO as GPIO
import time

LED1 = 29
LED2 = 31
LED3 = 33
LED4 = 35
LED5 = 37
LED6 = 36
LED7 = 38
LED8 = 40

GPIO.setmode(GPIO.BOARD)
GPIO.setwarnings(False)
GPIO.setup(LED1, GPIO.OUT)
GPIO.setup(LED2, GPIO.OUT)
GPIO.setup(LED3, GPIO.OUT)
GPIO.setup(LED4, GPIO.OUT)
GPIO.setup(LED5, GPIO.OUT)
GPIO.setup(LED6, GPIO.OUT)
GPIO.setup(LED7, GPIO.OUT)
GPIO.setup(LED8, GPIO.OUT)

def setLEDs(setValue):
    GPIO.output(LED1, setValue)
    GPIO.output(LED2, setValue)
    GPIO.output(LED3, setValue)
    GPIO.output(LED4, setValue)
    GPIO.output(LED5, setValue)
    GPIO.output(LED6, setValue)
    GPIO.output(LED7, setValue)
    GPIO.output(LED8, setValue)

while True:
    for flashCount in range(3):
        setLEDs(GPIO.HIGH)
        time.sleep(0.1)
        setLEDs(GPIO.LOW)
        time.sleep(0.1)
```

```
for flashCount in range(3):
    setLEDs(GPIO.HIGH)
    time.sleep(0.5)
    setLEDs(GPIO.LOW)
    time.sleep(0.5)
```

Looking at Python Lists

The last Python feature I want to show you in this chapter is perhaps one of Python's most interesting feature: the list feature. In Python, a *list* is a sequence of values that can be treated individually or as a whole. You can add any kind of data you want to a list.

Lists are commonly represented as a sequence of values separated by commas and enclosed in square brackets. For example, here is a list of string values:

```
names = ["Chapman", "Cleese", "Gilliam", "Idle", "Jones",
    "Palin"]
```

You can use a list in a for statement, like this:

```
for name in names:
    print(name)
```

This example prints the following output:

```
Chapman
Cleese
Gilliam
Idle
Jones
Palin
```

In fact, I should point out that the range function (see the preceding section) is actually a function that simply returns a list. For example, the range(3) function returns the list [0, 1, 2]. So the following two for loops are equivalent:

```
for x in range(3):
    print(x)

for x in [0, 1, 2]:
    print(x)
```

You can use lists to simplify your LED flashing program by using a list to store the pin numbers of the eight GPIO ports connected to the LEDs, like this:

```
LEDs = [29, 31, 33, 35, 37, 36, 38, 40]
```

Then you can replace the eight lines of code that set the GPIO ports to output mode with a simple for loop:

```
for LED in LEDs:
    GPIO.setup(LED, GPIO.OUT)
```

And you can rewrite your setLEDs function like this:

```
def setLEDs(setValue):
    for LED in LEDs:
        GPIO.output(LED, setValue)
```

Listing 2-9 shows a version of the LED Flasher program that uses lists to simplify the program.

| LISTING 2-9: | The LED Flasher Program with a List |

```
"""Listing 2-9: Blink all eight LEDs.

This version of the program uses a list to represent the
LED pins.
"""

import RPi.GPIO as GPIO
import time

LEDs = [29, 31, 33, 35, 37, 36, 38, 40]

GPIO.setmode(GPIO.BOARD)
GPIO.setwarnings(False)

for LED in LEDs:
    GPIO.setup(LED, GPIO.OUT)

def setLEDs(setValue):
    for LED in LEDs:
        GPIO.output(LED, setValue)

while True:
    for flashCount in range(3):
        setLEDs(GPIO.HIGH)
```

```
        time.sleep(0.1)
        setLEDs(GPIO.LOW)
        time.sleep(0.1)
    for flashCount in range(3):
        setLEDs(GPIO.HIGH)
        time.sleep(0.5)
        setLEDs(GPIO.LOW)
          time.sleep(0.5)
```

You can access items in a list individually by using an index number. In Python, this is called a *slice*, because it gives you just a slice of the entire list. For example, to access the first item in a list, slice the list using index 0:

```
print List1[0]
```

Here, the first item in a list named List1 is printed.

You can access items at the end of a list by using a negative index number. This example slices off just the last item in List1:

```
print List1[-1]
```

To access a range of items, specify the start and stop index values separated by a colon. For example, to slice the first three values in List1:

```
print List1[0:3]
```

Finally, you can specify a stride value as well. For example:

```
print[0:-1:2]
```

This example prints every other item in the list, starting with the first. Note that you can omit the start or stop indexes:

```
print[::2]
```

This example also prints every other item in the list, starting with the first. Note that if you omit a start or stop value, you must still include the colon for the omitted value.

To print the items in a list in reverse order, use –1 as the stride value:

```
print[::-1]
```

Listing 2-10 shows yet another version of the LED Flasher program. This one uses a slice in a for loop to flash the LEDs first in one direction and then in the opposite direction. This creates an effect similar to the spooky electronic eyes on the evil Cylons in the old TV series *Battlestar Galactica*. The first for loop flashes all eight LEDs. Then the second for loop flashes the middle six LEDs in reverse order by using the following slice:

```
for LED in LEDs[-2:1:-1]:
```

This slice selects the LEDs in reverse order, starting with the second-to-last item in the LEDs list and ending with the second item.

LISTING 2-10: **The Cylon LED Flasher**

```python
"""Listing 2-10: Cylon LED Flasher.

This version of the program uses a pair of for loops to
flash the LEDs one at a time, like a cylon.
"""

import RPi.GPIO as GPIO
import time

LEDs = [29, 31, 33, 35, 37, 36, 38, 40]

GPIO.setmode(GPIO.BOARD)
GPIO.setwarnings(False)

for LED in LEDs:
    GPIO.setup(LED, GPIO.OUT)

while True:
    for LED in LEDs:
        GPIO.output(LED,GPIO.HIGH)
        time.sleep(0.1)
        GPIO.output(LED,GPIO.LOW)
    for LED in LEDs[-2:1:-1]:
        GPIO.output(LED,GPIO.HIGH)
        time.sleep(0.1)
        GPIO.output(LED,GPIO.LOW)
```

Chapter **3**

Reading Digital and Analog Input

I n the previous chapter, I explain how you can use your Raspberry Pi's GPIO ports in output mode, sending digital HIGH and LOW signals to the output ports by using the GPIO.output function.

In this chapter, I turn the tables and use the GPIO ports as inputs. Here, I explain how to read a digital input from a GPIO port, which is easy because a digital input is simply HIGH or LOW. I also explain how to do something a bit more complicated: read an analog input through a digital port. This can be useful, for example, if you want to use a potentiometer as input to a Raspberry Pi. Unlike Arduino, Raspberry Pi does not have a built-in ability to read analog input. As a result, you'll have to construct some external circuitry to convert analog input — such as the voltage level flowing through a potentiometer — to a digital value that the Pi can use. I explain how to create and use an analog-to-digital converter in this chapter.

Using a GPIO Port for Digital Input

In Chapter 2 of this minibook, I explain how to connect an LED to a GPIO port and turn the LED on or off by using the GPIO.output function in a Python program. The GPIO.output function uses the GPIO ports as output pins by setting the status of a GPIO port to HIGH or LOW so that external circuitry (such as an LED) can react to the pin's status. Before you can use GPIO.output, you must configure the port as an output by calling GPIO.setup, specifying GPIO.OUT as the port's I/O mode.

But what if you want to use a GPIO port for input rather than for output? In other words, what if you want the Raspberry Pi to react to the status of an external circuit instead of the other way around? The easiest way to do that is to connect a push button to a GPIO port, and then use Python to program the port as an input port rather than as an output port. You can then use the GPIO.input function to read the status of the port.

WARNING

Unlike all the other digital circuits presented in this book, including Arduino circuits, the Raspberry Pi GPIO ports are designed to work at 3.3 V rather than 5 V. If you mistakenly connect a 5 V digital circuit to a GPIO input port, you're likely to damage the port — and possibly fry the entire Raspberry Pi. So be careful when working with input ports to make sure that you connect only 3.3 V inputs.

WARNING

Wow, two warnings in a row! I want to emphasize the importance of the 3.3 V versus 5 V warning by reminding you that the 40-pin GPIO header block on a Raspberry Pi has both +3.3 V and +5 V pins, and also by reminding you that none of the pins on the header block are labeled on the board. The +5 V power supplies are located on pins 2 and 4, so you must be careful to never use these pins in circuits that are connected to GPIO input ports. Instead, use pins 1 or 17, which provide +3.3 V each.

With all the warnings out of the way, let's look at two distinct ways of working with digital inputs connected to switches:

>> **Active-high:** This type of connection places +3.3 V on the input port when the push button is pressed. When the button is released, the input port sees 0 V.

>> **Active-low:** This type of connection sees +3.3 V at the input port when the push button is not pressed. When you press the push button, the +3.3 V is removed, and the input pin sees 0V.

In other words, an active-high circuit is HIGH when the button is pressed. An active-low circuit is LOW when the button is pressed. Neither option is better than the other, but a program that works with an input port needs to know whether the input is active-high or active-low.

Figure 3-1 shows examples of both active-high and active-low push buttons. In the active-high circuit, the input port is connected to ground through resistors R1 and R2 when the push button is not pressed. Thus, the voltage at the input port is 0. When the push button is pressed, the input port is connected to +3.3 V through R1, causing the input port to see +3.3 V. As a result, the input port is HIGH when the button is pressed and LOW when the button is not pressed.

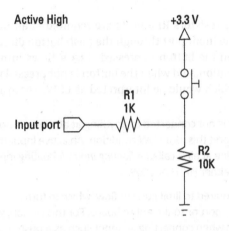

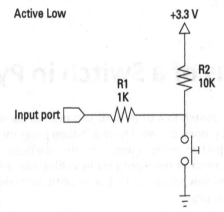

FIGURE 3-1:
Active-high
and active-low
input circuits.

In the active-low circuit, the input port is connected to +3.3 V through resistors R1 and R2, causing the input port to be HIGH. When the button is pressed, the current from the +3.3 V source is shorted to ground through R2, causing the voltage at the input port to drop to zero. Thus, the input pin is LOW when the button is pressed and HIGH when the button is not pressed.

TECHNICAL
STUFF

In an active-high circuit, R1 is called a *pull-down* resistor because it pulls the current from the I/O pin down to zero when the push button isn't depressed. In an active-low circuit, R1 is called a *pull-up resistor* because it pulls the voltage at the I/O pin up to +3.3 V when the push button isn't depressed.

WARNING

Note that in both circuits, R2 is connected between the +3.3 V and ground to prevent a short circuit when the button is pressed. Without this resistor, the Raspberry Pi would likely be damaged when the button is pressed.

TECHNICAL STUFF

You may be asking why resistors R1 and R2 are required at all in an active high. Why not just connect the input port through the push button directly to the +3.3 V supply? That way, when the button is pressed, +3.3 V flows into the input port, registering a HIGH condition, and when the button is not pressed nothing is present at the input port, which would be interpreted as LOW. There are two reasons:

» When an input port is not connected to anything at all, there is no guarantee that the Pi will interpret this as a LOW condition. An active input that is not connected to anything at all is called a *floating input.* A floating input cannot be definitively read as either HIGH or LOW.

» The resistors are required to limit current flow, which in turn prevents damage to the input port or to the entire board. For this reason, you should always use resistors when connecting an input such as a push button to a GPIO port.

Checking the Status of a Switch in Python

After you've connected a switch to a GPIO port, you need to know how to determine whether the switch is open or closed from a Python program. To do this, you must first designate the port as an input port by calling the GPIO.setup function. Then you can check the status of the input port by calling the GPIO.input function. The GPIO.input function returns GPIO.LOW or GPIO.HIGH depending on the current status of the input pin.

REMEMBER

In Python, you can designate a GPIO port using either its internal port number or its pin number. For example, GPIO port 2 is actually connected to pin 3. Throughout this chapter, I assume that you're using the pin numbers rather than the internal port numbers in your programs. To designate this option, you should place the following line in your program prior to any other statements that reference the ports:

```
GPIO.setmode(GPIO.BOARD)
```

To designate a port as an input port, use the GPIO.setup function, like this:

```
GPIO.setup(3, GPIO.IN)
```

Then, to get the status of the port, you can call `GPIO.input`, like this:

```
GPIO.input(3);
```

You can test the status of an input port in an `if` statement like this:

```
if GPIO.input(3) == GPIO.HIGH:
    print "Input is HIGH"
else:
    print "Input is LOW"
```

Here's an example that lights an LED on pin 5 (GPIO port 3) if the user presses the button on pin 3 (GPIO port 2) and turns the LED off when the button is released:

```
if GPIO.input(3) == GPIO.HIGH:
    GPIO.output(5, HIGH)
else:
    GPIO.output(5, LOW)
```

Listing 3-1 shows an interesting program that works with a Raspberry Pi that has an active-high push-button switch connected to pin 33 and LEDs connected to pins 35 and 37. The program flashes the LED connected to pin 35 on and off at half-second intervals until the push-button switch is depressed. Then it flashes the LED on pin 37 as long as the button is held. When you release the button, the program returns to flashing the LED on pin 35.

LISTING 3-1: **The Push-Button Program**

```
import RPi.GPIO as GPIO
import time


BTN1 = 33
LED1 = 35
LED2 = 37


GPIO.setmode(GPIO.BOARD)
GPIO.setup(BTN1, GPIO.IN)
GPIO.setup(LED1, GPIO.OUT)
GPIO.setup(LED2, GPIO.OUT)


while True:

    if GPIO.input(BTN1) == GPIO.HIGH:
        GPIO.output(LED1, GPIO.HIGH)
        time.sleep(0.5)
```

```
            GPIO.output(LED1, GPIO.LOW)
            time.sleep(0.5)
        else:
            GPIO.output(LED2, GPIO.HIGH)
            time.sleep(0.5)
            GPIO.output(LED2, GPIO.LOW)
            time.sleep(0.5)
```

Project 53 shows how to build a simple circuit you can use to test the program in Listing 3-1, and Figure 3-2 shows the completed circuit.

FIGURE 3-2:
A circuit for testing
an active-high
push-button
switch (Project 53).

Project 53: A Push-Button-Controlled Raspberry Pi LED Flasher

In this project, you connect a breadboard with two LEDs and a push button to a Raspberry Pi board.

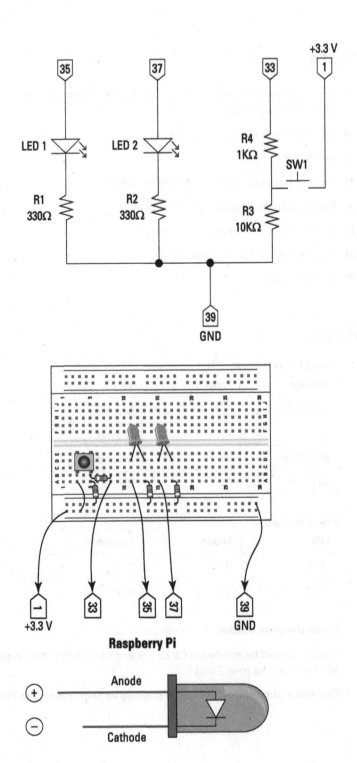

Raspberry Pi

Anode
+
−
Cathode

Parts

>> One Raspberry Pi with software installed, connected to a monitor, keyboard, and mouse

>> One small solderless breadboard

>> Two 5 mm red LEDs

>> One normally open DIP breadboard push button

>> Two 330 Ω resistors (yellow-violet-brown)

>> One 10 kΩ resistor (brown-black-orange)

>> One 1 kΩ resistor (brown-black-red)

>> 6 jumper wires

Steps

1. **Insert the resistors.**

Resistor	From	To
R1 (330 Ω)	A13	Ground
R2 (330 Ω)	A17	Ground
R3 (10 kΩ)	A5	Ground
R4 (1 kΩ)	B5	B8

2. **Insert the LEDs.**

LED	Anode	Cathode
LED1	D11	D13
LED2	D15	D17

3. **Insert the push button.**

   The pins should be inserted in C3, E3, C5, and E5 such that the switch opens and closes across rows 3 and 5.

4. **Connect a jumper from A3 to the positive voltage bus on the breadboard.**

5. Connect jumpers from the breadboard to the GPIO pins on the Raspberry Pi.

Breadboard	GPIO Raspberry Pi Pin
A8	33
A11	35
A15	37
Ground bus	39
Positive bus	1

6. Use Thonny to enter and run the Flashing LED program (refer to Listing 3-1).

LED2 will flash on and off at half-second intervals. If you press the push button, LED1 will flash instead.

Reading Analog Input

The Raspberry Pi is a purely digital machine. Unlike many microprocessors, including Arduino, Raspberry Pi has no built-in capability for reading analog input. As a result, there's no straightforward way to read something like a voltage level from a component such as a potentiometer, photocell, temperature sensor, or joystick.

To get around this limitation, you need to use a circuit that converts analog data to digital data. Such a circuit is called an analog-to-digital converter (ADC). There are many ways to build an ADC circuit, but the easiest way is to let someone else build the circuit for you. You can use an inexpensive IC called an MCP3008. This handy chip lets you convert as many as eight different analog inputs to a digital output, which can be read by your Raspberry Pi. You can purchase one from an online retailer for just a few dollars.

As shown in Figure 3-3, the MCP3008 is housed in a standard 16-pin package. Table 3-1 describes the 16 pins in greater detail.

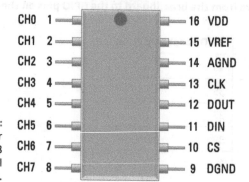

MCP3008

CH0	1		16	VDD
CH1	2		15	VREF
CH2	3		14	AGND
CH3	4		13	CLK
CH4	5		12	DOUT
CH5	6		11	DIN
CH6	7		10	CS
CH7	8		9	DGND

FIGURE 3-3:
Pinout for
the MCP3008
analog-to-digital
converter.

TABLE 3-1

Pinouts for the MCP3008 Analog-to-Digital Converter

Pin	Name	What it does
1	CH0	Analog input channel 0
2	CH1	Analog input channel 1
3	CH2	Analog input channel 2
4	CH3	Analog input channel 3
5	CH4	Analog input channel 4
6	CH5	Analog input channel 5
7	CH6	Analog input channel 6
8	CH7	Analog input channel 7
9	DGND	Ground for the digital circuits
10	CS	Chip Select
11	DIN	Digital Serial Data In
12	DOUT	Digital Serial Data Out
13	CLK	Serial Clock
14	AGND	Ground for the analog inputs
15	VREF	Voltage Reference Input
16	VDD	Power supply (+2.7 V to 5.5 V)

The function of each of the 16 pins described in Table 3-1 will make more sense when you understand the following points about what the MCP3008 does and how you can use it:

>> **You can connect as many as eight input voltages to the MCP3008, each on pins 1 through 8.** For example, suppose you want to use a potentiometer as the input for channel 0. To do that, connect one side of the potentiometer to any of the Pi's +3.3 V pins and the other side to any of the Pi's ground pins. Then connect the center terminal of the potentiometer to pin 1 on the MCP3008. As you turn the potentiometer's shaft, the voltage sent to the channel 0 of the MCP3008 will vary from zero to +3.3 V.

>> **Pin 15 provides a reference voltage, which determines the voltage range for the analog input channels.** You'll usually connect this to the +3.3 V power provided by the Pi. That way, the maximum voltage of the analog input will be the same as the reference voltage: +3.3 V.

>> **All the analog input channels share a common ground at pin 14.** This is generally connected to one of the ground pins in the Pi, as is the digital ground at pin 9 of the MCP3008.

>> **The MCP3008 converts the voltage level on the analog input pin to a 10-bit binary number.** A 10-bit binary number can have a maximum decimal value of 1023. Thus, the MCP3008 converts an analog input voltage to an integer between 0 and 1023.

>> **The MCP3008 uses a special form of serial data communication called *serial peripheral interface* (SPI) to tell the Raspberry Pi what voltage levels are present on the analog inputs.** SPI requires four digital ports to work, so the MCP3008 will tie up four GPIO ports on your Pi.

>> **The ARM processors in the Raspberry Pi versions 2, 3, and 4 include built-in SPI capabilities, so you don't have to worry about the details of making SPI work.** Instead, you just connect the MCP3008 to the right GPIO pins and use special functions in Python to read data from the MCP3008.

>> **By default, SPI communicates over the GPIO ports on pins 19, 21, 23, and 24.** Although it is possible to move the SPI communications to different pins, there's little reason to do so for most projects. So, if your project requires analog input, you'll want to reserve ports 19, 21, 23, and 24 as communication ports for the MCP3008.

>> **In case you're interested, here are the four pins used for SPI to perform the following functions:**

- Pin 24 is the *chip select* pin, which is used to initiate a conversation between the MCP3008 and the Pi.

- Pin 23 is the *clock* pin, which provides the timing used to send data between the MCP3008 and the Pi.

- Pin 19 connects to the MCP3008's *data input* pin and is used to transmit data one bit at a time from the Pi to the MCP3008.

- Pin 21 connects to the MCP3008's *data output* pin and is used to transmit data one bit at a time from the MCP3008 to the Pi.

» **The MCP3008 gives you an approximation of the actual voltage present on the analog input channel.** The accuracy of the estimate depends on a number of complex factors. Although the approximation is close enough for most purposes, I probably wouldn't use it to aim a laser that performs eye surgery!

Figure 3-4 shows a schematic of a typical circuit that uses an MCP3008 chip to connect a potentiometer to a Raspberry Pi. In this schematic, the potentiometer's wiper is connected to analog input channel 0 (pin 1), and the four SPI pins of the MCP3008 are connected to the four default SPI ports on the Pi. You see a variation of this circuit used in an actual project later in this chapter.

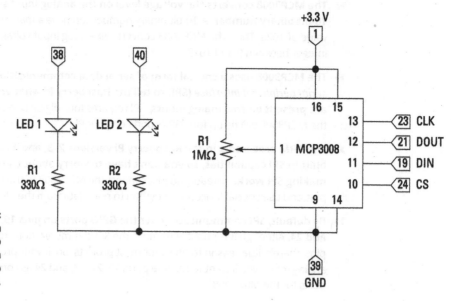

FIGURE 3-4:
Using an
MCP3008 to
connect a
potentiometer to
a Raspberry Pi.

Enabling SPI on Your Raspberry Pi

Although the Raspberry Pi has built-in support for the SPI interface used by the MCP3008 analog-to-digital converter chip, the Raspberry Pi OS disables this support by default. So, before you can use the MCP3008, you must first enable SPI support in the Raspberry Pi OS. Here are the steps:

1. **Click the Terminal icon (shown in the margin), found in the menu bar at the top of the Raspbian desktop.**

   A terminal window opens.

2. **Enter the command** sudo raspi-config.

   The Raspberry Pi Software Configuration Tool, shown in Figure 3-5, appears. (In this figure, I've pressed the down-arrow key twice to highlight Interface Options.)

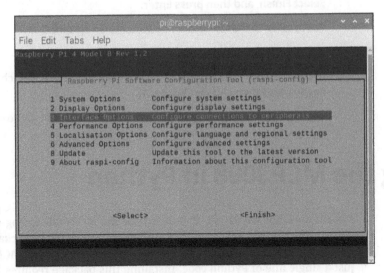

FIGURE 3-5:
The Raspberry
Pi Software
Configuration
Tool.

3. **Select Interface Options and press Enter.**

   The Interface Options page, shown in Figure 3-6, appears.

4. **Scroll down to SPI Enter.**

5. **When asked if you would like the SPI interface to be enabled, select Yes and press Enter.**

   A confirmation screen indicates that SPI has been enabled.

6. **Press Enter.**

Reading Digital
and Analog Input

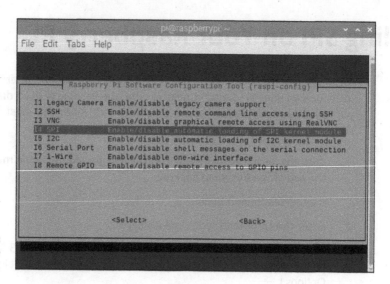

FIGURE 3-6:
The Interface
Options page.

7. **Select Finish, and then press Enter.**

   You're taken to the terminal window's command prompt.

8. **Close the terminal window.**

9. **In the Raspberry Pi OS Desktop, choose Menu⇨ Shutdown, choose Logout, and then choose Shutdown.**

   Your Pi restarts. When Raspberry Pi OS starts back up, SPI will be enabled.

Using the MCP3008 in Python

Python does not have built-in support for the MCP3008 analog-to-digital converter chip, but you can download and install a convenient package of functions named mcp3008 that will let you read an analog value from an MCP3008 chip with just a single line of Python code. Installing this package requires a few steps, but isn't too difficult.

First, you need to make sure that some key elements of your Python programming environment are up to date and properly installed. Ensure that your Pi is connected to the Internet; then open a terminal window by clicking the Terminal icon in the menu bar at the top of the Raspberry Pi OS Desktop (shown in the margin), and then enter the following commands:

```
sudo apt-get update

sudo apt-get dist-upgrade

sudo apt-get install build-essential

sudo apt-get install python-dev

sudo apt-get install python-smbus

sudo apt-get install python3-pip
```

Once the correct packages are installed, enter the following command to download and install the mcp3008 package:

```
pip install https://pypi.python.org/packages/source/m/mcp3008/mcp3008-0.1a2.
    tar.gz
```

Note that the above command must be entered all on one line, with no line breaks. (The last part of the command, which follows the final forward slash, should be mcp3008-0.1a2.tar.gz.)

After the mcp3008 package downloads and installs, you're ready to access the MCP3008 from your Python programs.

Using the mcp3008 Package

The mcp3008 package provides functions that let you determine the values of any or all of the input ports on an MCP3008 chip. To use these functions, you must include the following line near the top of your program:

```
import mcp3008
```

One more bit of housekeeping is required before you can read data from the MCP3008: You must create a variable that you'll use to access the MCP3008 chip, like this:

```
adc = mcp3008.MCP3008
```

This statement creates a variable named adc that you can use to access the MCP3008.

The `mcp3008` package defines eight constant values which you use to refer to the eight analog input channels:

```
mcp3008.CH0
mcp3008.CH1
mcp3008.CH2
mcp3008.CH3
mcp3008.CH4
mcp3008.CH5
mcp3008.CH6
mcp3008.CH7
```

You use the `read` function to read the value of one or more analog input ports. This function uses a list of ports to be read as an argument, which means that you must enclose the input channel constants in square brackets when you call the function. For example, to read the value of input channel 0:

```
analog = adc.read([mcp3008.CH0])
```

Here, the variable `analog` will be set to an integer between 0 and 1023, corresponding to the voltage read by the analog input channel relative to the reference voltage.

You can read more than one input channel, like this:

```
analog = adc.read([mcp3008.CH0, mcp3008.CH1])
```

In this example, the values of channels 0 and 1 are read. The result is a list of two values, which are then stored in the `analog` variable. To retrieve these values separately, you'd have to slice them. For example:

```
analogCH0 = analog[0]
analogCH1 = analog[1]
```

After these statements execute, the variable `analogCH0` will be set to the value of analog channel 0, and `analogCH1` will be set to the value of analog channel 1.

If you want, you can read all eight of the analog channels at once by calling the `read_all` function, like this:

```
analog = adc.read_all()
```

This simply returns a list that contains the value for all eight of the analog channels.

Listing 3-2 shows a simple program that alternately flashes LEDs connected to pins 38 and 40. The rate at which the LEDs flash is set by a potentiometer connected to analog channel 0 of an MCP3008. As you can see, the program divides the value returned by the read function by 1023, and then uses the result as the sleep interval between flashes. Thus, as the user turns the potentiometer's knob, the flash rate varies between 0 and one second.

LISTING 3-2: **An LED Flashing Program That Uses a Pot**

```
import RPi.GPIO as GPIO
import time
import mcp3008

LED1 = 38
LED2 = 40

GPIO.setmode(GPIO.BOARD)
GPIO.setup(LED1, GPIO.OUT)
GPIO.setup(LED2, GPIO.OUT)

adc = mcp3008.MCP3008

while True:

    pot = adc.read([mcp3008.CH0])
    pause = pot / 1023

    GPIO.output(LED1, GPIO.HIGH)
    GPIO.output(LED2, GPIO.LOW)

    time.sleep(pause)

    GPIO.output(LED1, GPIO.LOW)
    GPIO.output(LED2, GPIO.HIGH)

    time.sleep(pause)
```

Project 54 shows how to build a circuit that includes a 1 MΩ potentiometer so that you can test the code in Listing 3-2. Figure 3-7 shows the completed circuit.

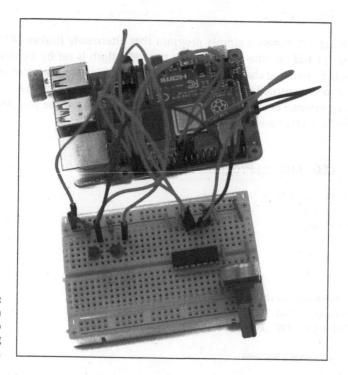

FIGURE 3-7:
Using a
potentiometer to
control flashing
LEDs (Project 54).

Project 54: A Variable-Rate LED Flasher

In this project, you connect a breadboard with two flashing LEDs and a potentiometer that controls the rate at which the LEDs flash. An MCP3008 analog-to-digital converter chip is used to interface the potentiometer to the Raspberry Pi.

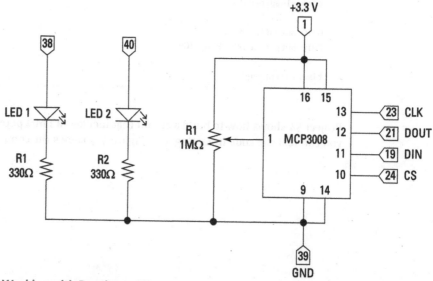

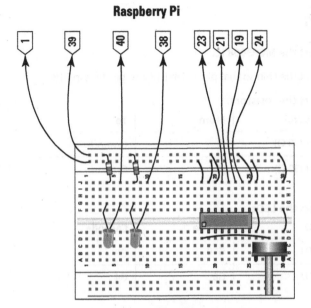

Raspberry Pi

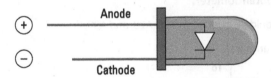

Anode

Cathode

Parts

» One Raspberry Pi with software installed, connected to a monitor, keyboard, and mouse

» One small solderless breadboard

» One MCP3008 analog-to-digital converter IC

» Two LEDs

» One 1 MΩ potentiometer

» Two 330 Ω resistors (yellow-violet-brown)

» Jumper wires

Steps

1. **Insert the MCP3008 IC.**

   Orient the chip so that pin 1 is in E18 and pin 16 is in F18.

2. **Insert the resistors.**

Resistor	From	To
R1 (330 Ω)	J8	Ground
R2 (330 Ω)	J4	Ground

3. **Insert the LEDs.**

LED	Anode	Cathode
LED1	G10	G8
LED2	G6	G4

4. **Insert the potentiometer.**

   The pins should be inserted in A26, A28, and A30.

5. **Insert the jumpers.**

From	To
D18	D28
E26	F26
E30	F30
J18	Positive bus
J19	Positive bus
J20	Ground bus
J25	Ground bus
J26	Positive bus
J30	Ground bus

6. **Connect jumpers from the breadboard to the GPIO pins on the Raspberry Pi.**

Breadboard	Raspberry Pi Pin
J6	40
J10	38
J21	23
J22	21
J23	19
J24	24
Ground bus	39
Positive bus	1

7. **Use Thonny to enter and run the Flashing LED program (see Listing 3-2).**

The LEDs will alternate on and off at a rate determined by the potentiometer. When you turn the knob on the potentiometer, the rate of flashing will increase or decrease.

8
Special Effects

Contents at a Glance

Chapter **1**

Building a Color Organ

One of the great things about Disneyland is that sometimes the long line you have to wait in to go on a particular ride is almost as good as the ride itself. One of the best examples of this is the famous *Indiana Jones Adventure: Temple of the Forbidden Eye.* Just outside of an ancient temple, you pass by a rickety steam-powered generator that is barely running. The clickity-clickity sound of the generator alternately grows louder and softer as the generator sputters and threatens. Once inside the temple, you pass through narrow tunnels and creepy caverns that are lit overhead by lights that appear to be powered by the rickety generator. The lights flicker and dim, then grow brighter for a moment, then flicker and dim again in sync with the laboring generator.

In this chapter, you learn how to build an electronic circuit that you can use to create this creepy lighting effect. I've used this circuit for many different purposes, including lighting the narrow passageways in my own haunted tomb at Halloween. I've also used the same circuit to create a thunderstorm in my front yard to add the right ambiance to my haunted Halloween graveyard. And the same circuit creates a spooky red heartbeat in the chest of a plastic skeleton that stands watch over the scene.

The circuit is called a *color organ,* and its operation is simple: It converts the volume of an audio input into an output voltage that gets higher as the sound source gets louder. If you connect a light to the output, the light will glow brighter when the audio input is louder and dimmer when the input is quieter.

WARNING
As with most of the chapters in this minibook, I need to start with a warning that this circuit requires that you work with line-level voltages (120 VAC), so it's potentially dangerous. The circuit is designed with safeguards, but you must be careful to not bypass them. You should inspect the project every time you use it to make sure none of the wiring has come loose or frayed, and you must never work on the circuit while it's plugged in.

Examining the Color Organ Project

The completed color organ described in this chapter is shown in Figure 1-1. The project is housed in a 2-x-3-x-6-inch project box that includes a power plug that connects the circuit to a wall outlet, a power outlet that you can connect a light to (maximum 120 W), an RCA-style audio input connector that you can connect a sound source to (up to 60 W), a knob for adjusting sensitivity, and a power switch.

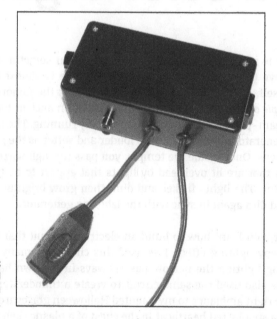

FIGURE 1-1: The completed color organ project.

To keep the project simple, you can build the electronics using an inexpensive kit — specifically, the Velleman MK110 Simple One Channel Light Organ kit. This kit is widely available on the Internet for under $10; just search for *Velleman MK110*, and you'll find several sources.

Figure 1-2 shows how the color organ project can be hooked up to create light that varies in brightness with a sound source. You'll need a source for the sound, such as a boom box or other sound system with external speaker outputs that you can tap into.

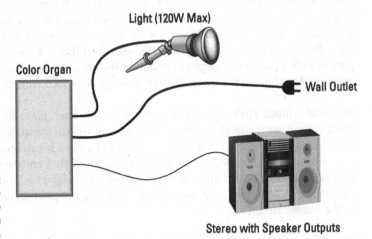

FIGURE 1-2:
Connecting the color organ to a light and a sound source.

Note that the diagram in Figure 1-2 doesn't show the details of how to connect the sound system to the color organ and the speakers. The easiest way to connect the color organ to the sound system is to simply replace one of the speakers with the color organ. That way, the color organ responds to one of the stereo channels while the speaker plays the sound coming from the other stereo channel. For this type of hookup, you'll need just a single cable that has an RCA connector on one end (to plug into the color organ) and the proper connector on the other end to connect to the sound system's speaker output.

You'll also need a suitable recording for your sound source. For example, if you want to use the color organ to create a thunderstorm effect, you'll need a recording of thunder. To make a red light flash in sync with a heartbeat, you'll need a recording of a heartbeat. The Internet is rife with such sound effects, so you shouldn't have too much trouble locating and downloading a sound effect to meet your needs. If you want to customize the sound effect, you can download a free audio editor called Audacity from www.sourceforge.net/projects/audacity.

Understanding How the Color Organ Works

There are several different ways to design a color organ circuit. Most of them rely on a special type of electronic component called a *triac*, which is essentially a transistor that's designed to work with alternating current. It has three terminals. Two are anodes, called A1 and A2, and the third is a gate. A voltage at the gate — either positive or negative — allows the anodes to conduct. The anodes are connected to the line load, and the gate voltage is derived from the audio input.

The audio input isn't connected directly to the triac gate, however. Instead, most color organs use one of two techniques to isolate the audio input from the line-voltage side of the circuit. One method is to use a transformer. The other is to use an *optoisolator*, which is a single component that consists of an infrared LED and a photodiode or other light-sensitive semiconductor. Voltage on the LED causes the LED to emit light, which is detected by the photodiode and passed on to the output circuit.

The Velleman MK110 kit uses an optoisolator triac, in which the photosensitive semiconductor is actually a triac whose gate is stimulated by light rather than by voltage. The optoisolator is an integrated circuit in a 6-pin DIP package.

Figure 1-3 shows a simplified schematic diagram for the circuit used by the Velleman MK110 kit. As you can see, the audio input is applied to the LED side of the optoisolator, controlled by a potentiometer, which lets you adjust the sensitivity of the circuit. The output from the optoisolator is applied to the gate of the triac, whose anodes are connected across the line voltage circuit. Thus, the volume of the audio input directly controls the voltage of the output circuit.

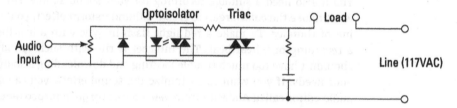

FIGURE 1-3: The schematic diagram for the color organ circuit.

Getting What You Need to Build the Color Organ

Other than the Velleman kit itself, most of the materials you need to build the color organ project can be purchased at your local RadioShack store or any other supplier of electronic components. The table below lists all the materials you'll need.

Quantity	Description
1	Velleman MK110 Simple One Channel Light Organ kit
1	2-x-3-x-6-inch plastic project box
1	20 mm PC board standoffs
1	RCA-style phono jack
1	¾-inch control knob
1	Chassis-type fuse holder for 1¼-x-¼-inch fuse
1	1 A, 250 V, fast-acting 1¼-x-¼-inch fuse
1	⅜-inch 4-40 machine screw and nut (for mounting the fuse holder)
1	SPST rocker switch
2	⅜-inch grommets
2	Screw-on wire connectors
5 inches	20-gauge stranded hook-up wire
1	Indoor extension cord

Assembling the Color Organ

Once you have gathered all the materials for the color organ, you're ready to assemble the project. You'll need the following tools:

>> Soldering iron, preferably with both 20 W and 40 W settings

>> Solder

TIP

Use thicker solder for the line-voltage wires and thin solder for assembling the MK110 kit.

- » Magnifying goggles
- » Phillips screwdriver
- » Small flat-edge jewelers screwdriver
- » Wire cutters
- » Wire strippers
- » Pliers
- » Hobby vise
- » Drill with ⅛-inch, ⁵⁄₃₂ -inch, ¼-inch, ⁵⁄₁₆ -inch, ⅜-inch, and ¾-inch bits

Here are the steps for constructing this project:

1. **Assemble the Velleman MK110 kit.**

   The kit comes with simple but accurate instructions. Basically, you just mount and solder all the components onto the circuit board. Pay special attention to the color codes for the resistors and the orientation of the diode.

   Figure 1-4 shows the completed MK110 kit.

   TIP

   When I build this kit, I find it best to mount the circuit board in a good hobby vise and use an alligator clip or masking tape to hold the components in place while soldering.

FIGURE 1-4:
The assembled
Velleman
MK110 kit.

2. **Drill all the mounting holes in the project box except the hole for the sensitivity control on the left side of the box.**

   Figure 1-5 shows the orientation of the approximate location of the mounting holes.

   Use the assembled circuit board to determine the exact drilling locations for the four holes that will mount the circuit board. The position of the other holes isn't critical, with the exception of the hole for the potentiometer knob. Don't drill that hole until Step 4.

3. **Mount the four standoffs in the four MK110 circuit board mounting holes.**

   Use four of the machine screws that came with the standoffs.

4. **Drill the hole for the circuit board's potentiometer.**

   Set the circuit board on top of the four standoffs to determine the exact location for this hole.

5. **Insert the two rubber grommets into the two ⅜-inch holes.**

   The grommets are difficult to squeeze into the hole, but work at it and you'll get them in. If necessary, use the small edge of a flat screwdriver to push the rubber edges into the holes.

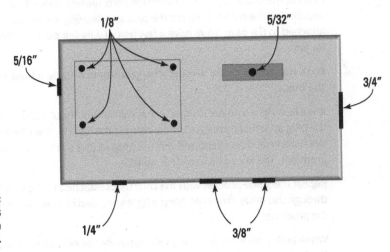

FIGURE 1-5:
Drill the holes
as indicated in
this diagram.

TIP In the steps that follow, you assemble all the parts into the project box. Use Figure 1-6 as a guide for the proper placement of each of the parts.

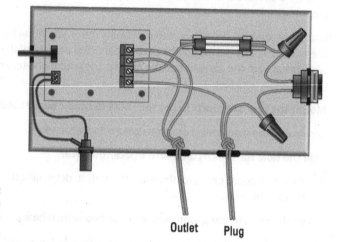

Outlet   Plug

FIGURE 1-6:
How the parts go
together inside
the project box.

1. **Cut the extension cord.**

   First cut the outlet end of the extension cord, leaving about 12 inches of wire attached to the outlet. Then cut the plug end, leaving about 3 or 4 feet of wire attached to the plug. You'll have a few feet of wire left over; set this wire aside for later.

2. **Push the power cords through the grommets and tie a knot inside the box.**

   It will be a tight squeeze, but the cords will fit. Pull about a foot of the cord with the plug attached through the hole nearest the switch. Then, tie it in a knot, cinch the knot down tight, and pull the plug so that the knot is snug against the grommet. The knot acts as a strain relief.

   Repeat the same process with the cord that's attached to the outlet, passing it through the other grommet, tying a tight knot, and pulling the knot up against the grommet.

   When both power cords are in place, separate the two wires of each cord inside the project box and strip about ⅜ inch of insulation from each wire.

3. **Cut two 1½-inch lengths of extension cord wire and solder them to the switch terminals.**

   You'll need to strip about ⅜ inch of insulation from each end of both wires. Put your soldering iron on its High setting and use thick solder. Set the switch aside when the solder sets.

4. **Cut two 1½-inch lengths of extension cord wire and solder them to the terminals on the fuse holder.**

   Again, you'll need to strip about ⅜ inch of insulation from each end of both wires and solder with high heat.

5. **Cut two 2½-inch lengths of the hookup wire and strip ⅜ inch of insulation from the ends.**

6. **Solder one of the hookup wires to the center terminal of the RCA-style phono jack and the other wire to the ground terminal.**

   At this point, you're done with the soldering iron, so you can turn it off.

7. **Mount the RCA-style phono jack in the ¼-inch hole in the project box.**

   To mount the jack, you'll first have to remove the nut, the ground terminal, and the lock washer from the jack. Then, pass the wire connected to the center terminal of the phono jack through the ¼-inch hole, and then insert the threaded end of the phono jack into the hole. Slip the lock washer, the ground terminal, and the nut over the wire connected to the center terminal, and then thread them onto the threaded part of the jack. Tighten with needle-nose pliers.

TIP

In the next few steps, you attach wires to the MK110 circuit card. Do *not* mount the circuit board to the standoffs quite yet. You'll have an easier time connecting the wires if the circuit board is loose. After the wires are all connected, you mount the board.

1. **Connect the separated wires of the cord that's attached to the outlet to the two terminals marked *Load* on the back of the MK110 circuit board.**

   Use a small, flat screwdriver to tighten the terminals. Make sure the wires are securely connected.

2. **Connect one of the wires attached to the fuse holder to one of the Mains terminals at the back of the MK110 circuit board.**

3. Connect one of the extension cord wires that's attached to the plug to the other Mains terminal at the back of the MK110 circuit board.

4. Connect the two hook-up wires from the phono jack to the input terminals at the front of the MK110 circuit board.

The input terminals are labeled *LS* on the board. You'll need a very small flat-blade screwdriver to tighten these terminals.

5. Mount the MK110 circuit board on the standoffs.

To mount the board, you'll need to tilt it a bit to slide the shaft of the potentiometer through the $\frac{5}{16}$-inch hole. Once the shaft is through, set the board down on the standoffs and secure it with the remaining four machine screws that came with the standoffs.

6. Use the $\frac{3}{8}$-inch 4-40 machine screw and nut to mount the fuse holder.

Slide the machine screw through the $\frac{5}{32}$-inch hole in the bottom of the project box. Then pass the machine screw through the hole in the center of the fuse holder and attach the nut. Tighten with a screwdriver.

7. Mount the switch.

To mount the switch, first remove the plastic nut on the threaded end of the switch. Then, pass the wires and the threaded end of the switch through the $\frac{3}{4}$-inch hole in the side of the project box. Finally, slip the nut over the wires and tighten it onto the switch.

8. Connect the switch to the power cord and the fuse.

Use one of the screw-on wire connectors to connect one of the switch wires to the unconnected wire on the fuse holder. Then, use the other wire connector to connect the other switch wire to the unconnected wire that leads to the power plug.

9. Insert the fuse in the fuse holder.

Guess what — you're almost done! Figure 1-7 shows the project with all the parts assembled.

10. Attach the knob to the potentiometer shaft protruding from the box.

Use a small, flat screwdriver to tighten the set screw on the knob.

11. Place the lid over the project box and secure it with the provided screws.

Now you really are done!

FIGURE 1-7:
The color organ is
almost finished.

Using the Color Organ

Having finished the color organ, it's time to put it to use making interesting sound and lighting effects. To do that, you'll have to connect both lights and a sound system to the color organ.

The procedure for actually using the color organ is quite simple:

1. Connect a light to the color organ's female extension cord connector.

2. Connect a speaker-level audio input to the RCA connector.

3. Plug the male extension cord connector into a power outlet.

4. Turn on the color organ.

5. Play the sound.

6. Turn the knob on the color organ to adjust the sensitivity.

7. If the light never comes on, try increasing the output volume on the stereo.

The color organ can handle just 120 W on the output circuit, so you need to be careful not to overload the circuit. You can use a single 100 W flood light or a couple of 60 W lamps. Or you can use a few strings of Christmas lights or other low-wattage lights.

As I mention earlier in this chapter, the easiest way to connect the color organ to the sound system is to simply replace one of the speakers with the color organ. The type of cable you'll need to do that depends on how the speakers connect to the sound system. If the speakers connect with simple post connectors, you'll need a cable with bare wire on one end and an RCA plug on the other end. If the speakers connect with RCA connectors, you'll need a cable with RCA plugs on both ends.

When you use the color organ in this way, it's important to realize that the color organ will respond to one channel of the stereo recording while the speaker plays the other channel. In most cases, you won't notice much difference. However, in some recordings, the sound on the left channel is very different from the sound on the right channel. This can affect the quality of the sound, and it can also prevent the light from flashing in perfect sync with the sound, since the light is responding to a different sound source from the one heard through the speakers.

In some cases, you can use this to improve the effect you're trying to achieve with the color organ. For example, in an actual thunderstorm, the lightning flashes well before the thunder is heard. To reproduce this effect, all you need is a sound recording of a thunderstorm in which the thunder is heard on the left channel before it's heard on the right channel. Then, if you connect the color organ to the left channel and the speaker to the right channel, the light will flash before the sound is heard.

Have fun with your color organ!

IN THIS CHAPTER

» Decorating your house with computer-controlled flashy holiday lights

» Looking at Light-O-Rama controllers

» Programming a light sequence with Light-O-Rama's Sequence Editor software

Chapter **2**

Animating Holiday Lights

Have you seen those crazy animated light shows some people put up for the holidays, in which thousands of lights (or maybe tens of thousands or in some cases hundreds of thousands) are sequenced in time with music to create dazzling displays? Usually the lights are sequenced to cool holiday music by Mannheim Steamroller or Trans-Siberian Orchestra. Often the music is broadcast over a low-wattage FM transmitter, so when you drive by the house to see the show, you can just tune your radio to the local broadcast to hear the music. If you're not sure what I'm talking about, just hop onto YouTube and search for *Christmas lights*. You'll find plenty of videos.

If you've ever wanted to create such a display yourself, this chapter will show you how. Although you can design the circuitry to control light shows like this from scratch, the easiest way to build a light show is to buy an inexpensive lighting controller. That way, you can focus your energies on the design of the show rather than on the design of the lighting controller.

There are several companies that sell lighting controllers for holiday displays. My favorite is Light-O-Rama, based in South Glens Falls, New York. What I like most about Light-O-Rama is that you can purchase preassembled lighting controllers, or you can purchase kits and assemble the controllers yourself. So if you're handy with a soldering iron (and I assume that you are, or you wouldn't be reading this book), you can save a few hundred dollars on a basic sequencer by building the circuit yourself.

For complete information about Light-O-Rama and its products, point your browser to www.lightorama.com.

TIP

Light-O-Rama controllers are useful for much more than just holiday displays; they're also useful for school events, carnivals, amusement parks, store and museum displays, theatrical productions, malls and shopping centers, and so on. In fact, the Light-O-Rama products shown in this chapter were donated by the good folks at Light-O-Rama for use at the Fresno Chaffee Zoo.

Introducing the ShowTime PC Controller

Light-O-Rama's most popular lighting controller for residential use is called the *ShowTime PC* controller, as shown in Figure 2-1. This controller (model number CTB16PC) offers the following basic features:

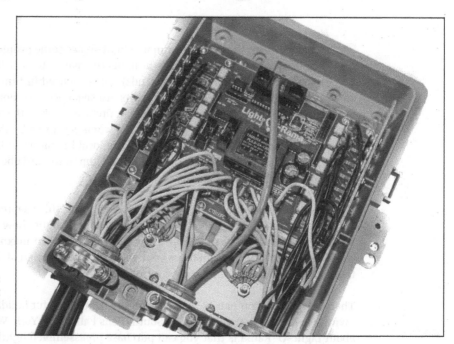

FIGURE 2-1:
The basic
ShowTime PC
controller from
Light-O-Rama.

>> **Sixteen separate *channels* of light:** Each channel is a separate 120 VAC circuit that can power up to 8 amps of lights. The power cords dangling from the bottom of the unit are where you plug the lights in; each cord connects the lights for one of the 16 channels. (The total current load for the controller should not exceed 30 amps.)

In most cases, you won't plug your lights directly into the power cords dangling from the bottom of the lighting controller. Instead, you'll use extension cords to reach from the lighting cords to where the lights are actually placed. If you plan on making a holiday lighting display, you'll need a lot of extension cords.

Sixteen channels is enough to get started with Light-O-Rama displays, but once you've set up your first show, you'll wish you had additional channels. Most Light-O-Rama setups include two, three, or four controllers for a total of 32, 48, or 64 channels.

>> **The ability to turn the lights on and off, separate for each channel:** The controller also has a few other special effects, such as fade up, fade down, different intensity levels, twinkling, and shimmering.

These special effects are what make the Light-O-Rama controller special. You can find much less expensive ways to simply turn lights on and off. For example, two Kit-74 relay controllers connected to a computer can turn 16 channels of lights on and off for less than half the price of a ShowTime PC controller. However, the Kit-74 can't fade the lights up or down, and the ability to gradually fade the lights adds a lot to the impact of the light show.

>> **Computer control of the light show:** You simply connect the ShowTime PC controller to the computer via a special USB adapter cable. Then, you run special software available from Light-O-Rama to control the show.

Although the ShowTime PC controller is mounted in a weatherproof container, your computer isn't. As a result, you'll want to place the computer inside your house or garage, then use a long cable to reach the ShowTime PC controller.

>> **Light show control without a computer:** If you don't want to drive the show from a computer, you can purchase a special device called an *MP3 Director* that lets you control the show without a computer.

>> **Expandability:** You can connect as many as 240 ShowTime PC controllers together to create a massive show with as many as 3,840 separate light circuits. You'd need a small nuclear reactor to power it all, but it's possible.

TIP

If you want to get started with Light-O-Rama, I suggest you purchase one of its starter kits. You can purchase a completely preassembled kit including a 16-channel ShowTime PC controller, USB adapter, and software for under $400, or you can purchase build-it-yourself kits for considerably less. These kits require you to solder the components to the board, but assembling the board yourself will give you the satisfaction of knowing you built it yourself.

Looking at a Basic Light-O-Rama Setup

Figure 2-2 shows a basic setup for using a single, 16-channel ShowTime PC controller to control up to 16 separate strands of lights. The following paragraphs describe each of the elements in this setup:

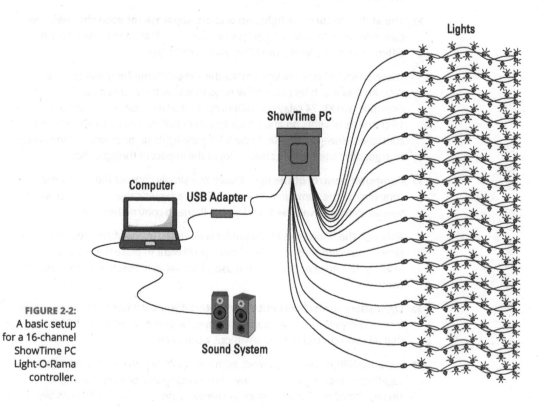

FIGURE 2-2:
A basic setup for a 16-channel ShowTime PC Light-O-Rama controller.

>> **ShowTime PC controller:** Usually located outside close to where your lights are. It comes with a weatherproof enclosure so you can place it outside.

TIP

Although it isn't shown in the figure, the ShowTime PC controller requires a source of AC power. As a result, you should locate the controller near an electrical outlet.

The controller has two electrical plugs; each provides power to half of the 16 light channels. If you plug both of these cords into the same electrical outlet, the total lighting capacity of the controller will be limited by the maximum amperage rating of the outlet you plug the cords into (typically 15 A). However, you can plug the two cords into separate outlets to double the lighting capacity of the controller to 30 A, provided that the two outlets you plug the controller into are located on separate electrical circuits.

>> **Lights:** Connected to the 16 power cords that hang from the bottom of the controller. For more information about choosing lights, see the section "Choosing Lights for Your Display" later in this chapter.

>> **Computer:** Runs the ShowTime software that controls the lights. You'll want to place the computer in a secure location that isn't exposed to weather.

REMEMBER

Unfortunately, Light-O-Rama's software isn't free. If you purchase one of Light-O-Rama's starter kits, the software is included in the purchase price, but if you purchase a do-it-yourself kit and assemble the circuit board yourself, you'll have to buy the software separately. (The cost is under $50.)

For more about using the Light-O-Rama software, see the section "Using the Light-O-Rama Sequence Editor" later in this chapter.

>> **USB adapter:** Required to connect the computer to the ShowTime PC controller. For more information about the USB adapter, see the section "Connecting the Controller to a Computer" later in this chapter.

>> **Sound system:** Plays or broadcasts the sound that is synchronized to the lights. The sound system connects to the computer's headphone output and either amplifies it for speakers or broadcasts it so it can be heard on an FM radio.

If you want to play the sound through speakers, you can use any amplifier that has an input jack and is powerful enough to play the sound at the volume you desire. I have a guitar amplifier that I usually use, but I have also used a boom box.

If you want to broadcast the music so that people can listen to it on their car radios as they drive by your house, you can purchase a low-power FM broadcaster from many sources on the Internet. Light-O-Rama sells one for about $125.

Understanding Channels and Sequences

In a ShowTime PC controller or other similar holiday lighting controller, lights are controlled via *channels*. Each channel is a separate electrical circuit to which you can connect one or more strings of lights. It's important to keep in mind that the controller can't control the individual lights that are connected to a channel; all the lights on a single channel operate together as a single unit.

A *sequence* is simply a recorded program that activates the lights connected to each of the controller's channels in a particular order, most often synchronized with music. The art of designing a good holiday light display lies in creating clever sequences that squeeze the most impact out of a limited number of channels.

When planning a light show, one of the first steps is to determine how many channels your show requires and what lighting elements will be controlled by each channel. For example, you might run several strings of icicle lights across the roofline of your house and connect those lights to one of the channels. Then, the controller can turn the icicle lights on or off together as a unit. Or you might put colored lights on a shrub connected to its own channel. Then, the controller can turn the shrub on or off.

There's no reason you can't physically overlap lights that are connected to different channels. For example, you might put a string of green lights connected to one channel on a shrub and place a string of red lights on the same shrub but connected to a different channel. Then, the controller can make the shrub glow green, glow red, or glow green and red at the same time. Or the controller might make the shrub alternately flash green and red, in rhythm with the music of course.

There's also nothing preventing you from connecting different colors or kinds of lights in different areas of your yard to the same controller. For example, you might use red lights on one shrub and green lights on another shrub on the same channel. Then, the controller can turn both shrubs on or off — one green, the other red.

You might even put a string of green lights and a string of red lights on both shrubs but connect the green string on the first shrub and the red string on the second shrub to one channel and the red string on the first shrub and the green string on the second shrub to another channel. Then, you can create a sequence that alternately flashes the shrub green and red.

Keep in mind that the controller can do more than just turn lights on and off. For example, the controller can cause each channel to fade up or down at any speed you want. If you have green and red lights on a single shrub, you can create a sequence that gradually changes the shrub from green to red, and then back to green.

TIP

You can also use channels to create the effect of motion in your light show. As a simple example, suppose you set up several strings of light radiating from a central point, like spokes on a wheel. If the lights on each spoke are connected to several channels, you can create the illusion that the wheel is spinning by activating each spoke in sequence.

The possibilities are limited only by your creativity — and the number of channels you have available to you. More channels are better because having more channels lets you create more complicated sequences. If you start with a 16-channel controller, it won't be long before you want to expand it to 32 channels so you can create more creative sequences.

Choosing Lights for Your Display

There are many different types of lights you can use with a Light-O-Rama or other lighting controller. Take a trip to any department store during the holiday season and you'll find a wide variety of lights to choose from, or you can purchase lights from online sources such as www.1000bulbs.com or www.christmaslightsetc.com.

TIP

The best time to buy lights for your display is the day after Christmas when most retail stores mark them down at least 50 percent.

The following paragraphs describe the most common types of lights used in holiday displays:

>> **Incandescent minilights:** Strings of 100 or more lights in various colors. These lights are readily available in white, red, green, blue, violet, and yellow. You can also find multicolored strings, which contain a mix of colors. If you hunt around, you may also find other colors. During the Halloween season, you can find orange and purple minilights.

Minilights are designed to run on 2.5 VAC. They're wired in series with 50 lights on each run, which enables them to operate on 120 VAC. The drawback of this arrangement is that because the lights are wired in series, if one of the lights pops out of its socket or comes loose, continuity is broken for the entire run, and all 50 lights will go out.

You can connect strings of minilights end-to-end on a single channel, but you shouldn't connect more than five strings of 100 lights together end-to-end or you'll blow the fuses that are built in to the plugs.

>> **Net lights:** These are minilights woven together into a meshlike net that can be spread over a shrub or other small bush. Usually, each net has 300 lights.

>> **LED lights:** Usually the same size as minilights but made with LEDs rather than incandescent light bulbs. LEDs consume much less power than incandescent lights; in fact, an entire string of LED lights consumes about the same amount of power as a single minilight, and you can connect as many as 20 strings together end-to-end. Thus, you can connect more LEDs to a single channel. However, LED lights are more expensive than incandescent minilights.

>> **C7 lights:** Incandescent lights that are the same size as night lights. Strings of C7 lights are usually made with 25 sockets per string. (The designation *C7* refers to the size of the base.)

C7 lights operate on 120 VAC, so they're wired together in parallel. If one goes out or if you remove one, the rest of the string will still light up. You can connect C7 strings end-to-end, but you should connect no more than three strings together to avoid overloading the circuit and blowing out the fuses.

>> **C9 lights:** Similar to C7 lights but bigger.

>> **Rope lights:** Strings of incandescent minilights or LEDs enclosed in a flexible transparent tube. You can purchase short lengths (typically 18 feet) of rope light at retail stores, or you can buy 150-foot commercial-grade spools of rope light online from sources such as www.1000bulbs.com or www.christmaslightsetc.com.

>> **Blow-mold figures:** Sculptures made of translucent plastic with a light bulb inside. Most blow-mold figures have a single incandescent light bulb inside, typically 40 W or 60 W. If you have several blow-mold figures in your display, you can connect them all to a single channel on your controller to turn them on or off together, or, if you have enough channels at your disposal, you can connect each to a separate channel to control each individually. (See why you need more channels?)

>> **Wire-frame sculptures:** Sculptures made from thick wires that are bent and welded into shapes such as deer, Christmas trees, presents, and so on. They are then lit by strings of incandescent minilights or LED lights.

>> **Megatree:** A popular type of light sculpture that is simple to make. All you need is a tall pole and several strings of lights. Attach one end of each string of lights to the top of the pole, and then use stakes to plant the other end of the light strings into the ground, forming a ring around the base of the pole. The more strings of light you use the better. If you use alternating colors of lights and connect each color to a separate channel, you can create some interesting animation effects.

Designing Your Layout

One of the most important steps for creating a holiday light display is designing the layout for your lights. To do that, first start with a diagram of your yard (or whatever space you'll be using for your display). It doesn't have to be exactly to scale, but it will help if the proportions are relatively accurate. If you want, grab a tape measure, make some rough measurements, and sketch out your space on a piece of graph paper. Then, sketch in where you'd like to place your lights and other decorations such as wire sculptures, blow-mold figures, or a megatree.

Once you've placed your lighting elements where you want them, start assigning channel numbers to each element. If you have a 16-channel controller, assign each lighting element a number from 1 to 16.

Assembling the ShowTime PC Controller

If you purchase the ShowTime PC controller from Light-O-Rama as a do-it-yourself kit, you'll have to assemble it yourself. That means you'll have to solder all the components to the main circuit board, install the board into the weatherproof container, and connect all the various cords that are required for the unit to work.

I'm not going to repeat the assembly instructions that come with the kit here because the instructions that come with the kit are excellent. But I would like to offer a few tips that may help ease the process:

>> **Allow plenty of time.** Even if you're skilled with a soldering iron, the circuit has a lot of components, and it may well take you two or three evenings to completely assemble the kit.

>> **Read through the entire assembly instructions before you start.** That way, there won't be any surprises.

>> **Clear off your workbench.** There are a million little parts in the kit, so you need plenty of space to spread your stuff out.

>> **Organize the parts.** Get some small containers to hold all the various parts and help keep them organized. I like to use small disposable plastic bowls.

>> **Secure parts on the board before you solder.** Get some blue painter's tape to hold parts to the board while you solder them in place.

>> **Turn off the TV.** The last thing you need while you're assembling a complicated kit is distraction.

» **Use lighted magnifying goggles.** They make your work a lot easier.

» **Secure the circuit board.** A third-hand tool or, better yet, a hobby vice will come in handy to hold the circuit board while you solder.

» **Use the proper soldering tools and techniques.** Make sure that you use a low-wattage soldering iron (25 W) and fine-gauge solder for the sensitive parts, especially the integrated circuits. The ICs are soldered directly to the circuit board and can be damaged if you heat them too much.

Figure 2-3 shows the circuit board being assembled.

FIGURE 2-3: Assembling the ShowTime PC circuit board.

After you've assembled the circuit board, you can install it in the weatherproof plastic box and connect the power cords. Again, Light-O-Rama provides excellent illustrated instructions for this task, so I won't duplicate them here.

One other bit of advice I will offer, however, is to pick up a set of numbered electrician's stickers so that you can properly label each of the 16 power cords. You can find these stickers at any home-improvement store that stocks electrical items. If you don't label the cords as you install them, you won't know which cord corresponds to each of the 16 channels.

Connecting the Controller to a Computer

The ShowTime PC controller communicates with the outside world using a digital protocol called RS-485. Most computers don't have an RS-485 port but do have a USB port. The USB adapter converts the USB signals to RS-485 signals so that the computer can communicate with the ShowTime PC controller. Figure 2-4 Shows the USB adapter.

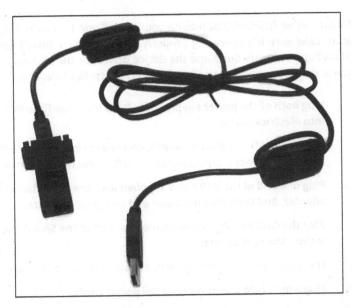

FIGURE 2-4:
A USB adapter converts the RS-485 protocol required by the ShowTime PC controller to USB.

The USB adapter comes with a short USB cable used to connect the adapter to the USB port on a computer as well as a Cat5 network cable used to connect the USB adapter to the ShowTime PC controller. Cat5 cable is the same type of cable used by most computer networks. The Cat5 cable can be as long as 4,000 feet. A cable that long might be suitable for the likes of Disneyland, but you'll probably want to keep your controller within a few dozen feet of the computer.

Light-O-Rama also sells a serial adapter that converts RS-485 signals to the standard serial port protocol that was once commonplace on PCs. Most computers don't come with serial ports anymore, so you'll probably use the USB adapter instead. (Note that if you use the serial adapter, the maximum length of the Cat5 cable that connects the adapter to the controller is 100 feet.)

TIP

It's very important that you follow the instructions that come with the USB adapter *before* you plug the adapter in to a USB port on your computer. Specifically, you should first run the driver installation software from the CD that came with the USB adapter. Plug the adapter into a USB port only after you've run the driver installation program. If you plug the adapter into a USB port before you run the driver program, there's a very good chance that Windows will install the wrong driver for the adapter, and that driver will be associated with the adapter, making it difficult to get the correct driver installed.

Testing the ShowTime PC Controller

When you've finished assembling your ShowTime PC controller, you should test it to make sure it's operating properly. You must first install the Light-O-Rama ShowTime Software Suite and the device driver for the USB adapter. Then, follow these steps to connect your ShowTime PC Controller to your computer:

1. **Plug both of the power supply cords from the ShowTime PC controller into electrical outlets.**

   The Status LED on the ShowTime PC controller's circuit board flashes, indicating that the power is on but the device isn't connected to a computer.

2. **Plug one end of the USB cable supplied with the USB adapter into the adapter, and then plug the other end into your computer.**

3. **Plug the Cat5 network cable that's attached to the ShowTime PC controller into the USB adapter.**

   The Status LED in the ShowTime PC controller continues blinking.

4. **Plug a light into one or more of the power cords.**

   I recommend using inexpensive night lights, which you can usually find at any dollar store for — you guessed it, $1. Since they're so inexpensive, you may want to purchase 16 of them and plug one into each of the channels. Otherwise, just get a few and plug them into the first few channels.

Once you have connected the controller to the computer, you can use the Light-O-Rama software to test the connection. Here are the steps:

1. **On the computer, run the Light-O-Rama Hardware program.**

   To run the program, click the Start button, choose All Programs, choose Light-O-Rama, and then choose Light-O-Rama Hardware. The Hardware program springs to life and displays the screen shown in Figure 2-5.

2. **Click the Auto Configure button near the top left of the Hardware program's window.**

   The Hardware program searches all the available COM ports on your computer until it finds the one that's associated with your USB adapter.

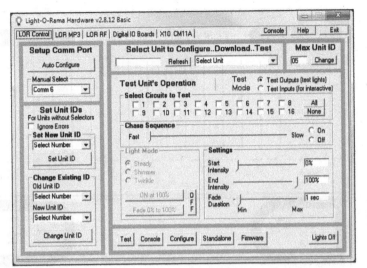

FIGURE 2-5:
The Light-O-Rama Hardware program.

3. **Change the maximum number of units to be searched to 5.**

To do so, click the Change button located in the Max Units section of the window (near the top right). This brings up a dialog box that has a slider control; drag this slider control all the way to the left, and then click Save.

This step saves you time in the next step, which searches for any available controllers. It takes a few seconds to look for controllers at every possible address, so changing the maximum from 240 to 5 can save you several minutes of your life that you would never get back as you waited for the program to search for controllers that don't exist.

4. **Click the Refresh button to look for ShowTime PC controllers.**

Assuming you completed Step 3, the Hardware program finds your controller within a few seconds. The display changes slightly to indicate that the controller has been found.

5. **Click the Console button located near the bottom of the window.**

This action brings up the fun console window shown in Figure 2-6.

6. **For each channel that you connected a light to, drag the corresponding slider up and down and verify that the light turns on and off.**

If a channel doesn't work, verify that the light bulb isn't burned out, that you've connected the light to the correct power cord, and that the light is turned on if it has a switch.

7. **Move the lights to other cords and repeat Step 6 to test all channels.**

You'll want to verify that all 16 of your controller's channels are operating correctly.

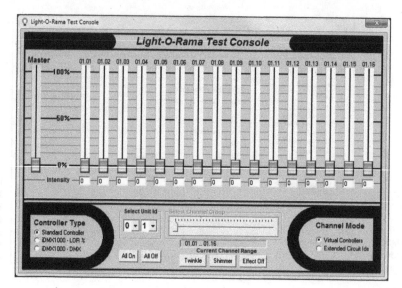

FIGURE 2-6:
The Hardware
program's
Console lets
you test your
controller
channels.

8. **Use the Twinkle and Shimmer buttons to verify that the twinkle and shimmer functions are working.**

   When you're tired of the twinkling and shimmering, click the All Off button.

   You're done!

Assuming all the lights came on, you've successfully tested your Light-O-Rama controller and are now ready to start creating sequences. (If any of the channels don't work, carefully double-check your wiring to make sure everything is connected properly. If you get stuck, contact Light-O-Rama's technical support department; they can help you get your controller working.)

Using the Light-O-Rama Sequence Editor

The meat of the Light-O-Rama Software Suite is the Sequence Editor, which lets you create light sequences that are synchronized with music. A *sequence* is a file that stores the information needed to synchronize lights with one selection of music. The music itself is stored in a separate audio file in the MP3 format.

You can also use the Sequence Editor to play a sequence. When it plays a sequence, the MP3 file is loaded and played through your computer's speakers or other audio output. In addition, the Sequence Editor sends instructions to the ShowTime PC controller to activate the lights in sequence with the music.

In Light-O-Rama, a *sequence* is usually associated with an individual song. Two or more sequences can be combined to create a *show*, which lets you cycle through a set of songs so that your audience doesn't have to watch just one song over and over again. You can also set up a *schedule*, which automatically starts and stops shows on particular times and days. Thus, by setting up sequences, shows, and a schedule, you can completely automate every aspect of your light show.

In the remainder of this chapter, I introduce you to some of the basic concepts for working with the Sequence Editor to create simple sequences. You should realize, however, that you can purchase preprogrammed sequences from Light-O-Rama's website for about $30 per song. Programming even a simple sequence for a single song can easily take 10 hours or more, and programming complicated sequences can quickly become a full-time job. So even at $30 per song, purchasing preprogrammed sequences can be worthwhile.

Understanding Sequences

In Light-O-Rama, a sequence is represented as a grid that's somewhat similar to the grid in a spreadsheet program. For example, Figure 2-7 shows part of a very simple sequence in which the lights on channels 1, 3, and 5 alternately turn on and off every half second.

Each row in the grid represents one of the channels available in the controller. Thus, a grid for a sequence played on a 16-channel controller will have 16 rows, one for each channel. Initially, these channels are named according to the control unit number (01 in the figure) and the channel number (1 through 16 in the figure). However, you can easily change the row names to something more meaningful, such as *Red Bush* or *Santa's Hat*.

Each column in the grid represents a time interval. When you first set up a sequence, you can designate the time intervals to use for the grid. For the sequence in Figure 2-7, I specified that each column should be one tenth of a second.

When a grid square is filled dark, the corresponding channel is turned on. When a grid square is empty, the corresponding channel is turned off. Thus, by following the grid cells from left to right, you can see that channels 1, 3, and 5 are on for 0.5 seconds, off for 0.5 seconds, on for 0.5 seconds, and so on.

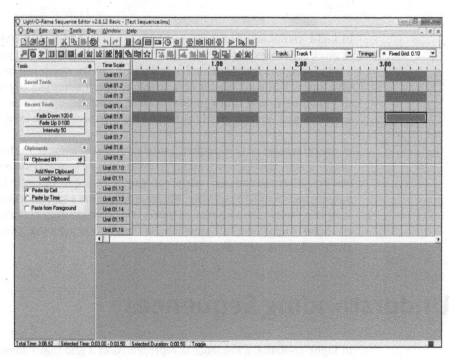

FIGURE 2-7:
A simple Light-O-
Rama sequence.

TIP

Light-O-Rama lets you create these two distinct types of sequences:

>> **Musical:** Lights are synchronized with music in this sequence. Musical sequences always have an MP3 file associated with them, and Light-O-Rama can play only one musical sequence at a time.

>> **Animation:** This sequence doesn't have a music file associated with it. Instead of synchronizing lights with music, animation sequences are used to create simple light animations, such as a waving Santa Claus or a snowman tossing a snowball. Light-O-Rama can play multiple animation sequences simultaneously. Thus, you can have your snowman throw his snowballs at the waving Santa Claus if you want.

Creating a Musical Sequence

To create a musical sequence using the Sequence Editor, follow these steps:

1. **In the Sequence Editor, choose File ⇨ New.**

   This brings up the dialog box shown in Figure 2-8.

FIGURE 2-8:
Choosing the
type of sequence
to create.

2. **Select New Musical Sequence, and then click OK.**

   You're prompted to select an MP3 file to use as the musical basis for your sequence.

3. **Locate and select the MP3 file to use for your sequence, and then click Open.**

   The dialog box shown in Figure 2-9 then appears.

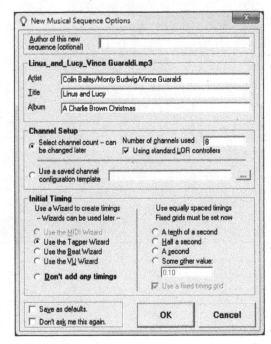

FIGURE 2-9:
Filling in
the musical
sequence options.

4. Stare at all the options for a while, and then change a few of them if you want.

You should familiarize yourself with the range of options that are available for creating new musical sequences. For your first few sequences, I suggest you choose the following options:

- *Number of Channels Used:* Set this to the number of channels in your controller. The default is 8. If you have a 16-channel controller, change this value to 16.

- *Initial Timing:* Change this value from the default choice (use the Tapper Wizard) to A Tenth of a Second. The Tapper Wizard is very useful, but you should create a few simple sequences without it before you try out the Tapper Wizard.

5. Click OK.

The Sequence Editor creates a blank sequence, as shown in Figure 2-10.

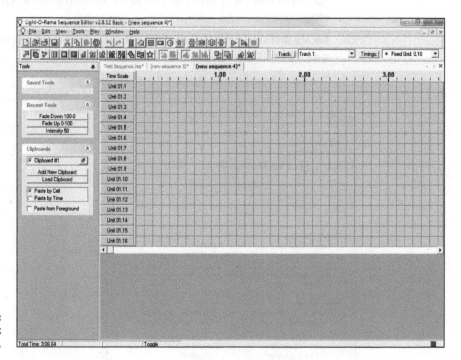

FIGURE 2-10:
A blank
sequence.

6. Fill in the sequence.

To do that, you use the buttons in the Tools toolbar to turn channels on and off in various ways. Here's the rundown on what the most popular of these buttons do:

Button	Name	What It Does
	On	Turns the light on.
	Off	Turns the light off.
	Toggle	Toggles the light: Turns it on if it's off or off if it's on.
	Set Intensity	Sets the light to any intensity you choose.
	Fade Up	Gradually fades the light from off to on. Click this tool and then drag over a range of cells to indicate how quickly to fade the light.
	Fade Down	Gradually fades the light from on to off. Click this tool and then drag over a range of cells to indicate how quickly to fade the light.
	Intelligent Fade	If you drag from left to right, fades the light up. If you drag from right to left, fades the light down.
	Twinkle	Causes the light to slowly twinkle.
	Shimmer	Causes the light to quickly shimmer.

7. To test a sequence, click the Play button in the toolbar.

When you've heard enough, click the Stop button to stop the playback.

8. Use the File ⇨ Save command to save your sequence.

As you work on your sequence, be sure to save it often so that if something goes wrong you won't lose much work.

Here are some additional tips and tidbits to ponder as you create your sequences:

>> You can use Undo (Ctrl+Z) to undo any mistakes you make.

>> If you select a tool and then drag over a range of cells, the tool will be applied to all cells within the range. Note that if you use the Toggle tool while dragging over a range of cells, each cell within the range will be individually changed from on to off or off to on.

>> When you use one of the fade tools, the coloration of the cells involved in the fade indicates how the lights will fade.

>> You can change the name of each row and the color it's rendered in by clicking the row button in the left margin of the row. This brings up the Channel Settings dialog box shown in Figure 2-11. Although it isn't a requirement, changing the color of the row to match the color of lights displayed by the channel can help you better visualize the show, and changing the name of the channel makes it a lot easier to keep track of how the channels are used in your show.

>> Another way to apply effects is to right-click a cell or range of cells. When you do, a pop-up menu listing all the various effects appears.

>> You can use the View ⇨ Zoom commands to zoom in or out to get finer control or to see more of the big picture.

>> If it helps, you can make the wave file visible by choosing View ⇨ WaveForm. Figure 2-12 shows a sequence with the waveform visible.

FIGURE 2-11: Changing the name or color of a row.

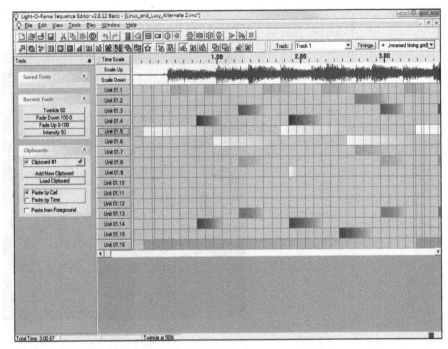

FIGURE 2-12:
Displaying the
waveform.

Visualizing Your Show

To help you picture how your show will appear when it is run, you can use the View ⇨ Animation command. This brings up the window shown in Figure 2-13, which lets you draw a crude representation of your lights. When you run the show within the Sequence Editor, the lights you draw in this window are turned on or off to simulate the appearance of the actual show.

Before you start drawing lights onto the Animation window, you should first import a photograph of your yard. To do so, you'll obviously first need to take a picture of your yard and upload it to your computer. Then, click the Select button, browse to the picture, and select it. You're asked to indicate the size of the grid you want to superimpose on the picture. Then, the picture is added to the Animation window, as shown in Figure 2-14.

To color in your lights, first select the channel you want to color in from the drop-down menu to the left of the picture. Then, select the Draw radio button and draw your lights on the picture roughly where they will appear in your yard. Draw the lights for each channel, one at a time. If you make a mistake, select the Erase radio button, and then click over your mistake to erase it. Figure 2-15 shows a picture of a house with some lights drawn on it.

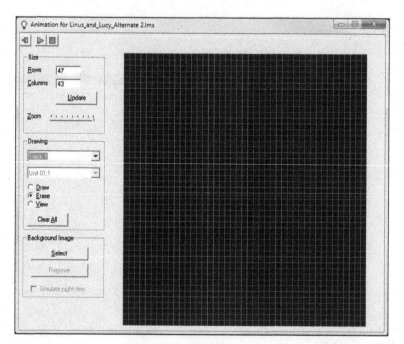

FIGURE 2-13:
The Animation window.

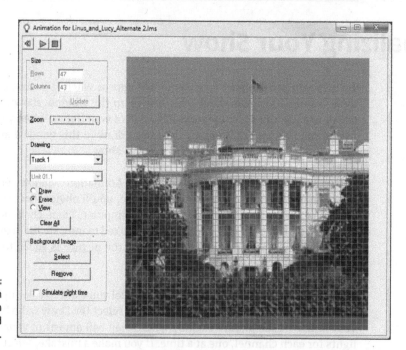

FIGURE 2-14:
The Animation window with a background picture.

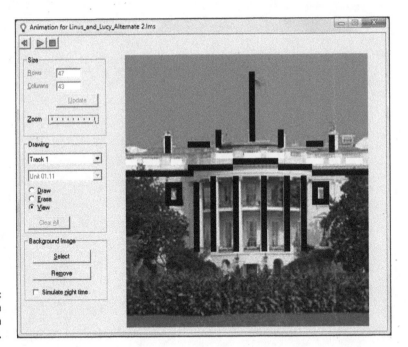

FIGURE 2-15:
The Animation
window with
lights drawn in.

Chapter **3**

Building an Animatronic Prop Controller

*A*nimatronics refers to the technology used to create automated puppets that move by themselves. The movements are provided by mechanical means such as motors, servos, compressed air, or pneumatics. Animatronic puppets are usually controlled by microprocessors.

My fascination with animatronics started at a very young age when I first visited Disneyland as a child and was mesmerized by three of the most amazing things I had ever seen: *Enchanted Tiki Room*, filled with marvelous robotic birds that sang and told jokes; *Pirates of the Caribbean*, with its swashbuckling pirates; and of course *The Haunted Mansion*, with its 999 ghosts and room for one more.

The Haunted Mansion was the *pièce de résistance*. I have been fascinated with all things Halloween ever since and have tinkered on and off over the years with building my own animatronic ghosts and goblins.

In this chapter, I show you how to build a microprocessor-based controller system that I've used in several of my best Halloween props. I've used this exact circuit to create a Frankenstein creature that pops up, a mummy that pops up, and a giant jack-in-the-box, but you can use this circuit to control just about any kind of animatronic creation you can imagine.

Note that Halloween isn't the only application for animatronics. You can use them in theatrical productions, school fundraisers, or for promotional or marketing purposes. For example, if you own a flower shop, you could create animatronic flowers for your storefront display, or if you own a tuxedo store, you could create an animatronic dancer dressed in your finest tuxedo to draw customers in. The only limit is your imagination.

WARNING

Animatronics should be built and used with care as they can be dangerous. You must ensure that no one who views your animatronic will be exposed to any danger, such as loose electrical connections or faulty compressed-air fittings. Make sure that you place your finished animatronic in a secure area where no one will be able to touch it, and make sure that all electrical and compressed-air fittings are completely secure.

If your animatronic prop uses compressed air, be especially careful! This chapter doesn't show you how to work with compressed air because that's a subject for an entire book. If you want to work with compressed air, always use professionally designed and manufactured components such as air cylinders and pop-up mechanisms. Don't try to build your own compressed air cylinders out of PVC pipe, bicycle pumps, or any other material that isn't specifically designed to work with compressed air. And beware of inexpensive compressed-air devices available on eBay or other online sites. A good, safe compressed air mechanism should cost you several hundred dollars. If it's selling online for $49, it's probably not safe.

Looking at the Requirements of Animatronic Prop Control

Whether it's used in a haunted house, trade show exhibit, store display, or museum, a typical animatronic prop requires the following elements to achieve the desired effect:

>> Most props require some sort of trigger to initiate the prop's routine. The trigger can be as simple as a button that you or a customer pushes to start the routine, or it could be a timer that causes the prop to start up

automatically every 5 or 10 minutes. A third option is to use a motion detector that causes the prop to start when motion is detected near the prop.

» All props require a way to control the prop's *actuators,* which are the components that cause the prop to move. The most common types of actuators are electric motors, servos, and compressed air cylinders. If you're using an electric motor, the prop controller usually turns the motor on or off directly by applying current to the motor. If you're using servos, the prop controller must send timing pulses to the servo to control the servo's position. And if you're using compressed air cylinders, the prop controller opens and closes electric valves that allow air into the cylinders.

» Most props require lighting effects that enable the viewer to see the prop in action. In some cases, this is as simple as one or more lights that come on when the prop starts its routine and go off when the routine ends. In other cases, a variety of lights are used during the routine to highlight different parts of the prop during different parts of the show.

» Some props use sound that is synchronized with the prop's movements. For example, an animated face might speak dialog, a scary creature might scream when it suddenly moves, or the prop may use music to enhance its routine.

The prop controller I show you how to build in this chapter meets all these needs. It's based on a microprocessor called a *BASIC Stamp,* which is similar in many ways to an Arduino. The BASIC Stamp provides more than enough computing power to control lights and sound and activate a few pneumatic valves and electric motors.

Instead of building the controller from a BASIC Stamp module from Parallax, the controller shown in this chapter uses components from a company named EFX-TEK, which specializes in microprocessor components for animatronic prop controls. The prop controller uses three products from EFX-TEK: a Prop-1 controller board, which includes a BASIC Stamp microprocessor; an RC-4 relay board that can hold up to four relays for switching 120 VAC circuits; and an AP-16+ sound board, which can be programmed to play sounds saved on a standard SD memory card.

For more information about these components and how to use them, see the section "Building the Prop Controller" later in this chapter.

TECHNICAL STUFF

EFX-TEK has recently replaced its RC-4 relay board with a new and improved version called the RC-4+. The RC-4+ is compatible with the RC-4, so you can build the prop controller using either board.

Examining a Typical Animatronic Prop

Figure 3-1 shows a typical animatronic prop that I made a few years ago to use in a Halloween haunted house. The prop is a large jack-in-the-box that benignly plays "Pop Goes the Weasel" while a crank on one side of the prop slowly spins until an unsuspecting visitor approaches it. Then — all in one frightening moment — the music stops, the lid flies open, a large clown head pops up, a bright light comes on to ensure that you can see the clown, and a ghastly voice screams, "Trick or treat!" A moment later, the light goes off, and the clown retracts back into the box. A few seconds after that, the crank starts spinning and the music plays again while the jack-in-the-box waits for another victim.

FIGURE 3-1: The jack-in-the-box prop.

The basic pop-up operation of the jack-in-the-box is accomplished with a compressed air mechanism that can be controlled with an electrically activated valve. When the valve is closed, the mechanism is retracted inside the box. When the valve opens, compressed air flows into an air cylinder that extends the mechanism upward. The clown head is mounted on top of the mechanism. The lid of the jack-in-the-box is positioned such that the clown's head pushes the lid open when the mechanism extends. Gravity takes care of closing the lid when the mechanism retracts and the clown goes back into the box.

Note that the point of this chapter is to show you how to build the electronic part of this prop — the prop controller that controls the prop's action. This chapter doesn't show you how to build the prop itself. The prop controller is generic enough that it can control just about any prop that includes simple actions such as pneumatic pop-ups, motors, lights, and sound. In fact I've used this very same design for half a dozen different Halloween props.

Nevertheless, it's beneficial to see how the prop controller works in a specific prop, so the following paragraphs describe in general terms how I built the jack-in-the-box prop. This will give you a general idea of how it operates so that you can see how the prop controller works with it:

>> The most important component of this prop is the pneumatic lifting mechanism. I purchased it for about $250 from an online distributor of Halloween prop mechanisms. If you search the web, you'll find several distributors that make safe, professional-quality compressed-air mechanisms. Just search for *Halloween prop lifter* and you'll find several suppliers. You can also find them for sale on eBay.

Figure 3-2 shows a close-up photograph of the lift mechanism.

FIGURE 3-2:
The lift
mechanism.

The lift mechanism I purchased included the welded steel frame, a pneumatic cylinder that provides the motion needed to lift the frame, and an electrical valve that can be connected to a compressed air hose. When the valve is connected to a source of compressed air and voltage is applied, the valve opens to allow compressed air into the cylinder.

WARNING

Do not, under any circumstances, attempt to cut corners by building your own lifter mechanism unless you are an experienced metalworker. The Internet has plans for similar devices made out of sprinkler pipe or bicycle pumps, but putting compressed air into plastic or metal parts that were not designed to handle compressed air is an excellent way to maim or even kill someone. PVC pipe, even the heavy schedule-40, is not strong enough to survive the pressure of an air compressor.

» The prop is housed in a 22-inch square box I constructed from ⅝-inch plywood, reinforced inside with 2-x-2-inch lumber. I was able to cut all six sides and the lid out of a single 4-x-8-foot sheet of plywood.

» To create the lid, I used a sabre saw to cut out the opening. Then, I cut a separate piece of plywood about 1 inch larger than the opening to use as the lid. I used a pair of small gate hinges to connect the lid to the box.

» I mounted the lift mechanism directly to the bottom of the box. Depending on the design of the mechanism you use, you may have to mount the lift on the back of the box instead of the bottom.

» I drilled two 1-inch diameter holes in the back of the box to pass the electrical power cord and the compressed air hose into the box.

» The clown consists of a small foam wig head that I bought at a thrift store and a mask that I purchased at a seasonal Halloween costume shop.

» I used a surplus automotive windshield wiper motor to turn the crank. You can find them for sale on the Internet for about $15. I fabricated the crank itself from a few pieces of scrap metal I had lying around.

» The prop controller is mounted inside the box, secured to one of the side panels with four screws.

» To trigger the prop, an infrared motion detector called a *PIR* is used. The motion detector is mounted to the outside of the box and connected via a 24-inch cable to the prop controller. When the motion detector detects nearby motion, the programming in the prop controller triggers the prop to activate the valve so that the clown pops up.

» To light up the clown's face when he pops up, I used a 75 W white flood lamp mounted in a plastic flood lamp holder I got from a home improvement store.

» For the sound, I used a set of inexpensive speakers that I had lying around. The speakers are connected to the audio output from the prop controller's sound board. This sound board includes a 20 W amplifier, which is more than loud enough to give a good scare when the clown pops up.

>> Figure 3-3 shows a wiring diagram of the prop so you can get an idea of the prop's electrical requirements. As you can see, the prop requires a total of three 12 VDC power adapters, commonly known as *wall warts*. The first is to power the prop controller, the second provides 12 VDC for the windshield wiper motor, and the third provides 12 VDC for the pneumatic valve.

To simplify the wiring, I used three six-outlet power strips and an eight-terminal barrier strip, which aren't shown on this diagram. One of the power strips plugs directly into a wall outlet to supply the main 120 VAC power for the prop. The other two power strips are connected to the RC-4 relay control board. Using these power strips allows you to simply plug in the 12 VDC power adapters and the flood lamp holder.

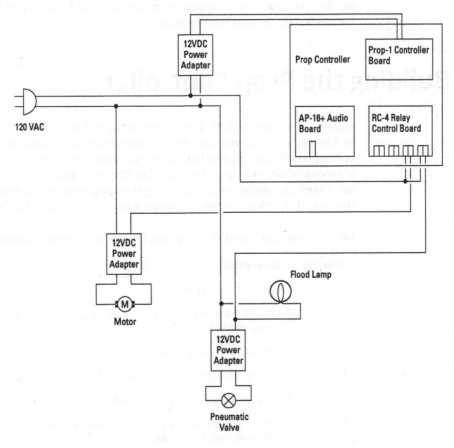

FIGURE 3-3:
Schematic
diagram for the
jack-in-the-
box prop.

WARNING

The prop runs on 120 VAC household power, and there are several exposed electrical connections inside the box. Therefore, it's imperative that the inside of the box is inaccessible to humans or pets.

It's especially important that the wiring is secure enough to withstand the vibrations that the box will experience every time the prop is activated. When the valve closes, the cylinder extends, the lid swings open, and the clown pops up his head, the box will be jolted more than a little. To insure that the frequent shaking doesn't cause any electrical connects to come lose, all electrical wires are fastened down with wire holders, and all connections are made with soldered terminals that are tightly screwed to the barrier strip.

That's about it for the jack-in-the-box prop itself. As I've already mentioned, the primary focus of this chapter is building the prop controller that provides the brains for this prop. The prop controller can be used for any other type of prop with similar requirements. And, if your prop is more complex than the jack-in-the-box, you can easily expand the prop controller by including additional relay control boards to control more devices.

Building the Prop Controller

Figure 3-4 shows the fully assembled prop controller used to control the jack-in-the-box prop. As you can see, the prop controller is made from three circuit boards, which you can purchase from EFX-TEK (www.efx-tek.com). The boards are mounted on an 8-x-10-inch sheet of Plexiglas, which you can purchase from most hardware stores. Before mounting the components on Plexiglas, spray-paint the back of the Plexiglas black to create a nice, professional-looking board.

Here's a complete list of the parts you'll need to build the prop controller:

Quantity	Description
1	EFX-TEK Prop-1 controller
1	EFX-TEK PIR sensor (includes a 14-inch extension cable for connecting to the Prop-1)
1	EFX-TEK AP-16+ audio player
1	EFX-TEK RC-4 relay control board
2	Crydom D2W203F solid-state relays
3	EFX-TEK stand-off kits (each kit includes four ⅝-inch stand-offs and eight 4-40 machine screws)

Quantity	Description
2	14-inch servo-style extension cables
2	6-inch lengths of 16-gauge stranded wire
1	8-x-10-inch sheet of $\frac{3}{16}$-inch Plexiglas
1	Can of black spray paint

FIGURE 3-4:
The assembled
animatronic
prop controller.

All these components except the wire, Plexiglas, and the paint can be ordered from EFX-TEK at www.efx-tek.com. The total price for all components should be around $275.

The following paragraphs describe the four main components of the prop controller, which are listed first in the parts list:

>> **Prop-1 controller:** The brains of the prop controller. This inexpensive controller board ($39.95) contains a BASIC Stamp microcontroller and two I/O busses that give you direct access to the BASIC Stamp's eight I/O pins:

- The first bus is a series of three-pin headers that you can connect standard servo cables to. These connectors provide a TTL interface to the BASIC

Stamp and are used for low-current applications such as communicating with other prop-control elements.

- The second bus is for high-current applications that can handle up to 500 mA. These outputs can be used to directly drive small motors, relays, solenoid valves, and so on.

The Prop-1 also includes a 5 V regulated power supply and a programming interface so you can connect it directly to your computer to download programs.

>> **PIR sensor:** An inexpensive (under $10) motion detector that can be connected via a three-pin servo cable to any of the Prop-1's low-current I/O pins. This device is used to trigger the prop's action.

>> **AP-16+ audio player:** At $129.95, the audio player is the most expensive component of the prop controller. However, sound is an essential element of any good prop, and the AP-16+ is an extremely versatile sound player. It can play sounds in WAV format directly from a micro-SD card and includes a built-in 20 W amplifier, so you can connect speakers directly to the AP-16+ without using a separate amplifier. The AP-16+ also connects to the Prop-1 via a three-pin servo cable. This connection allows the Prop-1 to send commands to the AP-16+ to tell the AP-16+ to play specific sound files on the micro-SD card.

>> **RC-4 relay control:** This module lets you control up to four line-voltage (120 VAC) circuits by using solid-state relays. It connects to the Prop-1 via a three-pin servo cable, enabling the Prop-1 to send commands to the RC-4 to tell it to turn its relays on or off under program control.

WARNING

I can't say it too many times in this chapter: The prop controller uses line-level voltages that can hurt or even kill you if you aren't careful. Make sure that the line-level wiring is properly secured and enclosed, and *never* work on the controller board when the line-level circuits are plugged in.

Here are the steps to assemble the prop controller:

1. **Paint one side of the Plexiglas with the black spray paint and allow the paint to dry.**

2. **Drill the necessary mounting holes in the Plexiglas.**

   You'll need to drill four mounting holes for each of the three boards that will be mounted to the Plexiglas (Prop-1, RC-4, and AP-16+. Use a $\frac{3}{16}$-inch drill bit. Figure 3-5 is a rough drilling guide, but note that this diagram isn't to scale. You'll need to lay your actual components out on the Plexiglas to determine the exact drilling locations.

3. **Mount the Prop-1 controller on the Plexiglas.**

   Use four of the stand-offs and eight of the 4/40 machine screws. Position the Prop-1 controller so that the row of screw terminal connectors is on the left side of the controller, as oriented in Figure 3-5.

4. **Mount the RC-4 board on the Plexiglas.**

   Use four of the stand-offs and eight of the 4-40 machine screws. Position the RC-4 board so that the four screw-terminal connectors are at the bottom of the board, as oriented in Figure 3-5.

5. **Mount the AP-16+ board on the Plexiglas.**

   Use four of the stand-offs and eight of the 4-40 machine screws. Position the AP-16+ board so that the audio output connectors are at the bottom of the board, as oriented in Figure 3-5.

6. **Use the two wires to connect the V+ and GND terminals on the Prop-1 to the V+ and GND terminals on the AP-16+ board.**

   These wires are used to provide power for the AP-16+ board. After test-fitting the wires, cut them to the correct length and strip ⅜ inch of insulation off each end. Then insert each end of the wire into the appropriate screw terminal and tighten the screw to ensure a solid connection.

7. **Use one of the extension cables to connect the P7 header on the Prop-1 controller to the SER header on the RC-4.**

   This cable enables the Prop-1 controller to send commands to the RC-4 board to cause the RC-4 board to turn its relays on or off.

   Eight three-pin headers are found near the center of the Prop-1 controller board. These headers provide access to the eight I/O ports of the BASIC Stamp microprocessor. The headers are labeled P0 through P7, so you shouldn't have trouble finding the correct header.

   There are two three-pin headers on the RC-4 board, both labeled SER. These two headers are connected to one another, so it doesn't matter which one you use as they're electrically the same.

   The only trick to connecting the extension cable is making sure that you get them oriented correctly. The extension cable has three wires: white, red, and black. The cables must be inserted into the header correctly or the Prop-1 won't be able to communicate with the RC-4.

   Orienting the cable correctly on the Prop-1 is easy because the three-pin headers are labeled to indicate which pin is white, which is red, and which is black. On the RC-4, the cable should be inserted into the header with the black wire facing the outside edge of the board and the white wire facing the center of the board.

8. **Use the second extension cable to connect the empty SER header on the RC-4 board to one of the SERIAL headers on the AP-16+ board.**

The purpose of this connection is to enable the Prop-1 controller to communicate with the AP-16+ audio player to tell it when to play its sounds.

As oriented in Figure 3-5, the SERIAL header pins are located near the top-right corner of the AP-16+ board. The pins on the AP-16+ headers are labeled W, R, and B, so you should have no trouble orienting the wire colors. As with the RC-4, there are two SERIAL headers on the AP-16+; it doesn't matter which of them you use.

9. **Use the third extension cable to connect the PIR sensor to the P6 header on the Prop-1 controller.**

The P6 header is right next to the P7 header.

10. **Insert the two Crydom solid-state relays in the K1 and K2 positions on the RC-4 board.**

Note that if your prop needs to control more than two 120 VAC circuits, you can add up to four Crydom relays to the RC-4 board. And you can add additional RC-4 boards if you need to control more than four circuits.

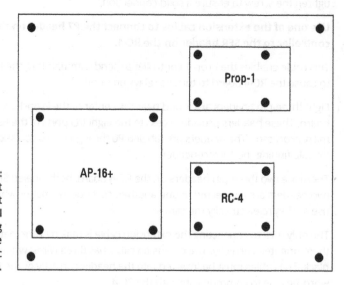

FIGURE 3-5: Where to mount the circuit boards and drill the mounting holes for the animatronic prop controller.

Programming the Prop-1 Controller

The Prop-1 controller from EFX-TEK is based on a BASIC Stamp microcontroller made by Parallax. Programming a BASIC Stamp is much like programming an Arduino: You install an integrated development environment (IDE) on your computer, use a USB port to connect the BASIC Stamp to your computer, compose a program in the IDE, and then download the program to the BASIC Stamp.

Really, the biggest difference between an Arduino and a BASIC Stamp is the programming language. The Arduino uses a variation of the C programming language. In contrast, the BASIC Stamp uses a variation of the BASIC programming language known as PBASIC.

Rather than provide a complete BASIC tutorial here, I'll just show you the coding needed to program the Prop-1 controller, the AP-16+ audio board, and the RC-4 relay board. If you're familiar with programming an Arduino in C or a Raspberry Pi in Python, you shouldn't have too much trouble understanding the code in this chapter.

Here's a short snippet of PBASIC that flashes an LED that is connected to pin 15:

```
' ($STAMP BS)
' ($PBASIC 2.5)
Main:
  HIGH 15
  PAUSE 500
  LOW 15
  PAUSE 500
  GOTO Main
```

Note that the preceding program is not a part of the prop controller. It's just a short sample program to show you how PBASIC works. Here are the salient points of this program:

>> The comments at the beginning (marked by apostrophes) are required to indicate the version of the BASIC Stamp and the version of the PBASIC language being used.

>> Main: is a *label* that marks a location in the program that can be referred to by other statements. In this example, a GOTO statement is used to cause the program's main loop to execute repeatedly as long as the controller is turned on.

>> HIGH 15 sets the value of pin 15 to HIGH (+5 V). Assuming an LED is connected to pin 15, the LED will light when the output is set to HIGH. (Note that you don't need to include a resistor to protect the LED, because all I/O pins on the Prop-1 board are protected by built-in 220 Ω resistors.)

>> PAUSE 500 pauses the program for 500 milliseconds.

>> LOW 15 sets the value of pin 15 to LOW (0 V). This turns off the LED.

>> PAUSE 500 pauses the program for another 500 milliseconds.

>> GOTO Main redirects the program back to the location marked by the label Main. This causes the program to repeat itself, over and over again, until the BASIC Stamp is turned off.

Sending Commands to the RC-4 or AP-16+ Modules

The RC-4 and AP-16+ modules are connected with three-pin jumpers to port 7. This enables the Prop-1 to send messages to the RC-4 and AP-16 modules using a simple serial communications protocol. Fortunately, you don't need to worry about the details of how the serial communication protocol works to send messages.

You use PBASIC's SEROUT command to send messages via an I/O pin. The SEROUT command has a simple syntax:

```
SEROUT pin, mode, (data)
```

The first parameter of the SEROUT command, *pin*, is simply the pin number that the data is to be sent to. Because the RC-4 and AP-16+ modules are connected to pin 7 in the animatronic prop controller, you'll simply specify pin 7 for this parameter.

The second parameter, *mode*, indicates the communication mode that will be used to send the data. For all EFX-TEK I/O modules, you should specify the built-in constant OT2400 for this parameter.

TECHNICAL STUFF

In case you're interested, OT2400 means that the transmission speed is 2400 bits per second, and the output polarity is open-drain driven high — which sounds pretty technical to me, so I wouldn't worry about it. Just use OT2400 for this parameter.

The third parameter, *data*, is the one that actually specifies the data to be sent to the external device. You must enclose the data in parentheses, and within the parentheses you can list as many string constants or variables as you want. For example:

```
SEROUT PIN7, OT2400, ("Hello World!")
```

This statement sends the string constant "Hello World!" to the device on pin 7.

Here's an example that accomplishes the same thing but with three string constants instead of one:

```
SEROUT PIN7, OT2400, ("Hello ", "World", "!")
```

And here's an example that uses a variable:

```
SYMBOL Bam = B0
Bam = "!"
SEROUT PIN7, OT2400, ("Hello, World", Bam)
```

You might reasonably ask why you would use several string constants or variables instead of just one. The answer to that excellent question will become apparent when you read the following two sections about programming the RC-4 and AP-16+ modules.

TIP

It's a good idea to assign a symbol to the I/O pin to which the RC-4 and AP-16+ modules are attached and to the communication mode. For example:

```
SYMBOL Sio = 7
SEROUT Sio, OT2400, ("Hello, World!")
```

That way, if you decide to change the I/O pin to which the external device is attached, you have to make only one change to your program — the symbol definition — instead of multiple changes in each SEROUT statement.

Though not as important, it's also a good idea to use a symbol for the mode parameter, like this:

```
SYMBOL Sio = 7
SYMBOL Baud = OT2400
SEROUT Sio, Baud, ("Hello, World!")
```

Why? Because someday you may decide to move your program over to a different controller, such as the more capable Prop-2 controller. When you do, you'll discover that the Prop-2 controller has additional mode settings that enable faster and more reliable serial communications. You'll be able to change the serial communications mode for all your program's SEROUT commands simply by changing the SYMBOL statement.

Programming the RC-4 Relay Control Module

The RC-4 relay control module supports up to four high-voltage solid-state relays that can turn line-voltage circuits on and off. EFX-TEK sells the board and relays separately, so you can equip it with just the number of relays your project needs.

A pair of jumper blocks on the RC-4 board lets you give it an address value of 0, 1, 2, or 3. This allows you to daisy-chain up to four RC-4 boards together, allowing you to control up to 16 separate relay circuits. The examples in this section assume that you have removed these two jumpers so that the address of the RC-4 is 0.

To control an RC-4 board from a Prop-1 controller, you use the SEROUT command to send data to the pin that the RC-4 board is connected to. The data that you send must follow this format:

```
"!RC4", address, command {,data}
```

The first data item is the string constant "!RC4". This string constant is a preamble that indicates that the data that follows is intended for an RC-4 relay control card. This allows you to chain several different kinds of EFX-TEK cards together and send commands only to the RC-4.

The second data item is the address of the RC-4 card you want to send the data to. The possible values are 0, 1, 2, or 3.

The third data item is a command, which is a single character that represents one of several commands that the RC-4 can respond to. Depending on the command used, you may also need to provide one or two additional bytes of data after the command. The possible commands are the following:

Command	What It Does
X	Resets all relays. No additional data is needed.
R	Sets or resets an individual relay to a particular value. Two data bytes are required: one to indicate which relay to set or reset, and a second to indicate whether the relay should be closed (set) or opened (reset).
S	Sets all relays. The S command is followed by a single byte of binary data that indicates which relays should be open and which should be closed.
V	Causes the RC-4 to send its firmware version information back to the prop controller. The version information can be read with a SERIN command.
G	Gets the current status of all relays. Use the SERIN command to read the data from the RC-4 after sending this command.

The three RC-4 commands you'll use most often are X, to reset all relays; R, to set an individual relay; and S, to set each of the four relays to a particular status (on or off). These commands are described in the following sections.

Turning all relays off

To turn all the relays on an RC-4 module off, use the X command. For example:

```
SEROUT PIN7, OT2400, ("!RC4", 0, "X")
```

It's a good idea to use this command near the beginning of your program to make sure that the program starts with all relays off.

Turning an individual relay on or off

To turn an individual relay on or off, you send an R command to the RC-4. The R command requires that you send five separate data items:

>> The preamble: "!RC4"

>> The RC-4 board's address

>> The command "R"

>> The relay number: 1 through 4

>> The relays status: 1 to turn the relay on, 0 to turn the relay off

Here's a SEROUT statement to turn the first relay (number 1) on:

```
SEROUT PIN7, OT2400, ("!RC4", 0, "R", 1, 1)
```

And here's a SEROUT statement to turn the first relay off:

```
SEROUT PIN7, OT2400, ("!RC4", 0, "R", 1, 0)
```

Setting all four relays at once

To set the status of all four relays at one time, you can use the S command. This command requires four data items:

>> The preamble: "!RC4"

>> The RC-4 board's address

>> The command "S"

>> A single byte value that represents the status of the four relays

The value that controls the relay status is usually written as a percent sign followed by four binary bits representing the status of each relay — 1 for on and 0 for off. The only trick is that the four binary bits are written in reverse order: The first is relay 4, the second is relay 3, the third is relay 2, and the fourth is relay 1.

For example, to turn relay 1 on and the other three relays off, you would use %0001 as the relay status. To turn relays 1 and 3 on and relays 2 and 4 off, you would use %0101 as the status.

Here's a SEROUT command that sends an S command to turn all four relays on:

```
SEROUT PIN7, OT2400, ("!RC4", 0, "S", %1111)
```

And here's a SEROUT command that turns relays 1 and 3 on and simultaneously turns relays 2 and 4 off:

```
SEROUT PIN7, OT2400, ("!RC4", 0, "S", %0101)
```

The following SEROUT command turns all relays off, which is equivalent to sending an "X" command:

```
SEROUT PIN7, OT2400, ("!RC4", 0, "S", %0000)
```

Using symbols to make RC-4 commands more readable

As I mention earlier in this chapter, it's a good idea to use symbols for the I/O pin and output mode. It's also a good idea to use symbols for the RC-4 address, for the relay number, and for the relay status (on or off). That can make your program more readable.

For example, here are some typical symbol declarations:

```
SYMBOL Sio = 7
SYMBOL Baud = OT2400
SYMBOL RC4 = 0
SYMBOL Relay1 = 1
SYMBOL Relay2 = 2
SYMBOL Relay3 = 3
SYMBOL Relay4 = 4
SYMBOL RelayOn = 1
SYMBOL RelayOff = 0
```

With these symbol declarations in place, you can turn on relay 1 with this statement:

```
SEROUT Sio, Baud, ("!RC4", RC4, "R", Relay1, RelayOn)
```

And here's a statement that turns relay 1 off:

```
SEROUT Sio, Baud, ("!RC4", RC4, "R", Relay1, RelayOff)
```

Here's a SEROUT statement that uses the symbols to send an S command that turns relays 2 and 4 on and relays 1 and 3 off:

```
SEROUT Sio, Baud, ("!RC4", RC4, "S",%1010)
```

A sample program for controlling all four RC-4 relays

Listing 3-1 shows a complete PBASIC program that gives the RC-4 card a bit of a workout by performing the following sequence of steps:

1. Turn off all relays with an X command.
2. Pause 1 second.

3. **Use a series of S commands to turn on relays 1, 2, 3, and 4 in sequence at 1-second intervals.**

   As each relay is turned on, the other three are turned off.

4. **Pause 1 second, and then turn off all relays.**

5. **Use a series of R commands to turn on relays 1, 2, 3, and 4 in sequence at 1-second intervals.**

   As each relay is turned on, the status of the other relays are left unchanged. As a result, this part of the program first turns on relay 1, then adds relay 2, then adds relay 3, and then adds relay 4. At this point, all four relays are on.

6. **Wait 1 second, and then repeat the entire program.**

You can download this program from this book's companion website.

LISTING 3-1:	An RC-4 Control Program

```
' {$STAMP BS1}
' {$PBASIC 1.0}

SYMBOL Sio = 7
SYMBOL Baud = OT2400
SYMBOL RC4 = 0
SYMBOL Relay1 = 1
SYMBOL Relay2 = 2
SYMBOL Relay3 = 3
SYMBOL Relay4 = 4
SYMBOL RelayOn = 1
SYMBOL RelayOff = 0

Main:

  'Turn off all relays
  SEROUT Sio, Baud, ("!RC4", RC4, "X")
  PAUSE 1000

  'Cycle through the relays one at a time
  SEROUT Sio, Baud, ("!RC4", RC4, "S", %0001)
  PAUSE 1000
  SEROUT Sio, Baud, ("!RC4", RC4, "S", %0010)
  PAUSE 1000
  SEROUT Sio, Baud, ("!RC4", RC4, "S", %0100)
  PAUSE 1000
  SEROUT Sio, Baud, ("!RC4", RC4, "S", %1000)
  PAUSE 1000
```

```
'Turn off all relays
SEROUT Sio, Baud, ("!RC4", RC4, "S", %0000)
PAUSE 1000


'Turn on relay 1
SEROUT Sio, Baud, ("!RC4", RC4, "R", Relay1, RelayOn)
PAUSE 1000


'Add relay 2
SEROUT Sio, Baud, ("!RC4", RC4, "R", Relay2, RelayOn)
PAUSE 1000


'Add relay 3
SEROUT Sio, Baud, ("!RC4", RC4, "R", Relay3, RelayOn)
PAUSE 1000


'Add relay 4
SEROUT Sio, Baud, ("!RC4", RC4, "R", Relay4, RelayOn)
PAUSE 1000


'Repeat!
GOTO Main
```

Programming the AP-16+ Audio Player Module

The AP-16+ audio player module is an amazingly versatile way to add sound to your animatronic prop. Like the RC-4, you communicate with the AP-16+ module using serial communication via the SEROUT statement. Some commands cause the AP-16+ module to return data, which you can retrieve with the SERIN statement. But for most props, all you have to do is send a few commands using SEROUT, so you can dispense with the SERIN statement.

The AP-16+ board can play sounds recorded in WAV format stored on a micro-SD card. So before you program your prop, you'll need to prepare a micro-SD card with the files you want the prop to play. You can simply use your computer to copy the files to the micro-SD card; the only restriction is that the filenames must be no longer than eight characters, and the filename extension must be .wav. (The AP-16+ card can't play MP3 files.)

For the jack-in-the-box prop, just two sound files are needed. The file that contains the "Pop Goes the Weasel" song is named weasel.wav. I created this recording myself; you can download it from this book's companion website.

The other file is a horrific scream that is played when the clown pops up. This file is named scream.wav. It is also available on the companion website.

TECHNICAL STUFF

The AP-16+ is an extremely versatile card that's designed to be usable in stand-alone mode — that is, not controlled by a microcontroller. Many of its features are useful primarily in stand-alone mode. For example, you can store files on the micro-SD card with predefined filenames such as SFX*nn*.wav and AUX*nn*.wav, where *nn* is a two-digit number (for example, SFX01.wav or AUX04.wav). The AP-16+ can be configured to automatically play these files in random ways in response to direct trigger inputs on the card. The AP-16+ can also be configured to play an ambient background sound whose filename must be ambient.wav. These are all very useful features but not relevant to the prop controller presented in this chapter. So although I ignore those features in this section, I suggest you carefully read the AP-16+ documentation supplied by EFX-TEK to see how these features work.

To control an AP-16+ module from a Prop-1 controller, you use the SEROUT command to send data to the pin that the AP-16+ is connected to. The data that you send must follow this format:

```
"!AP16", address, command {,data}
```

The string constant "!AP16" is the preamble that indicates that the command is intended for an AP-16+ module. This allows you to chain different kinds of modules together with the AP-16+, making it possible to control all the modules used in a project from a single I/O pin.

The second data item is the address of the AP-16+ module you want to send the command to. The possible values are 0, 1, 2, or 3. The address is set by a pair of jumpers on the AP-16+ board. For the examples in this section, I assume the AP-16+ address is 0.

The third data item is the command. Depending on the command used, you may also need to provide additional data after the command. The possible commands are:

Command	What It Does
X	Resets the AP-16+, which stops any sound that may be playing.
L	Sets the output volume.
PW	Plays a named file. Additional data provides the name of the file and the number of times to loop the file. (Specify 0 for the loop value to repeat the file indefinitely.)

Command	What It Does
G	Gets the current status of the AP-16+ card. Use the SERIN command to read a single byte that indicates whether the AP-16+ is currently playing a sound file and, if so, which sound file is being played.
PS	Plays one of the SFXnn.wav files. Additional data provides the SFX number (that is, the nn) and the number of times the file should be looped. (Specify 0 for the loop value to repeat the file indefinitely.)
P?	Randomly selects one of the SFXnn.wav files to play.
PA	Plays one of the AUXnn.wav files. Additional data provides the SFX number (that is, the nn) and the number of times the file should be looped. (Specify 0 for the loop value to repeat the file indefinitely.)
S	Changes the playback speed.
V	Gets the AP-16+ version information. The version information can be read with a SERIN command.

The four AP-16+ commands you'll use most are X, to reset the sound player and stop any sound that may be playing; L, to set the output volume; PW, to play a specific file; and G, to get the current status of the AP-16+ so you can find out if it has finished playing a sound file. These commands are described in the following sections.

Note that throughout these sections, I assume that the following symbols have been defined near the top of the program:

```
SYMBOL Sio = 7
SYMBOL Baud = OT2400
SYMBOL AP16 = 0
```

Resetting the AP-16+

To reset the AP-16+, send an X command like this:

```
SEROUT Sio, Baud, ("!AP16", AP16, "X")
```

This causes any WAV file that may be playing to immediately stop. Not only is it a good idea to send this command at the beginning of your program to make sure that the AP-16+ is ready to receive commands, but this command is also the best way to silence the AP-16+ if you want to stop it before the current file finishes playing.

Changing the volume

If you want, you can change the output volume from your program. This is useful if you want to play a file at less than full volume. It's also useful if you want the sound output to come from one side of the room or the other.

To change the output volume, use the L command. The syntax of the L command is this:

```
"!AP16", address, "L", left, right
```

Where *left* and *right* are values between 0 and 100 that indicate the percentage volume for the left and right outputs, respectively.

For example, to set the output volume to 50% for both left and right, use this statement:

```
SEROUT Sio, Baud, ("!AP16", AP16, "L", 50, 50)
```

To make the sound come exclusively from the left channel, use this:

```
SEROUT Sio, Baud, ("!AP16", AP16, "L", 100, 0)
```

To restore the sound to full volume, use this statement:

```
SEROUT Sio, Baud, ("!AP16", AP16, "L", 100, 100)
```

Playing a specific file

To play a specific WAV file, use the PW command. Here's the syntax:

```
"!AP16", address, "PW", name, 13, loops
```

The *name* parameter is the name of the file, without the .wav extension. The 13 is required because of a quirk in the way the AP-16+ programming works. Think of it as a good luck charm!

The *loops* parameter indicates how many times the file should be repeated. Normally, you'll just specify 1 to play the file once. The maximum is 255, but specify 0 if you want the file to keep playing forever — or at least until you reset the AP-16+ with an X command or use another PW command to play a different file.

Here's the command to play a file named scream1.wav just once:

```
SEROUT Sio, Baud, ("!AP16", AP16, "PW", "SCREAM1", 13, 1)
```

Here's a command that plays the file weasel.wav indefinitely:

```
SEROUT Sio, Baud, ("!AP16", AP16, "PW", "WEASEL", 13, 0)
```

Note that whenever you play a sound file, any other sound file that happens to be playing is immediately stopped. Thus, you don't have to wait for one sound file to end to play another.

Waiting for a file to finish playing

When you play a sound file, you will often want your program to wait until the sound file finishes playing. You could do that with a PAUSE command if you know exactly how long the sound is. But what if you aren't sure?

Fortunately, the AP-16+ includes a G command that gets the current status of the card. To use this command, you must first send the G command to the AP-16+ with a SEROUT statement. Then, you must retrieve the status result by using a SERIN statement.

The SERIN statement is pretty straightforward: Its syntax looks like this:

```
SERIN pin, mode, variable
```

The *pin* and *mode* parameters are the same as for the SEROUT statement. The *variable* parameter is simply the name of a variable you have set aside to receive the status result.

Here's how you can use the SEROUT and SERIN statements together to get the status of an AP-16+:

```
SYMBOL ap16Status B0
SEROUT Sio, AP16, ("!AP16", AP16, "G")
SERIN Sio, AP16, ap16Status
```

In this example, the SYMBOL statement equates the name ap16Status with the variable B0. Then, the SEROUT statement sends a G command to the AP-16+ and the SERIN statement retrieves the status data in the variable named ap16Status.

The status data includes a variety of useful information, but in nearly every case the only thing you really want to know is whether the AP-16+ is currently playing a sound. You can find that out by testing bit 7 of the status byte. The easiest way to do that is to define a symbol using the BASIC Stamp's predefined bit-variables, as in this example:

```
SYMBOL ap16Status = B0
SYMBOL ap16Playing = BIT7
```

Then, you can simply test the `ap16Playing` variable, like this:

```
IF ap16Playing = 0 THEN SoundDone
```

This statement will branch to the label `SoundDone` if the AP-16+ has finished playing the sound.

Here is a complete snippet of code that plays a sound and then locks itself in a loop until the sound is finished playing:

```
SEROUT Sio, Baud, ("!AP16", AP16, "PW", "SCREAM", 13, 1)
StillPlaying:
SEROUT Sio, Baud, ("!AP16", AP16, "G")
SERIN Sio, Baud, ap16Status
IF ap16Playing = 0 THEN SoundDone
GOTO StillPlaying
SoundDone:
'The sound is done!
```

A Sample AP-16+ Program

Listing 3-2 shows a complete PBASIC program that plays sounds on an AP-16+ card. The program plays the sound `weasel.wav` for 10 seconds, and then plays the `scream.wav` sound until that sound ends. Then it repeats. You can download this program from this book's companion website.

LISTING 3-2:	An AP-16+ Program

```
' {$STAMP BS1}
' {$PBASIC 1.0}

SYMBOL Sio = 7
SYMBOL Baud = OT2400
SYMBOL AP16 = 0
SYMBOL ap16Status = B0
SYMBOL ap16Playing = BIT7

'Reset the card
SEROUT Sio, Baud, ("!AP16", AP16, "X")
```

```
Main:
  'Play Pop Goes the Weasel
  SEROUT Sio, Baud, ("!AP16", AP16, "PW", "WEASEL", 13, 0)

  'Wait 10 seconds
  PAUSE 10000

  'Play the scream
  SEROUT Sio, Baud, ("!AP16", AP16, "PW", "SCREAM", 13, 1)

  'Wait for the scream to end
  StillScreaming:
  SEROUT Sio, Baud, ("!AP16", AP16, "G")
  SERIN Sio, Baud, ap16Status
  IF ap16Playing = 1 THEN StillScreaming

  'Do it again
  GOTO Main
```

Programming the PIR Motion Detector

The PIR sensor uses infrared light to detect motion. It sends a 1 to the microcontroller when such motion is detected.

At first glance, then, you would think that the PIR sensor is pretty easy to use. Just plug it into one of the Basic Stamp's I/O pins and wait for that pin to have a value of 1 to trigger your prop. The code would look something like this, assuming that Pir is a symbol assigned to the I/O pin that the PIR is connected to:

```
WaitForMovement:
IF Pir = 0 THEN WaitForMovement
MovementDetected:
'Activate your prop now!
```

Unfortunately, life isn't that easy where PIR motion detection is concerned. For starters, the PIR won't be reliable for a while when power is first applied. Thus, any program that uses a PIR to detect motion should start with an immediate 60-second pause. This gives the PIR time to settle down so it can reliably detect motion.

Second, the PIR will often send very brief false triggers when there's no actual motion present in the room. The usual way to get around this problem is to wait until the PIR's value is 1 for a quarter of a second before triggering your prop.

You can do this with a tricky bit of code:

```
SYMBOL Pir = PIN6
SYMBOL waitTimer = B2
PAUSE 60000
Main:
waitTimer = 0
WaitForMotion:
PAUSE 10
waitTimer = waitTimer + Pir * Pir
IF waitTimer &lt; 25 THEN WaitForMotion
'The PIR has detected motion for 0.25 seconds
'so the prop can now be triggered
```

The cleverness in this bit of code is that the WaitForMotion loop pauses for 10 milliseconds, then it adds the value of the PIR input status to the waitTimer variable while at the same time multiplying the result by the status of the PIR input. If the PIR input is 0, multiplying by the input status resets the waitTimer variable to 0 because any number times zero is zero. So this loop won't exit until the PIR has detected motion consistently for a quarter of a second.

Looking at Complete Jack-in-the-Box Program

Now that you've seen the techniques for programming the individual components of the prop controller, put them together to create a program for an actual prop. Listing 3-3 shows the complete program for the jack-in-the-box prop. I made liberal use of comments throughout this program, so you should have no trouble following its basic operation. (The program is available for download at this book's companion website.)

LISTING 3-3:	The Jack-in-the-Box Program

```
' {$STAMP BS1}
' {$PBASIC 1.0}
' Jack-in-the-Box Program
'
' The following devices are connected to the prop controller for this prop
'
' Pin 6 PIR Sensor
'
' Pin 7 Serial communication for:
```

```
'    RC-4      Address %00
'    AP-16+    Address %00
'
' RC-4 Relay 1:  Pneumatic Solenoid Valve to open prop
'                Light to illuminate clown
'
' RC-4 Relay 2:  Motor to turn crank handle
'
'
' AP-16+ Sound Files
'
' WEASEL.WAV Pop Goes the Weasel to be played while motor is running
' SCREAM.WAV Scream to play when clown pops up
'
' The action of the prop is as follows:
'
'     1. Turn the crank arm motor and play "Pop Goes the Weasel" until
'        motion is detected in the room.
'
'     2. When motion is detected, pop up the clown and play the scream sound.
'
'     3. When the scream sound ends, retract the clown, wait 2 seconds,
'        and then resume the crank motor and "Pop Goes the Weasel."
'
'     4. Wait at least 10 seconds before triggering the prop again.
'
SYMBOL Sio = 7
SYMBOL Baud = OT2400
SYMBOL RC4 = 3
SYMBOL RelayValve = 1
SYMBOL RelayMotor = 2
SYMBOL RelayOn = 1
SYMBOL RelayOff = 0
SYMBOL AP16 = 0
SYMBOL ap16Status = B0
SYMBOL ap16Playing = BIT7
SYMBOL Pir = PIN6
SYMBOL waitTimer = B2


'Reset everything & pause one minute to let the PIR stabilize.
SEROUT Sio, Baud, ("!AP16", AP16, "X")
SEROUT Sio, Baud, ("!RC4", RC4, "X")
PAUSE 50000

Main:

'Play Pop Goes the Weasel in an indefinite loop.
SEROUT Sio, Baud, ("!AP16", AP16, "PW", "WEASEL", 13, 0)
```

(continued)

LISTING 3-3: **(continued)**

```
'Start the motor.
SEROUT Sio, Baud, ("!RC4", RC4, "R", RelayMotor, RelayOn)

'Wait 10 seconds for the room to clear.
PAUSE 10000

'Wait for the PIR to trigger for 0.25 seconds.
waitTimer = 0
WaitForMotion:
PAUSE 10
waitTimer = waitTimer + Pir * Pir
IF waitTimer &lt; 25 THEN WaitForMotion

'At this point the trigger has been activated.

'Stop the motor.
SEROUT Sio, Baud, ("!RC4", RC4, "R", RelayMotor, RelayOff)

'Play the scream.
SEROUT Sio, Baud, ("!AP16", AP16, "PW", "SCREAM", 13, 1)

'Pop up the clown.
SEROUT Sio, Baud, ("!RC4", RC4, "R", RelayValve, RelayOn)

'Wait for the scream to end.
StillScreaming:
SEROUT Sio, Baud, ("!AP16", AP16, "G")
SERIN Sio, Baud, ap16Status
IF ap16Playing = 1 THEN StillScreaming

'Retract the clown.
SEROUT Sio, Baud, ("!RC4", RC4, "R", RelayValve, RelayOff)

'Wait 2 seconds.
PAUSE 2000

'Do it all again.
GOTO Main
```

Chapter **4**

Re-Creating a Retro Science-Fiction Robot Head

My love of science fiction movies began when I was a young boy in the 1960s. Yes, I really am that old. I watched *Star Trek* when it was originally broadcast (fortunately, my dad had just bought our first color television set), and I saw *2001: A Space Odyssey* on its original release.

I especially loved movies and TV shows that featured robots, whether the robot was good or evil. The '50s and '60s had plenty of them. Two of the most popular robots were Robby the Robot from the 1950s classic movie *Forbidden Planet* and B-9 (most often simply called "The Robot") from the classic TV series *Lost in Space*.

One defining characteristic of both robots is that they featured interesting heads that had cool-looking instruments inside a glass bubble. In this chapter, I fulfill one of my own life-long fantasies — creating my own vintage, glass-bubbled robot head. For lack of a better name, I call the vintage robot head "VIN-e." As a shout-out to one of my favorite Kurt Vonnegut novels, *VIN-e* stands for *virtual infundibuled extenuator*. His primary function is to make things seem not as bad as they really are and convince you that even if things are that bad, it's not your fault. After all, there are extenuating circumstances, and VIN-e is the ultimate extenuator.

VIN-e is a purely fantastical creation designed for nothing more than amusement. It's assembled from bits and pieces, most of which I had lying around. If you want to build your own version of Vin-E, please don't follow my version as if it's a blueprint. Instead, use your own creativity along with whatever bits and pieces you happen to have lying around or can find at a local discount store or even thrift store. Design it so it tickles your *own* funny bone, and most of all, have fun in the building!

In fact, I don't offer detailed construction plans in this chapter. Instead, I just show you what inspired me to create VIN-e, a few of the bits and pieces I used to build him, and a photo of VIN-e in his current form. VIN-e is a work in progress — I've been tinkering with him for a while, and I'll probably continue to tinker with him for a few more years.

Looking to Robby and B-9 For Inspiration

Forbidden Planet is a landmark science fiction movie that was released in 1956. There is much that is noteworthy about this movie:

>> It's the first science fiction movie to depict a starship that travels faster than light.

>> It stars Leslie Nielsen when he was still a serious leading man, not the comic genius we saw decades later in movies like *Airplane!* and *The Naked Gun* movies.

>> It's very loosely based on Shakespeare's play *The Tempest*.

>> It was one of the main influences on Gene Roddenberry as he conceived of *Star Trek*. To wit, the original pilot for *Star Trek* (called "The Cage") is remarkably similar to *Forbidden Planet*. And *Forbidden Planet* takes place on the planet Altair IV. The Altair system is featured heavily in *Star Trek*. Altair IV's sister planet Altair VI is mentioned in the original series episode "Amok Time," and Altair VI plays a role in the classic Kobayashi Maru simulation from the 1982 movie *Star Trek II: The Wrath of Khan*.

But the most memorable thing about *Forbidden Planet* is the introduction of a fantastic robot named Robby, affectionately known as Robby the Robot. Robby was a mechanical costume designed and built primarily by Robert Kinoshita. Kinoshita graduated from the University of Southern California with a degree in architecture in 1940, but like so many Japanese Americans, he was held by the U.S. government in a Japanese internment camp during World War II.

Robby's design was ambitious and very expensive to build — adjusted for inflation, the costume cost more than $1 million! The head consisted of numerous gadgets that moved and blinked, representing his electronic brain, all enclosed in a large dome made of Plexiglas. There are elements that resemble eyes, a nose, and ears, as well as a "mouth" that lights up when Robby talks.

Figure 4-1 shows a model (part of my collection) of Robby. I've zoomed in on the head so you can see its key components.

FIGURE 4-1:
Robby the
Robot from
Forbidden Planet

A decade after designing Robby, Kinoshita was hired by producer Irwin Allen to design another robot for his new TV series *Lost in Space*. This new robot would become every bit as famous as Robby. His technical designation is General Utility Non-Theorizing Environmental Control Robot (GUNTER), but he's almost always referred to simply as "The Robot." He's occasionally referred to by his model number, B-9. Irwin Allen called him Blinky.

Like Robby, B-9 is a costume with amazing mechanical and electrical detail. Rather than a huge parabolic dome head filled with mechanical parts, B-9 has a glass-encased sensor unit that contains more modern-looking electronics. I've always thought the sensor unit resembled a fluted ceiling-mounted bathroom or closet light. Like Robby, B-9 has a "mouth" that lights up when the robot talks.

Figure 4-2 shows a model (also part of my collection) of B-9.

FIGURE 4-2:
Robot B-9 from
Lost in Space.

Conceiving VIN-e

The idea for VIN-e came a long time ago, when I was watching *Forbidden Planet* for the umpteenth time. It occurred to me that it would be fun to design and build something like that — not an entire robot costume, just the head, because the heads of old science fiction robots always seem to be the most interesting parts.

I sat on the idea for a few years and got distracted by many other projects. But I kept thinking about it and finally decided to do it when the COVID-19 pandemic hit in 2020. Like many people, I needed a distraction.

I decided I wanted my robot head to have the following elements, mostly inspired by Robby the Robot and B-9:

- A glass dome that encloses the head

- Electronic or mechanical instruments (or both) that resemble eyes, nose, and ears

- A voice that plays random robot sayings, including the best utterances of Robby the Robot ("Sorry, miss, I was giving myself an oil job" or "The whole thing hardly comes to ten tons") or B-9 ("Danger, Will Robinson" or "It does not compute")

- A "mouth" that lights up when VEN-e talks

- At least one bit that moves, most likely a sensor or antenna outside the dome

- Blinking lights (duh)

- As much as possible, parts that look vintage, resembling vacuum tubes and the like, with all modern electronics hidden

Introducing VIN-e 1.0

Figure 4-3 shows VIN-e in his current form. I think he looks pretty good, but there's still room for a lot of improvement. It'll be fun to see what VIN-e looks like a year or two from now!

The following sections provide a brief rundown on the various bits and pieces that make up version 1.0 of VIN-e.

The dome

The dome that encloses the head is a plastic serving bowl I found at Target for about $12. I realized that plastic would be important so I could easily cut it. And it needed to be something I could find off the shelf, because I don't have the skills to mold my own bowl from Plexiglas.

The bowl is 11 inches in diameter at the top and 6 inches in diameter at the bottom, and it stands 6 inches tall. Of course, I flipped it upside down for VIN-e.

I made just one cut in the dome — a 2½-inch rectangle at the top to accommodate the central pillar. The bowl was easy enough to cut with a Dremel rotary tool. I bought two of the bowls just in case I ruined one, but my first attempt at cutting it was successful.

FIGURE 4-3:
VIN-e, version 1.

The base

The base that the head rests on is a plastic serving tray that I found at Big Lots for about $10. The tray was actually 12 x 18 inches, which is a bit large. So I cut it down to a 12-inch square with my Dremel tool. I also spray-painted it black.

The base stands 1¾ inches tall, which is large enough to accommodate all the electronics that control VIN-e.

The central pillar

The central pillar that rises from the base up through the dome and then above the top of the dome is an interesting solar yard lamp I found at Harbor Freight for $10. The lamp came apart into three pieces, as shown in Figure 4-4.

The top part is the actual lamp. It has a solar panel on top, a tiny circuit card inside, a battery, and an LED. I tore it all apart and kept just the ribbed clear part, which also has a cool reflector at the bottom to scatter light from the LED. I replaced the white LED that came with the lamp with a red LED, because that makes for a more interesting display.

FIGURE 4-4:
The solar-
powered yard
lamp I used
as the central
pillar in VIN-e.

I kept the center part of the lamp to use as a pillar in the center of VIN-e's head. It holds the eyes and nose. I also drilled some holes in it to run cables through.

The bottom piece is a stake for planting the lamp in the ground. I used my Dremel tool to cut off the actual stake portion, but I kept the round peg that jams into the pillar. I used that to mount the pillar to the base.

The eyes and nose

The eyes are made of the lids from two containers of spackle that I've had for years (the spackle dried out long ago). I painted them with metallic bronze paint.

The lights in the center of the eyes, as well as the nose light, are LED arcade game console lights I purchased from Amazon. I bought a few dozen of these a few years ago for another project and had a few left over. The LEDs are rated for 12 VDC and have a built-in current-limiting resistor, so you don't have to supply your own resistor.

I was hoping that the LEDs would work just fine at the 5 V sourced by the digital I/O pins on an Arduino board, but they didn't, so I used relays to switch the proper voltage for the LEDs.

The mouth

VIN-e's mouth is a 6-x-1-inch decorative LED lamp, the kind where the LED is designed to look like an old-fashioned filament. This particular lamp has a filament that twists from top to bottom. I connected this lamp to the very same Velleman MK110 Simple One Channel Light Organ kit that I show in Book 8, Chapter 1, with the line-out signal from the MP3 shield as its input.

I found some very cool plastic conduit holders on sale at Home Depot for 17¢ each. I mounted two of them to the base about 4 inches apart and simply stuck the lamp in between them. Figure 4-5 shows these pieces. I used an inexpensive lamp base, connected to the output of the Velleman color organ circuit.

FIGURE 4-5:
Bits and pieces used to support VIN-e's mouth and "vacuum tubes."

If I ever figure out a way to create a fluted clear plastic shield that looks like Robby or B-9's mouth, I'll switch to that. But I've never found anything that would work, so for now, the vintage-looking LED lamp provides VIN-e's mouth.

The "vacuum tubes"

I wanted something that would look really vintage, like old vacuum tubes. I came across a set of candelabra-base LED lights that have vintage-looking filaments that remind me of vacuum tubes, so I picked up a few along with a few candelabra lamp bases. I also found a low-profile plastic electrical box, which I used to mount the lights an inch or so above the base.

The lamp in front of the vintage LED lamps are 1-inch-square arcade game lamps purchase from Amazon. They're similar to the LEDs I used for the eyes and nose. One is red and the other is green.

The antenna

The antenna atop the pedestal is one of my favorite parts. It's mounted on a two-servo arm that I purchased as a kit from Amazon for about $20. The servo is mounted to the salvaged top of the solar yard lamp with a couple of screws.

The antenna itself came from a $5 flashlight I bought at Rite Aid. I just took it apart and replaced the super-bright white LED with a red LED.

Each of the servos has a three-wire lead, and the resistor itself has a two-wire lead. I drilled a hole in the back of the pillar to run the wires to the base so they can be connected to the appropriate pins on the Arduino.

Looking at VIN-e's Electronics

All the electronics required to control VIN-e are mounted under the base. The following components are used:

>> **An Arduino MEGA board:** The MEGA board is similar to the UNO board, but it has additional digital I/O pins so that you can control additional circuits, as shown in Figure 4-6. VIN-e doesn't really need these additional pins, but I chose to use a MEGA board because it's compatible with shields designed for the UNO but provides digital pins that aren't covered up by the shield.

>> **An Adafruit MP3 shield:** I scoured the Internet for things I wanted VIN-e to say and copied them to the micro-SD drive.

>> **An eight-channel relay card (see Figure 4-7):** This card provides eight relay circuits that you can drive from digital output pins on an Arduino. All the 12-volt lamps are switched on and off through the relay board.

>> **A Velleman MK110 Simple One Channel Light Organ kit:** This is used to synchronize VIN-e's mouth with his voice.

>> **A small audio amplifier:** I harvested one from an old cheap set of computer speakers. I simply took the speakers apart and discarded the cases, keeping just the amp circuit and the speakers.

FIGURE 4-6:
An Arduino
MEGA board.

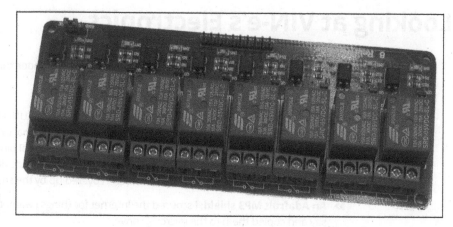

FIGURE 4-7:
An eight-channel
relay board.

>> **Two power sources:** VIN-e uses the following:

- *120 VAC directly from a wall outlet:* This source is used to power the vintage
  LED lamps. It's also used to power the other two power supplies. To safely
  manage the distribution of 120 VAC in the project, I used a covered terminal
  strip, as shown in Figure 4-8. In this figure, you can see the black incoming
  power cord connected to the terminal strip, with additional wiring to connect
  the 120 VAC power to the other 120 VAC components of the project.

- *5 V:* This is used to power the Arduino and the audio amplifier. I used a
  small wall-adapter, plugged into the 120 VAC circuit.

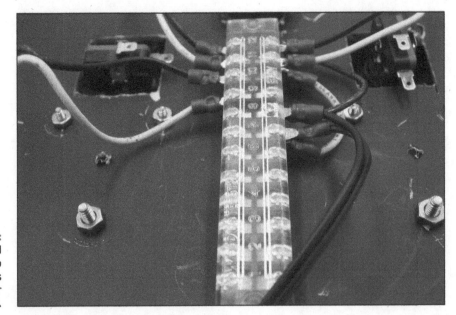

FIGURE 4-8:
A covered
terminal strip
safely routes
120 VAC power
for the project.

A Final Word about Safety

I want to end this chapter — and this book — with a final word about safety.

Like many of the projects you'll work on, VIN-e uses 120 VAC household current.
Working with household current can be extremely dangerous, so take all the necessary precautions. If you aren't sure how to build a circuit safely, *don't build it.*

Be sure to review Book 1, Chapter 4 for general tips on working safely with electronics. And review Book 1, Chapter 1 for additional safety information for working
with household current.

Finally, when you build a project like a vintage robot head, don't neglect other
aspects of safety. All power tools, even little ones like Dremel rotary tools, can be
dangerous. Always wear eye and face protection, keep your work area clean, and
know what you're cutting or drilling before you cut or drill.

And have fun!

Index

noninverting, 392

open-loop, 388–390

origin of, 387

overview, 383–384

symbol for in schematic diagrams, 384–385

terminology, 384

unity gain amplifiers

overview, 393–394

unity follower, 394

unity inverter, 394

as voltage comparator, 395–398

operations, logical, 493, 504

operators

increment, 639

math, 631–632

opt directory (Linux), 747

optoisolators, 818

OR gate

building, 499–500

CMOS, 552–555

description of, 505

four-input, 510

general discussion, 509–511

NAND gate, constructing from, 520

NOR gate, constructing from, 521

overview, 495–496

in sensor circuits, 511

truth table for, 509–510

two-input, 509–510

OR operation, 493

OR transistor circuit, 534

organs, color. See color organs

oscillator mode, 555 timer chip

description of, 352

typical circuit, 353–354

oscillators, 323, 445

oscilloscopes

calibrating, 155–157

connecting probe to audio plug, 159

display, 150–151

displaying signals with, 157–158

overview, 149–150

trace, 152, 157

waveforms, 152–154

OUTPUT constant, 617

output pin, 347

P

P? command, 873

PA command, 873

pace variable, 691

Parallax, 853

parallel circuits, 179–181

parallel inductors, 272

parallel lamp circuits, 183–186

parallel resistors, 222

parallel switch circuits, 190–193

parallel switches, 186–187

part number, 81

Parthian Empire, 14

PCBs (printed circuit boards)

coin-toss circuit, 108–113

custom, creating, 107

overview, 105

preprinted, 106–107

surface-mount, 105–106

through-hole, 105

PCBWay, 107

peak inverse voltage, 282–283

peak reverse voltage, 282–283

peak voltage, 407, 408

peak-to-peak voltage, 408

phases, 412

phenomenon, 22

Phillips-head screwdriver, 460

phone, 69

phototransistors, 469

Pi (Raspberry)

analog input, reading, 799–802

versus Arduino, 737

components, 738–739

connecting LED to GPIO port, 753–754

Cylon LED flasher, 790

description of, 739

desktop, 746

file system, 746–747

flashing the LEDs, 765–767

GPIO, digital input, using for, 791–794

About the Author

When **Doug Lowe** was 10 years old, his father gave him an electronics experimenter's kit that started a lifelong love of all things electronic. As a child, Doug dreamed of becoming an electrical engineer. But he soon discovered that he loved writing as much as he loved electronics, and his first technical book was published in 1981. Since then, he has written more than 100. All those books have been about one specific corner of the electronics world: computers. But now he has finally written a book about his first love, the joy of designing and building electronic circuits from scratch.

Doug lives in sunny Fresno, California, where the motto is "Fres-YES!" (unfortunately, he's not making that up). The fruits of his electronic passions can often be seen around the Halloween season in the form of animated computer-controlled Halloween decorations that rival Disney's Haunted Mansion.

Dedication

My grandpa had a wonderful workshop behind his home. It was cramped and cold in the winter and smelled of machine oil, stale coffee, aged rubber, and damp wood — like an ancient hardware store clinging to life down the road from a giant big-box store. It was there that I began a lifelong love of electronics, as my grandpa tried to explain the mystery of electricity and taught me how to use a soldering iron, read a voltmeter, and test a questionable vacuum tube.

In the spring of 1970, he showed up at my house with a trailer full of everything an 11-year-old boy could possibly need to create his own electronics lab. The trailer was stacked high with broken radios and box after box filled with tools and spare parts: vacuum tubes, soldering irons, condensers, voltmeters, resistors, tube testers, and tuning capacitors.

I can still remember the look on my dad's face when he saw that trailer full of junk pull up in front of our house, hitched to the back of Grandpa's old Buick.

I was in heaven. I think my dad may have been somewhere else.

But he helped unload the trailer, and then he built me an amazing workbench with enough shelves and drawers to store everything and more.

This book is dedicated to the memory of my grandpa, Kenneth D. Lowe, Sr., and to my dad, Kenneth D. Lowe, Jr.

To my grandfather, for giving me an amazing gift buried in a trailer full of junk. And to my dad, for letting me keep it.

Author's Acknowledgments

First, I'd like to thank Elizabeth Stilwell for giving me the opportunity to do a third edition of this book, as well as the Katies: Katie Mohr and Katie Feltman for green-lighting the two previous editions.

And special thanks to Elizabeth Kuball for editing this third edition, as well as the second edition before it and all the other projects we've done over the past few years. Let's do it again, please!

Thanks also to Kirk Kleinschmidt, who gave the book a great technical review and offered many excellent suggestions along the way. And, of course, thanks to all the behind-the-scenes people who chipped in with help that I'm not even aware of.

Publisher's Acknowledgments

Associate Editor: Elizabeth Stilwell

Project Editor: Elizabeth Kuball

Copy Editor: Elizabeth Kuball

Technical Editor: Kirk Kleinschmidt

Production Editor: Tamilmani Varadharaj

Cover Photos: © kilukilu/Shutterstock

Take dummies with you everywhere you go!

Whether you are excited about e-books, want more from the web, must have your mobile apps, or are swept up in social media, dummies makes everything easier.

Find us online!

dummies.com

Leverage the power

Dummies is the global leader in the reference category and one of the most trusted and highly regarded brands in the world. No longer just focused on books, customers now have access to the dummies content they need in the format they want. Together we'll craft a solution that engages your customers, stands out from the competition, and helps you meet your goals.

Advertising & Sponsorships

Connect with an engaged audience on a powerful multimedia site, and position your message alongside expert how-to content. Dummies.com is a one-stop shop for free, online information and know-how curated by a team of experts.

- Targeted ads
- Video
- Email Marketing
- Microsites
- Sweepstakes sponsorship

20 MILLION PAGE VIEWS EVERY SINGLE MONTH

15 MILLION UNIQUE VISITORS PER MONTH

43% OF ALL VISITORS ACCESS THE SITE VIA THEIR MOBILE DEVICES

700,000 NEWSLETTER SUBSCRIPTIONS TO THE INBOXES OF **300,000** UNIQUE INDIVIDUALS EVERY WEEK

# of dummies

Custom Publishing

Reach a global audience in any language by creating a solution that will differentiate you from competitors, amplify your message, and encourage customers to make a buying decision.

- Apps
- Books
- eBooks
- Video
- Audio
- Webinars

Brand Licensing & Content

Leverage the strength of the world's most popular reference brand to reach new audiences and channels of distribution.

For more information, visit **dummies.com/biz**

PERSONAL ENRICHMENT

Staying Sharp	**Facebook**	**Guitar**	**Investing**	**Beekeeping**	**Digital Photography**
9781119187790	9781119179030	9781119293354	9781119293347	9781119310068	9781119235606
USA $26.00	USA $21.99	USA $24.99	USA $22.99	USA $22.99	USA $24.99
CAN $31.99	CAN $25.99	CAN $29.99	CAN $27.99	CAN $27.99	CAN $29.99
UK £19.99	UK £16.99	UK £17.99	UK £16.99	UK £16.99	UK £17.99

Meditation	**Pregnancy**	**Samsung Galaxy S7**	**iPhone**	**Crocheting**	**Nutrition**
9781119251163	9781119235491	9781119279952	9781119283133	9781119287117	9781119130246
USA $24.99	USA $26.99	USA $24.99	USA $24.99	USA $24.99	USA $22.99
CAN $29.99	CAN $31.99	CAN $29.99	CAN $29.99	CAN $29.99	CAN $27.99
UK £17.99	UK £19.99	UK £17.99	UK £17.99	UK £16.99	UK £16.99

PROFESSIONAL DEVELOPMENT

Windows 10	**AutoCAD**	**Excel 2016**	**QuickBooks 2017**	**macOS Sierra**	**LinkedIn**	**Windows 10**
9781119311041	9781119255796	9781119293439	9781119281467	9781119280651	9781119251132	9781119310563
USA $24.99	USA $39.99	USA $26.99	USA $26.99	USA $29.99	USA $24.99	USA $34.00
CAN $29.99	CAN $47.99	CAN $31.99	CAN $31.99	CAN $35.99	CAN $29.99	CAN $41.99
UK £17.99	UK £27.99	UK £19.99	UK £19.99	UK £21.99	UK £17.99	UK £24.99

SharePoint 2016	**Fundamental Analysis**	**Networking**	**Office 2016**	**Office 365**	**Salesforce.com**	**Coding**
9781119181705	9781119263593	9781119257769	9781119293477	9781119265313	9781119239314	9781119293323
USA $29.99	USA $26.99	USA $29.99	USA $26.99	USA $24.99	USA $29.99	USA $29.99
CAN $35.99	CAN $31.99	CAN $35.99	CAN $31.99	CAN $29.99	CAN $35.99	CAN $35.99
UK £21.99	UK £19.99	UK £21.99	UK £19.99	UK £17.99	UK £21.99	UK £21.99

dummies.com

dummies®
A Wiley Brand